Gene Therapy for Viral Infections

Gene Therapy for Viral Infections

Patrick Arbuthnot

Wits and South African Medical Research Council,
Antiviral Gene Therapy Research Unit,
School of Pathology, Faculty of Health Sciences,
University of the Witwatersrand, South Africa

ELSEVIER

AMSTERDAM • BOSTON • HEIDELBERG • LONDON
NEW YORK • OXFORD • PARIS • SAN DIEGO
SAN FRANCISCO • SINGAPORE • SYDNEY • TOKYO
Academic Press is an imprint of Elsevier

Academic Press is an imprint of Elsevier
125 London Wall, London EC2Y 5AS, UK
525 B Street, Suite 1800, San Diego, CA 92101-4495, USA
225 Wyman Street, Waltham, MA 02451, USA
The Boulevard, Langford Lane, Kidlington, Oxford OX5 1GB, UK

Library of Congress Cataloging-in-Publication Data
A catalog record for this book is available from the Library of Congress

British Library Cataloguing-in-Publication Data
A catalogue record for this book is available from the British Library

ISBN: 978-0-12-410518-8

Publisher: Janice Audett
Acquisition Editor: Linda Versteeg-Buschman
Editorial Project Manager: Halima Williams
Production Project Manager: Chris Wortley
Designer: Maria Inês, Cruz

Typeset by TNQ Books and Journals
www.tnq.co.in

Dedication

To Pamela

Contents

Acknowledgments

I am indebted to the many colleagues, students, and friends who often unwittingly made significant contributions to this book. Support for the research in our laboratory has been vital to enabling me to engage with the challenging field of antiviral gene therapy. Funding received over a period of several years from the South African National Research Foundation, Medical Research Council, Cancer Association, Poliomyelitis Research Foundation, European Commission, l'ANRS of France and from the German Research Foundation (DFG) is gratefully acknowledged. I am also thankful to Elizabeth Gibson, Halima Williams, Chris Wortley, and others from Elsevier Press for their patience while putting the manuscripts together.

CHAPTER 1

Essentials of Viruses and their Suitability for Treatment Using Gene Therapy

1.1 GENE THERAPY

The term "gene therapy" was coined in 1972 to explain the use of procedures that are intended to treat or alleviate disease by genetically modifying the cells of a patient [1]. The concept of gene therapy was developed after publication of the first reports demonstrating that it was possible to alter gene expression in cultured cells. These early studies showed that gene expression in murine or human cells could be modified by transfection with DNA expressing herpes simplex virus thymidine kinase [2] or hexose-1-phosphate uridylyltransferase [3], respectively. Initial definitions of gene therapy referred exclusively to the use of genes to treat disease, but the meaning has now become broader [4–6]. Currently, gene therapy is defined by the use of nucleic acids, which may include DNA, RNA, or chemically modified derivatives, to alter gene function and treat disease. Because abnormalities of gene function underlie many disease processes, including those caused by viral infections, interventions using gene therapy potentially have wide-ranging applicability.

Gene therapy has many applications and may be used for restoring the health of diseased cells, killing of malignant tissue, and induction of immune responses to gene-encoded proteins. To treat diseased cells, gene therapy may entail repairing damaged genes or silencing "rogue" genetic elements that are expressed by viral pathogens. With the advent of recombinant DNA technology, polymerase chain reaction, and sophisticated nucleic acid sequencing procedures, insights into molecular biology and the fundamental mechanisms causing disease processes have greatly progressed. These developments have had a profound enabling effect on the rational design of gene therapy approaches.

Different methods of therapeutic inhibition of gene function have been used to counter viral infections. These include silencing of virus-encoded genes (Chapter 2) and introduction of targeted disabling mutations into viral genes or host factors (HFs) (Chapter 3). Using gene transfer to augment patients' immune responses to virus infections has been another way of achieving preventative or therapeutic antiviral therapeutic effects (Chapter 11).

Gene Therapy for Viral Infections. http://dx.doi.org/10.1016/B978-0-12-410518-8.00001-6

Exploiting gene therapy to counter virus replication has advanced considerably, and several viruses have now been shown to be candidates for treatment using this approach. However, there is no universal method of using gene therapy for viral infections. Individual viruses have particular characteristics, and this necessitates that viral gene therapy be tailored to specific infections. In developing viral gene therapy, tissue tropism, the acute or chronic nature of an infection, and the efficiency with which antiviral sequences can be delivered to infected tissues are important considerations (Table 1.1). Both synthetic nucleic acids and DNA expression cassettes are being developed for viral gene therapy. Expressed antiviral sequences may be more useful for countering chronic viral infection whereas synthetic antiviral nucleic acids are better suited to inhibiting acute viral infections. Preventing viral escape from gene therapy is important, and overcoming this problem by simultaneous targeting of multiple viral sites or suppressing HFs has shown promise.

1.2 ESSENTIALS OF VIRUSES

Viruses are the simplest life forms; not surprisingly, they are also very plentiful. Estimations place the number of viral particles in the biosphere to be between 10^{31} and 10^{32} [7–9]. In natural waters of the Earth, they are estimated to outnumber bacteria by an order of magnitude. Viruses are a major cause of disease and constitute a reservoir of enormous genetic diversity. They are highly varied with respect to their structure, genome replication mechanisms, and modes of interacting with their host organisms. Common and interrelated defining features of viruses are the following:

- They are obligate intracellular parasites that only reproduce within host cells and are incapable of independent replication.
- Viruses do not have the machinery required for translation of proteins. They use host protein synthesis mechanisms, with their own genetic material as template, to produce the components constituting intact infectious viral particles (virions).
- Viruses lack the mechanisms for generating the energy required to drive the biochemical processes required for their existence.

Viruses may essentially be considered as nucleic acid parasites that use virions to introduce their own genetic material into cells. Thereafter, the host cellular machinery is reprogramed for copying the viral genome and the formation of more virions to result in completion of the viral replication cycle. Viruses have evolved efficient mechanisms for introducing DNA or RNA genomes into cells, and this property has been exploited for development of viral gene therapy vectors. Ironically, in some cases these recombinant vectors are being developed as therapeutics to counter virus infections.

Table 1.1 Viral Characteristics That Influence the Gene Therapy Strategy

Viral characteristics	Implications for gene therapy of viral infections
DNA or RNA virus	Both DNA and RNA viruses are susceptible to RNA silencing mechanisms (e.g., by RNAi activators). DNA targeting, such as by sequence-specific nucleases, is only possible with DNA viruses.
Virus tropism	Delivery of gene therapies remains a challenging task for successfully implementing gene therapy of viral infections. Therefore, accessibility of infected tissues to gene therapy vectors is an important factor that currently influences success. For example, delivery of small interfering RNAs targeting respiratory syncytial virus after inhalation is easier than delivery to hepatocytes after systemic administration of gene therapy formulations.
Acute or chronic nature of infection	Acute infection may require a single dose of a therapeutic whereas chronic infection may require repeated administrations. Approaches that achieve sustained inhibition of viral replication may also be necessary for treating chronic infections.
Viral replication rate	Very high replication rate may overwhelm efficacy of antiviral gene therapy. Moreover, viral dormancy may also evade nucleic acid antivirals.
Predisposition to mutation	RNA viruses or viruses that use reverse transcription during genome duplication are prone to error and may introduce mutations that evade sequence-specific antivirals. Combinatorial approaches or host factor targeting may be necessary to provide a higher barrier to resistance.
Host immune response to virus	Host immunity may augment efficacy of antivirals. Although generally undesirable, induction of the innate immune response by viral gene therapies may augment their antiviral efficacy (e.g., 5′ triphosphate-containing small interfering RNAs against HBV infection).
Reliance on host factors (HFs) that may be silenced	Disabling HFs (e.g., CCR5 for HIV-1 infection and micro RNA 122 for HCV replication) that are required for viral replication may be used to inhibit viral replication. Successful use of this approach requires that inhibition of HF function is not toxic to cells.
Virus-encoded mechanisms of countering gene silencing	Viruses that are capable of inhibiting the RNAi pathway (e.g., adenovirus virus-associated RNAs) may result in attenuated inhibition of silencing efficacy.
Similarity of viral targets to host cellular sequences	Sequence homology between viral targets and cellular sequences may result in unintended harmful off-target effects of a gene therapy.

Whether viruses meet the basic requirements of what constitutes life has been a subject of debate. Definitions of life are themselves fraught [10], but viruses display at least some of the traits that characterize living entities. Key attributes of living entities include the ability to reproduce and evolve in response to external influences. Because viruses reproduce, albeit in a parasitic manner, they may be considered as living organisms. Moreover, adaptation of viruses to their environments through evolution, an important and clinically relevant property, is another argument in favor of viruses being classed as living entities. However, because viruses are obligate intracellular parasites that are incapable of independent replication, some have argued that they do not qualify as living organisms.

Defining what makes a virus a virus has also been the subject of some debate [11–13]. It has been argued that the disappearance and appearance of virions (disintegration and reconstitution) is the fundamental characteristic feature of viruses [13]. Disappearance of the virion occurs when the particle breaks down to release the viral genome into the cell, then reappearance happens when the intracellular machinery is used to propagate virion progeny. Although understanding the phenomenon of disappearance and appearance of viruses is useful, defining viruses on the basis of such phenotypic features may be problematic because they may not be particular to viruses. Therefore, the specifics of genome coding capacity have been preferred as the essential attribute of viruses [11,12]. Raoult and Forterre proposed the existence of genes that encode a viral capsid as the unique feature of viruses, and a paraphrased version of their widely accepted definition of viruses is the following: "Viruses are capsid-encoding organisms that are composed of proteins and nucleic acids. They self-assemble in nucleocapsids and use ribosome-encoding organisms for the completion of their life cycles." According to this description, all living organisms may be classified as either capsid-encoding viruses or organisms that have translational capacity (Bacteria, Archaea, and Eukarya). The definition of capsids themselves becomes important for distinguishing viruses from other organisms. It is well known that capsid-encoding genes may be integrated into other organisms. A good example is the existence in human immunodeficiency virus (HIV)-1-infected humans of capsid-encoding sequences located in the integrated provirus. Raoult and Forterre cleverly address the possible designation of host organisms as viruses by classifying a capsid as a "structure that is used to disseminate a genome that encodes the capsid proteins" [11].

1.2.1 Viral Origins

Identifying the origin of viruses has been contentious. The dependence of viruses on cellular life forms led to the idea that cellular life preceded the evolution of viruses. However, there is some evidence that viruses co-evolved with their cellular hosts. A particularly interesting recent insight into virus evolution has come from structural analysis of virion architecture and coat topology (reviewed in refs [14,15]). Unexpected similarities were found in viruses that infect organisms of different kingdoms (Bacteria, Archea, and Eukarya), which was interpreted as indicating a common ancestry to all viruses. The three main propositions to explain the origins of viruses are the following [16]:

1. The *progressive hypothesis* purports that mobile genetic elements gained an ability to be transmitted between cells and then gave rise to viruses. The similarity between eukaryotic retrotransposons, commonly found in eukaryotic cells [17], and retroviruses gives support to this notion. Among other common features, retroviruses and retrotransposons possess flanking long terminal repeats and encode reverse transcriptase and integrase. Both entities convert RNA into DNA, which is then integrated into the host genome by similar mechanisms. Although the progressive hypothesis postulates that retrotransposons gave rise to retroviruses, it is also possible that the converse is true: Infection of host cells with retroviruses gave rise to infection-deficient retrotransposons.
2. The *regressive hypothesis* states that viruses may have evolved from free-living, more complex intracellular parasites. According to this theory, an initial symbiotic relationship between the virus precursors and their hosts turned parasitic during evolution of the viruses. Characteristics of the nucleocytoplasmic large DNA viruses (NCLDVs) back this hypothesis. NCLDVs are large and have complex genomes. The Mimivirus member of this family, which is also the largest known virus, has a genome size of 1.18 million base pairs. Sequencing of the genome interestingly revealed evidence of remnants of translational machinery [18]. These included amino acyl transfer RNA synthetases, translation factors, and tRNA-encoding sequences. These observations suggest that NCLDVs lost translational capabilities as they evolved into obligate intracellular parasites.
3. The *virus-first hypothesis* proposes that viruses were the first replicating entities that gave rise to cellular life [19,20]. According to this theory, self-replicating viral units acquired membranes that gave rise to bacterial, archaeal, and eukaryotic cells. Furthermore, while parasitizing their cellular descendants, the hypothesis proposes that ancestral viruses evolved into present day viruses. Detailed analysis of the DNA polymerase genes of phycodnaviruses and other organisms supported this idea. Reconstruction of a phylogenetic tree demonstrated that the viral genes are exclusively located at the root of the clade containing all eukaryotic DNA polymerase delta genes [20]. The observation was reasonably interpreted as indicating that the polarity of the flow of genetic information was from viruses to primitive eukaryotic hosts, and not the reverse. An important underlying concept of this theory is that simpler viral genomes evolve more rapidly than their more complex host genomes. Thus, the resultant enhanced capacity for genetic novelty would be capable of providing a rich source of molecular complexity to the host.

There are compelling arguments in support of very different hypotheses to explain the origin of viruses. A possibility is that various underlying processes gave rise to different viruses, and that these events may also have taken place several times during the evolution of viruses. However, this notion may not reconcile with the structural evidence supporting a common viral origin [14].

1.2.2 Basics of Virus Replication

There is considerable variation in the molecular mechanisms that viruses use for their propagation. However, because all viruses are parasitic and their replication involves disintegration and reconstitution of virions after infection of host cells, there are some essential features that are common to all virus life cycles (Figure 1.1):

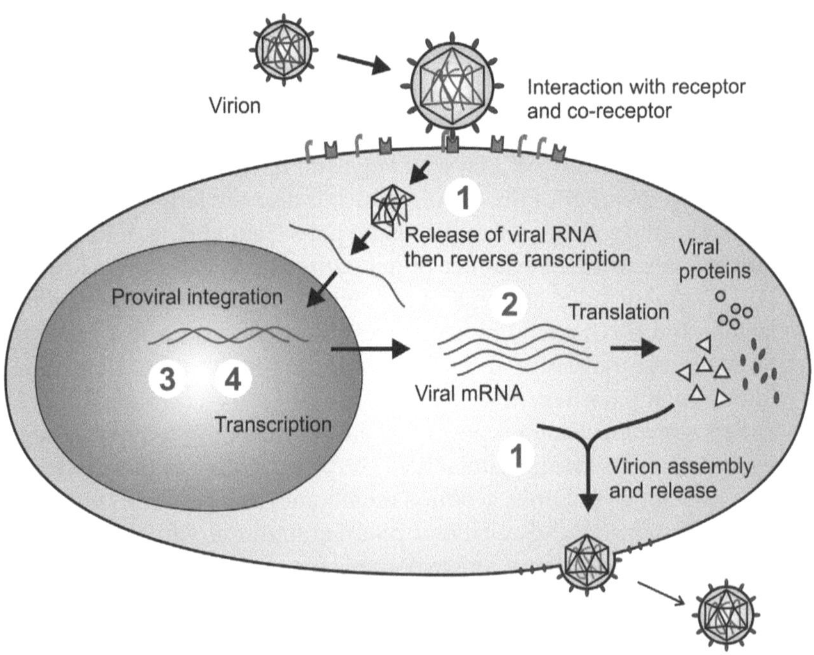

FIGURE 1.1 Schematic illustration of infection of a cell by a retrovirus with potential targets for gene therapy.
Attachment of the virion to the target cell involves interaction with a receptor (e.g., CD4) and co-receptor (e.g., CCR5). Release of viral RNA is followed by reverse transcription and integration of proviral DNA into the host genome. Transcription and then translation of viral mRNA result in the formation of viral proteins. Release of the newly formed virions involves assembly of the capsid, incorporation of viral genomic RNA, and budding from the cell membrane to form the viral envelope. Examples of potential mechanisms of gene therapy are (1) suppression of expression of the viral co-receptor and other host factors, (2) post-transcriptional gene silencing, (3) mutation of viral DNA, and (4) transcriptional silencing of viral genes.

1. The initial step is typically considered to be the interaction of virions with the host cell membrane. This involves binding by molecules on the surface of virions to receptors on the recipient cells. The specificity of this interaction is a major factor that defines the tissue tropism and the particular species that a virus infects. Some viruses have broad tropism and infect different species or several different tissues (e.g., cytomegalovirus). Others may be limited to particular cell types in one organism (e.g., hepatitis B virus (HBV) infection of hepatocytes of humans, certain primates, and the Asian tree shrew). In addition to susceptibility to virus infection conferred by interaction of virion and cell surface receptors, target cell infection range is also determined by permissiveness to infection. This permissiveness involves specific interactions of viral and HFs that are required for viral replication during subsequent stages of the virus infection cycle. Inhibiting the function of these HFs is an approach that is being used to counter viral replication using gene therapy. An important example is inactivation of the C–C chemokine receptor 5 (CCR5), an HIV-1 co-receptor, to make CD4 cells resistant to infection with the virus. Another example is the use of chemically modified oligonucleotides to hybridize and inhibit the function of micro RNA-122, a hepatitis C virus (HCV) HF. Using this method to treat HCV infection has reached an advanced stage of clinical testing [21].
2. After interaction with cell surface receptors, viruses penetrate target cells. The process may involve translocation; endocytosis; or, in the case of enveloped viruses, fusion of the viral envelope with the target cell membrane. Internalization of the particles is followed by breakdown of the viral particles and release of their genetic material.
3. Production of viral proteins follows, and this essential step exploits the host cell translational machinery. Viral protein production may involve direct use of virion-released nucleic acid, and an example is the use of poliovirus plus strand viral RNA as translation template (reviewed in ref. [14]). Processing of released viral nucleic acid may be required before translation of viral proteins occurs. An example is the reverse transcription of the HIV-1 RNA to form proviral DNA, which is in turn transcribed before being translated into viral proteins. Viruses have also evolved noncanonical mechanisms to ensure efficient protein expression in host cells. Three examples are the following:
   a. Use of an internal ribosomal entry site by viruses such as HCV and poliovirus to guide translation initiation in the absence of a 5′ cap on viral mRNA [22].
   b. Production of precursor polyproteins that form mature viral proteins after proteolysis. An example is the HCV polyprotein that generates 10 structural and functional protein elements of the virus (reviewed in ref. [23]; Chapter 7).

c. Exploiting RNA secondary structures to induce reading frame shifts. HIV-1 utilizes this mechanism to increase mRNA protein coding versatility and allow formation of required ratios of Gag and Pol protein components from a single mRNA template.

4. Interaction of viral proteins with host cellular factors leads to replication of viral genomes and assembly of new viral particles. This step is complex and highly variable among different viruses. It is important to note that it is largely the way in which cellular functions are subverted that determines whether a virus infection is manifested as disease. For example, the killing of $CD4^+$ T helper cells that is effected by HIV-1 results in compromised immunity and attendant complications leading to the acquired immunodeficiency syndrome (AIDS).
5. Release of newly formed virions is required to spread the infection between cells from the same individual or from one person to another. The process of releasing nascent virions is also variable and depends on properties of the host and viral elements. Viral capsid assembly and packaging of the genome is an essential step before release of newly formed virions. This process occurs by different mechanisms, but it generally requires interaction of a packaging signal on the viral genome with viral and cellular proteins to introduce the viral nucleic acid into the capsid particle. Enveloped viruses, such as HIV-1, may be budded from a cell and carry the cell membrane with them to form the envelope. Nonenveloped viruses (e.g., poliovirus) lack an outer membranous layer and may be released after cell death and lysis. After release from cells, virions are free to start a new round of infection by interacting with receptors on the surface of susceptible cells.

1.2.3 Virus Structures and Classification

The classification of viruses is based on their genome composition, size, shape, host range, and mode of replication [24]. Descriptive nomenclature has been helpful to assist workers in understanding pathogenic and structural properties of groups of viruses. For example, the Hepadnaviridae family, of which HBV is a member, includes viruses that have DNA genomes and infects the liver to cause hepatitis.

Viral genomes may be RNA or DNA, single or double stranded, linear or circular, and monopartite or multipartite. In addition, the genomes of single-stranded RNA viruses may have a sense (+) or antisense (−) orientation. The sense-stranded RNA genomes may serve as translational templates. However, antisense single-stranded viruses require conversion to plus stand sequences before translation is possible. Ambisense translation is also possible with some RNA viruses and is an interesting mechanism of increasing coding capacity.

With these viruses, protein translation is possible from genomic RNA or its complement. The term is derived from the ambiguous coding properties of each RNA strand. That is, each strand may have both sense and antisense polarity. Examples of viruses with ambisense genomes are found in the members of the Bunyaviridae family, such as Rift Valley Fever virus [25]. DNA viruses generally have less variability in their genome structures. Most of them contain a single DNA molecule that may be single stranded (e.g., parvovirus), double stranded (e.g., adenovirus), or circular (e.g., HBV). More viruses have RNA genomes than DNA genomes. Because the error rate of RNA replication is high, these viruses frequently generate mutations, which in turn may provide fitness advantages or disadvantages to the viruses.

Capsids, which provide a protective shell for viral genomes, are generally classified as being either helical or icosahedral in structure, although complex and pleomorphic alternatives may also occur. An example of a capsid with variant helical structure is that of HIV-1. This viral component has been described as a "fullerene cone." It comprises a helical arrangement of hexamers of capsid proteins that is sealed by 12 pentamers of the capsid protein [26–28]. Capsids themselves may be surrounded by a lipid membrane that is derived from the host cells. Enveloped virions typically have glycoproteins embedded in the lipid bilayer. These function as ligands for the receptors on target cells and serve as viral antigenic determinants.

1.2.4 Spread of Virus Infections

Modes of viral transmission, seasonal variation of infections, incubation periods, and phases of communicability are important to understand the epidemiology of infections and to implement measures to prevent spread. When transmission occurs between individuals it is referred to as "horizontal." Vertical transmission results from a mother infecting her unborn baby. Some viruses (e.g., HBV) may be passed from mother to baby during childbirth. In such cases in which transplacental transmission of the virus is rare, virus transmission is referred to as "perinatal." Horizontal transmission may occur across epithelial surfaces, which include intestinal, respiratory, genitourinary, conjunctival mucosa, and the skin. In addition, blood–blood (parenteral) contact between individuals and zoonotic spread between arthropods and different species may cause viral spread.

1.3 VIRAL PATHOGENESIS OF DISEASE

Because viruses are obligate intracellular parasites, it is not surprising that some disruption to cellular functions occurs during their propagation. Virulence, or the capacity of viruses to cause pathology, varies and is dependent on several host- and virus-derived factors. It is important to note that the severe cytotoxicity associated with high virulence potentially incapacitates virus propagation; therefore, it would be an evolutionary disadvantage to a virus. Consequently, viruses that have less virulent effects are more common. Disruptive effects may vary considerably and are largely responsible for the pathogenic effects of virus-related disease. Direct cytotoxic effects of viruses may result from disruption to essential host cellular functions, such as the maintenance of normal cell membrane ion permeability and synthesis of macromolecules. So-called "genotoxic effects" may occur after mutational effects of viruses on host genomes. Indirect toxicity is another potentially serious consequence of viral infection and typically results from effects of the host's immune response to a virus infection. HBV is an example of a virus that causes indirect toxicity. The virus has minimal direct cytopathic effects and symptoms of acute infection occur as a result of cell-mediated attack on infected hepatocytes. Ironically, severe symptoms indicate a good long-term prognosis with low risk for chronic infection. Conversely, asymptomatic infection indicates a poor immune response and high risk for viral persistence. Cytopathic effects of viruses manifest in different ways, such as the induction of programmed cell death, formation of inclusion bodies, changes to cell morphology, and syncytium formation.

Persistent infection occurs when a virus is not cleared and remains in infected cells. Naturally persistent virus infections have been characterized as being latent or chronic [29]. An example of a latent virus infection is that caused by herpes viruses (Chapter 11). The replication occurs during bouts of disease manifestation, but the viral genome lies dormant between such episodes. HBV and HCV may cause chronic infections, and these viruses are detectable during the periods of persistence (Chapters 6 and 7). Inadequacy of host immune responses to the infections plays a major part in determining the persistence of these virus infections [30]. However, latent and chronic viral infections are not mutually exclusive. For example, HIV-1 manifests latent and chronic characteristics. This has important implications for therapy of the infection. Current antiretrovirals are capable of suppressing HIV-1 replication, but they do not eradicate the reservoir of quiescent proviral integrants.

The mechanisms by which viruses become persistent are largely a result of immune- or gene expression-related effects. Viruses may have immunoevasive or immunomodulatory effects to limit or attenuate efficacy of host immune responses. Variation in viral antigens by HIV-1 is the classic example of avoidance of neutralizing effects of a host immune response. Reduced expression of major histocompatibility complex (MHC) class I molecules by cytomegalovirus [31] and modulation of monocytes and macrophages by Epstein Barr virus [32] are other examples of modulatory effects that viruses use to evade host immune responses. Stability and quiescence of viral replication intermediates may also add to persistence of viral infections. The enduring nature of integrated proviral DNA of HIV-1 together with suppressed viral gene expression are clinically important examples of viral mechanisms that enable avoidance of immunodetection by a host's antiviral immune response.

1.4 IMMUNE RESPONSES TO VIRUS INFECTIONS

Viral infections in vivo result in the stimulation of innate and adaptive immune responses. The innate response is activated during the initial stages of an infection and is triggered by pattern recognition receptors (PRRs) that distinguish particular pathogen-associated molecular patterns (PAMPs) that are found in various microbial pathogens such as viruses [33–36]. The adaptive immune response, with humoral and cell-mediated arms, occurs later during a virus infection and is a more recent evolutionary development. The innate system uses germ-line-encoded PRRs that identify groups of viral pathogens whereas the adaptive immune response entails selection of clonally expressed pathogen-specific receptors. Adequate stimulation of the innate immune response contributes significantly to the effectiveness of the adaptive response. Mounting of an antiviral immune response is understandably critically important for elimination of viral infections. Many antiviral therapies, including candidate viral gene therapies, require augmentation from the host's pathogen-specific immune response to be effective. Furthermore, because some antiviral gene therapies are based on use of recombinant viral vectors, antiviral immunity may also be a factor that affects the efficiency of delivery of antiviral sequences. In addition, some antiviral nucleic acids, particularly activators of RNA interference (RNAi), may be perceived as foreign; therefore, they induce an immunostimulatory effect. Therefore, the antiviral immune response has an important influence on the efficacy of gene therapy at several levels. The essentials of antiviral immune responses are described below. For comprehensive accounts of the topic, the reader is referred to the many excellent reviews in the field.

1.4.1 Innate Immunity

The components that are recognized by PRRs include viral double-stranded RNA, single-stranded RNA, RNA with 5′ triphosphates, proteins, and DNA [36]. PRRs comprise three main groups:

1. Retinoic acid-induced gene I (RIG-I)-like proteins (RLPs),
2. Toll-like receptor (TLR) proteins, and
3. Nucleotide oligomerization domain (NOD)-like receptors.

Each group is responsible for recognizing particular viral motifs. In addition, the downstream activation of pathways that counter viral replication differs. Activation of TLR and RLP pathways are schematically illustrated in Figure 1.2. RLPs, which are essential for mounting an innate immune response to RNA

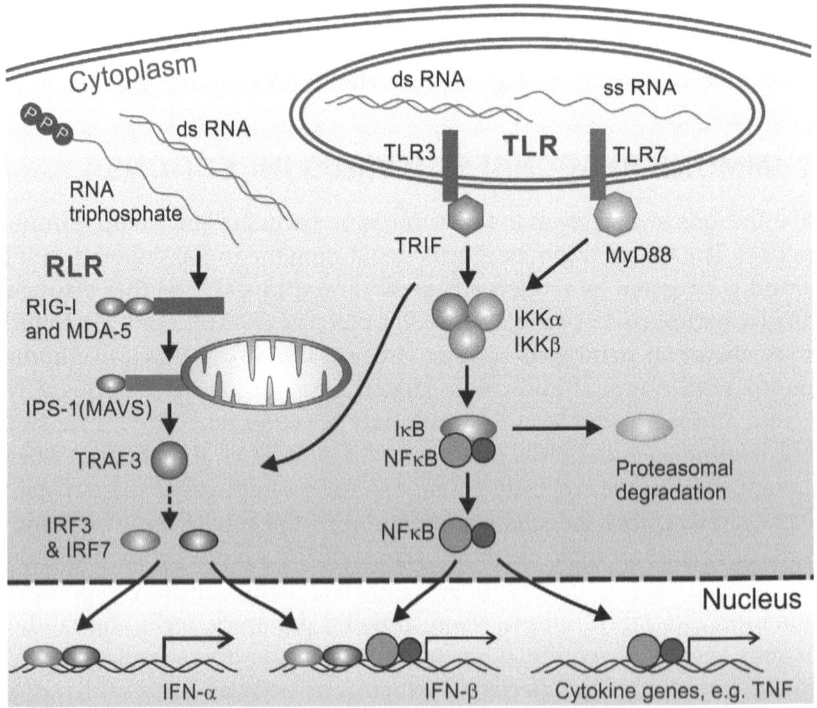

FIGURE 1.2 Schematic illustration of essential processes of the arms of the innate immune response to viral infection.

PRRs may be membrane bound (e.g., endosomal TLR3 and TLR7) or cytoplasmic (e.g., RIG-I and MDA5). Engagement with PAMPs such as double-stranded RNA or triphosphorylated RNA activates downstream signaling. Activation of RLRs involves downstream stimulation of IPS-1 and TARFs (e.g., TARF3) and then induction of IRF3 and IRF7 transcription factors. TLR activation is mediated by TRIF (TLR3) and MyD88 (TLR7). Subsequent phosphorylation of IκB by members of the IKKs releases NFκB to enter the nucleus and effect transcriptional activation. Crosstalk between the pathways may occur, such as through the activation of IRF3 and IRF7 by TRIF. Culmination of the effects of TLR and RLR pathway activation is expression of IFN-α, IFN-β, and cytokine gene expression.

viruses, contain helicase and C-terminal repressor domains. Caspase recruitment domains (CARDs), found in RIG-I and melanoma differentiation-associated gene 5 (MDA5) proteins but not in the laboratory of genetics and physiology-2 RLP, affect downstream signaling [37,38]. Members of the RIG-I-like receptor (RLR) family of PRRs respond differently to virus infections. For example, RIG-I is important for mediating responses to HCV [39], paramyxovirus, and influenza virus infections [40,41] whereas MDA5 activates an innate immune response to infections with picornavirus [42]. However, functioning of RLPs is not mutually exclusive and they may act in concert to recognize a viral infection. An example is the mounting of an innate immune response to reovirus, which requires RIG-I and MDA5 [43]. After RNA binding by RIG-I and MDA5, these proteins associate with the CARD-containing adapter termed IFN-β promoter stimulator-1 (IPS-1) or mitochondrial antiviral signaling (MAVS) protein. Subsequent activation of intermediates of the pathway (e.g., TNF-receptor-associated factor [TNAF]-3), which are common to other innate immunostimulatory mechanisms, result in stimulation of type I interferon (IFN) genes [33–36]. Ultimately, there is increased secretion of IFN-α and IFN-β, which have a major antiviral effect. Release of cytokines such as tumor necrosis factor (TNF)-α, interleukin (IL)-1β, and IL-6 recruit immune cells to the sites of infection and induce inflammation [33]. There is also release of proinflammatory cytokines and chemokines, with increased synthesis of co-stimulatory molecules, such as CD40, CD80, and CD86, which contribute to T cell responses. As may be expected, viruses have evolved mechanisms of subverting functioning of PRRs to evade effects of the innate immune response. Examples are the disabling of MDA5 by a poliovirus-encoded protease [44] and the proteolytic inactivation of MAVS by the HCV NS3/4A protease [45].

TLRs, which acquired their name because of similarity to the protein product of the *Drosophila* Toll gene, are membrane-associated PRRs. Functions of the TLRs are governed by their structure and their intracellular localization. At least 10 TLRs have been identified in humans [36]. These PRRs are capable of recognizing components of viruses that are located outside of cells and within cytoplasmic vacuoles. TLR2, TLR3, TLR4, TLR7, and TLR9 are particularly involved with recognizing viral motifs. TLR2 and TLR4 are located on plasma membranes and recognize surface viral proteins. Intracellular viral patterns, located within cytoplasmic vesicles, are recognized by TLR3, TLR7, and TLR9. Duplex RNA activates TLR3. Viral single-stranded RNA and DNA with unmethylated CpG elements activate signaling by TLR7 and TLR9. After activation, downstream signaling occurs when these PRRs interact with myeloid differentiation factor 88 (MyD88) or Toll IL-1 receptor homology domain-containing adapter inducing IFN-β (TRIF) in the case of TLR3. The signaling cascade culminates in activation of expression of inflammatory response genes, which is largely mediated by inhibitor of nuclear factor-κB (IκB), IκB kinases (IKKs), nuclear factor (NF)-κB, IFN regulatory factor (IRF)-3, and IRF7 [33–36]. There are several cross-talking intermediates in the pathway, such as the IL-1R-associated kinases (IRAK1, 2, 3, and 4), TNF receptor-associated factors, and mitogen-activated protein kinases. In addition, TLR- and RLR-activated pathways share molecular intermediates.

IL-1β and IL-18 mediate another important innate immunostimulatory mechanism during viral infection [46]. Unlike the TLR and RLR pathways, IL-1β and IL-18 production is activated by caspase-1-mediated cleavage of the cytokine precursors [47,48]. The inflammasome subcellular organelle provides a platform for this cleavage reaction [49]. Several inflammasome proteins, including the NOD leucine-rich repeat receptors and downstream apoptosis-associated speck-like protein containing a CARD mediate the activation of caspase-1 to process pro-IL-1β to form mature and biologically active pro IL-1β [36,50]. The precise role of the inflammasome in mediating innate immune responses to viral infections is not entirely clear. However, evidence indicates that double-stranded RNA and adenoviral DNA are capable of activating pro IL-1β processing. Moreover, evidence is beginning to emerge that DNA viruses may modulate inflammasome function to evade innate immune response mechanisms [51].

1.4.2 Adaptive Immunity

Virus-mediated activation of the adaptive immune response is largely dependent on adequate stimulation of an innate response [33]. The link between innate and adaptive immune responses has been confirmed by studies using various live and killed viral preparations as immunogenic vaccines [52]. Activation of the innate immune response using adjuvants is important to enhance the immunogenicity of vaccines (Chapter 11). Moreover, highly purified recombinant viral antigens, which do not activate innate immunity, are poor immunogens. However, it is difficult to generalize about the mechanisms. The innate immunostimulatory pathways are redundant, and there is a lack of selectivity of particular viral components for the activation mechanisms. Different viruses activate different components of the innate immune response. In addition, because some viruses use mechanisms of subverting the innate response, effects on adaptive immunostimulation will also be affected.

The adaptive immune response is initiated when an immature dendritic cell ingests a pathogen component, such as a viral protein [53,54] (Figure 1.3). This happens at the site of an infection and may occur simultaneously with innate immunostimulation of an immature dendritic cell. Dendritic cells' maturation ensues during their transport to a draining lymph node. The mature dendritic cell processes the antigenic peptide and displays it during the process of antigen presentation to naïve T cells. Therefore, mature dendritic cells are often referred to as antigen-presenting cells (APCs). The presentation of the antigens on the surface of APCs occurs through loading onto MHC class I or II molecules. When the naïve T cell recognizes a presented antigen through interaction with its own T cell antigen receptor, the cell responds by clonal proliferation and differentiation into an effector T cell. The type of T cell response that is elicited is dependent on the MHC molecule that is presenting the antigen. The two main types of T cell are $CD4^+$ and $CD8^+$. MHC class I is responsible for stimulating $CD8^+$ T cells whereas $CD4^+$ T cells are activated by peptides presented on MHC class II molecules. In addition to the interaction that takes place between the MHC molecules and the T cell antigen receptor, other molecules on the surface of APCs are also important to achieve optimal naïve T cell stimulation. These include co-stimulatory molecules, such as CD40,

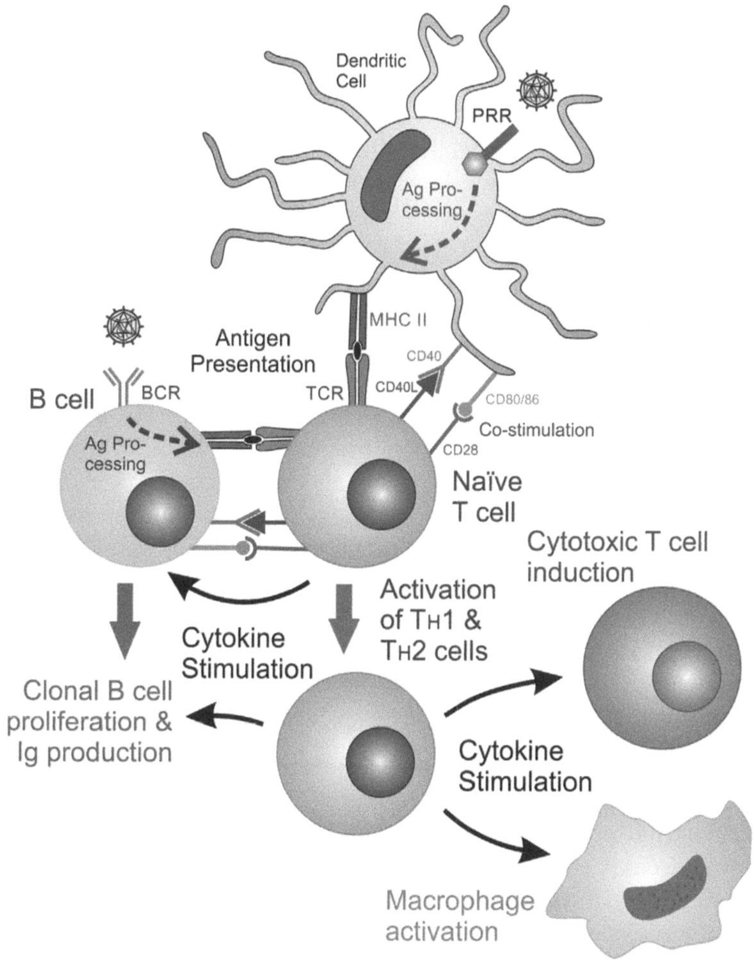

FIGURE 1.3 Schematic illustration of essential processes of the arms of the adaptive immune response to viral infection.

Adaptive immunity is initiated by the processing of pathogen-derived antigens by the dendritic cell. The mature dendritic cells, or APCs, then present a peptide on MHC class II molecules. Naïve T cells recognize the presented antigens through interaction with their own TCRs. Co-stimulation also occurs and involves binding of CD40, CD80, and CD86 on the dendritic cells with CD40L and CD28 on the T cell. Activation of Th1 and Th2 cells stimulates cytokine release with resultant cytotoxic T cell induction, macrophage activation, and clonal B cell proliferation with Ig production. B cells may themselves function as APCs by processing viral antigens, derived from internalization after binding to BCRs, and presentation on MHC class II molecules. Additional effects of Th1 and Th2 cells (e.g., as mediators of allergy) are not depicted in this diagram.

CD80, and CD86, which are induced during activation of the innate immune response (see above, section 1.4.1, and reviewed in ref. [55]). Receptors for these co-stimulatory molecules, such as CD40L for CD40 and CD28 for CD 80/86, are located on the surface of the naïve T cells (Figure 1.3).

$CD4^+$ T helper cells, which may be Th1 or Th2, induce $CD8^+$ cytotoxic T cells and maturation of B cells to form clones of antibody- (immunoglobulin [Ig]-) producing plasma cells and they improve macrophage and neutrophil phagocytic function [56]. The Th1 and Th2 subclasses of $CD4^+$ helper cells differ according to the cytokines that they secrete. Th1 cells produce IL-2 and IFN-γ whereas Th2 cells secrete IL-4, IL-10, and IL-13. Th1 cells are required for effective cell-mediated immunity against intracellular viral pathogens and for mounting of antibody-mediated immunity. Th2 cells differ in that their effects are mainly against extracellular pathogens through stimulating the production of IgM, IgG, and IgE subclasses of antibodies, and they are responsible for triggering allergic responses. HIV-1 infection occurs in $CD4^+$ cells and leads to depletion of the immunological helper cell function that ultimately leads to AIDS. Activation of $CD8^+$ cytotoxic T cells leads to their migration to a site of infection where they are capable of killing pathogen-infected cells. Memory T and B cells remain after an infection has resolved. These memory cells do not contribute to the initial adaptive immune response, but they are responsible for maintaining memory for particular pathogens. When re-exposure occurs, memory T and B cells are activated more rapidly than naïve lymphocytes. Consequently, a more effective immune response is mounted, which prevents establishment of a repeat infection. This secondary immune response, which is dependent on memory cells, underlies the use of vaccines to prevent infections.

Cytotoxic effects of T cells on pathogen-infected cells are mediated by interaction of the selected surface T cell receptor (TCR) molecules with MHC class I-associated viral peptides present on the surface of infected cells [57]. CD8 is required for this interaction, which induces release of cytotoxins, such as granzymes and perforins from the cytotoxic T cells. After penetrating the target cells, the caspase pathway is typically activated to induce programmed cell death [58].

Activation of the humoral immune response to viral infections occurs as a result of B cell activation (reviewed in ref. [54]). These cells are produced in the bone marrow and display B cell receptors (BCRs) on their surfaces. Triggering of B cell maturation occurs when the BCR encounters an antigen on an APC that it recognizes. The antigen is then engulfed and digested. Thereafter, antigenic peptides are exposed by display on MHC class II molecules. Recognition of antigens by TCRs on Th cells [56], together with cytokine release and interaction of co-stimulatory molecules with their ligands on B and Th cells, results in stimulation of the B cells to mature into antibody-producing plasma cells. Thus, B cells function as APCs and as antibody-producing cells. Antibodies interact with their cognate viral antigens to block infection of cells, and in concert with cell-mediated immunity, they generally provide an effective mechanism of countering viral infection.

The process of generating antibodies with specific antigen-binding properties, which is similar to the case with forming antigen-specific TCRs, involves somatic recombination of antibody-coding genes (reviewed in ref. [59]). This mechanism enables the generation of an enormous repertoire of antigen-binding proteins, which is estimated to be approximately 3×10^{11}. It is from this library that molecules that bind to specific antigens are selected. Segments encoding the Ig heavy chains comprise Variable (V), Diversity (D), Joining (J), and Constant (C) regions (Figure 1.4) whereas κ and λ light chains are formed by V and J segments. In maturing B cells, somatic recombination between VDJ heavy chain and κ and λ light chain segments give rise to the heavy and light chains of an antibody. Recombination-activating gene-1 (RAG1) and RAG2 proteins, together with several other proteins including ligases and DNA kinases, facilitate the process by making double-strand breaks at specific recognition sequences (recombination signal sequences [RSSs]) [60]. One copy of each type of recombination is present in a given lymphocyte, and the Ig is

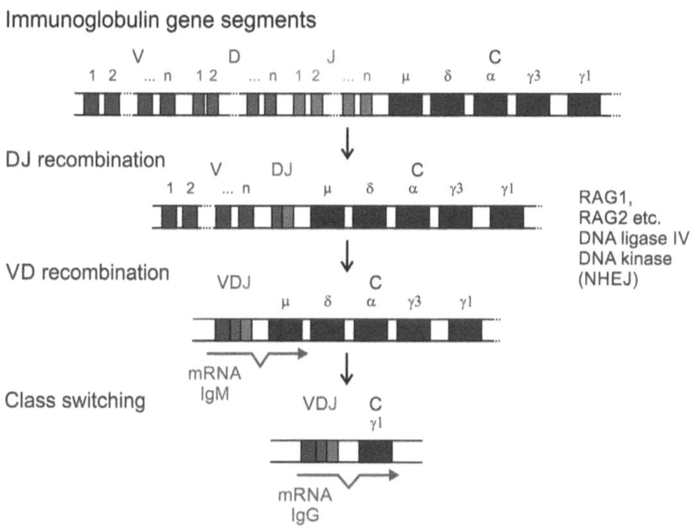

FIGURE 1.4 Rearrangement of Ig heavy-chain gene segments during assembly of Igs that bind viral proteins.

The V, D, J, and C segments recombine to form Ig heavy-chain sequences that contribute to antibody binding to specific viral epitopes. Recombination involves formation of double strand breaks at RSSs in the DNA between the gene segments, and then joining of blunt ends of the DNA by the nonhomologous end-joining (NHEJ) repair mechanism. RAG1 and RAG2 are essential for the process, and their action is complemented by other cellular enzymes that form part of the NHEJ machinery (e.g., DNA ligase IV). Initial recombination is between D and J segments and is then followed by recombination between V and D elements. IgM, derived from the C μ segment, is typically produced after the initial eliciting of an antibody response. Thereafter, switching to other antibody isotypes (e.g., IgG from the C γ1 segment) occurs and results from recombination with excision of intervening isotype-encoding C sequences.

expressed on the surface of the immature B cells. Pentameric IgM is typically produced early after exposure to a virus, and IgG is formed later during an infection. Switching in the type of antibody produced by a B cell is mediated by excision of sequences encoding the M chain and ligation to G constant heavy chain DNA. IgA plays a role in providing immunity at mucosal surfaces, such as in the genitourinary tract. TCRs comprise α and β chains, which are analogous to the Ig light and heavy chains, respectively. The α chains undergo VJ recombination whereas β chains undergo VDJ recombination to generate a range of TCRs with diverse antigen binding properties.

1.4.2.1 Antibody-Mediated Neutralization of Viral Infections

The host's humoral response to viral exposure plays an important role in neutralizing infections and occurs through interaction of antibodies with proteins located on the surface of viral particles [67]. In the case of enveloped viruses, the proteins are embedded within the surrounding lipid bilayer and on the surface protein shell of nonenveloped viral particles. Binding of the antibodies inhibits fusion to target cells in the case of enveloped particles and cellular penetration by nonenveloped virions. Blocking of surface proteins' interaction with cellular receptors and disrupting the mechanisms of fusion of enveloped viral particles with the target cells counter the viral infection. Another mechanism of neutralizing viral infection, particularly relevant to nonenveloped particles, is by antibody-directed activation of tripartite motif-containing-21-mediated intracellular proteasomal degradation of the viral capsids [68].

Variability of the surface proteins, which is conferred by error-prone replication of viral nucleic acids (see below, section 1.5), may enable virions to escape neutralization by the host's humoral immune response. Although this is a serious obstacle to eliminating viral infections, identification of broadly neutralizing antibodies (bNAbs) has been a significant and interesting development [67]. bNAbs are capable of inhibiting viral infection through an interaction with conserved regions of a broad range of different viral isolates. Characterization of bNAbs targeting HIV-1 is facilitating advances in therapeutic and prophylactic immune-based antiviral strategies (Chapters 8 and 11). The ability of a virus to mutate successfully and escape the humoral response is dependent on maintenance of fitness and essential structural elements of the surface proteins. An interesting consideration pertaining to viral escape concerns the differences in sizes that may exist between antibodies' complementarity determining regions (CDRs) and the virions' receptor binding sites [67]. The CDRs are typically larger than the receptor binding domains. As a result, the efficacy of an antibody may be compromised by its interaction with more variable viral sequences that are peripheral to the essential and vulnerable viral particles' receptor binding sites. Transient exposure of the fusion machinery or shielding by surface glycoproteins, such as is the case with HIV-1, further limits the ability of antibodies to neutralize viral infections.

1.4.2.2 *Modulation of the Adaptive Immune Response by Regulatory T Cells*

Regulatory T cells, or Tregs, play a role in suppressing peripheral immune responses, which are normally important to induce tolerance to self and modulate the antiviral immune responses [61,62]. Tregs influence the immune response through various mechanisms, which include cytokine-mediated disruption of function and maturation of APCs, induction of apoptosis, and suppression of effector T cells. Tregs are characterized by expression of the forkhead box protein-3 transcription factor. They may be derived from two lines: natural Tregs (nTregs) or induced Tregs (iTregs) [63]. iTregs are produced after differentiation of $CD4^+$ cells and are particularly important in the modulation of immune responses to viral infections. nTregs are derived from the thymus and have their main function in modulating recognition of self. Direct interaction of viruses with Tregs may influence the pathogenesis of an infection, and manipulating the function of Tregs has therapeutic potential [64]. When viruses interact directly with Tregs, as is the case with HCV and HIV-1, regulatory function may be modulated to influence the outcome of infection. Conflicting inhibitory and stimulatory effects of HIV-1 on immune response to the virus are examples. By attenuating $CD4^+$ T cell activation, Tregs may diminish the numbers of HIV-1-infectable cells [65]. Conversely, inhibition of the T cell response reduces the antiviral immune response [66].

1.5 MUTATION OF VIRUSES

The occurrence of mutations in viral genomes is an important feature that has a bearing on the success of antiviral treatments and immune-based elimination of infections. RNA viruses are more prone to mutations than are DNA viruses. This is largely a result of the highly error-prone nature of their replicating enzymes. The error rate of DNA viruses is similar to that of eukaryotic cells and has been calculated to be approximately 10^{-8} to 10^{-11} errors per nucleotide [69]. Because replicating enzymes of RNA viruses do not have proofreading capability, their error rates are considerably higher, and they have been reported to range between 10^{-3} and 10^{-4} errors per incorporated nucleotide. Mutations may also arise after recombination events that entail the exchange of genetic material. When a genetic change causes a deleterious effect on viral replication, these mutants are usually lost. However, mutations that permit replication may lead to the selection of variants that evade therapeutic and immunological pressure.

The plasticity of viral genomes is a factor that influences the frequency of emergence of mutations. For example, HBV has an unusually compact genome with most regions encoding multiple functions, such as the sequence encoding the envelope proteins, which overlaps with the polymerase open reading frame

(ORF; Chapter 6). Therefore, mutations in the surface ORF that permit viable replication of HBV are rare because they typically also compromise viral nucleic acid replication by Pol. Consequently, the vaccine against HBV, which comprises Surface protein, is very effective. Conversely, HIV-1, which has a highly plastic genome, readily generates immunoevasive replication competent variants, and vaccination against this virus has proven notoriously difficult to achieve (Chapters 8 and 11).

1.6 CONTROL OF VIRAL INFECTION

There are several interventions that may be used to limit viral spread. Methods depend on specific characteristics of viral infections, such as the epidemiology, mode of transmission, susceptibility to antiviral drugs, and whether effective vaccines are available. Each of these factors has a bearing on the most useful strategy that should be employed to limit spread of a viral infection.

1.6.1 Preventing Spread of Viral Infection

Insights into understanding the modes of transmission and epidemiology of particular viral infections may be used effectively to prevent spread of viral infections. Using practices of safe sex to prevent HIV-1 transmission is a widely understood important preventative measure. Likewise, avoidance of contaminated needles among intravenous drug users is important to prevent spread of HCV and HIV-1.

Vaccination as a means of immunoprophylaxis against viral infections is widely and successfully used. Administration of an immunogen to induce active immunization of individuals is a highly effective means of evoking T and B cell responses. Prevention of infection with measles, polio, rubella, mumps, and HBV is used in many countries within their expanded programs of immunization. These vaccines may be in the form of recombinant viral proteins (e.g., HBV surface proteins) or live attenuated or killed viral viruses (e.g., polio). Gene-based immunization is now emerging as a useful means of prophylaxis or treatment of several different viral infections (Chapter 11). After exposure to a nonpathogenic vaccine, memory T and B cells are formed. If an individual is subsequently exposed to the pathogenic virus, then a rapid immune response is mounted. This reaction is more effective than that which occurs after exposure of an unimmunized individual, and it prevents establishment of the infection.

Passive immunity involves administration of antibodies, derived from a different host, to prevent viral infection. An example is the use of hyperimmune globulin from individuals who are immune to HBV to prevent mother-to-child transmission of the virus during childbirth. The method is more complicated, short lived, and expensive than conventional vaccination; therefore, it is only used with specific indications. An interesting development that may have relevance for passive immunization using gene therapy has been based on the identification of bNAbs to HIV-1. Generation of an immunogen that would be suitable for prevention of HIV-1 infection has been notoriously difficult. The feasibility of using vectors to deliver DNA encoding bNAbs has interesting possibilities for prevention of HIV-1 transmission (Chapters 8 and 11). Adequate delivery of the antibody-encoding sequences and their sustained expression will of course be a requirement for the approach to be successful.

1.6.2 Antiviral Drugs

Developing selective treatment of viral infections using small molecules is a highly active field of work. The essential underlying principle governing the design of small-molecule antivirals is that these drugs should interact specifically with viral targets to disrupt replication without influencing the host cell. In addition, they should pose a high genetic barrier to resistance to prevent emergence of viral escape mutants that negate the drug's efficacy. To address problems of a low genetic barrier to resistance, combinations of drugs that have different mechanisms of action, such as is used in combination antiretroviral therapy for HIV-1 treatment, may be used.

The first small-molecule antivirals became available approximately 50 years ago when idoxuridine and methisazone were licensed, respectively, for the treatment of keratitis caused by herpes virus and prevention of smallpox virus infection (reviewed in refs [70,71]). Idoxuridine is a thymidine analog that has the methyl group on C5 substituted for an iodine atom. Disruption to gene coding is thought to occur after incorporation of idoxuridine into herpes virus DNA. Subsequent to licensing of idoxuridine, there has been rapid development in advancing nucleoside and nucleotide analogs to treat antivirals. These drugs are currently a major group of antiviral chemotherapeutics and are widely used in the treatment of viral infections of considerable public health importance, such as HBV [72] and HIV-1 [73]. Nucleoside and nucleotide analogs function by taking advantage of the highly error-prone nature of viral polymerases, particularly reverse transcriptases and RNA polymerases, and their lack of proofreading capabilities. Drug incorporation during viral replication results in disruption of viral gene expression.

Advances in developing small molecule antiviral drugs have been facilitated by improvement of high-resolution structural analysis of viral proteins. This burgeoning field is now assuming greater importance and has enabled rational design of new drugs. Knowledge of the three-dimensional structures of viral proteins enables identification of candidate drugs that have affinity for specific sites within the target, and it expands the range of potential viral targets. A particularly important example has been the development of small molecules that target HIV-1 protease. This viral enzyme is responsible for the maturation of Gag and Pol polypeptide precursors to form mature proteins required for virion formation [74]. The competitive inhibitors of HIV-1 protease comprise cleavage-resistant peptide mimics of the transition state of the viral protein substrates [75]. Since discovery of the first HIV-1 protease inhibitor, saquinavir, many other effective drug candidates have been developed. Similar approaches have been used successfully to identify protease inhibitors for other viruses such as HCV [23]. The mechanisms of action of small-molecule antivirals are also continually broadening to include new and different sites of action. For example, in addition to inhibition of viral protease and reverse transcriptase, HIV-1 treatment regimens may now include drugs that act as inhibitors of viral entry and proviral integration.

Gene therapy for viral infection is an emerging treatment approach that should add significantly to the available means of controlling viral infections. Because viral gene therapies are based on rational design that is essentially based on knowledge of viral nucleic acid sequences, the approach may be particularly useful to countering new emerging viral pathogens. Moreover, gene therapy broadens the range of antiviral strategies by enabling the inactivation of viral targets that have previously been inaccessible.

1.6.3 Immunostimulatory Drugs

In their seminal study that was published in 1957, Isaacs and Lindenmann described the identification and antiviral efficacy of IFN [76]. At the time, it had been known that treatment of cells with inactivated influenza virus was capable of interfering with growth of the live virus. Studies performed on chick chorio-allantoic membrane led Isaacs and Lindenmann to identify a released factor, termed "the interferon," that was responsible for inhibition of live influenza virus replication. It was subsequently shown that other microorganisms could stimulate release of IFN. In addition, there are different types of IFN, and they have been classified as type I (e.g., IFN-α and IFN-β) or type II (IFN-γ), which each interact with specific receptor types. Binding to the receptor results in activation of the JAK/STAT pathway to cause downstream signaling to activate genes of the innate immune response [33,34,36]. There is variation in the antiviral efficacy of different IFNs, and IFN-α has proved to be the most effective as an antiviral therapeutic. Currently, IFN-α is produced in large amounts using recombinant techniques and has been widely used for treatment of HBV and HCV infections. Modification by conjugation with polyethylene glycol has successfully been used to prolong the half-life of IFN-α and thereby prolong the duration of the antiviral action.

Although derivatives of IFN-α are the standard treatment for HBV [72] and HCV [77] infections, antiviral efficacy of other cytokines may also have utility. An example is IL-7, which promotes maturation of precursor B and T cells and may have use in the treatment of various viral infections [78–80].

1.7 METHODS OF USING GENE THERAPY TO TREAT VIRAL INFECTIONS

As with the development of most new therapies, advancing gene therapy for viral infections involves a stepwise series of investigations. The process requires preliminary analysis in simple models and then progression to more complex disease simulations before embarking on trials in humans (Figure 1.5). Candidate drugs that show good efficacy in the simple models are often not effective in humans. It is only a small proportion of candidate drugs that meet all requirements of safety and efficacy to proceed to clinical trial. Analysis may be started using computer simulations, but it is typically initiated on cells in culture. From there, lead compounds are taken forward for testing in various disease models in vivo and then in clinical trial.

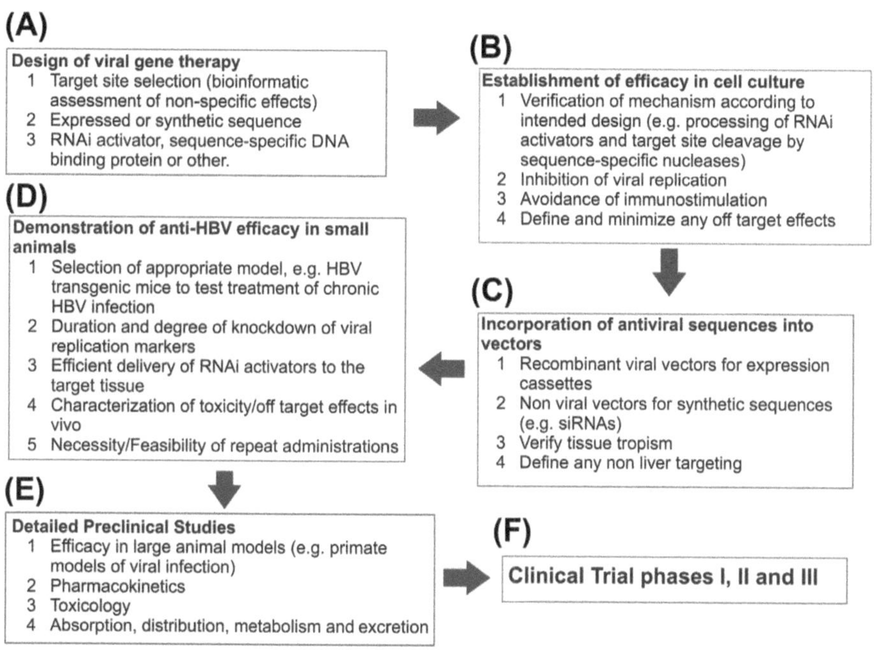

FIGURE 1.5 Stages in development of gene therapy for viral infections.
The essential stepwise investigations (A–F) that are required to develop viral gene therapies are indicated together with considerations that are important for advancement at each stage.

Understandably, most approaches to gene therapy of virus infections have entailed inhibiting the function of viral gene targets (Table 1.1). Achieving effective, sustained, and specific silencing is vitally important for therapeutic benefit and overcoming the challenges associated with meeting these objectives is the major focus of research in the field. In addition to direct inhibition of viral gene targets, disabling of HFs that are required for viral infection has also been used to counter viral infections. Absence of a toxic effect of HF inhibition is crucially important for the success of this approach. Using gene therapy to introduce targeted mutation of the CCR5 co-receptor for HIV-1 is an example of therapeutic inhibition of an HF [81,82]. Naturally occurring *CCR5* mutations do not appear to cause deleterious effects, and the resistance of these individuals to HIV-1 infection has prompted investigation of targeted *CCR5* disruption as a gene therapy for HIV-1 infection. Use of recombinant viruses as vaccine vectors is another rapidly developing field in the use of gene therapy for viral infections [83–85] (Chapter 11). These recombinant viral vectors have interesting features that contribute to inducing more effective immune responses, particularly through induction of a T-cell immune response.

Recent improvements in technologies enabling rational design of inhibitors of gene function have contributed significantly to developing viral gene therapy. Particularly important have been advances of RNAi (Chapter 2) and the targeted cleavage of DNA sequences using so-called designer nucleases (Chapter 3). Targets that were previously thought to be "undruggable" have been shown to be amenable to potentially therapeutic functional inhibition. Although significant challenges remain before viral gene therapy becomes a reality, the field is poised at an exciting stage when exciting developments are current and are likely to be imminent.

REFERENCES

[1] Friedmann T, Roblin R. Gene therapy for human genetic disease? Science 1972;175(4025):949–55.

[2] Munyon W, Kraiselburd E, Davis D, Mann J. Transfer of thymidine kinase to thymidine kinaseless L cells by infection with ultraviolet-irradiated herpes simplex virus. J Virol 1971;7(6):813–20.

[3] Merril CR, Geier MR, Petricciani JC. Bacterial virus gene expression in human cells. Nature 1971;233(5319):398–400.

[4] Crystal RG. Transfer of genes to humans: early lessons and obstacles to success. Science 1995;270(5235):404–10.

[5] Miller AD. Human gene therapy comes of age. Nature 1992;357(6378):455–60.

[6] Mulligan RC. The basic science of gene therapy. Science 1993;260(5110):926–32.

[7] Bergh O, Borsheim KY, Bratbak G, Heldal M. High abundance of viruses found in aquatic environments. Nature 1989;340(6233):467–8.

[8] Wommack KE, Colwell RR. Virioplankton: viruses in aquatic ecosystems. Microbiol Mol Biol Rev 2000;64(1):69–114.

[9] Suttle CA. Viruses in the sea. Nature 2005;437(7057):356–61.

[10] Benner SA. Defining life. Astrobiology 2010;10(10):1021–30.

[11] Raoult D. What makes a virus a virus: reply from Raoult and Forterre. Nat Rev Microbiol 2008;6(8):643.

[12] Raoult D, Forterre P. Redefining viruses: lessons from Mimivirus. Nat Rev Microbiol 2008;6(4):315–9.

[13] Wolkowicz R, Schaechter M. What makes a virus a virus?. Nat Rev Microbiol 2008;6(8):643. [author reply].

[14] Bamford DH, Grimes JM, Stuart DI. What does structure tell us about virus evolution? Curr Opin Struct Biol 2005;15(6):655–63.

[15] Abrescia NG, Bamford DH, Grimes JM, Stuart DI. Structure unifies the viral universe. Annu Rev Biochem 2012;81:795–822.

[16] Wessner D. The origins of viruses. Nat Educ 2010;3:37.

[17] Lander ES, Linton LM, Birren B, Nusbaum C, Zody MC, Baldwin J, et al. Initial sequencing and analysis of the human genome. Nature 2001;409(6822):860–921.

[18] Raoult D, Audic S, Robert C, Abergel C, Renesto P, Ogata H, et al. The 1.2-megabase genome sequence of Mimivirus. Science 2004;306(5700):1344–50.

[19] Prangishvili D, Forterre P, Garrett RA. Viruses of the Archaea: a unifying view. Nat Rev Microbiol 2006;4(11):837–48.

[20] Villarreal LP, DeFilippis VR. A hypothesis for DNA viruses as the origin of eukaryotic replication proteins. J Virol 2000;74(15):7079–84.

[21] Janssen HL, Reesink HW, Lawitz EJ, Zeuzem S, Rodriguez-Torres M, Patel K, et al. Treatment of HCV infection by targeting microRNA. N Engl J Med 2013;368(18):1685–94.

[22] Kieft JS. Viral IRES RNA structures and ribosome interactions. Trends Biochem Sci 2008;33(6):274–83.

[23] Scheel TK, Rice CM. Understanding the hepatitis C virus life cycle paves the way for highly effective therapies. Nat Med 2013;19(7):837–49.

[24] Gelderblom HR. Structure and classification of viruses. In: Baron S, editor. Medical microbiology. Galveston, TX, USA: University of Texas Medical Branch at Galveston; 1996.

[25] Muller R, Poch O, Delarue M, Bishop DH, Bouloy M. Rift Valley fever virus L segment: correction of the sequence and possible functional role of newly identified regions conserved in RNA-dependent polymerases. J Gen Virol 1994;75(Pt 6):1345–52.

[26] Ganser BK, Li S, Klishko VY, Finch JT, Sundquist WI. Assembly and analysis of conical models for the HIV-1 core. Science 1999;283(5398):80–3.

[27] Li S, Hill CP, Sundquist WI, Finch JT. Image reconstructions of helical assemblies of the HIV-1 CA protein. Nature 2000;407(6802):409–13.

[28] Zhao G, Perilla JR, Yufenyuy EL, Meng X, Chen B, Ning J, et al. Mature HIV-1 capsid structure by cryo-electron microscopy and all-atom molecular dynamics. Nature 2013;497(7451):643–6.

[29] Boldogh I, Albrecht T, Porter DD. Persistent viral infections. In: Baron S, editor. Medical microbiology. Galveston, TX, USA: University of Texas Medical Branch at Galveston; 1996.

[30] Rehermann B. Pathogenesis of chronic viral hepatitis: differential roles of T cells and NK cells. Nat Med 2013;19(7):859–68.

[31] del Val M, Hengel H, Hacker H, Hartlaub U, Ruppert T, Lucin P, et al. Cytomegalovirus prevents antigen presentation by blocking the transport of peptide-loaded major histocompatibility complex class I molecules into the medial-Golgi compartment. J Exp Med 1992;176(3):729–38.

[32] Savard M, Bélanger C, Tardif M, Gourde P, Flamand L, Gosselin J. Infection of primary human monocytes by Epstein-Barr virus. J Virol 2000;74(6):2612–9.

[33] Aoshi T, Koyama S, Kobiyama K, Akira S, Ishii KJ. Innate and adaptive immune responses to viral infection and vaccination. Curr Opin Virol 2011;1(4):226–32.

[34] Iwasaki A, Medzhitov R. Regulation of adaptive immunity by the innate immune system. Science 2010;327(5963):291–5.

[35] Palm NW, Medzhitov R. Pattern recognition receptors and control of adaptive immunity. Immunol Rev 2009;227(1):221–33.

[36] Takeuchi O, Akira S. Innate immunity to virus infection. Immunol Rev 2009;227(1):75–86.

[37] Saito T, Hirai R, Loo YM, Owen D, Johnson CL, Sinha SC, et al. Regulation of innate antiviral defenses through a shared repressor domain in RIG-I and LGP2. Proc Natl Acad Sci USA 2007;104(2):582–7.

[38] Yoneyama M, Kikuchi M, Natsukawa T, Shinobu N, Imaizumi T, Miyagishi M, et al. The RNA helicase RIG-I has an essential function in double-stranded RNA-induced innate antiviral responses. Nat Immunol 2004;5(7):730–7.

[39] Sumpter Jr R, Loo YM, Foy E, Li K, Yoneyama M, Fujita T, et al. Regulating intracellular antiviral defense and permissiveness to hepatitis C virus RNA replication through a cellular RNA helicase, RIG-I. J Virol 2005;79(5):2689–99.

[40] Kato H, Takeuchi O, Sato S, Yoneyama M, Yamamoto M, Matsui K, et al. Differential roles of MDA5 and RIG-I helicases in the recognition of RNA viruses. Nature 2006;441(7089):101–5.

[41] Melchjorsen J, Jensen SB, Malmgaard L, Rasmussen SB, Weber F, Bowie AG, et al. Activation of innate defense against a paramyxovirus is mediated by RIG-I and TLR7 and TLR8 in a cell-type-specific manner. J Virol 2005;79(20):12944–51.

[42] Gitlin L, Barchet W, Gilfillan S, Cella M, Beutler B, Flavell RA, et al. Essential role of mda-5 in type I IFN responses to polyriboinosinic:polyribocytidylic acid and encephalomyocarditis picornavirus. Proc Natl Acad Sci USA 2006;103(22):8459–64.

[43] Kato H, Takeuchi O, Mikamo-Satoh E, Hirai R, Kawai T, Matsushita K, et al. Length-dependent recognition of double-stranded ribonucleic acids by retinoic acid-inducible gene-I and melanoma differentiation-associated gene 5. J Exp Med 2008;205(7):1601–10.

[44] Barral PM, Morrison JM, Drahos J, Gupta P, Sarkar D, Fisher PB, et al. MDA-5 is cleaved in poliovirus-infected cells. J Virol 2007;81(8):3677–84.

[45] Li XD, Sun L, Seth RB, Pineda G, Chen ZJ. Hepatitis C virus protease NS3/4A cleaves mitochondrial antiviral signaling protein off the mitochondria to evade innate immunity. Proc Natl Acad Sci USA 2005;102(49):17717–22.

[46] van de Veerdonk FL, Netea MG, Dinarello CA, Joosten LA. Inflammasome activation and IL-1beta and IL-18 processing during infection. Trends Immunol 2011;32(3):110–6.

[47] Cerretti DP, Kozlosky CJ, Mosley B, Nelson N, Van Ness K, Greenstreet TA, et al. Molecular cloning of the interleukin-1 beta converting enzyme. Science 1992;256(5053):97–100.

[48] Thornberry NA, Bull HG, Calaycay JR, Chapman KT, Howard AD, Kostura MJ, et al. A novel heterodimeric cysteine protease is required for interleukin-1 beta processing in monocytes. Nature 1992;356(6372):768–74.

[49] Martinon F, Burns K, Tschopp J. The inflammasome: a molecular platform triggering activation of inflammatory caspases and processing of proIL-beta. Mol Cell 2002;10(2):417–26.

[50] Kanneganti TD, Body-Malapel M, Amer A, Park JH, Whitfield J, Franchi L, et al. Critical role for cryopyrin/Nalp3 in activation of caspase-1 in response to viral infection and double-stranded RNA. J Biol Chem 2006;281(48):36560–8.

[51] Amsler L, Malouli D, DeFilippis V. The inflammasome as a target of modulation by DNA viruses. Future Virol 2013;8(4):357–70.

[52] Coffman RL, Sher A, Seder RA. Vaccine adjuvants: putting innate immunity to work. Immunity 2010;33(4):492–503.

[53] Banchereau J, Steinman RM. Dendritic cells and the control of immunity. Nature 1998;392(6673):245–52.

[54] Zabel F, Kündig TM, Bachmann MF. Virus-induced humoral immunity: on how B cell responses are initiated. Curr Opin Virol 2013;3(3):357–62.

[55] Gool SW, Vandenberghe P, Boer Md, Ceuppens JL. CD80, CD86 and CD40 provide accessory signals in a multiple-step T-cell activation model. Immunol Rev 1996;153(1):47–83.

[56] Mosmann T, Coffman R. TH1 and TH2 cells: different patterns of lymphokine secretion lead to different functional properties. Annu Rev Immunol 1989;7(1):145–73.

[57] Yewdell J, Anton LC, Bacik I, Schubert U, Snyder HL, Bennink JR. Generating MHC class I ligands from viral gene products. Immunol Rev 1999;172(1):97–108.

[58] Trapani JA, Smyth MJ. Functional significance of the perforin/granzyme cell death pathway. Nat Rev Immunol 2002;2(10):735–47.

[59] Schatz DG, Oettinger MA, Schlissel MS. V(D)J recombination: molecular biology and regulation. Annu Rev Immunol 1992;10(1):359–83.

[60] Gellert M. V(D)J recombination: RAG proteins, repair factors, and regulation*. Annu Rev Biochem 2002;71(1):101–32.

[61] Scheffold A, Murphy KM, Hofer T. Competition for cytokines: T(reg) cells take all. Nat Immunol 2007;8(12):1285–7.

[62] Vignali DA, Collison LW, Workman CJ. How regulatory T cells work. Nat Rev Immunol 2008;8(7):523–32.

[63] Curotto de Lafaille MA, Lafaille JJ. Natural and adaptive foxp3+ regulatory T cells: more of the same or a division of labor? Immunity 2009;30(5):626–35.

[64] Veiga-Parga T, Sehrawat S, Rouse BT. Role of regulatory T cells during virus infection. Immunol Rev 2013;255(1):182–96.

[65] Eggena MP, Barugahare B, Jones N, Okello M, Mutalya S, Kityo C, et al. Depletion of regulatory T cells in HIV infection is associated with immune activation. J Immunol 2005;174(7):4407–14.

[66] Kinter A, McNally J, Riggin L, Jackson R, Roby G, Fauci AS. Suppression of HIV-specific T cell activity by lymph node CD25+ regulatory T cells from HIV-infected individuals. Proc Natl Acad Sci USA 2007;104(9):3390–5.

[67] Corti D, Lanzavecchia A. Broadly neutralizing antiviral antibodies. Annu Rev Immunol 2013;31:705–42.

[68] Mallery DL, McEwan WA, Bidgood SR, Towers GJ, Johnson CM, James LC. Antibodies mediate intracellular immunity through tripartite motif-containing 21 (TRIM21). Proc Natl Acad Sci USA 2010;107(46):19985–90.

[69] Fleischmann WRJ. Viral genetics. In: Baron S, editor. Medical microbiology. 4th ed. Galveston, TX, USA: University of Texas Medical Branch at Galveston; 1996.

[70] Prusoff WH. A review of some aspects of 5-iododeoxyuridine and azauridine. Cancer Res 1963;23:1246–59.

[71] Whitley RJ, Alford CA. Developmental aspects of selected antiviral chemotherapeutic agents. Annu Rev Microbiol 1978;32:285–300.

[72] Halegoua-De Marzio D, Hann H-W. Then and now: the progress in hepatitis B treatment over the past 20 years. World J Gastroenterol 2014;20(2):401–13.

[73] Chiao SK, Romero DL, Johnson DE. Current HIV therapeutics: mechanistic and chemical determinants of toxicity. Curr Opin Drug Discov Dev 2009;12(1):53.

[74] Brik A, Wong CH. HIV-1 protease: mechanism and drug discovery. Org Biomol Chem 2003;1(1):5–14.

[75] De Clercq E. The history of antiretrovirals: key discoveries over the past 25 years. Rev Med Virol 2009;19(5):287–99.

[76] Isaacs A, Lindenmann J. Virus interference. I. The interferon. Proc R Soc Lond Ser B Biol Sci 1957;147(927):258–67.

[77] Zhu Y, Chen S. Antiviral treatment of hepatitis C virus infection and factors affecting efficacy. World J Gastroenterol 2013;19(47):8963–73.

[78] Guillemard E, Nugeyre MT, Chene L, Schmitt N, Jacquemot C, Barre-Sinoussi F, et al. Interleukin-7 and infection itself by human immunodeficiency virus 1 favor virus persistence in mature CD4(+)CD8(−)CD3(+) thymocytes through sustained induction of Bcl-2. Blood 2001;98(7):2166–74.

[79] Milne CD, Fleming HE, Zhang Y, Paige CJ. Mechanisms of selection mediated by interleukin-7, the preBCR, and hemokinin-1 during B-cell development. Immunol Rev 2004;197:75–88.

[80] Napolitano LA, Stoddart CA, Hanley MB, Wieder E, McCune JM. Effects of IL-7 on early human thymocyte progenitor cells in vitro and in SCID-hu Thy/Liv mice. J Immunol 2003;171(2):645–54.

[81] Holt N, Wang J, Kim K, Friedman G, Wang X, Taupin V, et al. Human hematopoietic stem/progenitor cells modified by zinc-finger nucleases targeted to CCR5 control HIV-1 in vivo. Nat Biotechnol 2010;28(8):839–47.

[82] Perez EE, Wang J, Miller JC, Jouvenot Y, Kim KA, Liu O, et al. Establishment of HIV-1 resistance in $CD4^+$ T cells by genome editing using zinc-finger nucleases. Nat Biotechnol 2008;26(7):808–16.

[83] Draper SJ, Heeney JL. Viruses as vaccine vectors for infectious diseases and cancer. Nat Rev Microbiol 2010;8(1):62–73.

[84] Berezikov E, Chung WJ, Willis J, Cuppen E, Lai EC. Mammalian mirtron genes. Mol Cell 2007;28(2):328–36.

[85] Arbuthnot P. Harnessing RNA interference for the treatment of viral infections. Drug News Perspect 2010;23(6):341–50.

CHAPTER 2

Harnessing RNAi to Silence Viral Gene Expression

2.1 INTRODUCTION

RNA interference (RNAi) is a process that operates in metazoan cells to regulate gene expression. The mechanism typically involves gene silencing by short RNAs that have their effects by base pairing to cognate complementary mRNA sequences. The report on RNAi action in *Caenorhabditis elegans* by Fire, Melo, and their colleagues was a significant development in the field [1]. The finding was made after observing that introducing double-stranded RNA (dsRNA) into nematodes resulted in highly effective silencing of genes with homologous sequences. To the authors' surprise, they noticed that the inhibitory effectiveness was considerably more pronounced than when using the antisense or sense strands alone. Quelling and co-suppression, seemingly unrelated processes that had been described in plants and animals, were shown to operate by the same posttranscriptional RNAi-based gene silencing mechanism. Since publication of Fire and Melo's article in 1998, RNAi research has advanced very rapidly. In addition to providing important insights into the complexity of gene regulation, it has been shown that RNAi may be exploited to achieve posttranscriptional silencing of almost any intended gene target. The landmark study by Elbashir, Tuschl, and colleagues was the first to prove this [2]. They demonstrated that artificial synthetic short duplex RNAs acted as exogenous activators of RNAi to reprogram the pathway in mammalian cells. Since then, many studies have described use of exogenous RNAi activators that are effective against a wide variety of pathology-causing genes, including those expressed by viruses. DNA expression cassettes and synthetic RNA sequences are being used to silence gene expression, and both classes have been shown to have potential therapeutic utility. microRNAs (miRs) are the prototype endogenous activators of RNAi, and their mechanisms of action have directed the design of potentially therapeutic exogenous silencers. Mature miRs comprise 19–24 nt of noncoding RNA that exert posttranscriptional mammalian gene inhibition by partial base pairing to target mRNA. miRs control most human genes, and individual miRs may be capable of targeting in excess of 300 different transcripts [3]. Bioinformatic analysis suggests that there are more than 45,000 miR target sites within the human genome. Almost all cellular processes, including cell division, immune responses, programmed cell death,

Gene Therapy for Viral Infections. http://dx.doi.org/10.1016/B978-0-12-410518-8.00002-8

differentiation, and development of tissue-specific phenotypes, are subject to control by miRs [4]. There are also other short RNA sequences that are capable of regulating gene expression. An example is the piwi-interacting RNAs (piRNAs) that are enriched in gonadal tissue where they function to inhibit the effects of transposons [5]. Recently, piRNAs have been shown to have broader and important regulatory functions [reviewed in ref. [6]]. These include control of genome rearrangement, epigenetic programming, and possibly cancer etiology. Effects of piRNAs are not restricted to germ cells and also manifest in somatic cells. Understandably, because RNAi plays such an important role in controlling cell functions, the mechanisms that are involved in regulating RNAi are complex and subtle. Moreover, disruption to RNAi may lead to the emergence of disease states such as cancer.

Viruses themselves are capable of expressing miRs to influence the expression or viral and host genes (reviewed in ref. [7]). In addition, viruses may alter the functioning of host miRs to affect their own proliferation. As viral gene controlling elements, miRs have several features that are useful for viruses. miRs are

- Small and require minimal coding capacity within the viral genome,
- Nonimmunogenic,
- Capable of evolving rapidly to adapt to changes in the environment, and
- Able to regulate different target sequences with varying effects.

Given the importance of miRs for viral replication, it is not surprising that many viral miRs have been described. As with miRs detected in metazoan genomes, the number of similar viral candidate regulatory sequences is constantly expanding (http://crdd.osdd.net/servers/virmirna/index.html).

Understanding the details of the processes involved in RNAi-based gene silencing is very important for using the pathway efficiently to treat viral infections. In addition to posttranscriptional silencing by RNAi activators, transcriptional regulation of gene expression by RNA has recently emerged as an important topic [8]. Exploiting this mechanism also has potential application to therapy of viral infections.

2.2 BIOGENESIS OF miRs IN MAMMALIAN CELLS

Formation of mature miRs involves a series of consecutive steps. The process is initiated by transcription of RNAs that include miR precursors and culminates with silencing that is mediated by sequence-specific interaction between mature miR guide strands and cognates on mRNA targets (Figure 2.1). Various mechanisms control each stage of the process and control of gene expression by mature miRs is subjected to regulation by various influences (reviewed in refs [9–11]; Table 2.1). Exogenous RNAi activators are subject to similar control;

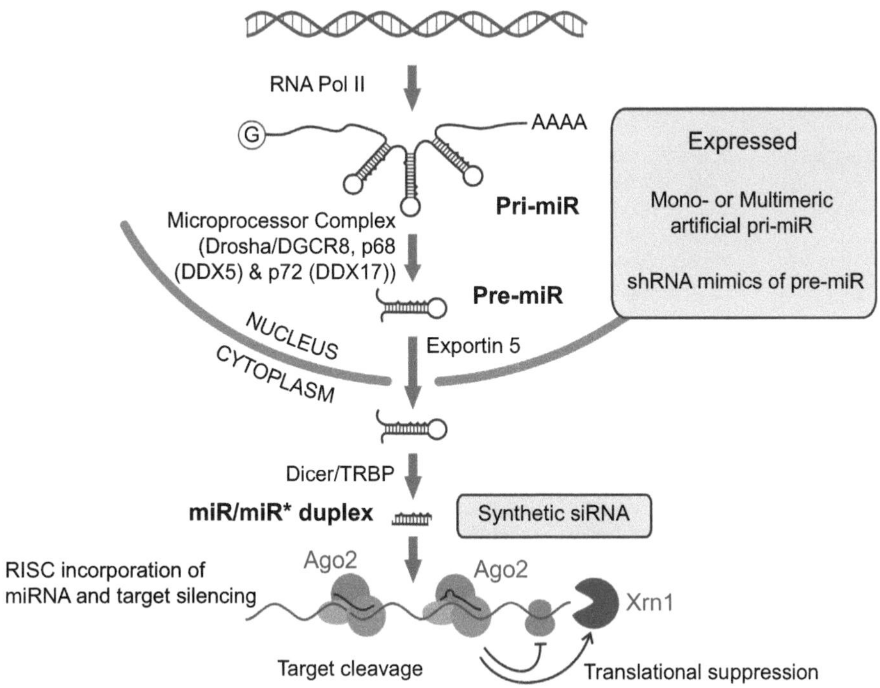

FIGURE 2.1 Natural biogenesis of miR.

Pol II transcripts, containing poly- or mono-cistronic miRs, are processed in the nucleus by the microprocessor complex to form pre-miR sequences of approximately 70 nt. After export to the cytoplasm, pre-miRs are cleaved by Dicer and TRBP to yield the mature miR duplex of approximately 22 bp. RISC activation follows incorporation of the duplex into the complex and removal of the passenger strand. The retained guide strand directs the complex to complementary mRNA targets. When complementarity between the target mRNA and guide strand is incomplete, translational suppression occurs, which involves mRNA degradation and impairment of ribosome function. Complete base pairing between the target and guide results in mRNA degradation. Expressed pri-miR and pre-miR sequences have been used to inhibit viral replication. Synthetic siRNAs, which enter the pathway at a more distal stage of the pathway, have also been used successfully against viruses.

therefore, understanding the mechanisms of natural miR biogenesis has relevance to antiviral therapeutic application of gene silencing.

2.2.1 Transcription of miR Precursors

Canonical miR biogenesis starts with Pol II-mediated transcription of sequences that include primary miRs (pri-miRs) [12]. These pri-miRs are typically found within 5′ capped and 3′ polyadenylated RNA and contain hairpin motifs with single-stranded flanking regions. Pri-miR sequences may be positioned within protein-coding mRNA, intergenic noncoding RNA, and within introns (reviewed in refs [13,14]). miRs may also be generated by noncanonical

TABLE 2.1 Mechanisms of controlling miR biogenesis and RNAi activity

Stage of RNAi pathway involved in canonical maturation of miRs	Examples and mechanisms of effects on miR biogenesis and RNAi activity
Transcription of pri-miR sequences	1. p53 activation of transcription of miR-34-containing sequences suppress malignant transformation and metastasis through effects on EMT. The mechanism involves Wnt and Snail1 signaling. 2. Nuclear presence of viral DNA that encodes pri-miRs per se may be a transcriptional mechanism that regulates miR production.
Microprocessor function	1. Drosha and DGCR8 regulate each other through protein stabilizing and posttranscriptional suppression. 2. Smad transcription factors regulate formation of mature miR-21 through sequence-specific interaction with pri-miR-21. The duplex binding motif resembles the DNA *cis*-elements required for transcription factor functioning. 3. p53 interacts with p68 to facilitate processing of tumor suppressor miRs such as miR-144. 4. Phosphorylated KSRP binds to pri-miRs in an apparently sequence-independent way to facilitate their processing. The mechanism is important for myoblast differentiation. 5. ER-α inhibits processing of pri-miRs that give rise to sequences which target genes that are also subject to transcriptional activation by ER-α. The mechanism involves interaction with p68 and p72. 6. The tertiary structure of pri-miRs may impose a physical barrier on processing. An example is the miR-17-92 cluster where the pri-miRs located at the surface are processed more efficiently than those at the center of the structure.
Nuclear export of pre-miRs	1. Function of exportin-5 may be limiting for transport of pre-miRs from the nucleus to the cytoplasm. This may be important for efficacy and possible toxicity of candidate therapeutic RNAi activators. 2. Nuclear accumulation of pre-miRs may play a role in cancers in which the C-terminal region of exportin-5 is truncated.
Dicer processing of pre-miRs	1. Dicer processing of let-7 may be inhibited by Lin-28. Let-7 targets several genes involved in regulating cell proliferation, and overexpression of Lin-28 is implicated in the ovarian carcinogenic process. 2. Lin-28 and let-7 are also involved in myoblast differentiation. 3. ADAR converts adenosine residues to inosines to inhibit miR biogenesis. This effect may occur at the microprocessor step (e.g., pri-miR-142) or during Dicer cleavage (e.g., miR-151). 4. Therapeutic inhibition of microprocessor and Dicer function may be affected by miR antisense oligonucleotides. An example is miravirsen, an HCV drug candidate, which inhibits miR-122 function.
Guide strand selection	1. Perfect pairing between miRs and their targets results in target cleavage. Seed region binding of a guide to the 3′ UTR of mRNA causes translational suppression, which is primarily caused by mRNA degradation. 2. The bias in favor of a guide strand's selection for incorporation into RISC influences efficiency of silencing. The strand of the miR duplex with lower thermodynamic stability at its 5′ end is generally preferred for inclusion in RISC.

Continued...

TABLE 2.1 Mechanisms of controlling miR biogenesis and RNAi activity *Continued*

Stage of RNAi pathway involved in canonical maturation of miRs	Examples and mechanisms of effects on miR biogenesis and RNAi activity
miR turnover	1. Exoribonucleases with 5′ to 3′ and 3′ to 5′ activity cause miR degradation. Specificity of the mechanism of action is likely to be determined by the miR sequences. 2. Cell division and growth factor stimulation affect miR-16 turnover in a manner that appears to reinforce the function of genes that are involved in controlling cell division. 3. mRNAs may also influence the stability of miRs in a reversal of the established silencing mechanism.
ceRNAs	1. Essential components of the ceRNA network include (1) miRs, (2) protein-encoding mRNA, (3) RNA transcribed from pseudogenes, and (4) lnc RNAs. Recent demonstration that circular RNA, formed as a result of nonlinear splicing, is an additional important component of the ceRNA system. The intricate network functions to control gene function by sequestering miRs and competing for binding to target mRNAs.

mechanisms. For example, pri-miRs sequences may be produced from primary tRNA transcripts that are synthesized by Pol III transcription [15]. Other non-canonical miR biogenesis involves debranching of lariat splice structures to give rise to mirtrons [16] as well as alternative folding of small nucleolar RNAs (snoRNAs) [17]. Each pri-miR sequence is variable and may comprise a few 100 nucleotides. Multiple pri-miR hairpins may be found in polycistronic miR transcripts, and each hairpin is responsible for generating a mature miR.

As expected pri-miR-containing sequences may be subject to transcriptional control, which affects formation of downstream mature miRs and target gene silencing. A well-characterized example is that of regulation of the endogenous miR-34 family by the p53 tumor suppressor protein [18–20]. p53 activates transcription of miR-34 sequences, which in turn act as tumor suppressors to influence cell growth regulatory pathways. One of the mechanisms involves suppression of canonical Wnt signaling [21]. The Wnt pathway is involved in regulation of a wide range of cell functions such as proliferation and induction of endothelial mesenchymal transition (EMT). Transformation of epithelial cells to mesenchymal cells and their migration after detachment from laminin arises through activation of EMT. Increased Wnt signaling is implicated in the development of cancer and metastasis of malignant cells. miR-34 interacts with conserved sites in the untranslated regions of several genes that are involved in Wnt signaling. Gene expression signatures have been found to correlate with p53 and miR-34 status in breast cancer and pediatric neuroblastoma malignancies. The effect of the miR-34 family on the Snail1 zinc finger transcription repressor is another example of the link between p53-mediated transcription of miR-34 and cancer progression [22]. Snail1 is involved in regulating EMT and invasive properties of cancer cells.

Viruses do not possess miR processing machinery, and the cellular RNAi pathway is essential for generating mature viral miRs. Thus, nuclear presence of viral DNA sequences that encode pri-miRs per se may be considered a transcription regulatory mechanism that controls biogenesis of exogenous miRs. miR-coding capacity is a feature that is thought to be particular to viruses that include a DNA intermediate during their replication [7]. Necessity for presence of viral pri-miR transcription in the nucleus, which is linked to essential subsequent nuclear processing, is the reason for this. Cytoplasmic RNA viruses that do not have access to the nucleus may not be able to initiate the early stages of miR biogenesis. The variability in cellular miR biogenesis mechanisms, including some that bypass nuclear processing, suggests that nuclear presence of viral RNA is not an absolute requirement for viral miR biogenesis.

2.2.2 Nuclear Processing of pri-miRs

The nuclear microprocessor is responsible for processing pri-miRs. Essential components of this complex comprise the Drosha RNase III enzyme and Di George Critical Region eight protein (DGCR8) [23,24]. DGCR8, a dsRNA binding protein, interacts with the pri-miR and directs Drosha to cleave the pri-miR within the duplex region at a position 11 bp from the stem base [25]. Precursor-miRs (pre-miRs) are formed as a result of this cleavage. These hairpin-containing structures comprise 70–100 nt and have characteristic 5′ phosphates and 2-nt overhangs at their 3′ ends. The DEAD box helicases, p68 (DDX5) and p72 (DDX17), also play a crucial role in the processing of certain pri-miRs [26]. The mechanism of this effect may involve provision of a supporting scaffold function to other proteins that regulate microprocessor function. Formation of some pre-miRs bypasses the microprocessor step through utilization of alternative processing steps. The mirtron group is an example, because pre-miRs of this set are formed directly following their splicing from introns [27]. Mirtrons appear to be considerably more common than originally thought, which suggests that they have significant functional importance [28].

Drosha and DGCR8 use target cleavage and protein stabilization mechanisms to achieve their own posttranscriptional regulation [29]. The microprocessor cleaves hairpin motifs located in *DGCR8* mRNA, and DGCR8/Drosha interactions stabilize Drosha. This interesting cross-regulatory mechanism contributes to homeostatic control of miR biogenesis. Phosphorylation of DGCR8, to increase stability of this protein, is another potential mechanism of regulating microprocessor function. This idea is based on the observation that microprocessors containing phosphomimetic DGCR8 were more stable than their unmodified counterparts [30].

The microprocessor complex has additional functions that are not part of pri-miR conversion to pre-miR [31]. Analysis using high-throughput sequencing and cross-linking immunoprecipitation (HITS-CLIP) revealed that DGCR8 binds several different classes of RNA, which include mRNAs, snoRNAs, and noncoding RNAs. These interactions are important for target cleavage and control of RNA abundance. In addition to Drosha, other endonucleases are involved in regulating mRNA abundance by DGCR8. In certain pathological states such as those that occur in cells of the fragile X-associated tremor/ataxia syndrome lineage, DGCR8 binds to expanded 5′ CGG 3′ repeats and is sequestered together with Drosha in 5′ CGG 3′ RNA aggregates [32]. As a result, functioning of the microprocessor is compromised. One of the best characterized regulatory effects on microprocessor function is that of the Smad transcription factors. Understanding this effect was initiated with the observation that treatment of cultured pulmonary artery smooth muscle cells (PASMCs) with transforming growth factor-beta (TGF-β), results in an increase in the formation of mature miR-21 [33]. However, the effect was not dependent on transcriptional activation of miR-21-encoding sequences because the concentration of pri-miR-21 was unaffected by the treatment. Thus, it appeared that the increased miR-21 production resulted from augmented processing of the precursor rather than transcriptional activation of pri-miR-containing sequences. This interpretation was confirmed by observing that mature miR-21 production was unaffected by administration of α-amanitin, a Pol II inhibitor, to cells. Nuclear translocation of Smad transcription factors occurs after TGF-β treatment of PASMCs and implicated a role for Smad proteins in the effect of TGF-β on miR-21 biogenesis. TGF-β mediates its effects primarily through the phosphorylation of receptor-specific Smads (R-Smads), which in turn leads to nuclear translocation and transcription activation through interaction with Smad4 [34]. The effect of TGF-β on miR-21 production was found to be independent of Smad4 interaction and rather involved in the formation of a complex with p68 and pri-miR-21 on the microprocessor [33]. The R-Smad interaction with the pri-miR-21 occurs through sequence-specific binding to dsRNA [35], and maturation of other pri-miRs may be influenced by Smads through this sequence-specific interaction. Interestingly, the binding motif (5′ CAGAC 3′) resembles the DNA binding sequence of Smads when acting as transcription factors. It seems that transcription factor binding to miR intermediates plays a more important regulatory role than has previously been appreciated. In support of this, it was observed that putative transcription factor binding sequences within duplex regions of miR precursors occurs more commonly than would be accounted for by chance [36].

In addition to its role in regulating transcription of pri-miR sequences, p53 regulates microprocessor function through contact with the p68 DEAD box helicase [37]. This interaction facilitates pri-miR processing to suppress potentially oncogenic genes, such as *c-myc*. miR-145 is one of the regulatory small RNAs that is involved in this antineoplastic effect. The mechanism for the specificity of action, although not yet elucidated, does not entail a sequence-specific interaction with pri-miR duplex regions, as with Smad proteins.

KH-type splicing regulatory protein (KSRP), a single-stranded RNA binding protein, is also capable of modulating microprocessor function [38]. This protein is known to regulate alternative splicing and degradation of mRNA in muscle cells [39]. KSRP also binds to the terminal loop of some pri-miR sequences to stimulate their conversion to pre-miRs [38]. Although initially thought to be based on specific interaction between 5′ GGG 3′ motifs and KSRP, the stimulatory mechanism appears to be broader and not particularly dependent on sequence [40–42]. Growth factor signaling that activates the Akt2 serine/threonine kinase is required for KSRP phosphorylation, which in turn leads to myoblast differentiation [40]. Increased pri-miR binding by phosphorylated KSRP leads to enhanced microprocessor function and generation of myogenic miRs. Heterogeneous nuclear ribonucleoprotein A1, another single-stranded RNA binding protein, antagonizes the effects of KSRP [43].

Stimulation of the estrogen receptor alpha (ER-α) by estradiol inhibits nuclear processing of a group of pri-miRs [44]. It appears that ER-α acts through specific contact with tertiary structural motifs as well as by binding to p68 and p72 helicases. Interestingly, the miR cognates are found in genes that are transcriptionally activated by ER-α. Therefore, ER-α acts at transcriptional and posttranscriptional levels to increase target gene expression.

Understanding the importance of particular miR sequences for their maturation comes from studies on the tertiary structure of the miR-17-92 cluster, which encodes six pri-miR elements [45]. RNA containing the pri-miRs folds tightly on itself to place a physical barrier on its processing. The pri-miRs located at the surface are processed more efficiently than those at the center of the structure. This structure/function relationship was further supported by the observation that disruption of the tertiary structure increased miR-92 processing with concomitant improved functional repression of an *integrin α5* mRNA target.

2.2.3 Nuclear Export of pre-miRs

After microprocessor cleavage of pri-miRs to form pre-miRs, export of pre-miRs to the cytoplasm occurs by a mechanism that is likely to have a role in regulating miR biogenesis. This function is performed by the exportin-5 karyopherin [46,47], which recognizes 2 nucleotide 3′ overhangs and the stem duplex region of pre-miRs [48]. Serious toxicity that resulted from saturation of the exportin-5-mediated transport of pre-miRs from hepatocyte nuclei indicates that this step is limiting [49]. The study reporting on this observation was aimed at developing therapy for hepatitis B virus (HBV) and utilized expressed pre-miR short hairpin RNA (shRNA) mimics to target the virus. Subsequent to this investigation, it was shown that Argonaute (Ago) proteins may play a more significant rate-limiting role in miR maturation [50]. Nevertheless, lethality observed in mice treated with vectors generating antiviral RNAi intermediates emphasizes the importance of understanding regulatory functions of miR processing when the pathway is being exploited for therapeutic use.

Additional evidence in support of the functional significance of pre-miR export from nuclei comes from the study of cancers with microsatellite instability. In these malignancies, C-terminal deletions were demonstrated in exportin-5 [51]. Resultant nuclear trapping of pre-miRs and global decrease in miR maturation are likely to contribute to malignant transformation.

2.2.4 Cytoplasmic Maturation of miRs

Dicer, another RNase III, cleaves pre-miR in the cytoplasm during the next miR maturation step [52,53]. This function is performed in conjunction with the human immunodeficiency virus (HIV)-1 *trans*-activation response element RNA binding protein (TRBP) in humans, which serves as a dsRNA binding protein. Dicer contains a PAZ (Piwi-Argonaute-Zwille) domain that anchors the 3′ end of pre-miR. The two RNase domains then usually cleave pre-miR 22 bp from the 3′ end of the pre-miR stem base. Dicer is also capable of binding to the 5′ end of pre-miR, which may be required to discriminate between pre-miRs and other RNAs [54]. Dicer cleavage of pre-miR generates a 22-nt duplex structure, which has 2-nt overhangs at each 3′ end.

The best understood mechanism of regulating pre-miR processing by Dicer is the inhibition caused by Lin-28 on lethal-7 (let-7) pre-miR maturation [55]. Interaction of Lin28 with Let-7 leads to the recruitment of terminal uridyl transferases and addition of uridyl residues to the 3′ end of let-7 [56]. The effects of this modification are inhibition of Dicer cleavage of the let-7 precursor and degradation of the pre-miR. Let-7 miR targets cell proliferation genes, such as c-*myc*, K-*ras*, and sequences encoding cyclin-dependent kinases, and it functions as a tumor suppressor. It is inactivated in several malignancies and is implicated in the neoplastic transformation process (reviewed in refs [57,58]). For example, overexpression of Lin-28, and the related lin-28b in ovarian cancer cells, correlates with diminished let-7 maturation and disease progression [59]. Derepression of growth regulatory genes, such as *cyclin-dependent kinase-2*, increases cell proliferation and contributes to the malignant phenotype [60].

Derangements of Lin-28 expression and disruption of pre-miR-1 maturation have also been implicated in dystrophic cardiac disease [58]. Sequestration of the RNA binding protein, muscleblind-like-1 protein (MBNL1) by expanded CUG or CCUG repeats contributes to the disease pathogenesis. Normally, MBNL1 competes with Lin-28 for binding to pre-miR-1 and counters the inhibitory effect of lin-28 on pre-miR-1 maturation. However, when MBNL1 is sequestered, lin-28 reduces formation of mature miR-1. As a result, there is disruption to the expression of genes encoding ion channel maintenance proteins, which in turn leads to abnormal cardiomyocyte contraction with dystrophy.

Conversion of adenosine to inosine residues by adenosine deaminase acting on RNA (ADAR) is an interesting possible mechanism of regulating miR processing. The effects may occur during Drosha/DGCR8 and Dicer cleavage steps. ADARs have been reported to act on pri-miR-142 to diminish its stability and processing by the microprocessor complex [61]. Tudor-SN, the nuclease responsible for degradation of inosine-containing RNA, then causes rapid decay of pri-miR-142. ADAR-mediated incorporation of inosine bases has a similar effect on pre-miR-151 processing, but at the Dicer cleavage step [62]. Although interesting, it is not yet clear whether this regulatory mechanism is widespread.

It has recently been demonstrated that antisense oligonucleotides may be used to inhibit the maturation of miRs that is effected by the Dicer cleavage step [63]. Miravirsen, the hepatitis C virus (HCV) oligonucleotide drug candidate that is complementary to sequences of mature miR-122 [64], is an example. The drug is also capable of binding to sequences within the stem region of pre-miR-122 and pri-miR-122 to inhibit both microprocessor- and Dicer-mediated maturation steps [63].

2.2.5 miR Silencing of Target mRNA

During the final step of miR-mediated gene silencing, one of the strands of the mature miR duplex is chosen to serve as a guide. The other strand, referred to as the passenger, miR star or miR*, is usually degraded, but it may cause some silencing. There is a bias in guide selection, which appears to be influenced by the helicase domain of Dicer [65], and preference is for the strand that has lower thermodynamic stability at its 5′ end. Loading of the RNA-induced silencing complex (RISC) with the selected strand is accomplished through the concerted action of TRBP, Ago proteins, and Dicer. Association with Ago proteins is fundamental to the silencing effect (reviewed in ref. [66]).

The mature miR guide typically binds to its target by complementary base pairing at the 3′ untranslated region (3′ UTR) of the mRNA cognate [67]. When complete base pairing between the guide and target occur, RISC causes mRNA cleavage. This effect is mediated by the slicer function of Ago2. In mammalian cells, it is uncommon for miRs to have a perfect match to their targets. Usually partial base pairing is responsible for target silencing and it is the seed region, comprising a minimum of six nucleotides from positions two to seven from the 5′ end of the guide, which is required for this effect [68]. A detailed study that entailed use of high-throughput sequencing to profile ribosome-bound RNA, mRNA sequencing, and proteomic analysis revealed that target mRNA degradation is primarily responsible for inhibition of target gene expression [69]. Approximately 85% of the silencing effect resulted from mRNA degradation, and only 15% was caused by translational suppression. mRNA degradation caused by interaction with a miR guide involves shortening of the polyA tail, decapping, and degradation by exonucleases such as Xrn1. Translational suppression is likely to result from reduced ribosomal translation initiation and increased dissociation of ribosomes from mRNA. Binding of RISC to mRNAs directs these complexes to cytoplasmic processing bodies (P bodies; reviewed in ref. [70]). Here, the mRNA is excluded from cellular translation and is inactivated by decapping, deadenyation, nuclease degradation, and effects of specific P body-associated components. However, P bodies do not appear to be essential for silencing by miRs [71]. Inhibitory effects are initiated in soluble cytoplasmic compartments, and RISC/mRNA location in the P bodies seems to be a consequence rather than a requirement for silencing.

An interesting alternative miR processing mechanism was revealed by study of miR-451. Analysis showed that miR-451 maturation occurred in a Dicer-independent but Ago2-dependent fashion in zebra fish [72] and mice [73]. Pre-miR-451 has an unusual and conserved secondary structure with a shortened duplex region of 19 bp. Ago2 cleavage, which occurs in the loop sequence of the pre-miR, generates a 30-nt intermediate that is uridylated and trimmed before final maturation. This unusual miR processing mechanism has practical application. Recent studies have demonstrated that it is possible to use a miR-451 scaffold to generate artificial HIV-1 gene silencing sequences [74]. As with the parental sequences, processing of these mimics is Dicer-independent but Ago2-dependent. An advantage of simulating miR-451 processing is that production of a passenger strand, which may cause off target silencing, is avoided.

2.2.6 miR Turnover as a Mechanism for Regulating Silencing Efficacy

Although most miRs are stable for at least 8 h in cultured human-derived cells [55], it has recently been reported that differential regulation of miR degradation may be controlled by several cellular processes [75]. Pathways that involve exoribonucleases with 5′ to 3′ and 3′ to 5′ activity have been implicated in influencing miR stability and silencing efficacy. In humans, these include Xrn1 [55], ribosomal RNA-processing protein-41, and polynucleotide phosphorylase [76]. Specificity of miR degradation by these ribonucleases would be important and suggests that sequence determinants of miRs play a role through activation of different miR degradation pathways. Physiological and pathological signals are also likely to have an effect on the mechanisms responsible for changing miR stability. Interesting insights are now emerging that confirm that miR turnover is important for the control of target gene silencing.

Recent evidence has corroborated the notion that sequence-specific mechanisms affect miRs' degradation. Analysis performed on miR-382 showed that seven nucleotides at the 3′ end of the mature miR, and positioned outside of the seed, control its stability [55]. Mutation in this region led to an increased half-life of miR-382 involving a mechanism that primarily influences the exosomal 3′ to 5′ exoribonuclease complex and Xrn-1 to a lesser extent. HIV-1 latency is thought to be regulated by miR-382, and rapid decay of this regulatory miR is an interesting way in which replication of the virus may be increased.

Influences of cell division and growth factor stimulation on miR stability have recently been studied in some detail. Analysis of turnover of mature sequences of the miR-16 cluster provided evidence in support of an effect of cell division on miR concentrations [77]. Endogenous miR-16 sequences accumulate when cells are arrested in G0, but their concentrations rapidly decrease upon stimulation of cells to start dividing. It appears that the intracellular miR-16 concentrations reinforce the function of genes involved in control of cell division and arrest. Stimulation of breast epithelial cells with epidermal growth factor causes a decrease in the concentrations of several miRs that interact with transcripts from growth response genes [78]. A similar effect was observed when cells were starved and suggests that the miR downregulation augments the proliferative effects of growth factor stimulation.

In a reversal of established silencing effects of miRs on mRNA, evidence has recently demonstrated that mRNA may influence stability of miRs (reviewed in ref. [75]). The degree of complementarity between mRNA and miR influences miR degradation. When there is near perfect matching, tailing with uridine or adenine residues coupled with miR trimming causes efficient miR degradation [79]. However, when pairing between miR and target mRNA involves the seed region alone, as is often the case with endogenous mRNA/miR interaction, minimal destabilization results. This mechanism has been exploited to influence miR efficacy in mice [80]. Delivering miR-complementary elements, so-called "tough decoys," with recombinant adeno-associated viral vectors (AAVs) caused specific inhibition of miR function. High-throughput sequencing confirmed that the mechanism entailed trimming and tailing. Viruses, particularly of the *Herpesviridae* family, produce RNAs that inhibit miR function by similar mechanisms. An example is the Herpesvirus saimiri U-rich noncoding RNA-1, which destabilizes miR-27 by a sequence-specific and binding-dependent mechanism [81]. However, the significance of the destabilization for viral replication remains to be established.

2.2.7 Regulation of miR Function by Competing Endogenous RNAs

The competing endogenous RNA (ceRNA) hypothesis describes control of gene function through regulating miR accessibility to targets by sequestering sequences. The mechanism involves an intricate interplay between diverse RNA species and provides a compelling explanation for the major role that nonprotein coding RNAs play in the regulating gene expression [82,83].

A complex network that involves competition of coding and noncoding RNAs for binding to a limited pool of miRs as well as communication between the different network components, is central to the mechanism. The essential constituents of the ceRNA network are (1) miRs, (2) protein-encoding mRNA, (3) RNA transcribed from pseudogenes, and (4) long noncoding (lnc) RNAs. As has been established (see above, section 2.2.5), miR response elements (MREs) located on protein coding sequences exert an effect on gene expression through translational suppression. Conversely, MREs found in nonprotein coding RNAs, such as lnc RNAs and pseudogenes, serve as decoys to diminish the available pool of miRs. The range of validated noncoding ceRNAs has recently been comprehensively reviewed [83]. Availability of inhibitory miRs to bind to protein-coding mRNAs is the ultimate determinant of the functional consequences of the ceRNA mechanism. To add complexity, mRNAs may also influence each other's function by acting as decoys or sponges for miRs that bind to other mRNAs [75].

Two recent reports on circular RNAs derived from the cerebellar degeneration-related protein 1 (CDR1) locus, and which function as miR sponges, have provided significant support for the ceRNA hypothesis [84,85]. Nonlinear mRNA splicing results in the formation of unusual circular RNA. The interesting characteristic of this circular RNA is that the sequence contains 73 miR-7 seed targets and a complete complement to miR-671. These targets are highly conserved in eutherial mammals; therefore, they suggest functional importance [84]. Sequence analysis revealed that the miR-7 targets were normally mismatched in the central regions. The circular RNA, termed ciRS-7 by Hansen et al. [84] and CDR1 by Memczak et al. [85], was resistant to endonucleolytic cleavage. This increased stability enhances the sponge effects of the nonlinear splice RNA. Functional importance for RNAi was confirmed by studies using HITS-CLIP, which showed that Ago2 associates with miR-7 at the nonlinear splice junction [84]. However, although Ago2 and miR-7 association directed ciRS-7 RNA to P bodies, the circular structure afforded resistance to degradation. Attenuation of expression of endogenous miR-7 targets and colocalization in vivo of ciRS-7 with miR-7 in the mouse brain reinforced the perception that a functional relationship exists between the two RNA species. Regulation of miR function by circular RNAs is likely to be more widespread than was originally thought and may also contribute to the pathogenesis of Parkinson's disease and brain tumor development. Another example of circular RNA influencing miR function is the sex determining region Y (Sry) [84]. In the case of Sry, nonlinear splicing generates a circular RNA that has multiple miR-138-sequestering sequences.

Artificial sequences were used experimentally to sequester and diminish function of specific miRs before they were shown to exist naturally [86]. These sequences were typically tandem repeats of intended miR targets. Interestingly, it was found that artificial sponges that were imperfectly matched to their target miRs, as is the case with natural sponges, resulted in more efficient and lasting inhibition of miR function. The explanation of this observation seems to be that degradation of miR/sponge duplexes that contain bulges is more delayed than the degradation that occurs with perfectly matched duplexes. Use of sponges for functional and therapeutic application is potentially very powerful. Miravirsen, the candidate HCV therapy, is an example of a drug that inhibits miR function by sequestration [64]. This antisense oligonucleotide therapy potentially acts at several steps of miR-122 biogenesis: mature miR-122 sequestration and pri-miR-122 and pre-miR-122 processing [63].

2.3 EXPLOITING RNAi TO SILENCE VIRAL GENE EXPRESSION

After it became apparent that RNAi could cause powerful and specific gene silencing, developing methods of harnessing the mechanism for therapeutic use was a logical progression. Because the specificity of the silencing mechanism is centered on base pairing, silencers may simply be generated using fundamental knowledge of the target gene sequences. The approach entails introduction of mimics of intermediates of the RNAi pathway into cells. These exogenous gene silencers reprogram RNAi and can theoretically be designed to target any sequence. Unlike the natural pairing that occurs between miR guides and their cognates, the mature guides that are generated from exogenous antivirals are usually designed to be completely complementary to their targets. This is intended to cause more efficient gene silencing by Ago2-mediated "slicing" of target mRNA rather than by translational suppression. Because viral replication results from intracellular expression of pathogen-specific genes, targeting these sequences has proven to be a useful approach to therapy.

There are several considerations for optimizing the therapeutic use of RNAi activators to treat viral infections. Particularly important are the following:

- Stability of the RNAi activator and durability of the antiviral silencing effect that is required,
- Location of viral infection and site of intended silencing action,
- Selection of a suitable vector to achieve efficient targeted delivery of the RNAi activators,
- Biodistribution and pharmacokinetics of naked or vector-delivered therapeutic sequences,
- Selection of potent RNAi activators that have antiviral effects at low dose,
- Accessibility of the viral target,
- Ability of a virus to mutate and escape silencing,
- Minimizing of off-target effects, and
- Elimination of immunostimulatory effects of the RNAi activators.

To date, a large variety of expressed and synthetic exogenous sequences has been used to reprogram RNAi to silence viral targets. Synthetic sequences typically resemble Dicer products and comprise central duplexes of 19 bp with 2-nt 3′ overhangs, which are termed short interfering RNAs (siRNAs). Expressed RNAi activators are produced by DNA templates that are engineered to transcribe mimics of pri-miRs or pre-miRs. There is no single set of rules that can be applied to the design of antiviral therapeutic sequences. Although useful algorithms have been developed to assist with conception of silencing sequences [87–90], empirical characterization of the properties of potentially therapeutic sequences remains important.

2.3.1 Synthetic RNAi Activators

siRNAs have been successfully used for viral gene silencing applications and have many useful properties. They have a cytoplasmic, not nuclear, site of action, which means that they may be delivered more easily than larger DNA expression cassettes. However, without a DNA template to provide a source of renewable silencing sequences, duration of silencing may be limited. As synthetic molecules, siRNAs may be chemically modified to confer desirable biological properties. This important property has been ingeniously exploited to improve efficacy of siRNAs. Chemical modification of siRNAs is a very active field of research, and several comprehensive recent reviews have been published on the topic [91–94]. Most synthetic siRNAs reported on to date have been designed as mimics of intermediates of RNAi that enter the pathway at the later stages, and function as Dicer products or Dicer substrates. Examples of representative synthetic siRNAs that have been successfully used are illustrated in Figure 2.2. The repertoire of structural variations in siRNAs has been comprehensively reviewed by Chang and colleagues [94].

2.3.1.1 Enhancing siRNA Processing and Potency

To be potent, an siRNA needs to be processed efficiently, have good access to the target site, and possess favorable sequence-specific properties. On the basis of the use of data from many gene silencers, tools have been developed to predict which are the most effective sequences against a particular target [87–90]. siRNAs typically comprise paired 21 nt oligonucleotides with a central duplex of 19 bp and 2-nt overhangs at their 3′ ends. This preference was based on the seminal studies of Elbashir and colleagues, who originally demonstrated activation of RNAi by synthetic sequences [2]. Incorporation of two deoxythymidine residues at the 3′ termini was initially favored. However, this feature has been shown to have no real benefit, and siRNAs that are complementary to their targets are presently more commonly used [93].

It seems that variation in the structures of synthetic exogenous gene silencers may be accommodated by the RNAi machinery. For example, many artificial

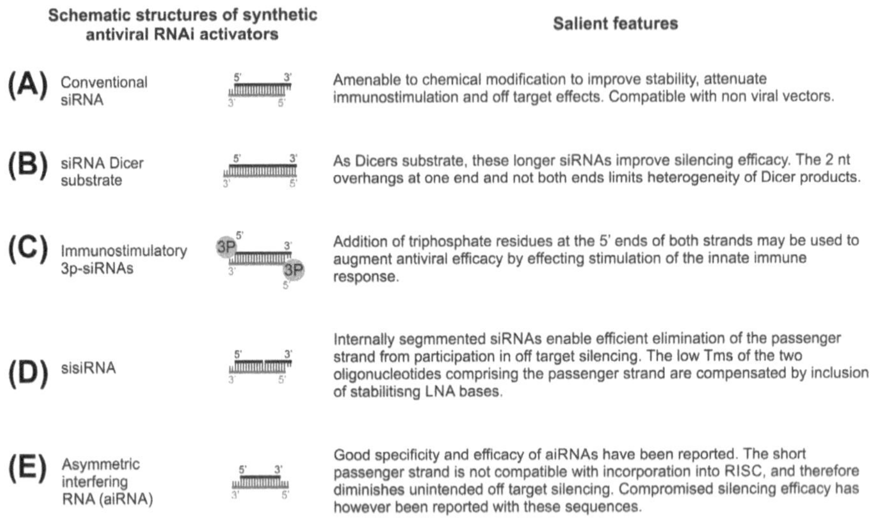

FIGURE 2.2 Illustration of synthetic RNAi activators that have been used to silence virus replication.

The schematic structures are illustrated of (A) standard siRNAs, (B) siRNA Dicer substrates, (C) immunostimulatory 3p-siRNAs, (D) sisiRNAs, and (E) asymmetric interfering RNAs (aiRNAs). All synthetic RNAi activators are compatible with use in nonviral vectors and are amenable to chemical modification to enhance specificity, efficacy, and stability and limit immunostimulation.

RNAi activators are now designed to serve as Dicer substrates, not only as Dicer products, and may have particular beneficial properties. In one of the first studies describing the effect of long duplexes on siRNA silencing efficacy, Kim et al. showed that siRNAs comprising 27-nt duplexes had silencing efficacy of up to 100 times that of conventional 21-mer siRNAs [95]. However, this silencing capability was not consistent, and in some cases the longer and standard length siRNAs demonstrated equivalent silencing efficacy. One of the difficulties of using longer siRNAs is that multiple mature siRNAs may be produced from a single substrate, and precision of the processing is not easy to predict. When many potential silencing sequences are generated, the risk of undesirable, nonspecific silencing effects is increased. To overcome this, long siRNAs have been designed to have one blunt end and the other with a 2-nt 3′ overhang [96]. Duplexes comprising a 25-nt sense and 27-nt antisense strand have been favored. Preferred binding of the PAZ domain of Dicer to the 3′ end with 2-nt overhangs limits the variability in the formation of mature Dicer products. Incorporation of two deoxynucleotides at the 3′ end of the sense passenger strand, at the blunt end of the duplexes, also appears to improve Dicer processing and silencing efficacy. Silencing by shorter synthetic siRNA duplexes with overhangs at both 5′ and 3′ ends of the guide, asymmetric interfering RNA (aiRNA), may also be effective [97,98]. Because the passenger sense strands of these siRNAs are too short to be incorporated into RISC, off-target silencing is diminished. However, good silencing efficacy of shorter siRNAs is not reliably attained [99].

Lower thermodynamic stability at the 5′ end of the intended guide is one of the most important predictors of siRNA efficacy [100], and helicase activity of Dicer is responsible for this sequence bias [65]. Therefore, target sites for siRNAs are selected such that the local melting temperature (Tm) at the 5′ end is lower (A:T rich) than at the 3′ end (G:C rich). Thermodynamic stability may also be influenced by chemical modification and incorporation of duplex-destabilizing mismatches.

2.3.1.2 Optimizing siRNA Stability

Because phosphodiester bond cleavage and resultant RNA breakdown typically occurs by a transesterification reaction involving the 2′-hydroxyl group of the ribose, chemical modification of this moiety within siRNAs has been a major method of improving their stability. Alterations to the phosphodiester backbone and substituting ribose for alternative sugars have also been used to extend the half-lives of siRNAs. Ribose modifications involving the 2′ -hydroxyl include incorporation of 2′-*O*-methyl (2′OMe), 2′-fluoro (2′-F), locked nucleic acids (LNAs), acyclic unlocked nucleic acids (UNAs), 2′-*O*-methoxyethyl (2′MOE), guanidinopropyl (GP), and deoxy residues. Backbone alterations to siRNAs have involved the use of phosphorothioate and boranophosphate residues [101], both of which have been reported to improve siRNA stability and silencing efficacy. Variable numbers of nucleotides may be modified within a siRNA and different chemical modifications may be added to single siRNAs. When formulated within protective nonviral vectors, chemical modification to protect siRNAs may not be as important as when these silencers are administered naked. In addition to improving siRNA stability, chemical modification may be used to influence siRNA duplex stability, target interaction, and immunostimulation.

The most widely studied ribose alteration is the 2′OMe substitution [93]. This nucleotide variant occurs naturally and is not toxic. Several studies have shown that siRNAs containing 2′OMe residues improve siRNA stability while retaining silencing efficacy [102]. The modifications may enhance resistance to both exo- and endonucleases. Positioning 2′OMe residues at different nucleotides of the siRNAs, alone or in combination, has been found to limit degradation. For example, strands with 2′OMe in nucleotides at the 3′ or 5′ ends of the guide strand, all nucleotides of the passenger strand, and at alternating ribose moieties of the guide may all contribute some improvement to stability [102]. Another approach entailing use of 2′OMe modification to stabilize siRNAs is particularly important for retaining guide strand efficacy. It is established that the lower thermodynamic stability at the 5′ end of a strand is important for biasing strand selection [100]. However, this property may also make the intended oligonucleotide more vulnerable to exonucleases with 5′ to 3′ activity that are found in the blood [103]. Incorporation of 2′OMe residues at the 5′ end of the guide has been shown to be useful to provide resistance to this degradation. UNAs have also been used successfully to confer nuclease resistance on siRNAs [104]. The ribose moieties in UNAs lack the C2–C3 bond, which results in opening of the sugar ring. The acyclic residues are poor substrates for nuclease degradation.

LNAs contain ribose molecules with a bicyclic structure. The linkage between the 2′ oxygen and 4′ carbon atoms results in a 3′-endo configuration, which increases Tm within duplexes and serum stability [105]. Oligonucleotides with LNA modifications are now in widespread use. However, excessive LNA residues within a siRNA may diminish silencing efficacy and also cause some toxicity [106].

2.3.1.3 Improving siRNA Specificity

Off-target effects of antiviral siRNAs may be caused by nonspecific guide strand interaction with targets or as a result of passenger strand selection and binding to cellular sequences. Therefore, searching for gene silencers that do not have complementary off-target sites is important to limit unintentional silencing of cellular genes. However, because complete pairing between guide and target is not required to cause nonspecific silencing, prediction of unplanned interaction between guide strands and targets is sometimes difficult to achieve. Although binding of the seed region is theoretically all that is necessary to mediate an off-target effect [107,108], using seed matches to predict risk of unintended silencing is not always reliable. The seven bases comprising a sequence complementary to the seed region of an antiviral siRNA guide may occur commonly throughout the human genome, but a silencing effect of consequence does not necessarily occur. However, evidence indicates that siRNAs with high seed frequencies in the 3′ UTR of human mRNAs does correlate with off-target silencing [109].

Chemical modification of nucleotides within the seed region of a guide strand may be used to diminish unintended off-target silencing. Incorporation of nucleotides with 2′OMe [110], deoxyribose [111], GP [112], UNA [113], and LNA [114] modifications at the ribose improved specificity. With these changes, potency of the siRNAs was variably affected and in some cases, minimal if any compromised silencing was observed. In one comparative study, inclusion of LNAs into siRNA seeds improved specificity more than did 2′OMe residues [114].

Blocking participation of the passenger strand in RISC function is useful to minimize off-target effects. In addition to designing siRNAs that favor guide strand selection, chemical modifications have been used to impede passenger strand incorporation into RISC. The presence of a phosphate group on the 5′ end of a candidate guide strand is required for loading onto RISC. Because the natural 5′ OH group is readily amenable to phosphorylation, the use of unphosphorylated passenger oligonucleotides is not adequate. Blocking phosphorylation of 5′-hydroxyl groups by inclusion of groups such as 2′OMe at the 5′ end is required [115]. Another ingenious approach to limiting silencing by passenger strands has employed the use of internally segmented siRNAs (sisiRNAs). With these silencing sequences, two contiguous LNA-modified oligonucleotides of 10–12 nt comprise the passenger strand. The lower Tms of the two oligonucleotides making up the passenger strand, with resultant destabilization of the siRNAs, are compensated by incorporation of the LNAs. Only the full-length guide strand, comprising 21 nt, is incorporated into RISC and functions highly effectively [116].

2.3.1.4 Minimizing Immunostimulation by siRNAs

Recognition of specific pathogen-associated RNA motifs by components of the innate immune response (Chapter 1) is highly relevant to the design of antiviral gene silencers. RNAi activators, particularly synthetic siRNAs, may be recognized as alien nucleic acids and therefore lead to innate immunostimulation. Nonspecific consequences of activating the innate immune response by RNAi activators is often underappreciated [93,117,118]. Many effects ascribed to specificity of siRNA-mediated silencing may in fact be artifact and caused by innate immunostimulation [119]. In addition to confounding interpretation of the specificity of gene silencing, innate immunostimulation may also cause toxicity. Membrane-bound Toll-like receptors (TLRs) distinguish exogenous RNA elements within the endosomal compartments. Therefore, this arm of innate immunity may interact with vector-delivered synthetic RNAi activators that are taken up by endocytosis. dsRNA of length greater than 30 bp is identified by TLR3, whereas TLR7 and TLR8 bind single-stranded RNA, which may include synthetic siRNAs. Ubiquitous retinoic acid-induced gene-I (RIG-I) and melanoma differentiation-associated gene-5 are cytoplasmic activators of the innate immune response and recognize 5′ triphosphates and blunt ends of dsRNA.

Although most opinion holds that activation of the innate immune response is an undesirable property of siRNAs, compelling evidence to the contrary has recently been presented [120–122]. In these studies, siRNAs containing 5′ triphosphates (3p-siRNAs) were used to inhibit HBV replication in cell culture and mice. These bifunctional 3p-siRNAs had direct antiviral effects by mediating cleavage of viral mRNA and also augmented the antiviral immune response by activating RIG-I to induce a strong type I IFN response. When tested in cell culture and in vivo, 3p-siRNAs had enhanced and more lasting antiviral efficacy.

Chemical modification has been used to enable synthetic siRNAs to evade innate immunostimulation. Inclusion of 2′OMe [123], LNA, and 2-′F residues has been successfully used to attenuate innate immunostimulation. A particularly interesting property of 2′OMe-modified oligonucleotides is that they are capable of competitive inhibition of TLR function [124]. Therefore, these oligonucleotides may achieve inhibition of innate immunostimulation in trans. Therefore, effects of 2′OMe-modified oligonucleotides on other molecules, conceivably other siRNAs, could be used to enhance specific effects of heterologous oligonucleotides.

2.3.2 Inhibiting Viral Gene Expression with Expressed RNAi Activators

Engineering DNA templates to express potentially therapeutic virus gene silencing sequences has also been successfully used (reviewed in ref. [125]). These expression cassettes have distinct features, some of which are particularly useful for therapeutic application to viral infections. Because DNA templates are more stable than synthetic siRNAs, and production of gene silencers from these templates is renewable, expressed RNAi activators may achieve more sustained viral gene knockdown. This makes DNA expression cassettes well suited to the silencing that is required to treat chronic viral infections. DNA cassettes may also be incorporated into efficient recombinant virus vectors. Although compatible with nonviral vectors (NVVs), efficient delivery of DNA templates with this type of vector is usually inadequate for therapeutic use. This difference between efficiency of NVV-delivered synthetic and DNA expression cassettes is largely ascribed to the large size of DNA cassettes. In addition, RNAi expression cassettes require more challenging nuclear and not cytoplasmic delivery, which is the case for siRNAs.

Using DNA templates to transcribe RNA sequences that contain hairpin motifs has been the main approach to activating RNAi with expression cassettes. Transcription of complementary RNAs from independent templates to form Dicer substrates may be used but has not gained wide favor. Cassettes that activate RNAi typically comprise a promoter with downstream sequences encoding a miR-like hairpin and transcription termination signal (Figure 2.3). Commonly, shRNA-encoding sequences have been used as pre-miR mimics and they comprise a duplex of approximately 22 bp with a single-stranded terminal loop. Mimics of pri-miRs comprise longer sequences and the hairpins are flanked by single-stranded regions that simulate natural pri-miR structures. Several variables within the hairpin motif-encoding sequences may change the efficacy of gene silencers. Influencing biological properties by introducing variations in duplex stem length, sequence composition of the stem and loop regions, incorporation of bulges and mismatches within the duplex region, and the number of hairpin motifs present in a transcript have all been assessed for possible improvements (Figure 2.4). In addition to shRNAs, single [126] or double long hairpin RNAs (lhRNAs) [127] have been incorporated into Pol III expression

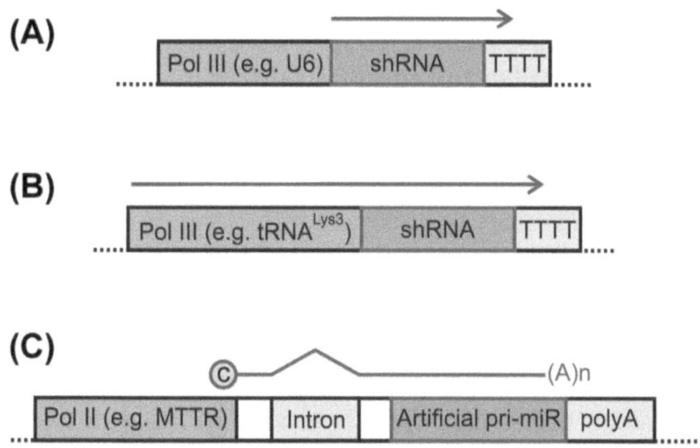

FIGURE 2.3 Types of miR-encoding DNA expression cassettes.
(A and B) Pol III or (C) Pol II regulatory sequences may be used to generate miR elements. miR mimics generated from tRNA promoters typically contain sequences that are derived from the promoter element. Artificial pri-miRs derived from Pol II transcripts may include an intron. As with their natural counterparts, they are capped on the 5′ end and have a 3′ poly(A) tail.

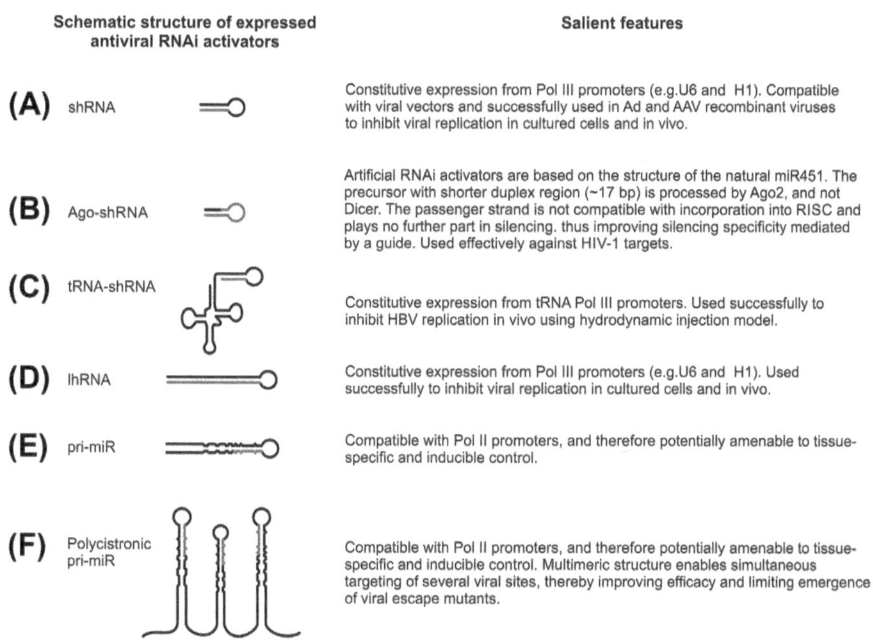

FIGURE 2.4 Illustration of expressed miR-like activators that have been used to silence virus replication.
The schematic structures of primary transcripts encoding (A) shRNAs, (B) Ago-shRNAs, (C) tRNA-shRNAs, (D) lhRNAs, (E) artificial pri-miRs, and (F) polycistronic pri-miR cassettes are illustrated. Compatibility of each type of transcript with types of Pol III and Pol II promoters is indicated. Intended guide strands are shown in color.

cassettes. The extended stem duplexes of these transcripts enable production of multiple siRNAs from a single RNA. This is important to prevent emergence of mutants that escape silencing effects of shRNAs, such as may occur during error-prone replication of HIV-1 [128–130] and HCV [131]. Expressed RNAi activators are transcribed within the nucleus; therefore, they do not traverse the endosomal compartments. As a result, TLR-mediated immunostimulation by expressed RNAi activators does not typically occur. However, the expressed sequences may stimulate cytoplasmic pattern recognition receptors and the DNA template may also have an effect on TLR7 and TLR9 during entry into cells.

As with synthetic RNAi activators, precisely defining the features of antiviral expression cassettes that confer optimal gene silencing is not always easy to achieve. Algorithms that take into account stem length and sequence have been described [132], but empirical assessment of efficacy remains important for use in miR expression cassettes. shRNA stems are typically 19–21 bp in length. The sequences at the ends of the predicted Dicer products may be changed to influence the bias of strand incorporation into RISC, but shRNA stem duplex length does not appear to correlate with silencing efficacy. As with synthetic siRNAs, this may be achieved by ensuring that the intended guide has a lower thermodynamic stability at its 5′ end. Another feature that is common to synthetic siRNAs and expressed shRNAs is that when the duplex region is longer, Dicer processing is less predictable. A heterologous population of Dicer products was generated from a panel of expressed shRNAs that had double-stranded regions comprising 24–29 bp [133]. Importantly, the variable processing generates a range of shRNA-derived siRNAs, which in turn may cause imprecise silencing and toxic off-target effects. This interpretation is in accordance with the observation by Grimm and colleagues, who showed that shRNAs with longer duplex regions were more toxic than their shorter counterparts [49]. Variable processing of expressed lhRNAs with duplexes comprising approximately 60 bp has also been described [126]. HBV-targeting lhRNAs were found to be processed more efficiently at the stem base than at the loop side of the sequences. Therefore, optimizing Dicer processing has logically been an objective in the field of generating expressed RNAi activators, and borrowing from endogenous miR processing has been useful to achieve this. Naturally, miRs have bulges within their duplexes and these play a role in regulating formation of shRNA-derived short RNA duplexes. By incorporating bulge sequences within shRNAs at a position 2 nt from a predicted site of Dicer cleavage, precision and predictability of processing of HCV-targeting shRNAs could be improved [133].

Insights from studying the processing of miR-451 have facilitated the design of other RNAi expression cassettes [72,73]. The stem region of pre-miR-451 comprises an unusually short duplex of 19 bp. Processing of pre-miR-451 occurs by a Dicer-independent mechanism that entails Ago2 cleavage within the loop region of the precursor (see above, section 2.2.5). The guide strand is formed after uridylation and trimming of the larger Ago2 cleavage product, whereas the shorter remaining strand is not incorporated into RISC and is degraded. The miR-451 backbone has been used to generate HIV-1-targeting alternatively processed exogenous Ago-shRNAs [74]. Cassettes expressing these sequences have the useful feature of eliminating problems that may arise from triggering RNAi by the passenger strand.

Both Pol II and Pol III promoters have been used in RNAi expression cassettes and each type of transcription regulatory element has advantages. Pol III promoters have been more widely used. Reasons for this are that Pol III promoters are (1) small in size, (2) capable of transcribing short RNA sequences of defined length, (3) constitutively active in most tissues, (4) limited in their composition to sequences that are almost all located upstream of the transcription initiation sites, and (5) amenable to convenient engineering using polymerase chain reaction to facilitate uncomplicated incorporation into expression cassettes.

Pol III promoters that have been commonly used to express RNAi activators are the U6 small nuclear RNA (snRNA) [134], 7SK snRNA [135], and RNase P H1 [136] transcription regulatory elements. DNA elements, each comprising a string of approximately six A residues on the antisense template, can be used conveniently to terminate transcription and incorporate 2–3 U residues at the 3′ end of the transcript. The protruding 3′ residues are recognized by the PAZ domain of Dicer to facilitate processing of the pre-miR mimics. The only sequence requirements downstream of the transcription initiation sites are for the initiating nucleotide of U6 and 7SK transcripts to be a G whereas for H1 the leading transcript nucleotides should be a G or A.

tRNA Pol III promoters may also be used to express RNAi activators that include antiviral silencers [137–139]. Processing of these tRNA-containing precursors is different to that of other Pol III transcripts. Because shRNA sequences are located on the same transcript as tRNA sequences, initial processing is likely to be effected by RNaseZ. This bypassing of Drosha/DGCR8 action has been shown to occur during processing of endogenous miR sequences [140].

A significant development in using RNAi to silence viral gene expression was the demonstration that production of HBV-targeting shRNAs from a U6 promoter may cause lethal toxicity resulting from saturation of the natural miR processing pathway [49]. Inhibition of exportin-5 function [49] and Ago proteins have been implicated in the toxicity [50]. Since this observation, one of the focuses of research has been development of improved methods of regulating the transcription of RNAi activators from expression cassettes. Tetracycline-responsive elements have been incorporated into Pol III promoters to improve dose regulation of shRNAs and achieve inducible transcription [141]. However, because Pol III promoters have a limited range of transcription control, Pol II promoters have been investigated for use in RNAi expression cassettes. Pol II transcriptional regulatory elements are more flexible; particularly important is that they are capable of tissue-specific regulation and responding to subtle regulatory signals. Therefore, their use should enable precise regulation of dose and targeted expression of RNAi effectors. Typically, Pol II promoter-containing cassettes have been engineered to transcribe pri-miR mimics rather than shRNAs. This design approach is based on the natural Pol II promoter transcription of pri-miRs within capped and polyadenylated RNA sequences [142–144]. Pol II promoter-derived pri-miRs are microprocessor substrates; therefore, they enter the RNAi pathway upstream of the point of entry of shRNAs. The view that each step of the miR biogenesis pathway is functionally coupled suggests that using RNAi activators that trigger the pathway at the most proximal steps should be more efficient. pri-miR mimics derived from Pol II transcripts reprogram RNAi at the earliest point of the pathway; therefore, they should also possess enhanced silencing efficacy. The polycistronic nature of natural pri-miR expression is an added beneficial feature. Therefore, incorporating multiple antiviral sequences into an expression cassette is possible and may further improve efficacy [142–144]. Simultaneous targeting of different viral sequences may augment target silencing and is desirable for suppressing emergence of viral escape mutants, which is particularly important for treating HCV and HIV-1 infections.

2.4 PERSPECTIVES ON USING RNAi ACTIVATORS TO COUNTER VIRAL INFECTIONS

As intracellular parasites, viruses are totally dependent on host factors and intracellular translation machinery for their propagation. Therefore, harnessing RNAi to inhibit expression of viral genes and host factors is potentially a useful method of treating viral infections. Although the approach is promising, silencing strategies need to be tailored to the specific characteristics of particular viral infections. Viral tissue tropism, whether an infection is acute or chronic, and ease of delivery of RNAi effectors to infected tissues all need to be considered when developing a gene silencing strategy. Topical administration of synthetic siRNAs is preferable if the sites of infection are readily accessible, such as is the case with acute respiratory syncytial virus (RSV) infection. Antiviral siRNAs delivered to infected respiratory epithelial tissue after inhalation of nebulized siRNA-containing formulations caused lower viral loads and diminished viral shedding [145,146], which is encouraging for treatment of the infection in pediatric and immunocompromised patients. Unlike with RSV, treatment of infections that are persistent and occur in tissues that are not readily accessible for delivery of RNAi effectors is more complicated. For example, to achieve the sustained virus gene silencing that is required to treat chronic infection caused by HBV, delivery of the RNAi activators with highly efficient hepatotropic vectors after their systemic administration is required.

Replication rate amongst different virus species varies considerably, and this factor is also likely to have an influence on the efficacy of RNAi-based antivirals. However, it is difficult to define the precise role of the rate of virus replication in contributing to the success or failure of candidate RNAi-based therapies. A high rate of viral replication with elevated transcription of viral genes may overwhelm the effects of gene silencing. On the other hand, viral replication dormancy, with low transcription of RNAi targets, may limit efficacy of RNAi-based viral gene silencing. Viral replication per se is also unlikely to be a factor that affects efficacy of RNAi-based antiviral efficacy alone. Host immune response, co-administration of antivirals, potency of RNAi activators, duration of silencing effects, and delivery efficiency are all likely to affect the efficiency of potentially therapeutic RNAi activators. Nevertheless, the observation that synthetic RNAi activators can be used to protect primates against hemorrhagic fever caused by Ebola virus [147] and Marburg virus [148] suggests that gene silencing may be used to counter highly replicative viruses.

Accumulation of viral target gene mutations is another important factor that may influence the efficacy of RNAi-based silencing of viral gene expression. The error-prone properties of certain viral RNA polymerases and reverse transcriptases may lead to incorporation of mutations during viral replication. When coupled to rapid viral replication and viral genome flexibility, selection of mutants that are resistant to effects of silencing may occur. Mutations that overcome silencing effects may result from changes to the viral target site, alterations in RNA secondary structure that diminish target site accessibility, and increased rate of viral replication [149]. Two main RNAi-based approaches to limiting viral escape have been used: simultaneous targeting of multiple targets using combinatorial methods and silencing of host factors. The principle underlying the combinatorial procedures is analogous to that of using combination antiretroviral therapy (cART) for the treatment of HIV-1 infection. That is, the ability of a virus to evade the silencing effects of multiple RNAi activators that simultaneously target different viral sequences should be lower than when single gene silencers are used. The efficacy of the approach is largely dependent on the plasticity of the viral genome that is being targeted. For example, HCV and HIV-1 have highly plastic genomes. For HIV-1 it has been calculated that four RNAi activators are required to prevent emergence of viral escape mutants, and multimeric RNAi activators targeting the virus have been designed accordingly [142,143,150]. However, the genome of HBV is very compact and mutations are not well tolerated without compromising viral replication fitness. Consequently, fewer individual gene silencing components would be required to prevent HBV escape. Inhibiting host factor function as a method of preventing viral escape is based on the notion that viruses are unlikely to develop mutations that would enable them to bypass dependence on an essential host factor. Use of miravirsen, although not an RNAi activator but an antisense molecule, to counter miR-122 function is an interesting example of inhibiting host factors for therapeutic use [64]. miR-122 is required for HCV replication and its functional inhibition with miravirsen is at an advanced stage of testing in clinical trial. Of course, host factors may have important cellular functions; therefore, their silencing may cause toxicity. Using host factor silencing for treatment of viral infections thus requires carefully controlled silencing and appropriate target selection.

Despite rapid and impressive advances in the use of RNAi against viral infections, successful implementation as a therapy for widespread clinical application is difficult. Problems with efficient and safe delivery, unpredictable pharmacokinetics, immunological effects, toxicity, durability of silencing, and in some cases uncertainty about mechanisms of action still need to be overcome. Challenges facing implementation of RNAi-based therapy for specific viral infections are discussed in more detail in subsequent chapters. Because many of the problems of RNAi-based virus therapy are not particular to this field, advances in related topics of basic and applied molecular biology are likely to be beneficial to antiviral gene silencing.

REFERENCES

[1] Fire A, Xu S, Montgomery MK, Kostas SA, Driver SE, Mello CC. Potent and specific genetic interference by double-stranded RNA in Caenorhabditis elegans. Nature 1998;391(6669):806–11.

[2] Elbashir SM, Harborth J, Lendeckel W, Yalcin A, Weber K, Tuschl T. Duplexes of 21-nucleotide RNAs mediate RNA interference in cultured mammalian cells. Nature 2001;411(6836):494–8.

[3] Bartel DP. MicroRNAs: target recognition and regulatory functions. Cell 2009;136(2):215–33.

[4] Bushati N, Cohen SM. microRNA functions. Annu Rev Cell Dev Biol 2007;23:175–205.

[5] Ishizu H, Siomi H, Siomi MC. Biology of PIWI-interacting RNAs: new insights into biogenesis and function inside and outside of germlines. Genes Dev 2012;26(21):2361–73.

[6] Ross RJ, Weiner MM, Lin H. PIWI proteins and PIWI-interacting RNAs in hte soma. Nature 2014;505:353–9.

[7] Skalsky RL, Cullen BR. Viruses, microRNAs, and host interactions. Annu Rev Microbiol 2010;64:123–41.

[8] Morris KV. The emerging role of RNA in the regulation of gene transcription in human cells. Seminars Cell Dev Biol 2011;22(4):351–8.

[9] Blahna MT, Hata A. Regulation of miRNA biogenesis as an integrated component of growth factor signaling. Curr Opin Cell Biol 2013;25(2):233–40.

[10] Tran N, Hutvagner G. Biogenesis and the regulation of the maturation of miRNAs. Essays Biochem 2013;54:17–28.

[11] Treiber T, Treiber N, Meister G. Regulation of microRNA biogenesis and function. Thromb Haemost 2012;107(4):605–10.

[12] Cullen BR. Transcription and processing of human microRNA precursors. Mol Cell 2004; 16(6):861–5.

[13] Miyoshi K, Miyoshi T, Siomi H. Many ways to generate microRNA-like small RNAs: non-canonical pathways for microRNA production. Mol Genet Genomics : MGG 2010;284(2): 95–103.

[14] Rother S, Meister G. Small RNAs derived from longer non-coding RNAs. Biochimie 2011;93(11):1905–15.

[15] Cole C, Sobala A, Lu C, Thatcher SR, Bowman A, Brown JW, et al. Filtering of deep sequencing data reveals the existence of abundant Dicer-dependent small RNAs derived from tRNAs. RNA 2009;15(12):2147–60.

[16] Ruby JG, Jan CH, Bartel DP. Intronic microRNA precursors that bypass Drosha processing. Nature 2007;448(7149):83–6.

[17] Ender C, Krek A, Friedlander MR, Beitzinger M, Weinmann L, Chen W, et al. A human snoRNA with microRNA-like functions. Mol Cell 2008;32(4):519–28.

[18] Bommer GT, Gerin I, Feng Y, Kaczorowski AJ, Kuick R, Love RE, et al. p53-mediated activation of miRNA34 candidate tumor-suppressor genes. Curr Biol 2007;17(15):1298–307.

[19] Chang TC, Wentzel EA, Kent OA, Ramachandran K, Mullendore M, Lee KH, et al. Transactivation of miR-34a by p53 broadly influences gene expression and promotes apoptosis. Mol Cell 2007;26(5):745–52.

[20] Raver-Shapira N, Marciano E, Meiri E, Spector Y, Rosenfeld N, Moskovits N, et al. Transcriptional activation of miR-34a contributes to p53-mediated apoptosis. Mol Cell 2007;26(5):731–43.

[21] Kim NH, Kim HS, Kim NG, Lee I, Choi HS, Li XY, et al. p53 and microRNA-34 are suppressors of canonical Wnt signaling. Sci Signal 2011;4(197):ra71.

[22] Kim NH, Kim HS, Li XY, Lee I, Choi HS, Kang SE, et al. A p53/miRNA-34 axis regulates Snail1-dependent cancer cell epithelial-mesenchymal transition. J Cell Biol 2011;195(3): 417–33.

[23] Denli AM, Tops BB, Plasterk RH, Ketting RF, Hannon GJ. Processing of primary microRNAs by the Microprocessor complex. Nature 2004;432(7014):231–5.

[24] Gregory RI, Yan KP, Amuthan G, Chendrimada T, Doratotaj B, Cooch N, et al. The Microprocessor complex mediates the genesis of microRNAs. Nature 2004;432(7014):235–40.

[25] Han J, Lee Y, Yeom KH, Nam JW, Heo I, Rhee JK, et al. Molecular basis for the recognition of primary microRNAs by the Drosha-DGCR8 complex. Cell 2006;125(5):887–901.

[26] Fukuda T, Yamagata K, Fujiyama S, Matsumoto T, Koshida I, Yoshimura K, et al. DEAD-box RNA helicase subunits of the Drosha complex are required for processing of rRNA and a subset of microRNAs. Nat Cell Biol 2007;9(5):604–11.

[27] Sibley CR, Seow Y, Saayman S, Dijkstra KK, El Andaloussi S, Weinberg MS, et al. The biogenesis and characterization of mammalian microRNAs of mirtron origin. Nucleic Acids Res 2012;40(1):438–48.

[28] Ladewig E, Okamura K, Flynt AS, Westholm JO, Lai EC. Discovery of hundreds of mirtrons in mouse and human small RNA data. Genome Res 2012;22(9):1634–45.

[29] Han J, Pedersen JS, Kwon SC, Belair CD, Kim YK, Yeom KH, et al. Posttranscriptional cross-regulation between Drosha and DGCR8. Cell 2009;136(1):75–84.

[30] Herbert KM, Pimienta G, Degregorio SJ, Alexandrov A, Steitz JA. Phosphorylation of DGCR8 increases its intracellular stability and induces a progrowth miRNA profile. Cell Reports 2013;5(4):1070–81.

[31] Macias S, Plass M, Stajuda A, Michlewski G, Eyras E, Caceres JF. DGCR8 HITS-CLIP reveals novel functions for the Microprocessor. Nat Struct Mol Biol 2012;19(8):760–6.

[32] Sellier C, Freyermuth F, Tabet R, Tran T, He F, Ruffenach F, et al. Sequestration of DROSHA and DGCR8 by expanded CGG RNA repeats alters microRNA processing in fragile X-associated tremor/ataxia syndrome. Cell Reports 2013;3(3):869–80.

[33] Davis BN, Hilyard AC, Lagna G, Hata A. SMAD proteins control DROSHA-mediated microRNA maturation. Nature 2008;454(7200):56–61.

[34] Shi Y, Massague J. Mechanisms of TGF-beta signaling from cell membrane to the nucleus. Cell 2003;113(6):685–700.

[35] Davis BN, Hilyard AC, Nguyen PH, Lagna G, Hata A. Smad proteins bind a conserved RNA sequence to promote microRNA maturation by Drosha. Mol Cell 2010;39(3):373–84.

[36] Piriyapongsa J, Jordan IK, Conley AB, Ronan T, Smalheiser NR. Transcription factor binding sites are highly enriched within microRNA precursor sequences. Biol Direct 2011;6:61.

[37] Suzuki HI, Yamagata K, Sugimoto K, Iwamoto T, Kato S, Miyazono K. Modulation of microRNA processing by p53. Nature 2009;460(7254):529–33.

[38] Trabucchi M, Briata P, Garcia-Mayoral M, Haase AD, Filipowicz W, Ramos A, et al. The RNA-binding protein KSRP promotes the biogenesis of a subset of microRNAs. Nature 2009;459(7249):1010–4.

[39] Apponi LH, Corbett AH, Pavlath GK. RNA-binding proteins and gene regulation in myogenesis. Trends Pharmacol Sci 2011;32(11):652–8.

[40] Briata P, Lin WJ, Giovarelli M, Pasero M, Chou CF, Trabucchi M, et al. PI3K/AKT signaling determines a dynamic switch between distinct KSRP functions favoring skeletal myogenesis. Cell Death Differ 2012;19(3):478–87.

[41] Ruggiero T, Trabucchi M, De Santa F, Zupo S, Harfe BD, McManus MT, et al. LPS induces KH-type splicing regulatory protein-dependent processing of microRNA-155 precursors in macrophages. FASEB J 2009;23(9):2898–908.

[42] Zhang X, Wan G, Berger FG, He X, Lu X. The ATM kinase induces microRNA biogenesis in the DNA damage response. Mol Cell 2011;41(4):371–83.

[43] Michlewski G, Caceres JF. Antagonistic role of hnRNP A1 and KSRP in the regulation of let-7a biogenesis. Nat Struct Mol Biol 2010;17(8):1011–8.

[44] Yamagata K, Fujiyama S, Ito S, Ueda T, Murata T, Naitou M, et al. Maturation of microRNA is hormonally regulated by a nuclear receptor. Mol Cell 2009;36(2):340–7.

[45] Chakraborty S, Mehtab S, Patwardhan A, Krishnan Y. Pri-miR-17-92a transcript folds into a tertiary structure and autoregulates its processing. RNA 2012;18(5):1014–28.

[46] Lund E, Guttinger S, Calado A, Dahlberg JE, Kutay U. Nuclear export of microRNA precursors. Science 2004;303(5654):95–8.

[47] Yi R, Qin Y, Macara IG, Cullen BR. Exportin-5 mediates the nuclear export of pre-microRNAs and short hairpin RNAs. Genes Dev 2003;17(24):3011–6.

[48] Okada C, Yamashita E, Lee SJ, Shibata S, Katahira J, Nakagawa A, et al. A high-resolution structure of the pre-microRNA nuclear export machinery. Science 2009;326(5957):1275–9.

[49] Grimm D, Streetz KL, Jopling CL, Storm TA, Pandey K, Davis CR, et al. Fatality in mice due to oversaturation of cellular microRNA/short hairpin RNA pathways. Nature 2006;441(7092):537–41.

[50] Grimm D, Wang L, Lee JS, Schurmann N, Gu S, Borner K, et al. Argonaute proteins are key determinants of RNAi efficacy, toxicity, and persistence in the adult mouse liver. J Clin Invest 2010;120(9):3106–19.

[51] Melo SA, Moutinho C, Ropero S, Calin GA, Rossi S, Spizzo R, et al. A genetic defect in exportin-5 traps precursor microRNAs in the nucleus of cancer cells. Cancer Cell 2010;18(4):303–15.

[52] Chendrimada TP, Gregory RI, Kumaraswamy E, Norman J, Cooch N, Nishikura K, et al. TRBP recruits the Dicer complex to Ago2 for microRNA processing and gene silencing. Nature 2005;436(7051):740–4.

[53] Zhang H, Kolb FA, Jaskiewicz L, Westhof E, Filipowicz W. Single processing center models for human Dicer and bacterial RNase III. Cell 2004;118(1):57–68.

[54] Park JE, Heo I, Tian Y, Simanshu DK, Chang H, Jee D, et al. Dicer recognizes the 5′ end of RNA for efficient and accurate processing. Nature 2011;475(7355):201–5.

[55] Bail S, Swerdel M, Liu H, Jiao X, Goff LA, Hart RP, et al. Differential regulation of microRNA stability. RNA 2010;16(5):1032–9.

[56] Heo I, Joo C, Cho J, Ha M, Han J, Kim VN. Lin28 mediates the terminal uridylation of let-7 precursor MicroRNA. Mol Cell 2008;32(2):276–84.

[57] Viswanathan SR, Daley GQ. Lin28: a MicroRNA regulator with a macro role. Cell 2010; 140(4):445–9.

[58] Rau F, Freyermuth F, Fugier C, Villemin JP, Fischer MC, Jost B, et al. Misregulation of miR-1 processing is associated with heart defects in myotonic dystrophy. Nat Struct Mol Biol 2011;18(7):840–5.

[59] Lu L, Katsaros D, Shaverdashvili K, Qian B, Wu Y, de la Longrais IA, et al. Pluripotent factor lin-28 and its homologue lin-28b in epithelial ovarian cancer and their associations with disease outcomes and expression of let-7a and IGF-II. Eur J Cancer 2009;45(12):2212–8.

[60] Li N, Zhong X, Lin X, Guo J, Zou L, Tanyi JL, et al. Lin-28 homologue A (LIN28A) promotes cell cycle progression via regulation of cyclin-dependent kinase 2 (CDK2), cyclin D1 (CCND1), and cell division cycle 25 homolog A (CDC25A) expression in cancer. J Biol Chem 2012;287(21):17386–97.

[61] Yang W, Chendrimada TP, Wang Q, Higuchi M, Seeburg PH, Shiekhattar R, et al. Modulation of microRNA processing and expression through RNA editing by ADAR deaminases. Nat Struct Mol Biol 2006;13(1):13–21.

[62] Kawahara Y, Zinshteyn B, Chendrimada TP, Shiekhattar R, Nishikura K. RNA editing of the microRNA-151 precursor blocks cleavage by the Dicer-TRBP complex. EMBO Reports 2007;8(8):763–9.

[63] Gebert LF, Rebhan MA, Crivelli SE, Denzler R, Stoffel M, Hall J. Miravirsen (SPC3649) can inhibit the biogenesis of miR-122. Nucleic Acids Res 2014;42(1):609–21.

[64] Janssen HL, Reesink HW, Lawitz EJ, Zeuzem S, Rodriguez-Torres M, Patel K, et al. Treatment of HCV infection by targeting microRNA. N Engl J Med 2013;368(18):1685–94.

[65] Noland CL, Ma E, Doudna JA. siRNA repositioning for guide strand selection by human Dicer complexes. Mol Cell 2011;43(1):110–21.

[66] Joshua-Tor L, Hannon GJ. Ancestral roles of small RNAs: an Ago-centric perspective. Cold Spring Harb Perspect Biol 2011;3(10):a003772.

[67] Friedman RC, Farh KK, Burge CB, Bartel DP. Most mammalian mRNAs are conserved targets of microRNAs. Genome Res 2009;19(1):92–105.

[68] Grimson A, Farh KK, Johnston WK, Garrett-Engele P, Lim LP, Bartel DP. MicroRNA targeting specificity in mammals: determinants beyond seed pairing. Mol Cell 2007;27(1):91–105.

[69] Guo H, Ingolia NT, Weissman JS, Bartel DP. Mammalian microRNAs predominantly act to decrease target mRNA levels. Nature 2010;466(7308):835–40.

[70] Eulalio A, Huntzinger E, Izaurralde E. Getting to the root of miRNA-mediated gene silencing. Cell 2008;132(1):9–14.

[71] Eulalio A, Behm-Ansmant I, Schweizer D, Izaurralde E. P-body formation is a consequence, not the cause, of RNA-mediated gene silencing. Mol Cell Biol 2007;27(11):3970–81.

[72] Cifuentes D, Xue H, Taylor DW, Patnode H, Mishima Y, Cheloufi S, et al. A novel miRNA processing pathway independent of Dicer requires Argonaute2 catalytic activity. Science 2010;328(5986):1694–8.

[73] Cheloufi S, Dos Santos CO, Chong MM, Hannon GJ. A dicer-independent miRNA biogenesis pathway that requires Ago catalysis. Nature 2010;465(7298):584–9.

[74] Liu YP, Schopman NC, Berkhout B. Dicer-independent processing of short hairpin RNAs. Nucleic Acids Res 2013;41(6):3723–33.

[75] Ruegger S, Grosshans H. MicroRNA turnover: when, how, and why. Trends Biochem Sci 2012;37(10):436–46.

[76] Das SK, Sokhi UK, Bhutia SK, Azab B, Su ZZ, Sarkar D, et al. Human polynucleotide phosphorylase selectively and preferentially degrades microRNA-221 in human melanoma cells. Proc Natl Acad Sci USA 2010;107(26):11948–53.

[77] Rissland OS, Hong SJ, Bartel DP. MicroRNA destabilization enables dynamic regulation of the miR-16 family in response to cell-cycle changes. Mol Cell 2011;43(6):993–1004.

[78] Avraham R, Sas-Chen A, Manor O, Steinfeld I, Shalgi R, Tarcic G, et al. EGF decreases the abundance of microRNAs that restrain oncogenic transcription factors. Sci Signal 2010;3(124):ra43.

[79] Ameres SL, Horwich MD, Hung JH, Xu J, Ghildiyal M, Weng Z, et al. Target RNA-directed trimming and tailing of small silencing RNAs. Science 2010;328(5985):1534–9.

[80] Xie J, Ameres SL, Friedline R, Hung JH, Zhang Y, Xie Q, et al. Long-term, efficient inhibition of microRNA function in mice using rAAV vectors. Nat Methods 2012;9(4):403–9.

[81] Cazalla D, Yario T, Steitz JA. Down-regulation of a host microRNA by a Herpesvirus saimiri noncoding RNA. Science 2010;328(5985):1563–6.

[82] Salmena L, Poliseno L, Tay Y, Kats L, Pandolfi PP. A ceRNA hypothesis: the Rosetta Stone of a hidden RNA language? Cell 2011;146(3):353–8.

[83] Tay Y, Pandolfi PP. The multilayered complexity of csRNA crosstalk and competition. Nature 2014;505:344–52.

[84] Hansen TB, Jensen TI, Clausen BH, Bramsen JB, Finsen B, Damgaard CK, et al. Natural RNA circles function as efficient microRNA sponges. Nature 2013;495(7441):384–8.

[85] Memczak S, Jens M, Elefsinioti A, Torti F, Krueger J, Rybak A, et al. Circular RNAs are a large class of animal RNAs with regulatory potency. Nature 2013;495(7441):333–8.

[86] Ebert MS, Neilson JR, Sharp PA. MicroRNA sponges: competitive inhibitors of small RNAs in mammalian cells. Nat Methods 2007;4(9):721–6.

[87] Fellman C, Lowe SW. Stable RNA interference rules for silencing. Nat Cell Biol 2014;16: 10–8.

[88] Huesken D, Lange J, Mickanin C, Weiler J, Asselbergs F, Warner J, et al. Design of a genome-wide siRNA library using an artificial neural network. Nat Biotechnol 2005;23(8):995–1001.

[89] Shabalina SA, Spiridonov AN, Ogurtsov AY. Computational models with thermodynamic and composition features improve siRNA design. BMC Bioinforma 2006;7:65.

[90] Wang X, Wang X, Varma RK, Beauchamp L, Magdaleno S, Sendera TJ. Selection of hyperfunctional siRNAs with improved potency and specificity. Nucleic Acids Res 2009;37(22):e152.

[91] Behlke MA. Chemical modification of siRNAs for in vivo use. Oligonucleotides 2008; 18(4):305–19.

[92] Engels JW. Gene silencing by chemically modified siRNAs. N Biotechnol 2012;30(3):302–7.

[93] Rettig GR, Behlke MA. Progress toward in vivo use of siRNAs-II. Mol Ther: J Am Soc Gene Ther 2012;20(3):483–512.

[94] Chang CI, Kim HA, Dua P, Kim S, Li CJ, Lee DK. Structural diversity repertoire of gene silencing small interfering RNAs. Nucleic Acid Ther 2011;21(3):125–31.

[95] Kim DH, Behlke MA, Rose SD, Chang MS, Choi S, Rossi JJ. Synthetic dsRNA Dicer substrates enhance RNAi potency and efficacy. Nat Biotechnol 2005;23(2):222–6.

[96] Rose SD, Kim DH, Amarzguioui M, Heidel JD, Collingwood MA, Davis ME, et al. Functional polarity is introduced by Dicer processing of short substrate RNAs. Nucleic Acids Res 2005;33(13):4140–56.

[97] Chang CI, Yoo JW, Hong SW, Lee SE, Kang HS, Sun X, et al. Asymmetric shorter-duplex siRNA structures trigger efficient gene silencing with reduced nonspecific effects. Mol Ther: J Am Soc Gene Ther 2009;17(4):725–32.

[98] Sun X, Rogoff HA, Li CJ. Asymmetric RNA duplexes mediate RNA interference in mammalian cells. Nat Biotechnol 2008;26(12):1379–82.

[99] Sierant M, Kazmierczak-Baranska J, Paduszynska A, Sobczak M, Pietkiewicz A, Nawrot B. Longer 19-base pair short interfering RNA duplexes rather than shorter duplexes trigger RNA interference. Oligonucleotides 2010;20(4):199–206.

[100] Reynolds A, Leake D, Boese Q, Scaringe S, Marshall WS, Khvorova A. Rational siRNA design for RNA interference. Nat Biotechnol 2004;22(3):326–30.

[101] Hall AH, Wan J, Shaughnessy EE, Ramsay Shaw B, Alexander KA. RNA interference using boranophosphate siRNAs: structure-activity relationships. Nucleic Acids Res 2004;32(20): 5991–6000.

[102] Choung S, Kim YJ, Kim S, Park HO, Choi YC. Chemical modification of siRNAs to improve serum stability without loss of efficacy. Biochem Biophys Res Commun 2006;342(3):919–27.

[103] Hoerter JA, Walter NG. Chemical modification resolves the asymmetry of siRNA strand degradation in human blood serum. RNA 2007;13(11):1887–93.

[104] Werk D, Wengel J, Wengel SL, Grunert HP, Zeichhardt H, Kurreck J. Application of small interfering RNAs modified by unlocked nucleic acid (UNA) to inhibit the heart-pathogenic coxsackievirus B3. FEBS Lett 2010;584(3):591–8.

[105] Mook OR, Baas F, de Wissel MB, Fluiter K. Evaluation of locked nucleic acid-modified small interfering RNA in vitro and in vivo. Mol Cancer Ther 2007;6(3):833–43.

[106] Swayze EE, Siwkowski AM, Wancewicz EV, Migawa MT, Wyrzykiewicz TK, Hung G, et al. Antisense oligonucleotides containing locked nucleic acid improve potency but cause significant hepatotoxicity in animals. Nucleic Acids Res 2007;35(2):687–700.

[107] Birmingham A, Anderson EM, Reynolds A, Ilsley-Tyree D, Leake D, Fedorov Y, et al. 3′ UTR seed matches, but not overall identity, are associated with RNAi off-targets. Nat Methods 2006;3(3):199–204.

[108] Jackson AL, Burchard J, Schelter J, Chau BN, Cleary M, Lim L, et al. Widespread siRNA "off-target" transcript silencing mediated by seed region sequence complementarity. RNA 2006;12(7):1179–87.

[109] Anderson EM, Birmingham A, Baskerville S, Reynolds A, Maksimova E, Leake D, et al. Experimental validation of the importance of seed complement frequency to siRNA specificity. RNA 2008;14(5):853–61.

[110] Jackson AL, Burchard J, Leake D, Reynolds A, Schelter J, Guo J, et al. Position-specific chemical modification of siRNAs reduces "off-target" transcript silencing. RNA 2006;12(7): 1197–205.

[111] Ui-Tei K, Naito Y, Zenno S, Nishi K, Yamato K, Takahashi F, et al. Functional dissection of siRNA sequence by systematic DNA substitution: modified siRNA with a DNA seed arm is a powerful tool for mammalian gene silencing with significantly reduced off-target effect. Nucleic Acids Res 2008;36(7):2136–51.

[112] Marimani MD, Ely A, Buff MC, Bernhardt S, Engels JW, Arbuthnot P. Inhibition of hepatitis B virus replication in cultured cells and in vivo using 2′-O-guanidinopropyl modified siRNAs. Bioorg Med Chem 2013;21(20):6145–55.

[113] Bramsen JB, Pakula MM, Hansen TB, Bus C, Langkjaer N, Odadzic D, et al. A screen of chemical modifications identifies position-specific modification by UNA to most potently reduce siRNA off-target effects. Nucleic Acids Res 2010;38(17):5761–73.

[114] Puri N, Wang X, Varma R, Burnett C, Beauchamp L, Batten DM, et al. LNA incorporated siRNAs exhibit lower off-target effects compared to 2′-OMethoxy in cell phenotypic assays and microarray analysis. Nucleic Acids Symp Ser 2008;52:25–6.

[115] Chen PY, Weinmann L, Gaidatzis D, Pei Y, Zavolan M, Tuschl T, et al. Strand-specific 5′-O-methylation of siRNA duplexes controls guide strand selection and targeting specificity. RNA 2008;14(2):263–74.

[116] Bramsen JB, Laursen MB, Damgaard CK, Lena SW, Babu BR, Wengel J, et al. Improved silencing properties using small internally segmented interfering RNAs. Nucleic Acids Res 2007;35(17):5886–97.

[117] Judge A, MacLachlan I. Overcoming the innate immune response to small interfering RNA. Hum Gene Ther 2008;19(2):111–24.

[118] Robbins M, Judge A, MacLachlan I. siRNA and innate immunity. Oligonucleotides 2009; 19(2):89–102.

[119] Robbins M, Judge A, Ambegia E, Choi C, Yaworski E, Palmer L, et al. Misinterpreting the therapeutic effects of small interfering RNA caused by immune stimulation. Hum Gene Ther 2008;19(10):991–9.

[120] Chen X, Qian Y, Yan F, Tu J, Yang X, Xing Y, et al. 5′-triphosphate-siRNA activates RIG-I-dependent type I interferon production and enhances inhibition of hepatitis B virus replication in HepG2.2.15 cells. Eur J Pharmacol 2013;721(1–3):86–95.

[121] Ebert G, Poeck H, Lucifora J, Baschuk N, Esser K, Esposito I, et al. 5′ Triphosphorylated small interfering RNAs control replication of hepatitis B virus and induce an interferon response in human liver cells and mice. Gastroenterology 2011;141(2):696–706. e1–3.

[122] Han Q, Zhang C, Zhang J, Tian Z. Reversal of hepatitis B virus-induced immune tolerance by an immunostimulatory 3p-HBx-siRNAs in a retinoic acid inducible gene I-dependent manner. Hepatology 2011;54(4):1179–89.

[123] Judge AD, Bola G, Lee AC, MacLachlan I. Design of noninflammatory synthetic siRNA mediating potent gene silencing in vivo. Mol Ther: J Am Soc Gene Ther 2006;13(3):494–505.

[124] Robbins M, Judge A, Liang L, McClintock K, Yaworski E, MacLachlan I. 2′-O-methyl-modified RNAs act as TLR7 antagonists. Mol Ther: J Am Soc Gene Ther 2007;15(9):1663–9.

[125] Arbuthnot P. MicroRNA-like antivirals. Biochim Biophys Acta 2011;1809(11–12):746–55.

[126] Weinberg MS, Ely A, Barichievy S, Crowther C, Mufamadi S, Carmona S, et al. Specific inhibition of HBV replication in vitro and in vivo with expressed long hairpin RNA. Mol Ther : J Am Soc Gene Ther 2007;15(3):534–41.

[127] Saayman S, Arbuthnot P, Weinberg MS. Deriving four functional anti-HIV siRNAs from a single Pol III-generated transcript comprising two adjacent long hairpin RNA precursors. Nucleic Acids Res 2010;38(19):6652–63.

[128] Boden D, Pusch O, Lee F, Tucker L, Ramratnam B. Human immunodeficiency virus type 1 escape from RNA interference. J Virol 2003;77(21):11531–5.

[129] Das AT, Brummelkamp TR, Westerhout EM, Vink M, Madiredjo M, Bernards R, et al. Human immunodeficiency virus type 1 escapes from RNA interference-mediated inhibition. J Virol 2004;78(5):2601–5.

[130] Westerhout EM, Ooms M, Vink M, Das AT, Berkhout B. HIV-1 can escape from RNA interference by evolving an alternative structure in its RNA genome. Nucleic Acids Res 2005;33(2):796–804.

[131] Wilson JA, Richardson CD. Hepatitis C virus replicons escape RNA interference induced by a short interfering RNA directed against the NS5b coding region. J Virol 2005;79(11):7050–8.

[132] Li L, Lin X, Khvorova A, Fesik SW, Shen Y. Defining the optimal parameters for hairpin-based knockdown constructs. RNA 2007;13(10):1765–74.

[133] Gu S, Jin L, Zhang Y, Huang Y, Zhang F, Valdmanis PN, et al. The loop position of shRNAs and pre-miRNAs is critical for the accuracy of dicer processing in vivo. Cell 2012;151(4):900–11.

[134] Bertrand E, Castanotto D, Zhou C, Carbonnelle C, Lee NS, Good P, et al. The expression cassette determines the functional activity of ribozymes in mammalian cells by controlling their intracellular localization. RNA 1997;3(1):75–88.

[135] Kruger W, Benecke BJ. Structural and functional analysis of a human 7 S K RNA gene. J Mol Biol 1987;195(1):31–41.

[136] Baer M, Nilsen TW, Costigan C, Altman S. Structure and transcription of a human gene for H1 RNA, the RNA component of human RNase P. Nucleic Acids Res 1990;18(1):97–103.

[137] Dyer V, Ely A, Bloom K, Weinberg M, Arbuthnot P. tRNA Lys3 promoter cassettes that efficiently express RNAi-activating antihepatitis B virus short hairpin RNAs. Biochem Biophys Res Commun 2010;398(4):640–6.

[138] Scherer LJ, Frank R, Rossi JJ. Optimization and characterization of tRNA-shRNA expression constructs. Nucleic Acids Res 2007;35(8):2620–8.

[139] Boden D, Pusch O, Lee F, Tucker L, Shank PR, Ramratnam B. Promoter choice affects the potency of HIV-1 specific RNA interference. Nucleic Acids Res 2003;31(17):5033–8.

[140] Bogerd HP, Karnowski HW, Cai X, Shin J, Pohlers M, Cullen BR. A mammalian herpesvirus uses noncanonical expression and processing mechanisms to generate viral MicroRNAs. Mol Cell 2010;37(1):135–42.

[141] Czauderna F, Santel A, Hinz M, Fechtner M, Durieux B, Fisch G, et al. Inducible shRNA expression for application in a prostate cancer mouse model. Nucleic Acids Res 2003;31(21):e127.

[142] Liu YP, Haasnoot J, ter Brake O, Berkhout B, Konstantinova P. Inhibition of HIV-1 by multiple siRNAs expressed from a single microRNA polycistron. Nucleic Acids Res 2008;36(9): 2811–24.

[143] Aagaard LA, Zhang J, von Eije KJ, Li H, Saetrom P, Amarzguioui M, et al. Engineering and optimization of the miR-106b cluster for ectopic expression of multiplexed anti-HIV RNAs. Gene Ther 2008;15(23):1536–49.

[144] Ely A, Naidoo T, Arbuthnot P. Efficient silencing of gene expression with modular trimeric Pol II expression cassettes comprising microRNA shuttles. Nucleic Acids Res 2009;37(13):e91.

[145] DeVincenzo J, Cehelsky JE, Alvarez R, Elbashir S, Harborth J, Toudjarska I, et al. Evaluation of the safety, tolerability and pharmacokinetics of ALN-RSV01, a novel RNAi antiviral therapeutic directed against respiratory syncytial virus (RSV). Antivir Res 2008;77(3):225–31.

[146] Simon AR, Karsten V, Meyers R, Cehelsky J, Gollob J, Vaishnaw A, et al. Preclinical and clinical studies employing RNA interference as a therapeutic for respiratory syncytial virus (RSV) infection in the Lung. In: RNA Interference from Biology to Therapeutics. Springer; 2013. p. 315–28.

[147] Geisbert TW, Lee AC, Robbins M, Geisbert JB, Honko AN, Sood V, et al. Postexposure protection of non-human primates against a lethal Ebola virus challenge with RNA interference: a proof-of-concept study. Lancet 2010;375(9729):1896–1905.

[148] Thi EP, Mire CE, Ursic-Bedoya R, Geisbert JB, Lee AC, Agans KN, et al. Marburg virus infection in nonhuman primates: therapeutic treatment by lipid-encapsulated siRNA. Sci Transl Med 2014;6(250):250ra116.

[149] Berkhout B. Toward a durable anti-HIV gene therapy based on RNA interference. Ann NY Acad Sci 2009;1175:3–14.

[150] Leonard JN, Schaffer DV. Computational design of antiviral RNA interference strategies that resist human immunodeficiency virus escape. J Virol 2005;79(3):1645–54.

CHAPTER 3

Engineering Sequence-Specific DNA Binding Proteins for Antiviral Gene Editing

3.1 INTRODUCTION

Human immunodeficiency virus-1 (HIV-1), hepatitis B virus (HBV), herpes simplex virus-2 (HSV-2), and human papilloma virus (HPV) are collectively very significant causes of global mortality and morbidity. A characteristic of these viruses is that they cause chronic infection and replicate using DNA intermediates. The DNA templates serve as latent reservoirs for the virus and are capable of reactivation of replication. Available therapies do not affect latent DNA directly, and the ability to inactivate viral DNA by targeted gene editing would present a very powerful means of permanently disabling viral replication. Therefore, the significant developments in engineering customized DNA-binding proteins have been of considerable interest to advancing gene therapy for the treatment of chronic viral infections. There are currently four main classes of sequence-specific binding proteins (Table 3.1):

1. zinc finger proteins (ZFPs),
2. transcription activator-like effectors (TALEs),
3. homing endonucleases (HEs) or meganucleases, and
4. clustered regularly interspaced palindromic repeats (CRISPR) with CRISPR-associated nuclease (Cas) systems.

Each of these groups has provided a platform to enable generation of binding proteins with an intended specificity. The CRISPR/Cas system is different from the other groups in that base pairing of an RNA guide determines sequence specificity. Interaction of ZFPs, TALEs, and HEs with their targets is determined by protein–DNA target binding specificity. The capability of DNA-binding specificity has been exploited to engineer additional functionality through coupling of nuclease, transcriptional activating, and transcriptional suppressing motifs.

Ensuring specificity of interaction with targets, and minimizing off-target effects, is critically important for clinical use of these DNA-binding proteins. This topic is a major focus of research and will no doubt be the main determinant of therapeutic success of targeted gene editing. The number of bases within a sequence of a viral target site that is required to keep off-target interaction to a minimum

Gene Therapy for Viral Infections. http://dx.doi.org/10.1016/B978-0-12-410518-8.00003-X

Table 3.1 Comparison of Different Types of Gene Editors That Have Been Developed for Use against Viral Infections

Type of gene editor	Mechanism of DNA sequence recognition	Ease of propagation of effective sequences	Versatility of functionalization	Target length and specificity	Size of DNA encoding editing sequence	Current stage of antiviral application
ZFP derivatives	Protein–DNA interaction	Difficult because of context-dependent effects on DNA binding by individual fingers	Versatile	Variable target length of 9–18 bp for monomers. Good specificity	Variable, but typically ~2 kbp	Clinical trial application against HIV-1 infection (Chapter 8)
TALE derivatives	Protein–DNA interaction	Easy, although preparation of tandemly arranged monomers may be time consuming	Versatile	Usually 18 bp for TALE and 36 bp for dimeric TALEN. Good specificity and some evidence to suggest better specificity than ZFP derivatives [42]	Large. Cassettes may be larger than 4 kbp and doubled when dimeric nucleases are required	Preclinical assessment against HIV-1 and HBV in cell culture and small animal models (Chapters 6 and 8)
CRISPR/Cas derivatives	Guided by complementary RNA	Very easy. Requires the generation of a cassette that expresses the targeting guide	Versatile	sgRNA targets 17–20 bp. Initially specificity thought to be poor, but technology improving	Large. Cassettes may be larger than 4 kbp	Preclinical assessment against HIV-1 and HBV in cell culture and small animal models (Chapters 6 and 8)
HEs	Protein–DNA interaction	Highly difficult. Co-location of enzymatic and DNA-binding regions of the HEs complicates engineering of target-specific nuclease function	Largely restricted to nuclease function	Target sites may be ~14–40 bp in length. Specificity may be good but complicated by difficulty of engineering effective HEs	Small, ~1 kbp	Preclinical assessment against HSV in cell culture

CRISPR/Cas, Clustered regularly interspaced palindromic repeats (CRISPR) with CRISPR-associated nuclease (Cas); HEs, Homing endonucleases; sgRNA, short guide RNA; TALE, Transcription activator-like effector; ZFP, Zinc Finger protein.

is estimated to be at least 17 [1]. The probability of a random 17-bp target site occurring within the human genome is approximately 0.15, and for an 18-bp target the probability is approximately 0.02. Although this should provide adequate flexibility for design of elements that target viral DNA with good specificity, other factors may contribute to the efficiency of this class of candidate antivirals. These include the inherent incomplete specificity for targets and the presence of human sequences that have homology to viral targets. Avoiding targets that have partial homology to host exonic or transcriptional regulatory elements is also important. In addition to directly targeting viral sequences, disabling host factors that are required for viral infection is another potential therapeutic use. Preliminary trial data indicate that use of this approach to disrupt the CCR5 co-receptor in autologous $CD4^+$ T cells has potential for HIV-1 therapy [2].

3.1.1 Targeted Gene Editing with Engineered Nucleases

Formation of double-stranded breaks (DSBs) at intended sites provides the basis for gene editing with customized nucleases. Normally such DSBs are repaired by nonhomologous end joining (NHEJ). Ligation of the ends of the cleaved DNA restores the sequence, but with repeated target cleavage mutations such as insertions and deletions (indels) are introduced at the target sites. The resulting disruption to a viral sequence may disable replication, which has therapeutic potential. When sequences with homology to those flanking the cleavage target are also introduced into a cell, homology-directed repair (HDR) may be induced. HDR is considerably more efficient in the presence of DSBs, and this property has been widely exploited for targeted insertion during genome modification (reviewed in refs [3,4]). In addition to therapeutic uses, targeted genome editing has been used extensively to generate model organisms, such as "knock-in" murine lines.

3.2 ZINC FINGER PROTEINS

ZFPs naturally function as transcription factors, and DNA sequence-specific binding properties are conferred by the finger-like domains of the proteins [5]. ZFP-encoding sequences are abundant in eukaryotic genomes and are now known to have diverse functions that involve recognition of molecules other than DNA [6]. Cys_2His_2 motifs are a characteristic of ZFPs, and finger-like structures are formed by a Zn^{2+} atom that coordinates formation of protein folds through interaction with the two cysteine and two histidine residues (Figure 3.1(A)). Each finger comprises approximately 30 amino acids and has a $\beta\beta\alpha$ helical structure. Amino acids exposed on the tip of the α helix of each finger bind specifically to three bases within the major groove of DNA. Tandem arrangement of modular finger domains confers specificity of binding to longer DNA sequences. Understanding of the structural features of ZFPs has enabled the application of these proteins to regulating and editing gene function [7].

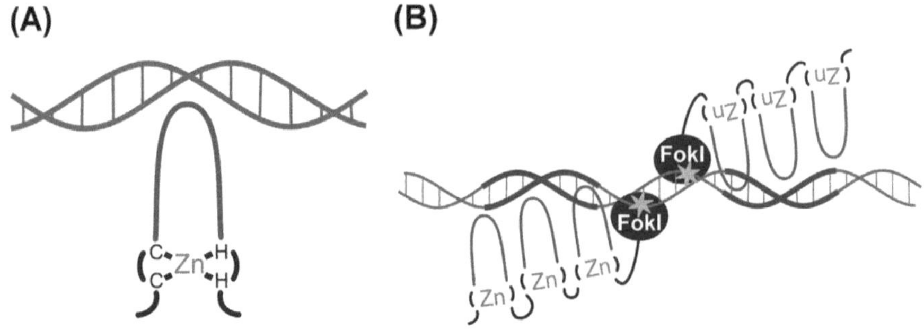

FIGURE 3.1 **Schematic illustration of a zinc finger motif and a dimeric zinc finger nuclease (ZFN).** (A) The zinc finger comprises a loop of $\beta\beta\alpha$ motifs that are formed by histidine and cysteine amino acids coordinated by a zinc atom. The amino acids at the tip of the finger interact with three bases in the major groove of the cognate DNA. (B) ZFNs comprise tandemly arranged zinc fingers, three in the illustrated example, which interact with a sequence of the intended target. Monomers of the complete ZFN bind to opposite strands of the target in a head-to-head arrangement. FokI nuclease domains, coupled to the arrays of zinc fingers, create nicks (stars) on opposite strands of the target within a spacer region between the monomer binding sites.

3.2.1 Engineering ZFPs with Custom DNA-Binding Properties

The modular arrangement of ZFPs, together with the identification of conserved linkers that occur between the fingers, has been used to engineer artificial sequence-specific binding proteins. Engineering dimeric ZFPs with coupled nicking nuclease function (Figure 3.1(B)) has been widely and successfully used. Although theoretically convenient, linking together fingers with individual defined specificity to constitute ZFPs does not reliably generate proteins with suitable DNA binding affinities. Particularly problematic are the following factors:

1. the efficiency of binding of individual fingers to their triplet cognates is influenced by the neighboring fingers,
2. individual fingers have variable specificity, and
3. certain nucleotide triplets are not recognized specifically by zinc finger elements.

To overcome the problems of context-dependent binding of fingers within a tandem array, elaborate and creative methods of selecting ZFPs with desired binding properties have been developed. One such approach, the oligomerized pool engineering (OPEN) bacterial two hybrid system, has been widely used for generating ZFNs [8]. After identification of suitable targets, bacterial strains are generated that harbor ZFN half sites within reporter gene constructs. This is followed by use of pool libraries of ZFP-encoding bacteriophages and selection of ZFPs with high-affinity target binding. Although the procedure entails a clever application of bacterial genetics to facilitate selection of the ZFNs, the procedures are time consuming and complicated. Therefore, refinements of the OPEN method have been devised that entail selection-free use of context-dependent assembly of three-finger arrays [9]. The method utilizes knowledge about functionality of N- and C-terminal fingers that have previously been shown to bind targets with good affinity when containing a common middle finger. Commercially available services have also been developed for generating artificial ZFPs. These are available through Sangamo Biosciences (Richmond, CA), which has developed the platform for generating designer ZFPs in partnership with Sigma-Aldrich (St. Louis, MO).

3.2.2 Adding Function to ZFPs

ZFPs and their derivatives have been the subject of research performed over a long period of time. The first demonstration that it is possible to design ZFPs to bind to specific target sites was reported 20 years ago by Choo, Sanchez-Garcia, and Klug [10]. In this study, a protein comprising three fingers was engineered to recognize a unique sequence of 9 bp occurring at the BCR-ABL fusion site that is commonly found in malignant cells of chronic myeloid leukemia. The artificial ZFP was capable of inhibiting gene transcription at the binding site. Since this seminal study, application of ZFPs to gene modifications has progressed significantly. Currently, there is a wealth of information on the topic, and engineered derivatives of ZFPs are now at advanced stages of development for clinical use. Particularly promising is the rendering of $CD4^+$ cells resistant to HIV infection after targeted mutation of the *CCR5* gene by ZFNs [2].

Fusion of the type IIS restriction enzyme, FokI, has commonly been used to generate ZFNs and was first reported in 1996 [11]. As a type IIS restriction enzyme, cleavage and DNA recognition sites of FokI are distinct. Normally, FokI functions as a dimer, and nuclease domains function as nickases by cleaving single strands of the target [12]. ZFNs have been generated by coupling DNA binding motifs of ZFPs, which comprise an engineered array of finger modules to the nickase of FokI. As with the natural FokI enzyme, ZFNs are engineered to form dimers. Two subunits of a ZFN bind to opposite strands of a duplex, and a nick in each strand is generated such that a DSB is formed at the targeted DNA. Repeated cleavage and error-prone repair by NHEJ leads to site-specific mutation. Improvement in specificity of ZFNs has been achieved by making use of nuclease domains that function as obligate heterodimers [13] (Figure 3.2). Complementary electrostatic and hydrophobic interactions at the interface between nuclease domains, engineered by site-directed mutagenesis, favors formation of functional heterodimeric nucleases. Repulsion between homodimers inhibits dimerization and therefore nuclease function.

It has long been known that direct binding of ZFPs to transcription regulatory elements is capable of impeding access by transcriptional factors to their regulatory elements [10]. This inhibitory effect has been demonstrated in several settings, and by showing suppression of duck HBV replication potential utility of the mechanism for treating HBV infection has been proposed [14]. Although

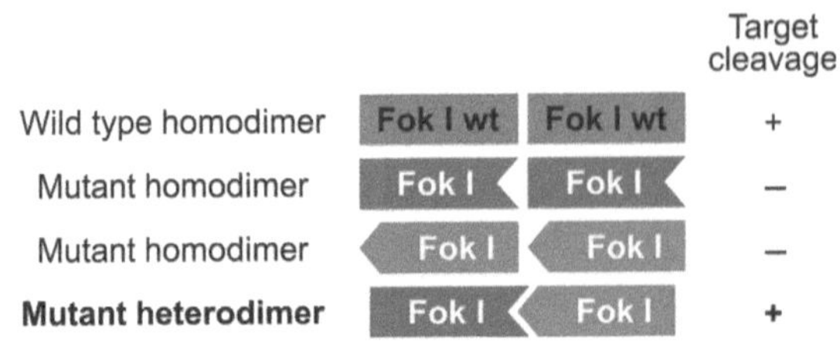

FIGURE 3.2 Rationale for the use of obligate heterodimers of FokI nuclease to improve cleavage specificity.
Juxtaposition of the wild-type FokI monomers results in digestion of both target strands. Homomonomers of the mutant forms of FokI are not compatible with target cleavage, whereas favorable interaction of the monomers in a heterodimer arrangement enables cleavage of both strands of the target DNA.

suppression may be effective, the inhibition may not be sufficiently durable for therapeutic use. To enhance transcriptional suppression, the Krüppel-associated box (KRAB) has commonly been fused to DNA-targeting ZFPs [15]. The KRAB domain is naturally found in ZFPs that perform transcription inhibition in tetrapods [16]. The potent inhibitory effect is caused by recruitment of KRAB-associated protein-1, which serves as a scaffold for other factors such as heterochromatin protein-1 and various histone-modifying proteins [17]. Resultant heterochromatin induction leads to potent transcriptional suppression, which may be sustained and effective over long ranges of DNA [18]. Other suppressor motifs have also been fused to ZFPs to achieve target gene suppression [15], but their use has not been as popular as that of KRAB.

To achieve transcription activation with engineered ZFPs, the viral protein 16 (VP16) of HSV, or the tetrameric combination of this protein (VP64), has been used successfully [15,19]. VP16 is a potent transcription activator and is essential for regulating expression of HSV genes [20]. Transcriptional activation with VP16 or VP64 has been used in many studies that have been aimed at functional gene analyses but few if any that use the technology for treatment of viral infections.

3.3 TRANSCRIPTION ACTIVATOR-LIKE EFFECTORS

TALEs are naturally found in the *Xanthomonas* genus of plant pathogenic bacteria. Species of *Xanthomonas* infect a wide variety of plants, and pathogenic forms that infect rice and citrus plants may be particularly destructive of agricultural crops [21]. TALEs enable bacteria to regulate gene expression in host cells and thereby enhance their survival. To have this effect, TALEs are secreted from the bacteria via a type III secretory system [22]. The mechanism involves injection of the bacterial protein through a needle-like structure to transport the TALEs from the *Xanthomonas* bacteria into the host plant cell. AvrBs3 is one of the best characterized TALEs. When introduced into host cells, it increases expression of UPA20 [23]. This helix-loop-helix-containing transcriptional

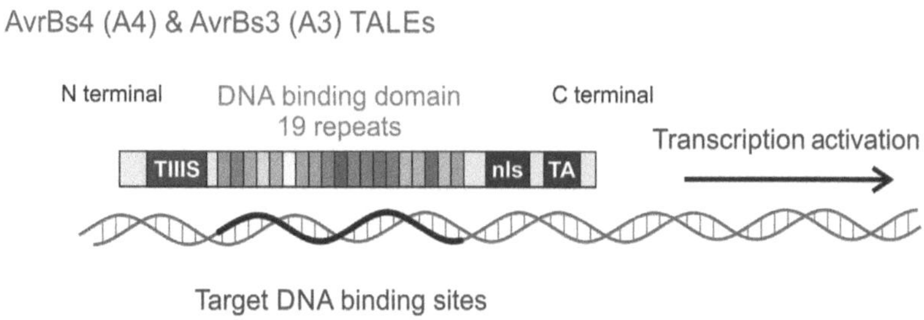

FIGURE 3.3 Structure of the natural transcription activator-like effectors (TALEs).
After infection of plant cells, Xanthomonas bacteria secrete TALEs via a type III secretory system, which requires the function of a signal peptide sequence (TIIIS). The DNA-binding domain comprises a tandem arrangement of monomeric units that each binds to a specific nucleotide on one strand of the target DNA. A nuclear localization signal (nls) directs the TALE to the plant cell nucleus, and a transcription activator (TA) motif is responsible for targeted gene activation.

factor is a master regulator of cell growth, and, after activation, it causes a cascade of cellular events that culminates in hypertrophy of plant cells.

Naturally, TALEs characteristically comprise three main domains (Figure 3.3):

1. a nuclear localization signal (nls) at the C terminal end;
2. an acid activation domain that is involved in transcription activation through the recruitment of host factors; and
3. a central region, which comprises a tandem arrangement of modules and constitutes the DNA binding component of the TALE.

Specificity of target gene regulation is achieved by interaction of the set of DNA-binding modules with their target sequences [24,25]. It was this property of TALEs that attracted the interest of researchers working on targeted gene editing. Each of the modules of the DNA binding domain comprises 33–35 amino acids (reviewed in refs [22,26,27]). Typically, there are 18.5 modular repeats within a DNA binding domain, and each recognizes a single nucleotide in a DNA sequence. The C-terminal repeat usually comprises 20 amino acids; therefore, it is referred to as a half module. The sequence of amino acids in each of the modules is similar except for the amino acids at positions 12 and 13. These two residues constitute the repeat variable diresidues (RVDs) and are essential for nucleotide binding specificity. The sequence of RVDs within the tandem of monomers determines the DNA to which the TALE will bind.

Structural analysis has shown that each monomer presents an RVD on an apical loop that is formed between two left-handed helical bundles [28,29]. The 13th amino acid forms a base-specific contact with the DNA, and the 12th amino acid stabilizes the protein–DNA interaction. The series of monomers is arranged in a right-handed superhelical structure, and TALE-DNA interactions occur within the major groove of the cognate DNA.

3.3.1 Engineering TALEs with Sequence-Specific DNA Binding Properties

Deciphering the specificity that RVDs have for nitrogen bases within DNA [24,25] was a significant development in advancing use of TALE derivatives for gene editing [23,24]. There is some degeneracy to the code. In some cases, more than one RVD may bind to the same base; conversely, the same RVD may bind to different bases. Typically, RVDs comprising NN (Asn–Asn), NI (Asn–Ile), NG (Asn–Gly), and HD (His–Asp) have been used to bind to G, A, T, and C, respectively. However, NN may also recognize an A residue. NK (Arg–Lys) may be more specific for G, but the overall activity of TALEs containing these RVDs is lower [30]. NH (Asn–His) seems to have high affinity for G, but it does not compromise TALE activity [31].

A significant advantage of engineering TALEs to bind to intended targets is that, unlike with ZFPs, individual monomers may be coupled without neighboring sequences significantly altering functionality. Developing the methodology required to piece together monomeric units with desired DNA binding affinity has been challenging, but ingenious strategies have been devised to overcome this technical difficulty [27]. The three main methods of generating the complete TALEs are:

1. restriction enzyme and ligation (REAL),
2. golden Gate, and
3. fast Ligation-based Automatable Solid-phase High-throughput (FLASH) procedures.

The REAL technique entails standard restriction enzyme digestion of monomers derived from a plasmid library. Individual monomers are thereafter ligated in a stepwise hierarchical manner to form dimers, tetramers, and octamers before assembling the complete TALE [32,33]. Assembling monomers using the Golden Gate methodology has been more popular and widely used [34–38]. The procedure involves use of type IIS restriction enzymes, which have the useful property of cutting DNA at positions that fall outside of their recognition sites. Using plasmid libraries comprising monomeric units with flanking type IIS recognition sites, it is possible to assemble the monomers in any required order and thereby generate a TALE with intended DNA binding affinity. After type IIS restriction digestion, the enzyme recognition sites are removed and sticky ends are formed that define the monomers' positioning within the complete TALE array. Typically the complete TALE is built after two or three steps. The FLASH procedure for generating TALEs involves solid phase synthesis [33,39,40]. Biotinylated DNA is immobilized on streptavidin-coated magnetic beads, and the TALE is built up by a stepwise ligation of DNA modules. The final TALE sequence is removed after restriction digestion and is ready for insertion into a destination plasmid. Solid phase synthesis bypasses the need for repeated electrophoresis and purification of fragments from gels.

The method has been modified to enable addition of multimers, rather than monomers, to speed up TALE synthesis. Moreover, FLASH is amenable to automation for high-throughput and cost-effective synthesis. Another innovative method of generating TALEs has entailed ligation-independent cloning [41]. The method makes use of the controllable 3' exonuclease activity of T4 DNA polymerase to generate nonpalindromic overhangs, which in turn may be used to assemble a complete TALE. The procedure is rapid and may be automated. When using libraries with modules containing up to 5 monomers, more than 600 complete TALE sequences may be generated within a day.

Superiority of the specificity of engineered transcription activator-like effector nucleases (TALENs) for their target DNA was confirmed in a study comparing ZFNs with TALENs [42,43]. Mussolino and colleagues demonstrated that TALENs targeted to *CCR5* caused very little off-target cleavage at the *CCR2* homologous sequence and also caused minimal cytotoxicity of transfected cells. However, corresponding ZFNs had significant mutagenic effects at the *CCR2* locus and caused more cytotoxicity in the targeted cells.

It was originally thought that TALEs were unable to bind to target DNA containing 5-methyl cytidine (5 mC) residues [44]. Because 5 mC has functional importance for gene regulation, particularly in CpG islands of promoter elements, lack of binding to this modified base was considered to be a shortcoming of TALEs. However, studies have recently shown that the NG RVD, which had previously been thought to be specific for T, also binds to 5 mC [45,46]. Structural analysis indicated that the lack of a side chain in the G^{13} allows the RVD to accommodate the methyl group of 5 mC. Removal of the G^{13} also allows for binding to 5 mC, which is likely to be possible because conserved G residues occur at positions 14 and 15 of TALE monomers. Importantly, van der Waals forces responsible for base-specific binding are not affected by the presence of the amino at position four of 5 mC, instead of the keto of T. The ability to design TALEs that are capable of binding specifically to sequences containing 5 mC residues extends their versatility [27].

3.3.2 Adding Nuclease and Other Functions to TALEs

As with ZFPs, TALEs may be used to target nuclease or gene regulatory functions to specific DNA sequences. Addition of the nuclease domain of FokI to generate TALENs has been widely used to engineer sequence-specific target cleavage (Figure 3.4; reviewed in refs [26,27]). The typical dimeric arrangement involves formation of two customized TALEN subunits that bind to opposite strands of a target DNA duplex and flank the cleavage site. Each FokI nuclease is capable of cutting one strand (nicking), and DSBs are formed when the nicks in the opposite strands are close together. The initial studies that generated TALENs used natural TALE sequences to form the TALENs [47,48]. It soon became clear that design of the TALEN scaffold was important for efficiency of cleavage and that natural

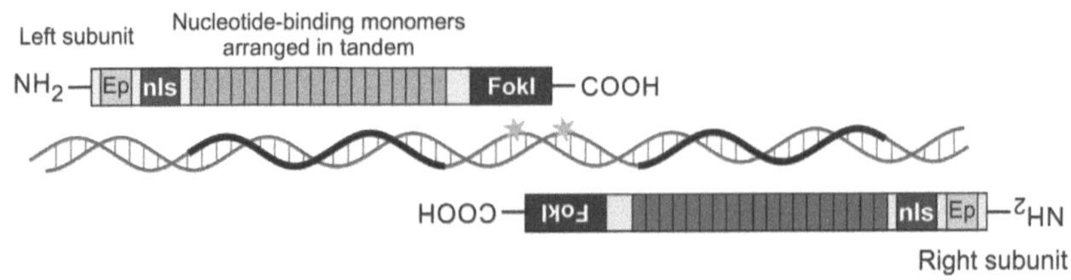

FIGURE 3.4 Schematic illustration of dimeric transcription activator-like effector nuclease (TALEN).

Left and right subunits of the TALEN each contain a tandem arrangement of monomers that is responsible for interacting with specific DNA sequences on opposite strands of the target DNA. Inclusion of FokI nuclease domains in the C-terminal region provides nicking functions to the subunits. When appropriately juxtaposed, the dimeric TALEN causes a double-stranded break in the target (stars). A nuclear localization signal (nls) is incorporated to ensure transfer of the TALEN subunits to the cell nucleus. An epitope (Ep), such as that of HA, may be included to facilitate immunological precipitation and detection of TALENs.

sequences, required for TALE secretion and host gene activation, are not necessary and in fact hamper functioning of engineered TALENs. Several different TALEN scaffolds have been generated in which N-terminal and C-terminal domains were shortened [42,47–50]. Additional elements, such as mammalian nls and epitope tags, have been included. The variety in TALEN backbone structures has resulted in slight differences in cleavage function. Particularly important has been the positioning of the TALEN subunits such that the intervening spacing is optimal for introducing DSBs at the target. Good double-stranded DNA cleavage with spaces ranging in length from 10 to 30 bp has been reported. Because optimal TALE interaction with DNA occurs when the base immediately 5′ of the cognate (position 0) is a T residue [24,25], the target site selection is important to ensure that both TALEN subunits are appropriately positioned on the cognate DNA. To improve specificity of target selection, as has been used with ZFNs (Figure 3.2), some investigations have used obligate heterodimers of FokI [51–53].

In addition to modification of TALEs by adding nuclease domains, transcription activating or suppressing domains may also be added to confer intended functions. Transcription activation of *frataxin*, suppression of which is implicated in Freidreich ataxia, has been successfully achieved by adding a tandem array of four VP16 peptides to a TALE targeting the *frataxin* promoter [54]. Targeted demethylation of DNA has also been used successfully to activate gene expression with TALE fusion proteins [55]. Introducing the KRAB sequence has been used to inhibit gene expression [56,57] and may be useful for inhibition of pathology-causing genes such as those of viruses.

3.4 HOMING ENDONUCLEASES

HEs, also known as meganucleases, are capable of cleaving double-stranded DNA at specific recognition sites comprising 14–40 bp [58]. Unlike restriction

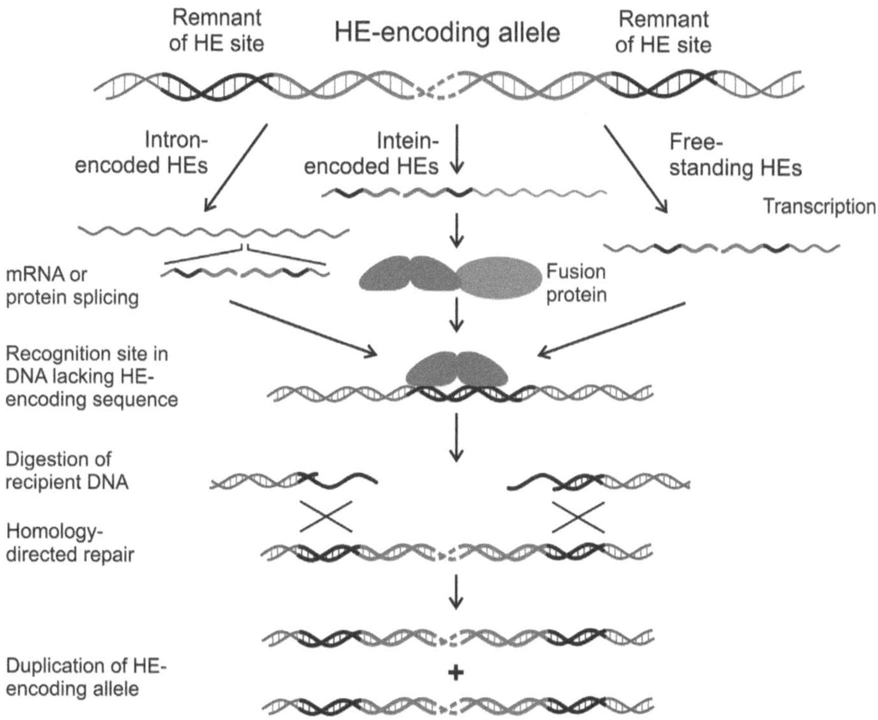

FIGURE 3.5 Natural production of homing endonucleases (HEs) and lateral transfer of sequences encoding these enzymes.

HEs are generated from elements that may be located within introns, intein elements, or from free-standing genes. By cleaving DNA at specific sites, double-strand breaks are generated that facilitate HDR and duplication of HE-encoding sequences through their incorporation at new sites. The HE recognition site is typically destroyed by the recombination event and is not susceptible to repeat digestion by the HE.

enzymes, which provide defenses to bacteria against invading DNA, HEs facilitate lateral mobility of genetic elements within an organism (Figure 3.5). This process is referred to as "homing" and gives the name to HEs. Interest in using HEs to treat viral infections is based on the specificity for the long target sites (reviewed in refs [1,59]). As with ZFNs, TALENs, and CRISPR/Cas-nucleases, repeated specific cleavage of viral DNA by HEs results in activation of repair pathways and targeted mutation. HEs may be engineered to cleave an intended DNA target, which provides the means to direct these enzymes to specific sites and their potential use for therapeutic antiviral application.

3.4.1 Natural Functions of HEs

HEs have been found in all three biological kingdoms (reviewed in ref. [60]). Earliest evidence of the existence of HEs came from studies performed in the 1970s on budding yeast [61]. The genetic marker omega (ω), which was transferred across yeast strains, was identified as a group I intron of the large ribosomal RNA. The mechanism of the transfer of the genetic element involved formation of DSBs at specific sites, and it was shown that the endonuclease responsible for the cleavage was encoded by sequences within the group I intron itself [62]. This HE, termed I-*Sce*I, was the first of many to be described. Cleavage was followed by HDR at the target, resulting in introduction of the intron sequence at the "intronless" target. There is some tolerance for sequence variation in the target sites, and this is thought to be important for HEs to accommodate genetic drift of the host organisms. Interestingly, the sites at which degeneracy is accommodated correspond to the wobble positions of the protein-coding sequences.

Sequences encoding HEs are typically found within group I and group II self-splicing introns, within protein inteins, and as free-standing sequences (Figure 3.5). The location within introns and inteins ensures diminished cost of fitness to the host during lateral transfer of genetic elements [60]. Classification of HEs has been based on the presence of characteristic identifying conserved amino acid motifs [63]. There are four main families of HEs, which each have LAGLIDADG, H-N-H, His–Cys box, and GIY-YIG motifs. Genomic locations (i.e., within group I introns, group II introns, or inteins) are often variable. For example, the LAGLIDADG HEs have been found in all three genomic niches.

Naming of HEs follows a similar convention to that which is used to describe restriction endonucleases. The first of three letters abbreviates the genus, and the second and third letters the species, from which the HE is derived. Numbering refers to the order of discovery. Origin from an intron (I), intein (PI), and free-standing (F) is indicated by a prefix. For example I-CreI is an intronic HE that was the first to be discovered from *Chlamydomonas reinhardtii*. When an HE is presumed to exist, which is typically based on bioinformatic analysis and homology to known HEs, then the name is proceeded by a "P." In the case of engineered HEs, an "H" prefix is used, and the three letter description is typically a combination of the source HEs and is devised according to the preference of the researchers who generated the hybrid HE.

3.4.2 Engineering HEs for Customized Gene Editing

Detailed understanding of the molecular mechanisms of duplex DNA cleavage by HEs has provided valuable insights into their modification for application to gene therapy. Specificity of target recognition occurs as a result of interaction between β-sheets, which bind at the curvature of the major groove of DNA [64,65]. The DNA binding is provided by contact between the protruding side chains of the HEs and the hydrogen donor or acceptors within the base pairs (Figure 3.6). Specific patterns of three or four hydrogen donors and acceptors are characteristic of each base pair. Contact of the HEs at two points is sufficient

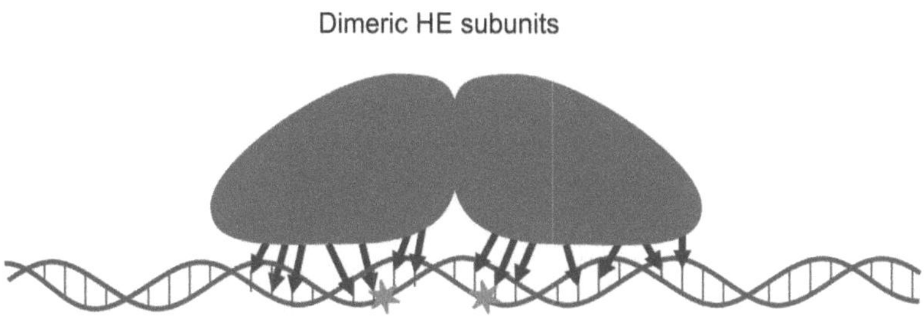

FIGURE 3.6 **Target DNA cleavage by homing endonucleases (HEs).**
The dimeric HE interacts with specific DNA sequences through interaction between motifs of the protein and DNA sequences exposed in the major groove of the nucleic acid. The DNA binding is provided by contact between amino acids protruded from the HEs and the hydrogen donor or acceptors within the DNA base pairs. Target cleavage (stars) typically results in the formation of staggered ends of DNA at the site of the double-strand break.

to confer binding specificity. The detailed structural analysis has demonstrated that contact with single hydrogen donor or acceptors sites often occurs, but interaction at all potential contact points is unusual. Contact at single points is less specific and accounts for the degeneracy of target recognition by HEs.

Understanding the specificity of the contact points between HEs and base pairs has been very useful. The enormous number of different HEs that exists naturally indicates that this family of proteins is flexible, and their modification may be used to alter binding specificity. Detailed structural characterization of different HEs has been, and will continue to be, important to inform the design of hybrid HEs (H-HEs) for gene therapy. A difficulty with engineering HEs is that functions of cleavage and target recognition are located within the same regions of the protein and altering sequences to confer binding specificity may compromise endonuclease efficiency. Moreover, the amino acids responsible for the formation of base pair-specific contacts are not modular and may function in a context-dependent manner.

In a comprehensive analysis of the specificity of HEs, Li and colleagues determined the binding affinity and cleavage by I-CreI and I-MsoI for all single base pair variants of the targets [66]. These two canonical LAGDIDADG enzymes have almost identical target sequences, with only 2 differences out of the 22 bp of the natural cognates. Interestingly, I-CreI and I-MsoI use significantly different DNA-protein contacts to mediate target recognition and cleavage. These variations suggest differences in the mechanisms of evolution of HEs and provide useful insights to enable rational design of customized HEs.

Creative approaches have been developed to generate and select HEs with intended DNA binding recognition and cleavage efficacy, and DNA shuffling is one such method [67,68]. Essentially, the procedure entails fusing DNA binding domains from different HEs to engineer new specificities of target binding.

Selection of HEs from libraries generated from randomized sequences encoding the DNA-binding domains has also been proposed as another method for propagating HEs with defined target specificity [58]. Although not yet widely tested, the extensive range of unique homing targets of naturally occurring HEs suggests that the selection procedure should be useful. In an attempt to simplify engineering of HEs, monomers instead of homodimers of the HEs have been engineered [69]. The monomeric versions of I-CreI and I-MsoI, mCreI and mMsoI, respectively, were capable of efficiently cleaving their cognates without the need to form homodimers. Two studies have reported on the usefulness of computer-based design of HEs to facilitate reprogramming of DNA binding and cleavage by these enzymes [70,71]. Ashworth and colleagues demonstrated that their algorithm was more reliable than methods that use standard substitution of DNA-binding regions to retarget HEs [70]. Using the approach, I-MsoI was initially reprogrammed to cleave a target that differed from the natural cognate by three contiguous base pairs. Efficiency was similar to that of the wild-type I-MsoI. The crystal structure of the engineered HE was then used successfully to inform computer-based design of an HE that cleaves at a target with a longer stretch of four contiguous substitutions. Importantly, unanticipated shifts in the conformation of protein interaction with the DNA target were observed. However, these alterations could be accommodated during iterative redesigning of HEs that acted on targets with greater numbers of substitutions within the wild-type target. Using different methods of computer-based structural design, Ulge et al. generated a panel of 3200 variants of mCre-I that targeted 16 of the 22 bp of the natural HE target [71]. Experimental analysis performed on 35 designs revealed that 24 (66%) of the selected engineered HEs had the intended altered site specificity. Although the results collectively demonstrate flexibility of engineering HEs to target novel sites, empirical evaluation of cleavage efficiency of modified HEs remains important. The main reason for this is that context-dependent effects of altering amino acid sequences in the DNA binding and catalytic domains of HEs are difficult to anticipate.

3.5 CRISPR/CAS

Rapid progress has recently been made in understanding RGNs derived from the CRISPR/Cas nuclease bacterial system. Many high-profile papers have been published on the topic, and the capabilities of these RGNs have generated considerable enthusiasm. Particularly important is the ease with which the CRISPR/Cas components can be directed to recognize specific intended target sites. Unlike with other DNA binding proteins, specificity of the CRISPR/Cas system is dependent on base pairing between an RNA guide and a strand of the DNA target. Engineering specificity by this mechanism is considerably easier to achieve than by construction of proteins that bind to particular DNA sequences, such as is required for the generation of HEs, TALEs, and ZFP derivatives.

3.5.1 Natural Function of CRISPR/Cas

The CRISPR/Cas system was originally described in 2002 after in silico analysis of prokaryotic repetitive DNA sequences of Archaea and Bacteria [72]. The DNA elements comprised an array of varied "spacer" sequences that were interspaced with repetitive DNA. The repeats and spacers were of similar length and comprised 21–37 bp. In this seminal study, *Cas* genes and conserved leader sequences of 300–500 bp were often found adjacent to the CRISPR regions. Analysis revealed helicase and exonuclease motifs in the Cas-encoded proteins, which suggested that they play a role in DNA metabolism or gene expression. A significant early observation was that the variable spacer sequences of the CRISPR had sequence homology to bacteriophages and plasmids [73,74]. This finding led to the idea, which was later confirmed, that the CRISPR/Cas system in prokaryotes is involved in adaptive defenses against invading nucleic acids of bacteriophages and plasmids.

Three types of CRISPR/Cas system have been described:

1. type I, which cleaves and degrades target DNA;
2. type II cleaves target DNA without causing degradation; and
3. type III, which cleaves DNA or RNA.

Most work on engineering DNA editing has entailed use of the type II system. Identifying the components of the type II CRISPR/Cas system was important to enable harnessing the mechanism for customized cleavage of DNA. The essential functional parts of the natural CRISPR/Cas are the following:

1. the CRISPR RNA (crRNA);
2. *trans*-encoded RNA (tracrRNA);
3. a protospacer adjacent motif (PAM); and
4. the Cas protein, which performs the endonuclease function of the CRISPR/Cas complex.

A good understanding of the mechanisms of CRISPR/Cas cleavage of target DNA has now emerged. Processing of invading DNA and integration of the sequences into the genomes of Bacteria and Archaea generate the CRISPR sequences that encode crRNAs (Figure 3.7). These short RNAs have spacer sequences that are complementary to the protospacer target DNA and serve as a guide to identify the DNA to be cleaved. tracrRNA plays a crucial role in the cleavage mechanism that was only recently fully appreciated. Deltcheva and colleagues showed that tracrRNA and Csn1, a widely conserved host RNase III, are responsible for the maturation of CRISPR pre-crRNAs to form mature crRNA spacers [75]. In *Streptococcus pyogenes*, tracrRNA contains sequences that are complementary to 24 nucleotides of the CRISPR repeat elements. Duplexes

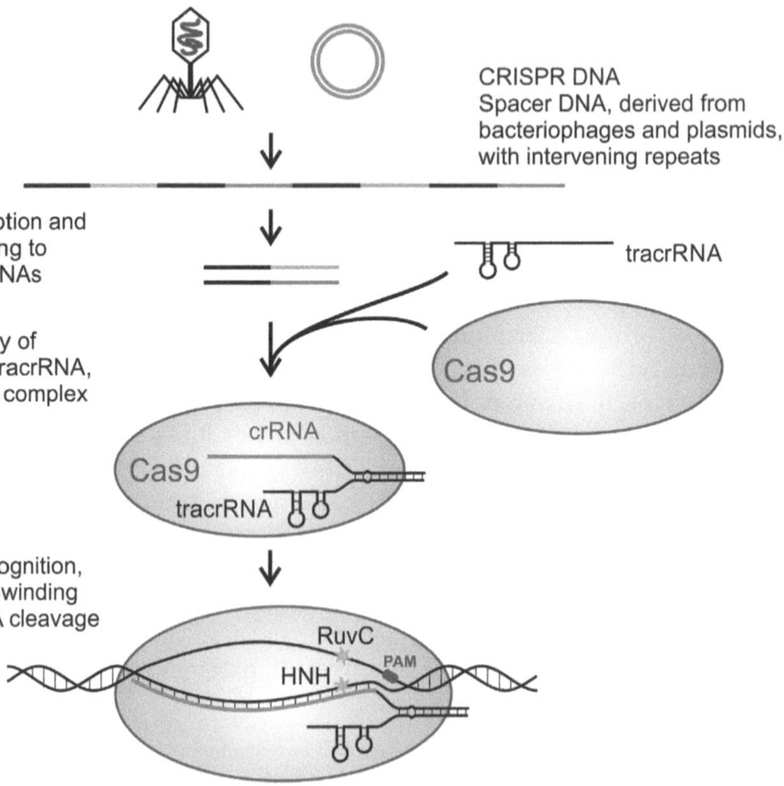

FIGURE 3.7 Natural function of the clustered regularly interspaced short palindromic repeats (CRISPR) with CRISPR-associated (Cas) system in bacteria.

Invading DNA from bacteriophages or plasmids is processed and incorporated between repeat elements of the CRISPR loci. These spacers are transcribed to form crRNA, which in turn interacts with tracrRNA through hybridizing sequences. The combined crRNA and tracrRNA form a complex with the Cas protein through interaction with secondary structures of the tracrRNA. The complex is directed to target DNA through interaction with the protospacer adjacent motifs (PAM). Target cleavage occurs after guiding hybridization of the crRNA to the target DNA and cleavage of each of the strands by RuvC and HNH domains of Cas. The Cas9 from *Streptococcus pyogenes* is most commonly used and is depicted here.

that are formed between the tracrRNA and pre-crRNA are recognized and diced by Csn1. Individual repeat-spacer-repeat units are released, which are subsequently further processed to form the shorter repeat-spacer sequences. tracrRNA is also crucial for establishing a complex between the crRNA, Cas, and the target DNA. It performs this function by retaining a linking duplex with the crRNA and binding to Cas9 through interaction of secondary structures with the targeting endonuclease (Figure 3.7). After defining the natural role of tracrRNA, another significant development was the demonstration that the crRNA and tracrRNA could artificially be combined into a short guide transcript, commonly termed sgRNA [76]. The importance of this observation was

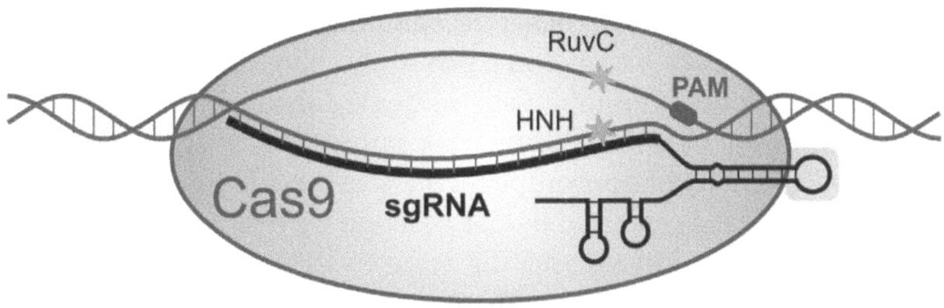

FIGURE 3.8 Schematic illustration of engineered clustered regularly interspaced short palindromic repeats (CRISPR) with CRISPR-associated (Cas) system.

Typically, the tracrRNA and crRNA are combined into one transcript, termed short guide RNA (sgRNA), through the engineering of a loop sequence (shaded) in an artificial expression cassette. This sgRNA performs the functions required for both Cas binding and for guiding of the complex to target DNA. As occurs during natural target cleavage, a PAM motif is essential and DNA strands are cleaved by RuvC and HNH domains of Cas.

that programming target DNA cleavage may be reduced to a requirement for only two components: Cas9 and the sgRNA (Figure 3.8). Combining crRNA and tracrRNA into the one sgRNA transcript has also enabled bypassing of the natural CRISPR RNA processing, which is needed to effect CRISPR/Cas targeting in prokaryotic cells.

The PAM comprises a triplet of base pairs and for Cas9 from *S. pyogenes* the canonical sequence is 5' NGG 3' in the DNA strand that is noncomplementary to the crRNA. Other PAM triplets (e.g., 5' NAG 3') have been described but support less efficient CRISPR/Cas function [76–78]. As an adjacent motif, the PAM is not complementary to the crRNA. The PAM is essential for Cas binding to its target [79], and absence of PAM sequences in the host CRISPR DNA ensures that self-cleavage does not occur. Insights into the dynamic mechanisms of Cas9 interaction with target DNA were provided by detailed studies using single molecule and bulk biochemical analyses [80]. Cas9 samples DNA through a mechanism that involves collision in three dimensions, rather than by moving along the target in one dimension. In the absence of a PAM, Cas9 binding to DNA is transient and weak. In the presence of a PAM, Cas9 binding affinity to the target increases and is further enhanced with greater length of the sequence complementarity between the guide and target DNA. Upon target binding by Cas9, the mechanism of R loop formation is thought to be initiated as a result of local DNA melting. This is followed by unwinding from the PAM-adjacent nucleation site, and RNA–DNA heteroduplex formation occurs through a Brownian ratchet mechanism.

Recent solving of the crystal structure of Cas9 has aided understanding of the mechanism of DNA cleavage [81]. Studies performed on Cas9 from *S. pyogenes* in association with target and guide RNA revealed that the protein comprises two lobes with nuclease and target recognition (REC) functions [82]. The RNA–DNA heteroduplex is bound in a positively charged groove that is located between the two lobes. The lobe with nuclease function contains HNH and RuvC domains, which are respectively responsible for cleavage of DNA strands that are complementary or noncomplementary to the crRNA guide (Figure 3.7). The DNA ends at DSBs generated by Cas cleavage are typically blunt, unlike the staggered ends that are formed after digestion by ZFNs, TALENs, and HEs. The PAM-interacting motif is located within the nuclease domain and is at the C terminal of Cas9. Further crystallographic analyses of related Cas9 proteins from *S. pyogenes* and *Actinomyces naeslundii* led to similar conclusions [83]. Studies performed in the absence of nucleic acids confirmed conservation of the HNH and RuvC nuclease domains and that a bilobed structure provides the nucleic acid-binding cleft. Some variation in the REC and PAM interacting domains were reported, which is in accordance with the differences in the crRNA and tracrRNA sequences with which the proteins from *S. pyogenes* and *A. naeslundii* interact.

3.5.2 Applying the CRISPR/Cas System for Gene Editing

The ease of generating expression cassettes that transcribe RNA to serve as sgRNA and tracrRNA to direct sequence-specific gene editing is the major advantage of the CRISPR/Cas system (Figure 3.8). Understanding the mechanism of CRISPR/Cas action on target DNA spurred many investigations that have been aimed at harnessing the system to modify target DNA of interest. During early 2013, the successful use of CRISPR/Cas to achieve gene editing in eukaryotic cells was a milestone development [84–88]. Exogenous CRISPR/Cas elements were shown to be capable of modifying gene function through application of principles that had been established from the mechanistic insights into the natural process of target cleavage by the RGN. Only minor modifications, such as codon optimization and incorporation of nuclear localization signals into Cas9, were required. Targeted genome editing was accomplished in many different cell types including those of human [84–87], zebrafish [88], roundworm [89], silkworm [90], tobacco [91], Arabidopsis [91], and yeast [92] origins. Versatility and ease of using the approach was further corroborated by using more than one guide to achieve multiplexed target editing [85,87]. Since these first studies, it rapidly became apparent that CRISPR/Cas has broad utility for genome editing of animals and plants. Application of CRISPR/Cas to genome-wide screens is a significant development [93]. Recent studies using large panels of sgRNAs that target most or all annotated human and mouse genes have been performed [94–96]. These studies have enabled identification of genes that are important for cell viability and resistance to chemotherapy or bacterial toxins. Using RGNs to edit genomes is also generating enthusiasm for the propagation of animal and cell models [93]. One particularly noteworthy recent development on this topic was the successful editing of genomes of cynomolgus monkeys [97]. Instead of introducing DNA expression cassettes into cells, Cas9 mRNA and guide RNAs were injected into single-cell stage embryos. Simultaneous disruption of two genes was achieved without evidence for off-target effects.

3.5.3 Improving Specificity CRISPR/Cas Gene Editing

Forming ZFNs, HEs, or TALENs with specificity for customized DNA targets requires somewhat complicated engineering of sequences that encode DNA-binding proteins. Expressing RNA to target Cas9 nuclease is considerably easier; therefore, it has gained widespread favor. However, specificity of the CRISPR/Cas system for target DNA has been a concern [78,98–101]. Studies have reported tolerance of up to five mismatches between guide and target sequences. Not only are mismatches compatible with Cas cleavage, but insertions and deletions that result in formation of bulges in the target or guide are also tolerated [102]. Understandably, considerable effort has therefore gone in to developing improved methods of achieving greater target specificity.

In a comprehensive assessment of Cas9 binding to targeted sequences, a genome-wide analysis was performed on mouse embryonic stem cells (mESCs) [103]. After stably transducing mESCs with DNA encoding a hemagglutinin (HA)-tagged dead Cas9 (dCas9), cells were transfected with one of a panel of four sgRNAs targeting *Phc1* or *Nanog*. ChIP-Seq was performed after pull down with an antibody to HA. Analysis revealed that each of the sgRNAs directed the dCas9 to as many as thousands of different genomic sites. A seed sequence of only five nucleotides, together with the PAM comprising 5' NGG 3', was commonly found at the target sites. Interestingly, U-rich seeds improved specificity of dCas9 binding to a target and also diminished sgRNA abundance. The importance of U residues in the PAM-proximal region is likely to relate to an influence that this base has on Cas9 interaction with the sgRNA and target DNA. In addition, multiple U residues may effect premature termination of sgRNA transcription by U6 Pol III to cause lower concentration of sgRNAs that have a high U content. Although the study found that binding of dCas9 to targets was promiscuous, mutation at off-target sites caused by Cas9-mediated DNA cleavage was rare. The authors propose a model in which Cas9 binding to a target requires only seed complementarity, but target cleavage only occurs when there is complete pairing between the sgRNA and cognate DNA. Interestingly, this proposed mechanism is reminiscent of the RNA-induced silencing complex-mediated cleavage of RNA, which only occurs when there is complete base pairing between the RNAi-activating guide and its target (see Chapter 2).

To address concerns about specificity of target DNA cleavage by Cas9, mutations have been introduced to disable either the HNH or RuvC nuclease domains [98,104,105]. The resultant mutants are capable of cutting either the noncomplementary or complementary strands of the target, respectively. Using nickases to achieve cleavage on opposite DNA strands requires paired DNA enzymes and improves specificity of cleavage, which has been reported to be augmented by 50- to 1500-fold [105]. This arrangement of Cas9 nickases is similar to that involving use of protein dimers to constitute functional ZFNs or TALENs. There is flexibility in the length of the sequence between nicking Cas9 pairs that permits target cleavage, and separation of Cas9 nickases' target sites by 4–100 bp has been reported to be capable of target gene editing [98,104,105]. Although specificity is markedly improved by use of paired nickases, off-target mutation may still occur, and it appears that the mechanism involves the formation of DSBs at a nick when a DNA replication fork passes through the site of action by the nickase [106]. To overcome this problem, systems have been developed in which cutting of DNA targets occurs only when nicking enzymes are bound to DNA in close proximity to each other [107,108]. The ingenious methods use nuclease-dCas9s that are fused to FokI nucleases. Cleavage is dependent on the precise juxtaposition of guides such that FokI domains are only active when the spacing between them is optimal. This configuration substantially diminishes the likelihood of unwanted off-target cleavage.

To direct Cas9 endonuclease activity to an intended target, the artificial sgRNAs have typically been engineered such that the guide corresponds to the 20 nt of the target that are 5′ of the PAM [106]. Shorter truncated transcripts encoding guides of 17–18 nucleotides, so-called "tru-gRNAs" or truncated guide RNAs, have been used successfully; importantly, they appear to confer greater specificity on the endonuclease [109]. Using this approach also has little if any effect on the efficiency of on-target cleavage. A useful advantage of the method is that it is easy to implement, and generating tru-gRNAs does not impose additional technical demands. The tru-gRNAs may also be used in conjunction with paired nickases to further improve specificity.

Because the promoters used to generate the guides often have sequence requirements at the transcription initiation site at the 5' ends, there are usually additional restrictions placed on target selection. For example, the popular U6 Pol III promoter, which is capable of highly efficient and precise transcription of short RNAs, requires that a G residue be the first nucleotide of the transcript. As a result, the target of a guide expressed from a U6 promoter should comprise 5' $GN_{16\text{-}19}NGG$ 3'. Recently, limitation of a G residue at the 5' end of U6 PolIII transcripts was obviated by co-transfecting of an expression cassette that produces the Csy4 RNase from *Pseudomonas aeruginosa* [107]. Engineered site-specific cleavage by Csy4 within the U6 transcript released multiple gRNAs that were functional and did not have any sequence restrictions to their 5′ ends.

In addition to use of tru-gRNAs and paired Cas9 nickases, methods that use reduced concentrations of Cas9 and gRNAs have been assessed. However, varied usefulness of the approaches has been reported. In one study, it was found that titration of the Cas9 and gRNA concentrations may be used to diminish off-target effects while maintaining on-target cleavage [101]. However, this observation was not corroborated in a study reported at the same time, which showed decreased on- and off-target effects when using lower concentrations of Cas9 and gRNA [100]. Other methods of improving target specificity have entailed alteration of tracrRNA sequences. An example is the use of synthetic gRNAs with two G residues at the 5′ ends that were generated from transcription in vitro with T7 polymerase [98]. Although improved specificity has been reported, overall target cleavage efficiency may be compromised [106].

3.5.4 Additional Functionality of CRISPR–Cas

In addition to using its natural nuclease activity for gene editing, the CRISPR/Cas system may also be adapted to effect other functions [106]. This property may be used to control gene expression and to mark particular target sites. Initial investigations aimed at using the system to control gene expression made use of dCas9 with D10A and H840A mutations in the RuvC and HNH nuclease domains, respectively. The protein may be directed to target DNA by guide RNAs, but it does not cleave the target. Studies performed in *Escherichia coli* [110] and human cells [111–114] revealed that inhibition of promoter function could be achieved by the dCas9. As with TALEs and ZFPs, it is also possible to couple transcription regulatory elements to dCas9. Examples include the use of the KRAB suppressor, VP64, and nuclear factor-κB transcriptional activators. In the absence of regulatory elements, the effect of dCas9 is likely to occur as a result of impedance to transcription factors' access to the regulatory elements on the DNA. In addition, use of multiple sgRNAs synergistically augments transcriptional effects of dCas fusions [111].

The CRISPR/Cas system has also been used for dynamic imaging in live cells [115]. Fusion of a reporter such as eGFP to dCas9 enables targeting of the reporter to a specific sequence of interest. This very useful feature was used to monitor spatiotemporal organization of repetitive genomic sequences of telomeres and coding regions. To detect nonrepetitive sequences, multiple sgRNAs that spanned the DNA target were used.

3.6 DELIVERY OF CUSTOMIZED SEQUENCE-SPECIFIC DNA BINDING PROTEINS FOR ANTIVIRAL THERAPEUTIC APPLICATION

As with most strategies aimed at achieving successful gene therapy, efficient delivery of customized DNA binding proteins poses challenges. Generating recombinant viral vectors that contain expression cassettes encoding the therapeutic sequences has generally been favored. Recombinant lentiviruses, adeno-associated viruses (AAVs), and adenoviral vectors (Ads) are the most widely used vectors and have features that are useful for certain applications (Chapter 4). Transient expression of engineered sequences may be all that is required to achieve gene editing. Persistent expression from constitutively active cassettes is unlikely to be necessary and may in fact be undesirable. Continuous expression over a long duration is likely to cause unintended off-target effects. For this reason, delivery of RNA or recombinant proteins has the advantage of causing a transient effect and enabling better dose regulation. Moreover, by using proteins or RNA to achieve a therapeutic effect, risk of mutagenic effects of introducing DNA into cells is avoided. Ex vivo delivery of therapeutic sequences, such as is feasible with targeted mutation of *CCR5* in $CD4^+$ T cells, overcomes some of the problems associated with systemic administration of vectors.

A particularly important consideration that is relevant to delivery of DNA-binding proteins is the size of the sequence(s) required to encode functional gene editors (Table 3.1). Each of the classes of editing proteins is encoded by different length sequences. HEs typically comprise approximately 330 amino acids and may be encoded by the shortest transgenes. ZFP derivatives are longer, usually approximately 650 amino acids, and may be accommodated by most viral vectors. With TALENs and Cas9, DNA cassettes encoding these proteins comprise approximately 4 kb. In the case of TALENs, in which two cassettes are required to constitute a nuclease that is capable of effecting DSB formation, DNA encoding the complete nuclease comprises approximately 8 kb. With engineered CRISPR/Cas, there is also a requirement for two cassettes: one encoding the Cas protein and the other to transcribe the gRNA. For both classes of DNA editing enzymes, the total length of DNA cannot easily be accommodated by the popular recombinant AAVs. To utilize AAVs for delivery of complete nucleases capable of effecting DSBs, in most cases it is necessary to generate two vectors, each encoding one of the subunits. Consequently, effective therapy would be reliant on delivery of each of the two vectors to the same virally infected cells. A toolbox of methods for using AAVs to deliver CRISPR/Cas systems has recently been described [116]. Another disadvantage of using AAVs for delivery of TALEN or nicking Cas9 is that single-stranded vectors, not the more efficient self-complementary (sc) AAVs, would need to be used. The scAAVs have a capacity for transgenes of approximately 2.4 kb, which is approximately half that of the transgene DNA length that may be incorporated into single-stranded AAVs [117,118]. HEs and ZFNs are considerably smaller than TALENs; therefore, they have the advantage of being compatible with incorporation onto AAVs, and in the case of HEs use of scAAVs is feasible.

Repetitive sequences within the monomers encoding the DNA-targeting domain of TALEs have been considered to be potentially problematic for maintenance of lentiviral vector integrity. This property of lentiviral vectors was recently confirmed, but it was not problematic when TALENs were delivered with Ads

[119]. A novel alternative approach, which entailed delivery of proteins fused to the Gag protein within lentiviral vectors, successfully accomplished delivery of pairs of ZFNs and TALENs [120]. Ads, particularly helper-dependent (HD) Ads, do have the capacity to accommodate large transgenes. With HD Ads, all viral vector protein-coding sequences are removed. In the presence of a helper virus, only inverted terminal repeats and the packaging signals are required to form the HD Ads. These "gutless" vectors are able to accommodate in excess of 20 kb of transgene, which is more than the total requirement for expression of complete engineered TALENs and CRISPR/Cas systems. Ads have been used successfully to deliver engineered CRISPR/Cas encoding sequences to effect targeted mutagenesis in a wide variety of cells [121]. The natural hepatotropism of Ads after systemic administration is useful for delivery of sequences to treat viral infections in the liver, such as are caused by HBV. However, a drawback of using Ads for therapy is the powerful and potentially toxic immunostimulatory effect of these vectors (Chapter 4).

Alternative approaches to delivery of TALENs have also been investigated and potentially have interesting possible applications. Recently, Liu and colleagues conjugated an arginine-rich cell penetrating peptide to TALENs that targeted the *CCR5* and *bone morphogenetic receptor protein type 1A* (*BMPR1A*) genes [122]. Successful targeted cleavage of intended sites was observed. This interesting approach has several advantages of safety for clinical application:

1. without the need for introducing of DNA expression cassettes into cells, the risk of insertional mutagenesis is decreased;
2. the effect of the TALENs is transient, which diminishes the likelihood of off-target effects that result from sustained production of the nucleases; and
3. regulation of dosage of the TALENs is more precise.

Studies have demonstrated that it is feasible to activate CRISPR/Cas gene editing after administration to cells of mRNA encoding Cas9 and sgRNA. Use of RNA has the advantage that potentially mutagenic integration of DNA is obviated, which would be important in a therapeutic setting. Many different methods of introducing the CRISPR/Cas nucleic acids into various cells have been used [106]. To date, most have been developed for functional rather than therapeutic application and have entailed direct microinjection, transfection, and use of polymer conjugates for delivery. Some studies have described generation of recombinant lentiviral vectors, but these have been used for genome-wide screening and not therapeutic use [94–96]. The development of vectors for therapeutic delivery of CRISPR–Cas9 components is at an early stage of testing for use against viral infections, but the generic considerations of vector toxicity, transgene size constraints, tissue targeting, and persistence of transgene expression are likely to apply to delivering exogenous CRISPR–Cas9 systems.

3.7 USE OF SITE-SPECIFIC DNA TARGETING FOR TREATMENT OF VIRAL INFECTIONS

Results from studies performed on particular viral targets have been explained in some detail in chapters dealing with gene therapy for specific viruses, and only a summarizing outline is provided here. Recent reports have shown that TALENs are very effective against HBV [123,124]. These initial studies used TALENs that were targeted to sequences distributed throughout the viral genome. Good efficacy of inhibition of replication was observed, and mutagenic gene editing at the intended sites was observed. The inhibition of HBV replication appeared to be better than that achieved with ZFP derivatives [14,125]. As with use of ZFNs for targeted disruption of the *CCR5* sequence encoding a co-receptor for HIV-1 infection, TALENs have also been developed to target this cellular sequence [42,122]. Importantly, side-by-side comparison of ZFNs with TALENs revealed that TALENs were as effective as ZFNs [42]. However, off-target mutations at the *CCR2* sequences, which have homology to the *CCR5* target, were lower after TALEN administration. Cytotoxicity, which was used as a surrogate indicator of nonspecific gene disruption, was lower in the TALEN-treated cells. These conclusions were recently corroborated by broader investigations into specific and off-target effects of ZFNs and TALENs on other human genes [43].

The number of studies published on the topic of using engineered CRISPR/Cas for treatment of viral infections is rapidly gaining momentum. Using a reporter gene assay to assess efficacy of HIV-1-targeting CRISPR–Cas, Ebina and colleagues showed disruption of the viral LTR function [126]. Disabling mutations were introduced at the target site of the integrated viral DNA, which indicates that the approach has utility for overcoming problematic latent HIV-1 infection. In another study, Hu et al. successfully engineered CRISPR/Cas RGNs to excise proviral sequences from HIV-1-infected cells [127]. CRISPR/Cas was used to introduce disruptive mutations in the CCR5 HIV-1 co-receptor [99]. Although mutations were detected in the targeted gene, nonspecific mutations in the related *CCR2* gene were also observed, which highlights the concerns about specificity when developing the CRISPR/Cas system for therapeutic application. Because improvements are being made in the specificity and therapeutic applicability of RGN-mediated gene editing technology, in the near future it is likely that several studies will be reported on the utility of gene editing using CRISPR–Cas.

3.8 CONCLUSIONS

Very impressive progress has been made in developing the use of customized sequence-specific DNA-binding proteins to disable viral replication. Nucleases and transcriptional suppressors both have potential therapeutic utility through direct action on viral gene expression as well by inactivating host factors required for viral replication. By introducing mutations at intended sites, nucleases have a significant advantage over other approaches, which has potential for achieving permanent inhibition of viral replication. There is currently substantial interest in use of CRISPR/Cas systems for gene editing. The simpler requirement for an RNA guide to target genes means that the technology is easy to implement and avoids difficulties associated with engineering of DNA-targeting proteins such as ZFPs, TALENs, and HEs. Use of RGNs against viral infections has not yet been widely explored, but it is highly likely that demonstrating utility of this application of RGNs will soon be forthcoming. Nevertheless, thorough comparisons of the different gene editing technologies, particularly with respect to ease of delivery to target tissues and specificity for intended cognates, will be important before a preference for one approach to treating viral infections can be established. As with most gene therapy strategies, the obstacles to advancing customized DNA binding proteins to a stage of clinical application are significant. Some of the challenges, which are generic to gene therapy, are the following:

1. establishing efficient delivery to target tissues,
2. controlling dose and expression of therapeutic sequences,
3. limiting off-target effects,
4. ensuring that host immune responses to the customized DNA binding proteins do not compromise antiviral efficacy, and
5. providing a high barrier against viral resistance.

Long-lasting expression may be complicated by increased risk of unintended off-target effects. The delivery of RNA or recombinant proteins, instead of protein-encoding DNA cassettes, may be preferable by ensuring that a transient and more regulatable effect is achieved. Ex vivo delivery of sequences encoding therapeutic DNA-binding proteins may be easier, but technically demanding procedures limits widespread use. Although use of gene editing against other viruses has been limited, the utility of the approach is likely to broaden. In a proof-of-principle study, direct cleavage of human T-cell leukemia virus type-1, a cause of T cell leukemia, has been achieved with ZFNs [128]. Successful use of TALENs to edit genes of HPV and ameliorate cervical cancer induced in a murine model of the infection has recently been demonstrated [129]. An important feature of this viral infection is that infected cells in the cervical epithelium are accessible. Thus, the TALEN-expressing plasmid DNA could be topically administered using a transfection reagent. Despite these significant challenges, use of customized and functionalized sequence-specific DNA binding proteins for treatment of viral infections is poised at an exciting stage, and significant developments are surely imminent.

REFERENCES

[1] Schiffer JT, Aubert M, Weber ND, Mintzer E, Stone D, Jerome KR. Targeted DNA mutagenesis for the cure of chronic viral infections. J Virol 2012;86(17):8920–36.

[2] Tebas P, Stein D, Tang WW, Frank I, Wang SQ, Lee G, et al. Gene editing of CCR5 in autologous CD4 T cells of persons infected with HIV. N Engl J Med 2014;370(10):901–10.

[3] Gaj T, Gersbach CA, Barbas III CF. ZFN, TALEN, and CRISPR/Cas-based methods for genome engineering. Trends Biotechnol 2013;31(7):397–405.

[4] Wei C, Liu J, Yu Z, Zhang B, Gao G, Jiao R. TALEN or Cas9-rapid, efficient and specific choices for genome modifications. J Genet Genomics 2013;40(6):281–9.

[5] Berg JM. Zinc finger domains: hypotheses and current knowledge. Annu Rev Biophys Biophys Chem 1990;19:405–21.

[6] Laity JH, Lee BM, Wright PE. Zinc finger proteins: new insights into structural and functional diversity. Curr Opin Struct Biol 2001;11(1):39–46.

[7] Klug A. The discovery of zinc fingers and their applications in gene regulation and genome manipulation. Annu Rev Biochem 2010;79:213–31.

[8] Maeder ML, Thibodeau-Beganny S, Sander JD, Voytas DF, Joung JK. Oligomerized pool engineering (OPEN): an "open-source" protocol for making customized zinc-finger arrays. Nat Protoc 2009;4(10):1471–501.

[9] Sander JD, Dahlborg EJ, Goodwin MJ, Cade L, Zhang F, Cifuentes D, et al. Selection-free zinc-finger-nuclease engineering by context-dependent assembly (CoDA). Nat Methods 2011;8(1):67–9.

[10] Choo Y, Sanchez-Garcia I, Klug A. In vivo repression by a site-specific DNA-binding protein designed against an oncogenic sequence. Nature 1994;372(6507):642–5.

[11] Kim YG, Cha J, Chandrasegaran S. Hybrid restriction enzymes: zinc finger fusions to Fok I cleavage domain. Proc Natl Acad Sci USA 1996;93(3):1156–60.

[12] Bitinaite J, Wah DA, Aggarwal AK, Schildkraut I. FokI dimerization is required for DNA cleavage. Proc Natl Acad Sci USA 1998;95(18):10570–5.

[13] Miller JC, Holmes MC, Wang J, Guschin DY, Lee YL, Rupniewski I, et al. An improved zinc-finger nuclease architecture for highly specific genome editing. Nat Biotechnol 2007;25(7):778–85.

[14] Zimmerman KA, Fischer KP, Joyce MA, Tyrrell DL. Zinc finger proteins designed to specifically target duck hepatitis B virus covalently closed circular DNA inhibit viral transcription in tissue culture. J Virol 2008;82(16):8013–21.

[15] Beerli RR, Segal DJ, Dreier B, Barbas 3rd CF. Toward controlling gene expression at will: specific regulation of the erbB-2/HER-2 promoter by using polydactyl zinc finger proteins constructed from modular building blocks. Proc Natl Acad Sci USA 1998;95(25):14628–33.

[16] Huntley S, Baggott DM, Hamilton AT, Tran-Gyamfi M, Yang S, Kim J, et al. A comprehensive catalog of human KRAB-associated zinc finger genes: insights into the evolutionary history of a large family of transcriptional repressors. Genome Res 2006;16(5):669–77.

[17] Friedman JR, Fredericks WJ, Jensen DE, Speicher DW, Huang XP, Neilson EG, et al. KAP-1, a novel corepressor for the highly conserved KRAB repression domain. Genes Dev 1996;10(16):2067–78.

[18] Groner AC, Meylan S, Ciuffi A, Zangger N, Ambrosini G, Denervaud N, et al. KRAB-zinc finger proteins and KAP1 can mediate long-range transcriptional repression through heterochromatin spreading. PLoS Genet 2010;6(3):e1000869.

[19] Beerli RR, Dreier B, Barbas 3rd CF. Positive and negative regulation of endogenous genes by designed transcription factors. Proc Natl Acad Sci USA 2000;97(4):1495–500.

[20] Triezenberg SJ, Kingsbury RC, McKnight SL. Functional dissection of VP16, the trans-activator of herpes simplex virus immediate early gene expression. Genes Dev 1988;2(6):718–29.

[21] Mew T, Alvarez A, Leach J, Swings J. Focus on bacterial blight of rice. Plant Dis 1993;77(1):5–12.

[22] Boch J, Bonas U. Xanthomonas AvrBs3 family-type III effectors: discovery and function. Annu Rev Phytopathol 2010;48:419–36.

[23] Kay S, Hahn S, Marois E, Hause G, Bonas U. A bacterial effector acts as a plant transcription factor and induces a cell size regulator. Science 2007;318(5850):648–51.

[24] Boch J, Scholze H, Schornack S, Landgraf A, Hahn S, Kay S, et al. Breaking the code of DNA binding specificity of TAL-type III effectors. Science 2009;326(5959):1509–12.

[25] Moscou MJ, Bogdanove AJ. A simple cipher governs DNA recognition by TAL effectors. Science 2009;326(5959):1501.

[26] Joung JK, Sander JD. TALENs: a widely applicable technology for targeted genome editing. Nat Rev Mol Cell Biol 2013;14(1):49–55.

[27] Sun N, Zhao H. Transcription activator-like effector nucleases (TALENs): a highly efficient and versatile tool for genome editing. Biotechnol Bioeng 2013;110(7):1811–21.

[28] Deng D, Yan C, Pan X, Mahfouz M, Wang J, Zhu JK, et al. Structural basis for sequence-specific recognition of DNA by TAL effectors. Science 2012;335(6069):720–3.

[29] Mak AN, Bradley P, Cernadas RA, Bogdanove AJ, Stoddard BL. The crystal structure of TAL effector PthXo1 bound to its DNA target. Science 2012;335(6069):716–9.

[30] Huang P, Xiao A, Zhou M, Zhu Z, Lin S, Zhang B. Heritable gene targeting in zebrafish using customized TALENs. Nat Biotechnol 2011;29(8):699–700.

[31] Streubel J, Blucher C, Landgraf A, Boch J. TAL effector RVD specificities and efficiencies. Nat Biotechnol 2012;30(7):593–5.

[32] Reyon D, Khayter C, Regan MR, Joung JK, Sander JD. Engineering designer transcription activator-like effector nucleases (TALENs) by REAL or REAL-Fast assembly. Curr Protoc Mol Biol 2012; Unit 12. 5.: http://dx.doi.org/10.1002/0471142727.mb1215s100.

[33] Reyon D, Tsai SQ, Khayter C, Foden JA, Sander JD, Joung JK. FLASH assembly of TALENs for high-throughput genome editing. Nat Biotechnol 2012;30(5):460–5.

[34] Cermak T, Doyle EL, Christian M, Wang L, Zhang Y, Schmidt C, et al. Efficient design and assembly of custom TALEN and other TAL effector-based constructs for DNA targeting. Nucleic Acids Res 2011;39(12):e82.

[35] Li L, Piatek MJ, Atef A, Piatek A, Wibowo A, Fang X, et al. Rapid and highly efficient construction of TALE-based transcriptional regulators and nucleases for genome modification. Plant Mol Biol 2012;78(4–5):407–16.

[36] Morbitzer R, Elsaesser J, Hausner J, Lahaye T. Assembly of custom TALE-type DNA binding domains by modular cloning. Nucleic Acids Res 2011;39(13):5790–9.

[37] Sanjana NE, Cong L, Zhou Y, Cunniff MM, Feng G, Zhang F. A transcription activator-like effector toolbox for genome engineering. Nat Protoc 2012;7(1):171–92.

[38] Weber E, Gruetzner R, Werner S, Engler C, Marillonnet S. Assembly of designer TAL effectors by Golden Gate cloning. PloS One 2011;6(5):e19722.

[39] Briggs AW, Rios X, Chari R, Yang L, Zhang F, Mali P, et al. Iterative capped assembly: rapid and scalable synthesis of repeat-module DNA such as TAL effectors from individual monomers. Nucleic Acids Res 2012;40(15):e117.

[40] Wang Z, Li J, Huang H, Wang G, Jiang M, Yin S, et al. An integrated chip for the high-throughput synthesis of transcription activator-like effectors. Angew Chem Int Ed Engl 2012;51(34):8505–8.

[41] Schmid-Burgk JL, Schmidt T, Kaiser V, Honing K, Hornung V. A ligation-independent cloning technique for high-throughput assembly of transcription activator-like effector genes. Nat Biotechnol 2013;31(1):76–81.

[42] Mussolino C, Morbitzer R, Lutge F, Dannemann N, Lahaye T, Cathomen T. A novel TALE nuclease scaffold enables high genome editing activity in combination with low toxicity. Nucleic Acids Res 2011;39(21):9283–93.

[43] Mussolino C, Alzubi J, Fine EJ, Morbitzer R, Cradick TJ, Lahaye T, et al. TALENs facilitate targeted genome editing in human cells with high specificity and low cytotoxicity. Nucleic Acids Res 2014;42(10):6762–73.

[44] Bultmann S, Morbitzer R, Schmidt CS, Thanisch K, Spada F, Elsaesser J, et al. Targeted transcriptional activation of silent oct4 pluripotency gene by combining designer TALEs and inhibition of epigenetic modifiers. Nucleic Acids Res 2012;40(12):5368–77.

[45] Deng D, Yin P, Yan C, Pan X, Gong X, Qi S, et al. Recognition of methylated DNA by TAL effectors. Cell Res 2012;22(10):1502–4.

[46] Valton J, Dupuy A, Daboussi F, Thomas S, Marechal A, Macmaster R, et al. Overcoming transcription activator-like effector (TALE) DNA binding domain sensitivity to cytosine methylation. J Biol Chem 2012;287(46):38427–32.

[47] Li T, Huang S, Jiang WZ, Wright D, Spalding MH, Weeks DP, et al. TAL nucleases (TALNs): hybrid proteins composed of TAL effectors and FokI DNA-cleavage domain. Nucleic Acids Res 2011;39(1):359–72.

[48] Miller JC, Tan S, Qiao G, Barlow KA, Wang J, Xia DF, et al. A TALE nuclease architecture for efficient genome editing. Nat Biotechnol 2011;29(2):143–8.

[49] Christian M, Cermak T, Doyle EL, Schmidt C, Zhang F, Hummel A, et al. Targeting DNA double-strand breaks with TAL effector nucleases. Genetics 2010;186(2):757–61.

[50] Sun N, Liang J, Abil Z, Zhao H. Optimized TAL effector nucleases (TALENs) for use in treatment of sickle cell disease. Mol Biosyst 2012;8(4):1255–63.

[51] Cade L, Reyon D, Hwang WY, Tsai SQ, Patel S, Khayter C, et al. Highly efficient generation of heritable zebrafish gene mutations using homo- and heterodimeric TALENs. Nucleic Acids Res 2012;40(16):8001–10.

[52] Dahlem TJ, Hoshijima K, Jurynec MJ, Gunther D, Starker CG, Locke AS, et al. Simple methods for generating and detecting locus-specific mutations induced with TALENs in the zebrafish genome. PLoS Genet 2012;8(8):e1002861.

[53] Lei Y, Guo X, Liu Y, Cao Y, Deng Y, Chen X, et al. Efficient targeted gene disruption in Xenopus embryos using engineered transcription activator-like effector nucleases (TALENs). Proc Natl Acad Sci USA 2012;109(43):17484–9.

[54] Tremblay JP, Chapdelaine P, Coulombe Z, Rousseau J. Transcription activator-like effector proteins induce the expression of the frataxin gene. Hum Gene Ther 2012;23(8):883–90.

[55] Maeder ML, Angstman JF, Richardson ME, Linder SJ, Cascio VM, Tsai SQ, et al. Targeted DNA demethylation and activation of endogenous genes using programmable TALE-TET1 fusion proteins. Nat Biotechnol 2013;31(12):1137–42.

[56] Cong L, Zhou R, Kuo YC, Cunniff M, Zhang F. Comprehensive interrogation of natural TALE DNA-binding modules and transcriptional repressor domains. Nat Commun 2012;3:968.

[57] Garg A, Lohmueller JJ, Silver PA, Armel TZ. Engineering synthetic TAL effectors with orthogonal target sites. Nucleic Acids Res 2012;40(15):7584–95.

[58] Chevalier BS, Stoddard BL. Homing endonucleases: structural and functional insight into the catalysts of intron/intein mobility. Nucleic Acids Res 2001;29(18):3757–74.

[59] Weber ND, Aubert M, Dang CH, Stone D, Jerome KR. DNA cleavage enzymes for treatment of persistent viral infections: recent advances and the pathway forward. Virology 2014: 454–5. 353–361.

[60] Gimble FS. Invasion of a multitude of genetic niches by mobile endonuclease genes. FEMS Microbiol Lett 2000;185(2):99–107.

[61] Coen D, Deutsch J, Netter P, Petrochilo E, Slonimski PP. Mitochondrial genetics. I. Methodology and phenomenology. Symp Soc Exp Biol 1970;23:449–96.

[62] Dujon B. Sequence of the intron and flanking exons of the mitochondrial 21S rRNA gene of yeast strains having different alleles at the omega and rib-1 loci. Cell 1980;20(1):185–97.

[63] Hafez M, Hausner G. Homing endonucleases: DNA scissors on a mission. Genome/Natl Res Counc Can 2012;55(8):553–69.

[64] Flick KE, Jurica MS, Monnat Jr RJ, Stoddard BL. DNA binding and cleavage by the nuclear intron-encoded homing endonuclease I-PpoI. Nature 1998;394(6688):96–101.

[65] Jurica MS, Monnat Jr RJ, Stoddard BL. DNA recognition and cleavage by the LAGLIDADG homing endonuclease I-CreI. Mol Cell 1998;2(4):469–76.

[66] Li H, Ulge UY, Hovde BT, Doyle LA, Monnat Jr RJ. Comprehensive homing endonuclease target site specificity profiling reveals evolutionary constraints and enables genome engineering applications. Nucleic Acids Res 2012;40(6):2587–98.

[67] Chevalier BS, Kortemme T, Chadsey MS, Baker D, Monnat RJ, Stoddard BL. Design, activity, and structure of a highly specific artificial endonuclease. Mol Cell 2002;10(4):895–905.

[68] Steuer S, Pingoud V, Pingoud A, Wende W. Chimeras of the homing endonuclease PI-SceI and the homologous Candida tropicalis intein: a study to explore the possibility of exchanging DNA-binding modules to obtain highly specific endonucleases with altered specificity. Chembiochem: A Eur J Chem Biol 2004;5(2):206–13.

[69] Li H, Pellenz S, Ulge U, Stoddard BL, Monnat Jr RJ. Generation of single-chain LAGLIDADG homing endonucleases from native homodimeric precursor proteins. Nucleic Acids Res 2009;37(5):1650–62.

[70] Ashworth J, Taylor GK, Havranek JJ, Quadri SA, Stoddard BL, Baker D. Computational reprogramming of homing endonuclease specificity at multiple adjacent base pairs. Nucleic Acids Res 2010;38(16):5601–8.

[71] Ulge UY, Baker DA, Monnat Jr RJ. Comprehensive computational design of mCreI homing endonuclease cleavage specificity for genome engineering. Nucleic Acids Res 2011;39(10):4330–9.

[72] Jansen R, Embden JD, Gaastra W, Schouls LM. Identification of genes that are associated with DNA repeats in prokaryotes. Mol Microbiol 2002;43(6):1565–75.

[73] Mojica FJ, Diez-Villasenor C, Garcia-Martinez J, Soria E. Intervening sequences of regularly spaced prokaryotic repeats derive from foreign genetic elements. J Mol Evol 2005;60(2):174–82.

[74] Pourcel C, Salvignol G, Vergnaud G. CRISPR elements in *Yersinia pestis* acquire new repeats by preferential uptake of bacteriophage DNA, and provide additional tools for evolutionary studies. Microbiology 2005;151(Pt 3):653–63.

[75] Deltcheva E, Chylinski K, Sharma CM, Gonzales K, Chao Y, Pirzada ZA, et al. CRISPR RNA maturation by trans-encoded small RNA and host factor RNase III. Nature 2011;471(7340):602–7.

[76] Jinek M, Chylinski K, Fonfara I, Hauer M, Doudna JA, Charpentier E. A programmable dual-RNA-guided DNA endonuclease in adaptive bacterial immunity. Science 2012; 337(6096):816–21.

[77] Jiang W, Bikard D, Cox D, Zhang F, Marraffini LA. RNA-guided editing of bacterial genomes using CRISPR-Cas systems. Nat Biotechnol 2013;31(3):233–9.

[78] Pattanayak V, Lin S, Guilinger JP, Ma E, Doudna JA, Liu DR. High-throughput profiling of off-target DNA cleavage reveals RNA-programmed Cas9 nuclease specificity. Nat Biotechnol 2013;31(9):839–43.

[79] Sapranauskas R, Gasiunas G, Fremaux C, Barrangou R, Horvath P, Siksnys V. The *Streptococcus thermophilus* CRISPR/Cas system provides immunity in *Escherichia coli*. Nucleic Acids Res 2011;39(21):9275–82.

[80] Sternberg SH, Redding S, Jinek M, Greene EC, Doudna JA. DNA interrogation by the CRISPR RNA-guided endonuclease Cas9. Nature 2014;507(7490):62–7.

[81] Garside EL, Macmillan AM. Cas9 in close-up. Nat Biotechnol 2014;32(4):338–40.

[82] Nishimasu H, Ran FA, Hsu PD, Konermann S, Shehata SI, Dohmae N, et al. Crystal structure of Cas9 in complex with guide RNA and target DNA. Cell 2014;156(5):935–49.

[83] Jinek M, Jiang F, Taylor DW, Sternberg SH, Kaya E, Ma E, et al. Structures of Cas9 endonucleases reveal RNA-mediated conformational activation. Science 2014;343(6176):1247997.

[84] Cho SW, Kim S, Kim JM, Kim JS. Targeted genome engineering in human cells with the Cas9 RNA-guided endonuclease. Nat Biotechnol 2013;31(3):230–2.

[85] Cong L, Ran FA, Cox D, Lin S, Barretto R, Habib N, et al. Multiplex genome engineering using CRISPR/Cas systems. Science 2013;339(6121):819–23.

[86] Jinek M, East A, Cheng A, Lin S, Ma E, Doudna J. RNA-programmed genome editing in human cells. eLife 2013;2:e00471.

[87] Mali P, Yang L, Esvelt KM, Aach J, Guell M, DiCarlo JE, et al. RNA-guided human genome engineering via Cas9. Science 2013;339(6121):823–6.

[88] Hwang WY, Fu Y, Reyon D, Maeder ML, Tsai SQ, Sander JD, et al. Efficient genome editing in zebrafish using a CRISPR-Cas system. Nat Biotechnol 2013;31(3):227–9.

[89] Friedland AE, Tzur YB, Esvelt KM, Colaiacovo MP, Church GM, Calarco JA. Heritable genome editing in *C. elegans* via a CRISPR-Cas9 system. Nat Methods 2013;10(8):741–3.

[90] Wang Y, Li Z, Xu J, Zeng B, Ling L, You L, et al. The CRISPR/Cas system mediates efficient genome engineering in *Bombyx mori*. Cell Res 2013;23(12):1414–6.

[91] Li JF, Norville JE, Aach J, McCormack M, Zhang D, Bush J, et al. Multiplex and homologous recombination-mediated genome editing in *Arabidopsis* and *Nicotiana benthamiana* using guide RNA and Cas9. Nat Biotechnol 2013;31(8):688–91.

[92] DiCarlo JE, Norville JE, Mali P, Rios X, Aach J, Church GM. Genome engineering in *Saccharomyces cerevisiae* using CRISPR-Cas systems. Nucleic Acids Res 2013;41(7):4336–43.

[93] Baker M. Gene editing at CRISPR speed. Nat Biotechnol 2014;32(4):309–12.

[94] Shalem O, Sanjana NE, Hartenian E, Shi X, Scott DA, Mikkelsen TS, et al. Genome-scale CRISPR-Cas9 knockout screening in human cells. Science 2014;343(6166):84–7.

[95] Wang T, Wei JJ, Sabatini DM, Lander ES. Genetic screens in human cells using the CRISPR-Cas9 system. Science 2014;343(6166):80–4.

[96] Koike-Yusa H, Li Y, Tan EP, Velasco-Herrera Mdel C, Yusa K. Genome-wide recessive genetic screening in mammalian cells with a lentiviral CRISPR-guide RNA library. Nat Biotechnol 2014;32(3):267–73.

[97] Niu Y, Shen B, Cui Y, Chen Y, Wang J, Wang L, et al. Generation of gene-modified cynomolgus monkey via Cas9/RNA-mediated gene targeting in one-cell embryos. Cell 2014;156(4):836–43.

[98] Cho SW, Kim S, Kim Y, Kweon J, Kim HS, Bae S, et al. Analysis of off-target effects of CRISPR/Cas-derived RNA-guided endonucleases and nickases. Genome Res 2014;24(1):132–41.

[99] Cradick TJ, Fine EJ, Antico CJ, Bao G. CRISPR/Cas9 systems targeting beta-globin and CCR5 genes have substantial off-target activity. Nucleic Acids Res 2013;41(20):9584–92.

[100] Fu Y, Foden JA, Khayter C, Maeder ML, Reyon D, Joung JK, et al. High-frequency off-target mutagenesis induced by CRISPR-Cas nucleases in human cells. Nat Biotechnol 2013;31(9):822–6.

[101] Hsu PD, Scott DA, Weinstein JA, Ran FA, Konermann S, Agarwala V, et al. DNA targeting specificity of RNA-guided Cas9 nucleases. Nat Biotechnol 2013;31(9):827–32.

[102] Lin Y, Cradick TJ, Brown MT, Deshmukh H, Ranjan P, Sarode N, et al. CRISPR/Cas9 systems have off-target activity with insertions or deletions between target DNA and guide RNA sequences. Nucleic Acids Res 2014;42(11):7473–85.

[103] Wu X, Scott DA, Kriz AJ, Chiu AC, Hsu PD, Dadon DB, et al. Genome-wide binding of the CRISPR endonuclease Cas9 in mammalian cells. Nat Biotechnol 2014;32(7):670–6.

[104] Mali P, Aach J, Stranges PB, Esvelt KM, Moosburner M, Kosuri S, et al. CAS9 transcriptional activators for target specificity screening and paired nickases for cooperative genome engineering. Nat Biotechnol 2013;31(9):833–8.

[105] Ran FA, Hsu PD, Lin CY, Gootenberg JS, Konermann S, Trevino AE, et al. Double nicking by RNA-guided CRISPR Cas9 for enhanced genome editing specificity. Cell 2013;154(6):1380–9.

[106] Sander JD, Joung JK. CRISPR-Cas systems for editing, regulating and targeting genomes. Nat Biotechnol 2014;32(4):347–55.

[107] Tsai SQ, Wyvekens N, Khayter C, Foden JA, Thapar V, Reyon D, et al. Dimeric CRISPR RNA-guided FokI nucleases for highly specific genome editing. Nat Biotechnol 2014;32(6):569–76.

[108] Guilinger JP, Thompson DB, Liu DR. Fusion of catalytically inactive Cas9 to FokI nuclease improves the specificity of genome modification. Nat Biotechnol 2014;32(6):577–82.

[109] Fu Y, Sander JD, Reyon D, Cascio VM, Joung JK. Improving CRISPR-Cas nuclease specificity using truncated guide RNAs. Nat Biotechnol 2014;32(3):279–84.

[110] Bikard D, Jiang W, Samai P, Hochschild A, Zhang F, Marraffini LA. Programmable repression and activation of bacterial gene expression using an engineered CRISPR-Cas system. Nucleic Acids Res 2013;41(15):7429–37.

[111] Cheng AW, Wang H, Yang H, Shi L, Katz Y, Theunissen TW, et al. Multiplexed activation of endogenous genes by CRISPR-on, an RNA-guided transcriptional activator system. Cell Res 2013;23(10):1163–71.

[112] Gilbert LA, Larson MH, Morsut L, Liu Z, Brar GA, Torres SE, et al. CRISPR-mediated modular RNA-guided regulation of transcription in eukaryotes. Cell 2013;154(2):442–51.

[113] Perez-Pinera P, Kocak DD, Vockley CM, Adler AF, Kabadi AM, Polstein LR, et al. RNA-guided gene activation by CRISPR-Cas9-based transcription factors. Nat Methods 2013;10(10):973–6.

[114] Qi LS, Larson MH, Gilbert LA, Doudna JA, Weissman JS, Arkin AP, et al. Repurposing CRISPR as an RNA-guided platform for sequence-specific control of gene expression. Cell 2013;152(5):1173–83.

[115] Chen B, Gilbert LA, Cimini BA, Schnitzbauer J, Zhang W, Li GW, et al. Dynamic imaging of genomic loci in living human cells by an optimized CRISPR/Cas system. Cell 2013;155(7):1479–91.

[116] Senis E, Fatouros C, Grosse S, Wiedtke E, Niopek D, Mueller AK, et al. CRISPR/Cas9-mediated genome engineering: an adeno-associated viral (AAV) vector toolbox. Biotechnol J 2014;9(11):1402–12.

[117] Grieger JC, Samulski RJ. Packaging capacity of adeno-associated virus serotypes: impact of larger genomes on infectivity and postentry steps. J Virol 2005;79(15):9933–44.

[118] Wu J, Zhao W, Zhong L, Han Z, Li B, Ma W, et al. Self-complementary recombinant adeno-associated viral vectors: packaging capacity and the role of rep proteins in vector purity. Hum Gene Ther 2007;18(2):171–82.

[119] Holkers M, Maggio I, Liu J, Janssen JM, Miselli F, Mussolino C, et al. Differential integrity of TALE nuclease genes following adenoviral and lentiviral vector gene transfer into human cells. Nucleic Acids Res 2013;41(5):e63.

[120] Cai Y, Bak RO, Mikkelsen JG. Targeted genome editing by lentiviral protein transduction of zinc-finger and TAL-effector nucleases. eLife 2014;3:e01911.

[121] Maggio I, Holkers M, Liu J, Janssen JM, Chen X, Goncalves MA. Adenoviral vector delivery of RNA-guided CRISPR/Cas9 nuclease complexes induces targeted mutagenesis in a diverse array of human cells. Sci Rep 2014;4:5105.

[122] Liu J, Gaj T, Patterson JT, Sirk SJ, Barbas 3rd CF. Cell-penetrating peptide-mediated delivery of TALEN proteins via bioconjugation for genome engineering. PloS One 2014;9(1):e85755.

[123] Bloom K, Ely A, Mussolino C, Cathomen T, Arbuthnot P. Inactivation of hepatitis B virus replication in cultured cells and in vivo with engineered transcription activator-like effector nucleases. Mol Ther: J Am Soc Gene Ther 2013;21(10):1889–97.

[124] Chen J, Zhang W, Lin J, Wang F, Wu M, Chen C, et al. An efficient antiviral strategy for targeting hepatitis B virus genome using transcription activator-like effector nucleases. Mol Ther: J Am Soc Gene Ther 2014;22(2):303–11.

[125] Cradick TJ, Keck K, Bradshaw S, Jamieson AC, McCaffrey AP. Zinc-finger nucleases as a novel therapeutic strategy for targeting hepatitis B virus DNAs. Mol Ther: J Am Soc Gene Ther 2010;18(5):947–54.

[126] Ebina H, Misawa N, Kanemura Y, Koyanagi Y. Harnessing the CRISPR/Cas9 system to disrupt latent HIV-1 provirus. Sci Rep 2013;3:2510.

[127] Hu W, Kaminski R, Yang F, Zhang Y, Cosentino L, Li F, et al. RNA-directed gene editing specifically eradicates latent and prevents new HIV-1 infection. Proc Natl Acad Sci USA 2014;111(31):11461–6.

[128] Tanaka A, Takeda S, Kariya R, Matsuda K, Urano E, Okada S, et al. A novel therapeutic molecule against HTLV-1 infection targeting provirus. Leukemia 2013;27(8):1621–7.

[129] Hu Z, Ding W, Zhu D, Yu L, Jiang X, Wang X, et al. TALEN-mediated targeting of HPV oncogenes ameliorates HPV-related cervical malignancy. J Clin Invest 2015;125(1):425–36.

CHAPTER 4

Viral Vectors for Delivery of Antiviral Sequences

4.1 INTRODUCTION

Because viruses are fundamentally nucleic acid parasites that are dependent on cellular functions for their replication, they have evolved sophisticated mechanisms for delivering their genomic material into cells. This fundamental property has been exploited to generate vectors that may be used for antiviral therapeutic application. The essential approach to generating replication-defective recombinant viruses entails replacing viral elements with therapeutic transgenes. Deficiencies required for proliferation of the vectors are typically provided in trans from packaging cells, helper viruses (HVs), or transfected plasmids. Removal of viral genes from a viral genome serves three important purposes:

1. rendering of the viral vector replication defective,
2. making space available within the vector genome to accommodate transgene sequences, and
3. limiting the risk of recombination that may give rise to formation of replication-competent vectors.

Several different vectors are being developed for applications to gene therapy, which include treatment of viral infections. The most commonly used viral vectors are derived from

1. adeno-associated viruses (AAVs),
2. adenoviruses (Ads), and
3. lentiviruses or retroviruses.

Recombinant poxvirus vectors have been used extensively as antiviral vaccines and are discussed in Chapter 11. Other recombinant viral vectors include derivatives of herpes viruses and Semliki Forest virus, but they have not been commonly used for gene therapy of viral infections. These vectors and other recombinant viruses, such as those that function by an oncolytic mechanism to treat cancer, are not discussed in detail.

The efficiency with which transgene delivery can be achieved with recombinant viruses has made them a popular choice for use as vectors to advance antiviral gene therapy. They are better than nonviral vectors for delivery of expression

Gene Therapy for Viral Infections. http://dx.doi.org/10.1016/B978-0-12-410518-8.00004-1

cassettes in vivo, but they are unsuited to delivery of synthetic RNA. Synthetic nonviral vectors have better utility for delivering short interfering RNAs (siRNAs) and synthetic mRNA (Chapter 5). As biological therapeutics, viral vectors are inherently complex, and several factors need to be considered in developing vectors from this class for antiviral application:

1. Generating viral vectors typically involves large-scale cell culture, which is generally complex and expensive. Therefore, with currently available technology, viral vectors may not be suited to widespread use for treatment of commonly occurring infections such as are caused by hepatitis B virus (HBV), hepatitis C virus (HCV), and human immunodeficiency virus (HIV)-1.
2. Viral vectors are capable of inducing powerful immune responses after in vivo administration. Innate and adaptive immune responses are both important. A strong innate immune response is potentially severely toxic, which has been shown to be the case after administration of recombinant Ads [1,2]. Adaptive immunity results in destruction of vectors and transduced cells with resultant compromised delivery efficiency. This is particularly important when repeat administration of viral vectors is required. In addition, when community-acquired infection with a natural parent of viral vectors occurs, then gene transfer efficiency will be compromised. Again, this may be problematic with Ads. To overcome the drawback, substitution of sequences with those that are derived from uncommon or zoonotic parent viruses may be used.
3. The duration of transgene expression that is required to achieve an intended therapeutic effect is also important. For treating chronic infections using a posttranscriptional RNA interference (RNAi)-based gene silencing approach, long-term expression of the therapeutic cassette is important. Conversely, transient expression for treating an acute infection may be preferable. In the case of using gene editing to directly disable viral sequences, transient expression may be desirable to limit problems of off-target effects. As derivatives of HIV-1, lentiviral vectors form proviral integrants in target cells. This capacity for stable integration is very useful to achieve sustained therapeutic transgene expression, but it may also increase the risks of off-target effects and mutation.
4. The size of the therapeutic transgene to be used for treatment of a viral infection is another important consideration. Large cassettes, such as those encoding transcription activator-like effector nucleases (TALENs) or clustered regularly interspaced palindromic repeats (CRISPR) with CRISPR-associated (Cas) sequences, are difficult to accommodate in the popular AAV class of recombinant viral vectors. To deliver TALENs

or CRISPR-Cas cassettes using AAVs, it may be necessary to administer more than one vector simultaneously to constitute the complete assembly of gene editors in the target cells. Helper-dependent (HD) Ads are capable of efficient delivery of large transgenes, but tissue tropism and immunity to vectors are significant issues.

5. Tropism of viral vectors is vital for therapeutic efficacy. As a result of interaction of the hexon protein of the vector with blood clotting factor X (FX) [3], Ads are naturally highly hepatotropic after systemic administration. This property may be useful for treatment of HBV and HCV infections, but it is undesirable for treatment of infections occurring in other tissues. When administered to cells in culture, Ads have a broad tropism. Thus, if ex vivo therapeutic gene transfer is feasible, such as is the case with modification to the CCR5 co-receptor of HIV-1 in $CD4^+$ cells [4], then the liver tropism of Ads is not problematic. Selection of variants of naturally occurring parent viral vectors is an approach that has been used successfully to engineer vector tropism. This has been used for the formation of AAVs with a wide range of tissue specificities.
6. Avoiding use of reagents that are potentially biohazardous, such as those of bovine origin, is important.
7. Instability of viral vectors necessitates appropriate storage to prevent decreases in efficiency.

4.2 ADENO-ASSOCIATED VIRAL VECTORS

Recombinant AAVs are very popular vectors for use in gene therapy. Their main advantages are that they are safe, amenable to engineering to change their biological properties, efficiently deliver transgenes to various cells, and achieve sustained transgene expression. The molecular biology of AAVs is now well understood, which has enabled improvement of the methods used for vector propagation and development of new approaches to adapting vector design [5].

4.2.1 AAV Biology

AAVs belong to the family *Parvoviridae*, and genus *Dependoviridae* [6–8]. They are small, nonenveloped viruses, which are not known to cause disease in humans. The capsid has icosahedral symmetry with a triangulation number of 1. It comprises 60 molecules of 3 capsid proteins, VP1, VP2, and VP3, which are encoded by the *cap* open reading frame (ORF) [9,10] (Figure 4.1). Translation of the three capsid proteins is influenced by viral mRNA splicing, and resulting variation in the translation efficiencies leads to an approximately 10-fold excess of VP3 relative to VP1 and VP2. Formation of the capsid is facilitated by the assembly-activating protein (AAP), which is also encoded by the

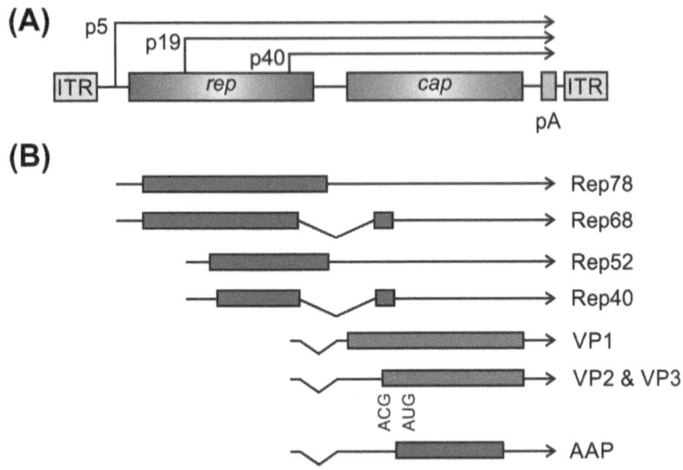

FIGURE 4.1 AAV genome and expression of the viral genes.
(A) The viral genome encodes *rep* and *cap* genes, which are flanked by ITRs. Three promoters, p5, p19, and p40, initiate transcription of mRNAs that terminate at a common transcription termination signal (pA). (B) The four Rep proteins, Rep78, Rep68, Rep52, and Rep40, are encoded by transcripts derived from the p5 and p19 promoters. Unspliced mRNAs generate Rep78 and Rep52 whereas spliced variants translate Rep68 and Rep40. VP1, VP2, and VP3 are encoded by spliced mRNAs initiated at the p40 promoter and translated from three in-phase start codons. The initiation codon of VP2 is unusual in that it comprises an ACG sequence. The regulatory AAP is also encoded within the *cap* sequence and is responsible for modulating assembly of the VP proteins to form the mature capsids.

cap sequence, but uses an alternate reading frame [11,12]. The *rep* ORF encodes four nonstructural proteins, Rep40, Rep52, Rep68, and Rep 78, which regulate AAV gene expression, DNA replication, and virion assembly.

Naturally, AAVs are dependent on helper functions of Ads or herpes viruses during co-infection of cells. The linear single-stranded DNA genome of AAVs, comprising approximately 5 kb, has two inverted terminal repeats (ITRs) and may be of positive or negative strand polarity (Figures 4.1 and 4.2). After infection of cells, AAV DNA traverses the endosomes and is transported to the nucleus, where the viral genome replicates. Site-specific integration of the genome on chromosome 19 may occur during latent infection and is mediated by the viral Rep proteins. AAV integration is a rare event after infection of cells with recombinant vectors because they are devoid of Rep-encoding sequences. The ITRs comprise 145 bases and are essential for AAV replication [13]. Intrastrand base pairing results in formation of characteristic T-shaped hairpin loop structures of ITRs, which prime synthesis of new DNA by host polymerase. When helper functions are provided in trans, the ITRs are the only *cis* elements that are required to propagate recombinant AAVs. During

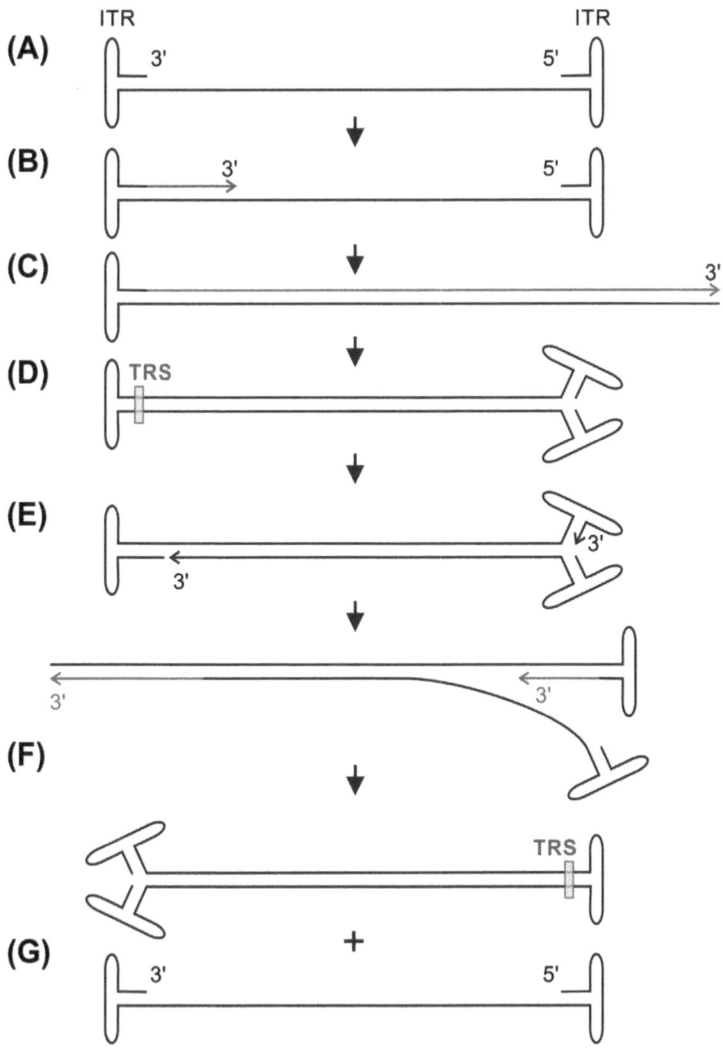

FIGURE 4.2 Replication of the AAV genome.
The single-stranded genome of AAVs, which may have positive or negative strand polarity, is flanked by palindromic ITRs that fold into characteristic hairpin structures (A). After infection of cells, and translocation to the nucleus, host DNA polymerase initiates DNA synthesis at the 3′ end (B). Extension to the 5′ end of the intramolecular template (C) copies the ITRs, which then refold into hairpins (D). Rep proteins introduce a nick at the TRS (E). Thereafter, both of the 3′ ends of the DNA strands are extended by the host polymerase to result in duplication of the ITRs and displacement of the nascent viral genomic DNA (F). Completion of the synthesis steps leads to formation of a new genomic strand and a template that again undergoes nicking, DNA synthesis, and formation of new genomic DNA of opposite strand polarity (G).

natural replication of AAVs, Rep proteins act in concert to facilitate nicking by Rep68 of viral DNA at the terminal resolution sites (TRSs) within the ITRs (Figure 4.2) [14,15]. This process is required for extension by host DNA polymerase at the nicked sites. When the single-stranded endonuclease function of Rep68 is compromised by a mutant TRS, a dimeric genome is formed, which comprises self-complementary sequences within the same strand of DNA (Figure 4.3).

4.2.2 Rationale for the Propagation of Recombinant AAVs

The typical method of propagating recombinant single-stranded AAVs (ssAAVs) entails generation of a plasmid containing the DNA cassette of interest, which is substituted for the *rep* and *cap* genes and inserted between ITRs. Co-transfection of HEK293 cells with this plasmid, and helper plasmids is then used to generate the recombinant AAVs (Figure 4.4). Co-infection with Ads, herpes viruses [16], or baculoviruses [17] may also be used to provide helper function, but these methods are not commonly used because of the simplicity and efficiency of plasmid co-transfection. There is a constraint on the sizes of inserts that may be incorporated into ssAAVs, and cassettes of up to approximately 4.7 kb in length may be used to replace the natural *rep* and *cap* genes. After transduction, the genomes of AAVs exist stably as circular monomeric or concatemeric episomes (reviewed in ref. [18]).

Efficiency of expression of transgenes delivered with AAVs is dependent on several factors. Vector uncoating and nuclear delivery of the DNA affect the efficiency of transduction of cells [19,20], but conversion of the single-stranded genome to a double-stranded transcription template is particularly important [21,22]. De novo synthesis of the complementary strand is usually required to form double-stranded viral DNA during natural infection. Alternatively, when base pairing occurs between complementary genomes of two viruses that simultaneously infect the same cell, duplex DNA may be formed [21]. Theoretically, the multiplicity of infection may be manipulated to favor formation of such DNA. However, because there are several cellular factors that play a role in annealing of the strands, it is difficult to predict the efficiency of duplex formation [20].

Understanding the importance of the TRSs in this natural viral replication mechanism led successfully to introduction of mutations at these sites as a means for generating self-complementary AAVs (scAAVs) [22]. By mutating one TRS, packaged vector DNA comprises a strand that folds back on itself to form a duplex that may serve as a template for transcription of the transgene. Thus, synthesis of the second strand of DNA is bypassed and transgene expression is enhanced as a result (Figure 4.3). Although transgene expression after transduction by scAAVs is superior to that achieved with ssAAVs, a drawback of the self-complementing vectors is that the size of the insert is further limited to approximately 2.5 kb. This may not be problematic for transduction of

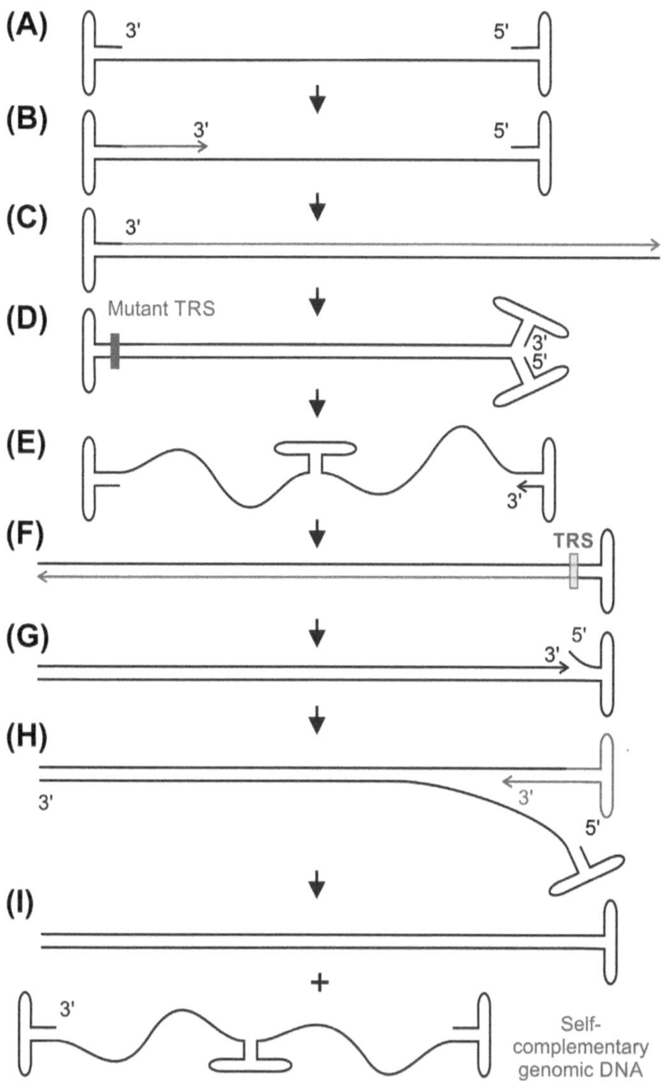

FIGURE 4.3 Replication of the AAV genome to form self-complementary sequences.
As with ssAAVs, the vector's genome is flanked by palindromic ITRs (A). After infection of cells, host DNA polymerase initiates DNA synthesis at the 3′ end (B). Extension to the 5′ end of the intramolecular template (C) copies the ITRs, which then refold into hairpins (D). When the TRS is mutated, Rep proteins are incapable of introducing a nick in the site (D). Thereafter, the 3′ end of the DNA is extended by the host polymerase to result in formation of dimeric DNA (E). Nicking at the opposite TRS (F) and strand displacement with DNA synthesis (G) leads to displacement of the self-complementary genomic DNA (H) and formation of a new dimeric template (I).

cells with virus-targeting RNAi expression cassettes that are small. However, the large size of many of the genome editing expression cassettes, such as those encoding TALENs and combined CRISPR-Cas elements, complicates their incorporation into scAAVs.

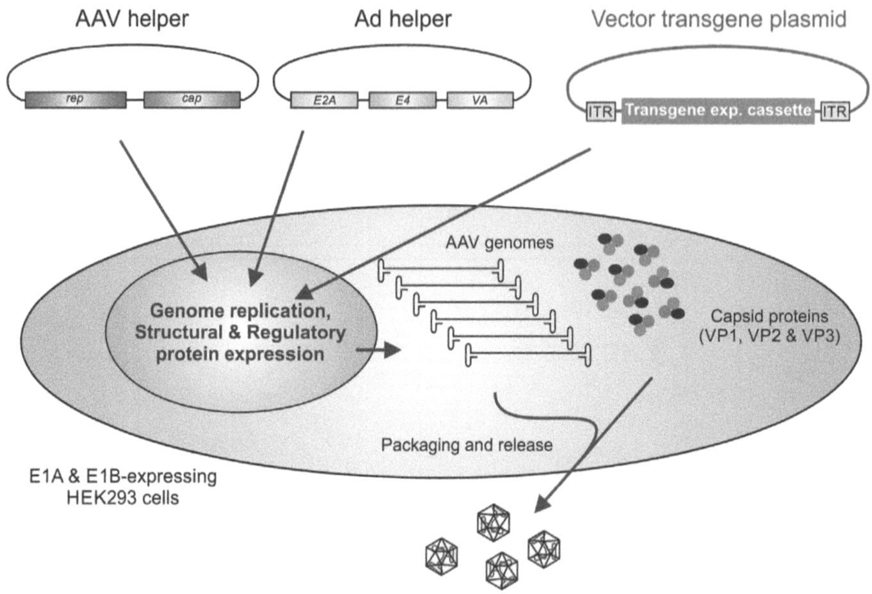

FIGURE 4.4 Production of recombinant AAVs.

HEK293 cells, which constitutively produce E1A and E1B proteins of Ads, are transfected with vector and helper plasmids. The vector plasmid contains the transgene of interest, which is flanked by ITRs. Helper plasmids encode AAV sequences (*rep* and *cap*) and the *E2A*, *E4*, and *VA* genes of Ads. After transfection of HEK293 cells with the three plasmids, genome replication and vector protein production occur. Packaging of genomes into the capsid particles is followed by release of the vectors.

4.2.3 Alteration of Sequences Encoding Recombinant AAVs to Modify their Biological Properties

The ability to package transgenes into recombinant AAVs generated from capsid proteins of different natural or artificial sources of AAVs has been particularly useful. This has enabled improvement of the versatility of the vectors by altering their tropism and immunogenicity (reviewed in refs [5,6]; Table 4.1). Capsids of different serotypes were initially reported to have no effect on the size of the DNA that may be packaged [23], but use of a capsid from serotype 5 was shown to be able to accommodate larger packaged DNA sequences of up to 8.9 kb [24]. Although significant, this property has not been widely exploited. Researchers have generally restricted the length of the DNA to be packaged to approximately 4.5 kb for ssAAVs and half of that for scAAVs. Directed evolution of AAVs, rational design of capsid proteins, and variation in capsid proteins of different serotypes have all been used successfully to match the properties of recombinant AAVs to their requirements for specific applications as vectors.

Exposure to AAVs is widespread in general populations, and neutralizing antibodies are commonly found [25]. Not surprisingly, this community-acquired

Table 4.1 Approaches to Modifying AAVs to Improve Their Clinical Utility

Approach	Method	Effect
Directed evolution of capsid-encoding sequences	DNA shuffling Error-prone PCR Site-specific randomization	1. Alteration of tissue tropism 2. Evasion of preexisting immunity 3. Improvement of transduction efficiency 4. Attenuation of adaptive immune response
Rational design of capsid proteins	Site-specific mutation to inhibit ubiquitination and proteolytic degradation	Attenuation of adaptive cell mediated immune response (MHC class I presentation)
	Inhibition of antibody interaction with capsid proteins	Attenuation of adaptive humoral immune response
	Capsid modification with ligands	Alteration of tissue tropism
Decoy capsids	Co-administration with transducing vectors	Immunoevasion

immunity diminishes gene transduction by recombinant AAVs. Immune responses to both wild-type viral epitopes and transgenes may interfere with the efficacy of vectors [26,27]. Activation of cell-mediated immunity after presentation of epitopes from AAV capsids on major histocompatibility complex (MHC) class I molecules in transduced cells also plays a role in diminishing efficiency of AAVs. Because $CD4^+$ and $CD8^+$ T cells recognize many epitopes of commonly used AAVs [28–30], evasion of immunity to the vectors has proven to be difficult. In one of the first clinical trials to treat hemophilia B using gene transfer with AAVs, declining expression of the blood clotting factor IX protein after vector administration was indeed a result of this MHC class I-mediated elimination of transduced cells [31]. In addition to altering capsid proteins to avoid interaction with B and T cell receptors, inhibition of the processing of the capsid proteins in the proteasome is also potentially useful to avoid immunostimulatory effects. Understanding the importance of phosphorylation of tyrosine residues within capsids for proteasomal degradation and antigen presentation led to engineering of capsid proteins in which the tyrosine residues were substituted with phenylalanines [32,33]. The resultant decreased phosphorylation attenuated immunostimulation because of impaired proteasomal degradation and epitope presentation. Other methods of enabling AAVs to evade the host immune response have entailed rational introduction of mutations at antibody docking sites of the capsid protein [34] and use of empty vectors to function as "decoys" for neutralizing antibodies [35].

Directed evolution of AAVs is a powerful method of engineering capsids to isolate vectors with desirable properties (reviewed in ref. [5]). The procedure entails generation of a library of viral particles that contain different mutations within the *cap* sequence, and they may be distributed throughout the ORF or be localized to specific sites. Three main methods have been used to engineer capsid libraries: error-prone polymerase chain reaction (PCR), DNA shuffling, and ligation of degenerate oligonucleotides into the *cap* ORF. Generation of viral particles occurs through a process in which a mutant capsid protein packages the particular gene encoding that capsid. This is followed by repeated iterative infection and selection to enrich particles that have desired properties, such as an intended tissue tropism. The procedure has recently been used for the isolation of AAVs that are tropic for human hepatocytes [36]. Behavior of AAVs in vivo in different animals cannot always be reliably extrapolated to predict the vectors' properties in humans. An example is the variable infection of murine and human hepatocytes by AAV8. The recombinant derivatives of AAV8 infect murine hepatocytes with high efficiency, but transduction of human hepatocytes is poor [36,37]. To address this problem, Lisowski and colleagues used a procedure that involved DNA shuffling and selection of replication-competent AAVs that were capable of efficiently transducing human hepatocytes in xenografted mice [36]. Serial selection of vectors capable of infecting the human cells was then performed. A potentially clinically useful chimeric capsid, made up of sequences from five different parental capsids and that had the desired properties, was isolated.

4.3 ADENOVIRAL VECTORS

Ads have several features that make them attractive for application to antiviral gene therapy. The vectors have been used to deliver directly acting antiviral sequences and immunostimulatory cassettes (Chapter 11). Useful properties of Ads include the following:

1. Ads have an ability to infect dividing and nondividing cells very efficiently.
2. The molecular biology of Ads is understood well, and as a result the methods of vector preparation may be adapted to particular applications.
3. After infection of cells, the genomes of Ads exist stably as episomes and rarely integrate into the host genomes.
4. Transgene expression in Ads is efficient, and there is minimal if any interference from transcriptional control elements of the vectors.
5. With development of HD Ads, the capacity of the vectors for accommodating large transgenes is good, which may be useful for delivery of large antiviral cassettes such as those encoding TALEN and CRISPR-Cas systems.
6. Ads are amenable to chemical modification, such as the addition of polymers, which is useful to attenuate immunostimulation and alter vector tropism.

Despite these significant advantages, a limitation of the vectors is that they are rapidly taken up by Kupffer cells, the resident macrophages of the liver [38]. Interaction of Ads with the Kupffer cells results in destruction of the viruses and the macrophages, with concomitant inflammation and immunostimulation [39]. Both innate and adaptive arms of the immune system are activated by Ads. Consequently, the vectors may be toxic and transduction efficiency may also be attenuated. A major focus of research on use of Ads has understandably been on attenuating these effects to improve their usefulness as vectors for gene therapy.

4.3.1 Ad Biology

Ads belong to the *Adenoviridae* family (reviewed in ref. [40]). Human Ad serotypes 5 and 2 have been most extensively studied, and insights into the biology of these viruses have provided the basis for Ad vector development. The virion consists of a nonenveloped capsid with a diameter of approximately 90 nm, which contains a genome of linear double-stranded DNA that comprises 27–44 kbp. The capsid is constituted from hexon, penton, and fiber proteins. The pentameric penton proteins are located at the 12 apices of the capsid icosahedrons. Trimeric fiber proteins interact with the pentons at their N-terminal ends. The globular knob is located at the C terminal and protrudes from the virion capsid to make it available for interaction with cell surface receptors. Trimeric hexon proteins, positioned between the apices of the capsid, are the most abundant protein of the virion. There are 720 hexon proteins within each capsid, and the 240 tower-like trimers are formed by winding of the monomers' protruding loop sequences around each other [41,42]. Hypervariable regions (HVRs), of which there are seven per hexon protein, are positioned at the surface of the capsid and are available for interaction with host proteins [43,44]. HVRs are highly charged and are important for interaction of Ads with scavenger receptors, particularly those of Kupffer cells [45]. Thus, modification of HVR sequences by substitution of amino acids [46] and polymer modification [47] has been used to attenuate binding of Ads to scavenger receptors and diminish immunostimulation by the vectors.

During natural infection, the fiber knob interacts with several different receptors (reviewed in ref. [48]). With the exception of hepatocytes, the Coxsackie and Adenovirus Receptor (CAR) is the primary binding site for Ads, and susceptibility to Ad infection correlates with CAR expression. Other receptors with lower affinity have been described and include heparan sulfate proteoglycans, CD46, CD80/86 MHC class I proteins, and vascular cell adhesion molecule-1. Ads bind to CAR and other receptors on several blood components, such as erythrocytes [49], macrophages [50], neutrophils [51], and platelets [52]. This promiscuous interaction may result in vector sequestration and compromised efficiency as a gene therapy vector. After binding to CAR, Ads are internalized by a mechanism that entails interaction of the penton proteins with cellular α_v integrins [53,54]. Interestingly, hepatocytes, which are the cells that are primarily infected by Ad5 after systemic administration, are not dependent on CAR expression for transduction by the vector [55]. The mechanism of transduction of these cells involves high-affinity binding of the viral hexon protein to blood clotting factor X (FX) [3]. When warfarin treatment is used to inhibit FX synthesis, Ad infection of hepatocytes is considerably diminished. There are 240 binding sites for FX per virion, and the clotting factor acts as a bridge receptor to facilitate liver cell transduction. FX interacts with the hexon protein at the zymogen's γ-carboxyglutamic acid domain, and the serine protease region binds to heparan sulfate on the hepatocyte surface [56]. It has recently been shown that by binding to Ads, FX also provides a protective shield to the virus [57]. This prevents immunoglobulin-M-mediated activation of the classical complement pathway that would otherwise destroy the virus.

The genes encoded by Ads are categorized as early or late, according to the stage at which they are transcribed during the viral life cycle (Figure 4.5). ITRs flank the genome, which also contains a packaging signal (ψ) that is located near the 5′ ITR. The early and late proteins are formed from transcription units that are active during the corresponding stages of the virus replication cycle (reviewed in ref. [58]). Early genes are involved in regulating processes that are important for initiating the viral replication cycle whereas late proteins are typically structural components of the viral particles. Functions of early proteins include the control of viral gene transcription, modulation of the host's response to infection, and promoting viral DNA synthesis. E1A and E1B are the first two genes to be expressed after virus entry into cells. Their role is to inhibit function of p53 and Rb in the host cell. Consequently, Ad replication is facilitated by inhibition of cell cycle arrest and programmed cell death. E3 plays a role in inhibiting immune-mediated cytotoxic destruction of infected cells [59].

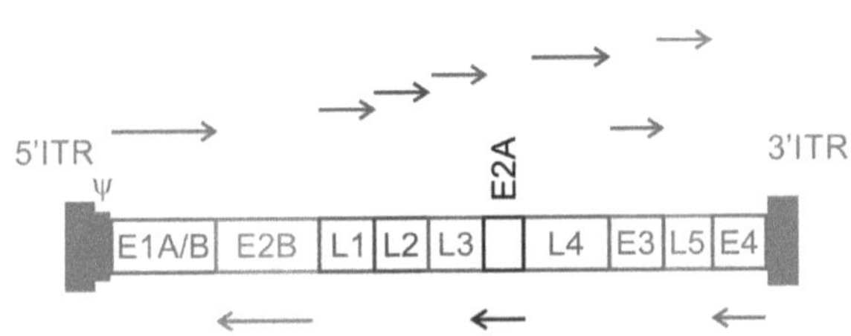

FIGURE 4.5 Schematic illustration of the Ad genome.
The viral genome comprises linear DNA, approximately 35 kb in the case of Ad5, which is flanked by 5′ and 3′ ITRs. The ψ sequence is essential for packaging of the genome into the viral capsid particles. Early (E) and late (L) genes are respectively required to encode regulatory or structural components of the virus. Multiple transcripts, which serve as translation templates for viral proteins, are depicted as arrows above and below the schematic of the genome. Direction of the arrows indicates 5′ to 3′ polarity.

There are several late proteins that are translated from the late transcriptional units [58]. Most of the proteins are involved with formation of the viral capsid and include hexon, penton base, and fiber proteins. The two late virus-associated (VA) transcripts, VAI and VAII, are highly expressed during the lytic phase of Ad replication and perform regulatory functions that are important for increasing translation of both early and late proteins (reviewed in ref. [60]). VAI is transcribed by Pol III and is a highly structured RNA that contains hairpin motifs that resemble precursor microRNAs. Part of their function, which is important for RNAi-mediated inhibition of viral replication, is competitive inhibition of exportin V, Dicer, and disruption of RNA-induced silencing complex (RISC) activity [61,62]. VAI may be processed by Dicer to form potentially functional guide sequences that are incorporated into RISC [63]. Interestingly, analysis of a panel of mutants with changes in the seed regions of either of the possible guides revealed no disruptive effects on the formation of viral progeny. This observation suggested that the regulatory function of VAI does not occur as a result of silencing of cellular genes. However, in an earlier study, VAI-encoded gene silencing was indeed found to be important for the inhibition of cellular growth regulatory genes [64]. Nevertheless, VAI and VAII RNAs seem to be multifunctional and have an important role in modulating the interferon-induced sensing of double-stranded RNA in infected cells [60].

Although expression of VA RNAs is potentially disruptive for therapeutic RNAi-mediated silencing of viral replication, these hairpin-containing transcripts play little if any role in vector-mediated transduction of cells. Because VA transcripts are produced late during infection, their formation is attenuated in recombinant Ads. This is supported by the observation that first-generation Ads do not disrupt miR biogenesis [65]. In addition, RNAi-activating expression cassettes are efficiently expressed and processed after delivery with recombinant first-generation and HD Ad vectors [66–70].

4.3.2 Propagation of Recombinant Adenoviral Vectors

The requirement of Ads for the expression of the E1 gene has been exploited to develop recombinant Ads. Thus, first-generation recombinant vectors were initially generated by deleting E1 from the viral genome and using HEK293 cells to provide the function of the protein in trans (reviewed in ref. [71]) (Figure 4.6). To incorporate transgenes into the Ad genome comprising approximately 35 kbp, methods initially entailed use of techniques of homologous recombination in cultured mammalian cells. Subsequently, this methodology was improved by development of approaches that entailed easier homologous recombination in bacteria [72]. Although it was demonstrated that the first-generation Ads transduced cells with high efficiency, the concomitant expression of several viral proteins was problematic. Understandably, a highly immunostimulatory effect resulted in increased toxicity and diminished duration of transgene expression.

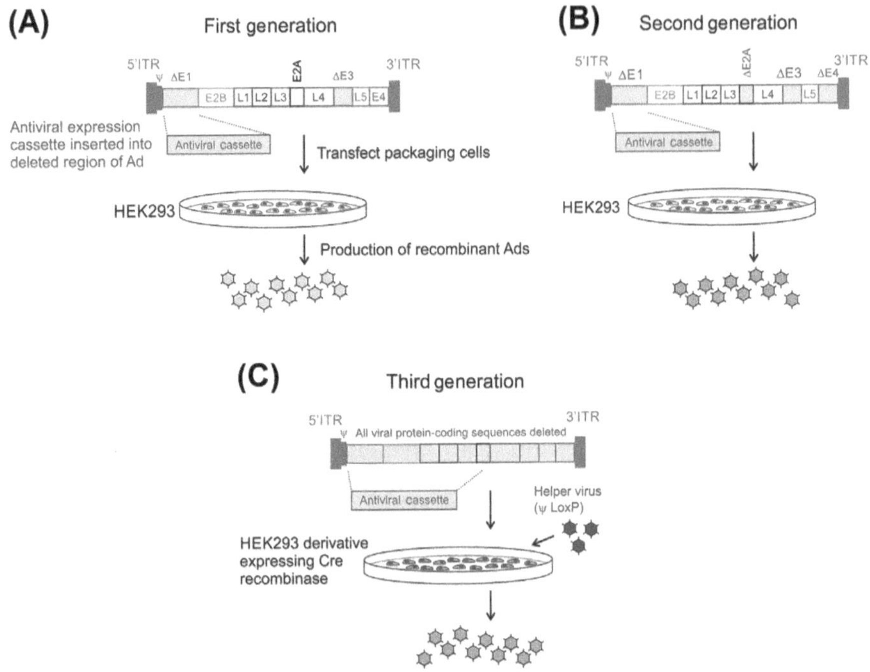

FIGURE 4.6 First-, second-, and third-generation recombinant Ad vectors.
(A) Antiviral expression cassettes (e.g., RNAi-activating cassettes) are inserted to replace the deleted E1 or E3 sequences in first-generation vectors. (B) E1, E2, E4, and/or E3 are deleted from second-generation vectors. First- and second-generation vectors may be propagated after introducing DNA that constitutes the recombinant vectors into cells that provide the missing essential genes (e.g., *E1A* and *E1B*) in trans. (C) Third-generation HD Ads have all of the viral sequences deleted, with the exception of the 5′ and 3′ LTRs and the ψ packaging element. To produce these gutless Ads, HVs are used to provide the viral proteins in trans. Typically, the HVs have their own ψ sequences flanked by targets of recombinases (e.g., LoxP cognates of Cre recombinase). As a result, preferential packaging of the HD Ads occurs and the HVs are eliminated during the propagation of the vectors.

These findings prompted development of second- and third-generation Ads, which contained fewer viral sequences than their first-generation counterparts (Figure 4.6). Second-generation Ads have E2 and E4 sequences removed whereas third-generation HD Ads are "gutless" and have all of the viral protein coding sequences removed. The only remaining viral sequences in HD Ads are the ITRs and the ψ sequence. An HV is required to enable propagation of the vectors. These HVs provide the functions required to generate the HD Ads in trans. A concern of preparing pure HD Ads is the possible contamination with HVs. A particularly successful method of eliminating HVs from vector preparations has been to flank the ψ sequences of the HVs with recognition sites for recombinase enzymes, such as LoxP [73] and FRT [74]. By using HEK293 cells that stably produce the recombinase, it is possible to eliminate almost all of the HVs by preventing their packaging. To avoid problems of homologous recombination that may arise when HVs and HD Ads are present in the same cells, Ng and Palmer reversed the orientation of the ψ sequence of the HVs [75].

The highly efficient hepatotropism of Ads after systemic administration is very useful for the treatment of viruses that infect hepatocytes. However, targeting the liver is often not desirable, and altering the vectors' tropism may be required for certain applications of antiviral gene therapy. The main methods that have been used to adapt capsid structures to change Ad tropism have included chemical modification, genetic alteration, and a combination of the two [76]. Addition of PEG moieties has been widely used and is convenient to improve stability on blood, prevent scavenging by Kupffer cells, evade immunostimulation, and retarget the vectors [77]. When conjugated to PEG 5000 (polyethylene glycol with molecular weight of 5000 Da), Ads retain their hepatotropism. However, when PEG variants with higher molecular weights are used, such as PEG 20,000 and PEG 35,000, infection of liver cells is diminished. Subsequent studies aimed to modify Ads through the conjugation of polymers to specific exposed cysteine residues of the hexon protein [47,78]. This strategy was capable of achieving greater precision of modification to diminish scavenging by Kupffer cells and retargeting the vectors. PEG molecules are also useful to use as spacers to couple targeting moieties onto the vectors [79]. Conjugating the integrin-binding RGD peptides to Ads through a PEG linker has successfully been used to enhance transduction of CAR-positive and CAR-negative cells [80].

4.3.3 Immunostimulation by Ads

Immunostimulation by Ads follows classical mechanisms [39,81–83]. After uptake by antigen-presenting cells (APCs), an innate immune response is mediated by activation of pathways involving mitogen activating protein kinases. There is associated release of proinflammatory cytokines that may also cause a toxic effect. Importantly, activation of the innate immune response does not depend on viral gene expression; therefore, HD Ads may also stimulate the pathway [82]. The APCs activate an adaptive immune response involving MHC class I presentation of processed antigens to $CD8^+$ T cells. Resultant differentiation to form cytotoxic T cells culminates in elimination of transduced cells. Ad antigens may also be presented to $CD4^+$ cells by mechanisms that involve MHC class II. Resultant cytokine release leads to destruction of Ad-infected cells, and activation of B cells induces a humoral immune response.

Although the immune activation mechanisms that occur during natural Ad infection are not unusual, the immunostimulatory effects of the viral vectors are more complex [76]. Several factors influence these effects, such as the dose of vectors, route of administration, and targeted tissue. When low dose and targeted administration of the vectors is feasible, then attenuation of immunostimulation is easier to achieve. However, high-dose administration via the systemic circulation is more complicated, and immunostimulatory effects are difficult to control. Use of immunosuppressive agents such as steroids may be useful [84]. Simultaneous delivery of cassettes that may silence key mediators of an immune response, such as the Cys-X3-Cys chemokine ligand, has been shown to have utility for attenuation of acute hepatotoxicity [85].

4.4 LENTIVIRAL AND RETROVIRAL VECTORS

Retroviridae are the family of viruses to which lentiviruses and gammaretroviruses belong. Interestingly, retroviruses were among the earliest viral infectious agents to be discovered [86]. The equine infectious anemia virus, a lentivirus, and murine leukemia virus (MLV), a gammaretrovirus, were isolated as filterable infectious agents during the early 1900s. Subsequent outbreak of the HIV-1 pandemic, and the serious complications associated with the infection, provided enormous impetus to research members of the *Retroviridae* family (Chapter 9) and their development for gene transfer. A feature of lentiviruses and retroviruses is that they have RNA genomes that serve as templates for reverse transcription. After infection of cells, the reverse-transcribed DNA is integrated into the genome of host cells to form a provirus. This fundamental feature is crucial for achieving sustained transgene expression after transduction with the vectors.

Vectors derived from lentiviruses have largely superseded their gammaretroviral counterparts because they have several advantages:

1. Lentiviruses are capable of infecting nondividing or slowly dividing cells [87], whereas retroviral vectors are only capable of transducing rapidly proliferating cells. This is as a result of the ability of the lentiviral preintegration complex (PIC) to traverse the intact nuclear membrane [88,89].
2. The considerable insights into the molecular biology of lentiviruses that have resulted from research on HIV-1 have led to improved understanding of the ways in which lentiviruses may be manipulated to improve the vectors' properties. This has facilitated propagation of safer vectors with disabled viral transcriptional elements (reviewed in ref. [90]). Improvement in the regulation of expression of transgenes and limiting interference with transcription of cellular genes after integration of the provirus has also resulted.
3. The gene encoding the lentiviral reverse transcriptase may be conveniently mutated to render the vectors integration defective [91]. This feature is useful to improve safety by preventing any possible insertional mutagenesis.

4. The favored integration of lentiviruses into actively transcribed host sequences, when compared with the preferred insertion of gammaretroviral proviruses at transcriptional regulatory elements, is less oncogenic.

4.4.1 Essential Biology of Lentiviruses and Retroviruses

Each retrovirion contains two identical copies of the RNA genome, which have sense polarity. Viral enzymes and accessory proteins are encapsidated together with the genomic RNA. An envelope membrane, which is derived from the cell of a previously infected cell, surrounds the capsid. Viral proteins are embedded in the envelope, and these give the virions their particular tropism. A well-documented example is the binding of the gp120 envelope protein of HIV-1 with the CD4 protein on T helper cells. Interaction of the viral envelope proteins with cognate receptors on cells' surfaces triggers membrane fusion and internalization of the viral core particle. After uncoating, the viral RNA is reverse transcribed to form DNA, which is then transported into the nucleus of the host cell. The ends of the viral DNA comprise two long terminal repeats (LTRs) which include repeat (R), unique 5′ (U5), and unique 3′ (U3) sequences, which are duplicated during the process of reverse transcription (Figure 4.7). The PIC, comprising viral and cellular proteins, is responsible for transporting viral RNA to the nucleus. Properties of the PIC are important for determining whether the vectors are capable of transducing replicating or nondividing cells. In the case of lentiviruses, the PIC traverses nuclear pores without the need for breakdown of the nuclear membrane. However, the PICs of gammaretroviruses require that the cells go through mitosis, with nuclear membrane breakdown, before integration of the viral DNA occurs. This important distinguishing feature enables lentiviral vectors to transduce nondividing cells, whereas gammaretroviral derivatives are dependent on cell division for transduction to take place.

After crossing the nuclear membrane, or after mitosis in the case of gammaretroviruses, the DNA within the PIC is integrated into the host genome to form the stable provirus. Integrase and reverse transcriptase functions are performed by the protein encoded by the *pol* gene. The stably integrated provirus is flanked by the LTRs. The viral genes and *cis* elements, such as the ψ packaging element, are located between these elements. The provirus serves as the template for transcription of viral genes (Figure 4.7). These include the structural, enzymatic, regulatory, and accessory proteins of the lentiviruses (see detailed discussion in Chapter 9). Gammaretroviruses have a simpler genome structure and lack the regulatory and accessory proteins that are characteristic of lentiviruses. All *retroviridae* contain *cap*, *env*, and *pol* sequences, which encode the capsid protein, envelope protein, integrase, and reverse transcriptase respectively. The *pol* gene encodes both integrase and reverse transcriptase enzymatic functions.

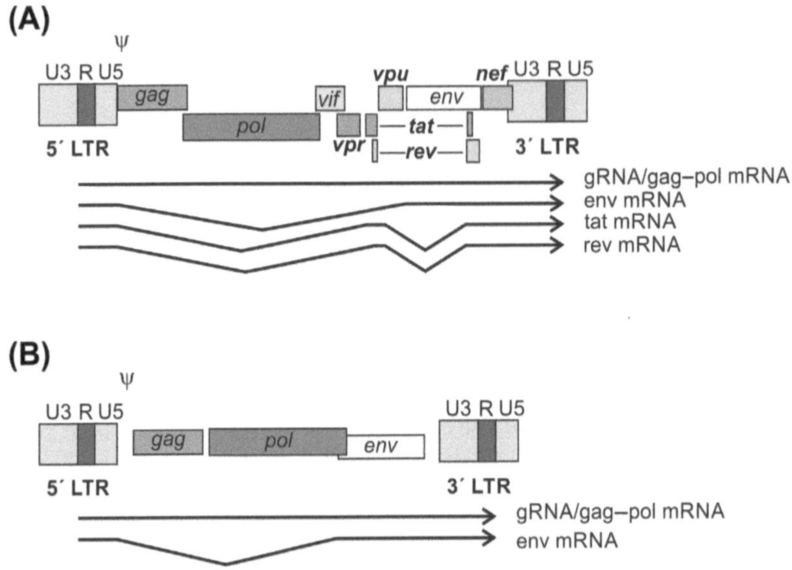

FIGURE 4.7 Arrangement of gammaretroviral and lentiviral proviruses.
(A) Organization of HIV-1 provirus indicating ORFs together with the 5′ and 3′ LTRs. Unspliced and some spliced transcripts are shown below the genome together with the proteins that they encode. (B) Organization of gammaretroviral provirus indicating ORFs together with the 5′ and 3′ LTRs. As in panel A, unspliced and spliced transcripts are shown below the genome.

LTRs regulate transcription of the proviral sequences. Promoter elements are only found within the U3 sequence and are responsible for initiating transcription of the viral RNA from the 5′ LTR. The accessory Tat protein augments the rate of transcription of viral genes by performing a facilitatory role in concert with host cellular sequences (reviewed in ref. [92]). The U5 sequences within the 3′ LTR serves as a transcription termination signal. Interestingly, and to enable transcription of viral genes, the U5 sequence within the 5′ LTR does not cause premature transcription termination. This appears to be as a result of the influence of additional sequences, located within the U3 region and not present in the 5′ end of the viral transcripts, acting in concert with the U5 element to stop transcription [93].

There are several transcription factor binding sites within the U3 sequence (reviewed in ref. [94]). Those that are responsible for interaction with the nuclear factor-κB family of transcription factors are among the most important, and they activate viral transcription when T cell activation occurs in HIV-1-infected cells [95]. Early during the transcription of lentiviral genes, short transcripts are predominantly synthesized [96], and a small amount of full-length transcripts is generated. These full-length transcripts are responsible for the production of the Tat regulatory protein, which then increases transcription of full-length viral transcripts (Chapter 9). The full-length lentiviral transcripts contain many splice donor and splice acceptor sites, and the total number of resultant mRNA variants is 30 for HIV-1 [97]. The transcripts may be mono- or polycistronic; thus, they are capable of encoding single or multiple proteins [94]. ORFs that are located toward the 5′ end of the mRNA are preferentially expressed. The sequence context of the initiation codon also influences the efficiency of the translation of particular proteins. Naturally, intron-containing cellular mRNAs are not efficiently exported from the nucleus, which is important to prevent aberrant cytoplasmic translation of intronic sequences. The regulatory Rev protein of lentiviruses is responsible for facilitating the export of incompletely spliced viral transcripts by a mechanism that enables translation of these transcripts. Because most transgenes that are included within lentiviral vectors are derived from unspliced transcripts, adaptation of these viruses for use as gene therapy vectors requires that the function of the Rev protein be retained.

In addition to functioning as a template for translation, the complete unspliced transcript constitutes the viral genome. Incorporation of this RNA into the viral capsids is dependent on the ψ sequence. The ψ element is removed from spliced transcripts, which prevents packaging of these incomplete viral sequences (Figure 4.7). The essential role of the packaging signal has been exploited for the design of retroviral and lentiviral vectors. By coupling a transgene-containing transcript with the ψ element, and removing this sequence from transcripts encoding *trans* viral elements required to constitute the virion, intended transgene-containing transcripts may be efficiently packaged.

4.4.2 Propagation of Recombinant Retro- and Lentiviruses

Vectors derived from the gammaretroviruses such as the MLV were studied extensively during the late 1980s to mid-1990s. This class of viral vector was the earliest to be developed for experimental and therapeutic use. The simple arrangement of the gammaretroviral genome may readily be manipulated using generic approaches to generate the recombinant viral vectors (Figure 4.8). Viral protein-coding sequences are substituted with a transgene in a plasmid containing flanking LTRs and ψ. After transfection of packaging cells, which express *gag*, *pol*, and *env* sequences, retroviral vectors are transiently formed. Subsequent infection of another packaging cell line with these recombinant viruses enables stable generation of the viral vectors. The procedure typically entails initial transfection of packaging cells that produce murine ecotropic vectors and then infection of murine cells that produce amphotropic recombinant retroviruses. The resulting amphotropic vectors have a broad tissue and host range.

Although understanding the principles of generating recombinant gammaretroviruses provided valuable insights into vectorology, it became apparent that this class of vectors has significant limitations for gene therapy. Recombinant

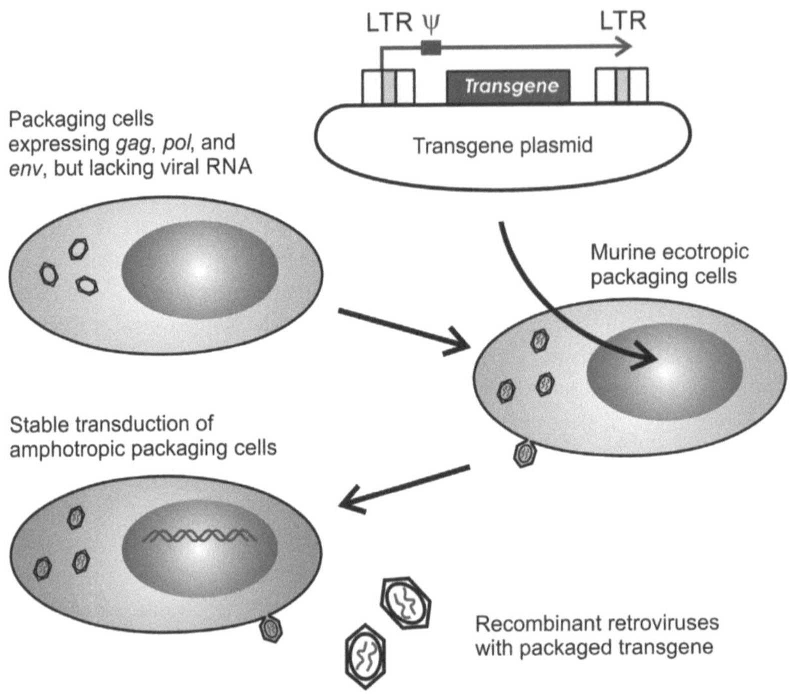

FIGURE 4.8 Scheme used for the preparation of recombinant gammaretroviral vectors.
A plasmid containing a transgene expression cassette, flanked by LTRs and containing a packaging signal (ψ), is used to transfect packaging cells. The packaging cells stably produce Gag, Pol, and Env proteins to generate recombinant viruses that are ecotropic for murine cells. These recombinant viruses are then used to transduce another packaging cell line, which is amphotropic, to result in the propagation of vectors with a broad host cell range.

HIV-1-derived lentiviruses have superior qualities and have now displaced engineered gammaretroviruses as the preferred vectors for stably transducing target cells. Again, propagation of replication-defective lentiviruses involves provision of structural and enzymatic factors in trans, whereas packaging and other *cis* elements are colocalized with the transgene (Figure 4.9) [92,99–109]. Standard procedures for generating lentiviruses involve transfection of HEK293 cells with four plasmids:

1. The transgene plasmid, which generates the vector genome. The plasmid contains essential *cis* elements such as LTRs, the central polypurine tract, the TAT-activating region, and ψ. The woodchuck hepatitis virus regulatory element may be included to enhance transgene expression. To minimize risk of recombination and transcriptional activation of cellular genes by integrated proviral LTRs, the 5′ LTR may be substituted with a different transcription control element such as the cytomegalovirus (CMV) immediate early promoter enhancer

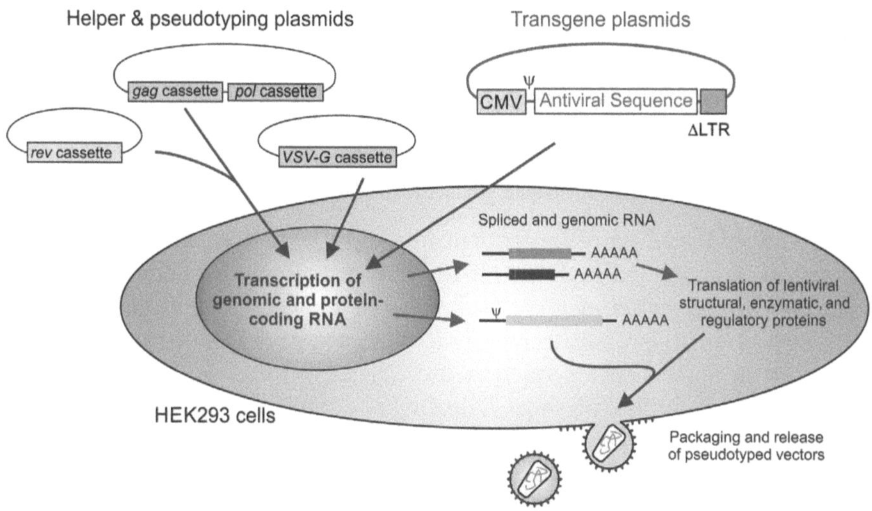

FIGURE 4.9 Outline of the scheme used for the preparation of recombinant lentiviral vectors. Typically, HEK293 cells are co-transfected with plasmids containing the transgene cassette together with plasmids encoding pseudotyping (VSV-G), viral structural, regulatory (Rev), and enzymatic components. The transgene cassette, encoding an antiviral sequence, contains the packaging signal (ψ), other *cis* elements, and the therapeutic sequence positioned downstream of a promoter. To improve safety and minimize risk of recombination, the transcriptional control elements contained within the LTRs are usually replaced with a promoter such as the CMV transcriptional control element. The 3′ LTR, containing the transcription termination signal, is mutated to render the sequence self-inactivating (SIN) when copied to the 5′ end of the provirus. The recombinant pregenomic RNA is selectively packaged and secreted from transfected cells as pseudotyped viral particles.

sequence. Further safety improvement may be achieved by deleting the enhancer and TATA elements within the 3′ LTR. Because the 3′ LTR is copied to the 5′ LTR during reverse transcription of lentiviral RNA, both LTRs will be disabled after integration of the provirus. Disabling the vectors' transcription controlling elements also has the advantage of minimizing interference with expression of transgene cassettes. Vectors with this structure are termed self-inactivating (SIN). Attenuation of transcription termination may occur in SIN vectors to cause increased read through into cellular sequences [98,99]. Consequently, disruption of neighboring genes may still occur with SIN vectors [100].

2. A helper plasmid that contains *gag–pol* sequences, which encode the capsid components, reverse transcriptase, and integrase functions. Nonintegrating lentiviral vectors are propagated using mutant *pol* sequences that are incapable of provirus insertion into the host cellular genome [91]. Viral DNA formed after reverse transcription by the mutant *pol* exists as a stable circular episome.

3. A Rev-encoding plasmid, which is essential to facilitate nuclear export of unspliced lentiviral RNA.
4. Lentiviral vectors are typically pseudotyped. Plasmids expressing the vesicular stomatitis virus G protein (VSV-G) are most commonly used for this purpose. Lentiviral vectors with the VSV-G protein embedded into the viral envelope have broad and efficient tropism.

The traditional method of incorporating a therapeutic transgene into lentiviral vectors has entailed insertion of the transgene-encoding sequence downstream of the promoter element at the 5′ end. However, other methods have also been used to propagate lentiviral vectors expressing therapeutic RNAi activators. Because 3′ LTR sequences are copied to the 5′ end during reverse transcription, insertion of short hairpin RNA (shRNA) expression cassettes within these elements results in their duplication in each LTR of the provirus, which potentially augments silencing efficacy [101]. Efficient production of vectors expressing polycistronic microRNA cassettes targeting HIV has also been reported [102,103]. This interesting observation indicates that lentivirus production may be achieved despite processing of miR intermediates within the vector pregenome.

Although stable integration of lentiviral proviral sequences is important for achieving sustained expression of transgenes, mutation at sites of insertion at cellular sequences has always been a concern. Sophisticated new technology has been developed to map the sites of vector integration (reviewed in refs [90,104]). Ligation-mediated [105] and linear amplification-mediated [106] PCR in conjunction with deep sequencing have been particularly useful. Lentiviral vectors preferentially integrate at actively transcribed regions, but they do so in a sequence-independent manner [104,105,107–109]. Integration occurs across the span of transcribed sequences, but elements that also have regulatory function are generally avoided. Retroviral vectors derived from MLV have a different integration pattern. Their preference is for integration at transcription regulatory elements, and in particular within the vicinity of transcription start sites [107]. There is evidence that gammaretroviral vectors are capable of discriminating between different types of transcriptional regulatory sequences and favor elements that are responsible for growth and differentiation of cells, rather than those found in housekeeping genes [107]. This interesting property may improve the fitness of gammaretroviruses, but it increases the risks for transformation of cells that are transduced with the vector derivatives. Malignancies caused by retroviruses have been observed in patients after administration of recombinant gammaretroviruses [110]. Insertional activation of *LMO-2*, a cellular oncogene, by MLV-derived vectors caused leukemia in some of the patients receiving gene therapy for X-linked severe combined immunodeficiency [111,112]. Although complications arising from insertional mutagenesis caused by lentiviral vectors are rarer, malignancies have been observed in mice after administration of the recombinant vectors [113,114]. Other methods of attenuating the effects of retroviral vectors on transcription have involved use of nonintegrating vectors [91] and, perhaps controversially, insulators that isolate the provirus from the neighboring cellular sequences [90,115,116]. Nonintegrating vectors are generated from sequences that have mutations in the integrase enzyme and form circular episomal DNA structures in transduced cells. Although transgene expression from these vectors is not as lasting as that of the integrating variants, they may have utility in the treatment of infectious diseases in which durable expression of the therapeutic sequences may not be essential. Duplicated LoxP sequences at either end of the integrants may be used for Cre-mediated excision of the proviral sequences and limit effects of insertional mutagenesis [117]. The approach has been used to achieve safer conditional RNAi [118,119].

The problems that may arise from formation of replication-competent vectors are potentially significant. In addition to affecting a treated patient, such derivatives of the vectors could also cause problems to untreated individuals. Therefore, considerable effort has gone into minimizing the risk of recombination, giving rise to replication-competent lentiviral vectors [90]. Splitting the genome to ensure that no or very little homology occurs between the vector genome containing the transgene and the sequences required for the formation of the intact virions has been useful. Pseudotyping of lentiviral vectors with envelope proteins, such as those of the vesicular stomatitis virus, further diminishes the risks of recombination and formation of replication-competent viruses.

4.5 USE OF VIRAL VECTORS FOR DEVELOPING TREATMENT VIRAL INFECTIONS

Utility of recombinant viral vectors has been demonstrated in several applications of gene therapy aimed at treating viral infections. One of the main approaches to achieving therapeutic inhibition of viral replication has been to harness the RNAi pathway (Chapter 2). Because artificial RNAi activating expression cassettes are generally small, scAAVs may readily be used to deliver these engineered sequences. Typically, an RNAi expression cassette comprises a Pol II or Pol III promoter that controls transcription of downstream sequences encoding an artificial shRNA sequence or artificial primary-microRNA (pri-miRNA) together with a transcription termination sequence. Commonly used Pol III promoters are the H1 and U6 transcription regulatory elements. These are short promoters, 0.35 kb for U6, and together with RNAi-activator-encoding sequences they can comfortably be accommodated by scAAVs. A disadvantage of Pol III promoters is that they are constitutively active, and tissue-specific expression requires regulation of vector tropism.

Pol II promoters offer the advantage of better transcription control and tissue-specific transcription, which are important to limit off-target effects. The size of Pol II-pri-miRNA cassettes is larger than that of equivalent Pol III transcription templates, which in some cases may limit their use in scAAVs. Lentiviral vectors are able to accommodate larger expression cassettes, but when used for therapeutic application, there are several important considerations: (1) promoter interference may compromise production of vectors; (2) when using repetitive elements, disruptive recombination of the transgenes may occur; and (3) RNAi precursors are naturally found in the vector pregenomic RNA, and processing may diminish titers. Ads, particularly HD Ads, have the advantage of being able to deliver large transgenes, which are expressed efficiently without influence of viral transcriptional elements. However, hepatotropism after systemic administration is not suited to therapy for extrahepatic viral infections. When a treatment strategy may be performed ex vivo, this is not problematic and the vectors have a broad host range under these conditions.

The variability in the sizes of gene editing cassettes has a bearing on the selection of viral vectors for therapeutic use. The large CRISPR-Cas and TALEN encoding sequences preclude use of scAAVs. Minimizing cassettes to incorporate them into ssAAVs is being investigated, but successful application is yet to be reported. The smaller size of zinc finger protein (ZFP) and homing endonuclease (HE) derivatives mean that choice of viral vector to deliver these gene editing sequences is more flexible. Although sequences encoding ZFP, HE, and CRISPR-Cas derivatives are small enough to be accommodated within lentiviral vectors, these vectors may not be ideally suited to a gene editing antiviral approach. Because vector-derived proviral sequences are stably integrated into target tissue, the risk of unintended off-target effects on cellular genes is markedly increased (Chapter 3). Ads are useful for delivery of all types of gene editing expression cassettes. However, immunostimulation and hepatotropism may be problematic after their systemic administration. Ad-mediated ex vivo delivery of ZFNs targeting the CCR5 gene has been successfully used to confer resistance of $CD4^+$ T cells to HIV-1 infection [4].

Application of lentiviral vectors to antiviral therapy is not widespread, but it is potentially useful for conferring durable resistance of cells to infection to chronic viral infections. Stable expression of antiviral sequences, such as artificial RNAi activators targeting HIV-1, may be useful to achieve this [120]. However, the derivation of lentiviral vectors from HIV-1 may complicate production of vectors that silence genes of HIV-1. Selecting targets that are absent from sequences required for lentiviral propagation or choosing target sequences such as *rev*, which are not required in abundance for vector propagation, may improve vector titers. With the recent impressive advances in the use of pluripotent stem cells, a potentially useful application of retro- and lentiviral vectors for treatment of viral infections is to transduce these cells before differentiation and reinfusion into patients. Such an approach would enable ex vivo propagation of cells that are resistant to viral replication and subsequent reinfusion of autologous differentiated cells into patients. Although the approach may be feasible in principle, the concerns relating to insertional mutagenesis and other potentially harmful off-target effects of sustained expression of antiviral sequences need to be established.

4.6 CONCLUSIONS

Recombinant viral vectors are highly efficient carriers of sequences encoding virus-disabling sequences. The features of the available viral vectors are varied. Appropriate viral vectors usually need to be selected and adapted for application to treatment of specific viral infections. However, their use is not without disadvantages and particularly important are the following:

1. With available technologies, the high cost of propagating recombinant viral vectors in sufficient quantities for treatment of commonly occurring viral infections is prohibitive.
2. The immunostimulatory effects of viral vectors, particularly Ads, may be toxic.
3. Presence within cells of DNA derived from viral vectors poses some risk for mutagenic events that may occur as a result of illegitimate recombination, and this potential problem is not limited to use of retro- and lentiviral vectors.
4. Immunity to a vector, which may be community acquired or follow prior administration of the recombinant virus, limits the vectors' efficacy.
5. Tropism of vectors after their systemic administration may not be suited to delivering antiviral sequences to intended target tissues.

Significant advances have been made in overcoming these challenges, which make their clinical use seem more feasible. However, despite the progress, it is unlikely that the available range of viral vectors will be applicable to treatment of all types of infection. For example, treatment of the world's HBV- or HCV-infected individuals would require enormous amounts of vector that would be impractical to propagate. Although recombinant viral vectors may be useful for achieving preclinical proof of concept, their ultimate clinical indications for delivery of therapeutic sequences are likely to be restricted. Future use of viral vectors may be particularly important for delivery of immunostimulatory or antibody-encoding sequences (Chapter 11).

REFERENCES

[1] Hollon T. Researchers and regulators reflect on first gene therapy death. Nat Med 2000;6(1):6.

[2] Raper SE, Chirmule N, Lee FS, Wivel NA, Bagg A, Gao GP, et al. Fatal systemic inflammatory response syndrome in a ornithine transcarbamylase deficient patient following adenoviral gene transfer. Mol Genet Metab 2003;80(1):148–58.

[3] Waddington SN, McVey JH, Bhella D, Parker AL, Barker K, Atoda H, et al. Adenovirus serotype 5 hexon mediates liver gene transfer. Cell 2008;132(3):397–409.

[4] Li L, Krymskaya L, Wang J, Henley J, Rao A, Cao LF, et al. Genomic editing of the HIV-1 coreceptor CCR5 in adult hematopoietic stem and progenitor cells using zinc finger nucleases. Mol Ther J Am Soc Gene Ther 2013;21(6):1259–69.

[5] Kotterman MA, Schaffer DV. Engineering adeno-associated viruses for clinical gene therapy. Nat Rev Genet 2014;15(7):445–51.

[6] Buning H, Perabo L, Coutelle O, Quadt-Humme S, Hallek M. Recent developments in adeno-associated virus vector technology. J Gene Med 2008;10(7):717–33.

[7] Daya S, Berns KI. Gene therapy using adeno-associated virus vectors. Clin Microbiol Rev 2008;21(4):583–93.

[8] Schultz BR, Chamberlain JS. Recombinant adeno-associated virus transduction and integration. Mol Ther J Am Soc Gene Ther 2008;16(7):1189–99.

[9] Muralidhar S, Becerra SP, Rose JA. Site-directed mutagenesis of adeno-associated virus type 2 structural protein initiation codons: effects on regulation of synthesis and biological activity. J Virol 1994;68(1):170–6.

[10] Cassinotti P, Weitzand M, Tratschin JD. Organization of the adeno-associated virus (AAV) capsid gene: mapping of a minor spliced mRNA coding for virus capsid protein. Virology 1988;167(1):176–84.

[11] Sonntag F, Kother K, Schmidt K, Weghofer M, Raupp C, Nieto K, et al. The assembly-activating protein promotes capsid assembly of different adeno-associated virus serotypes. J Virol 2011;85(23):12686–97.

[12] Sonntag F, Schmidt K, Kleinschmidt JA. A viral assembly factor promotes AAV2 capsid formation in the nucleolus. Proc Natl Acad Sci USA 2010;107(22):10220–5.

[13] Bohenzky RA, LeFebvre RB, Berns KI. Sequence and symmetry requirements within the internal palindromic sequences of the adeno-associated virus terminal repeat. Virology 1988;166(2):316–27.

[14] Im DS, Muzyczka N. The AAV origin binding protein Rep68 is an ATP-dependent site-specific endonuclease with DNA helicase activity. Cell 1990;61(3):447–57.

[15] Im DS, Muzyczka N. Partial purification of adeno-associated virus Rep78, Rep52, and Rep40 and their biochemical characterization. J Virol 1992;66(2):1119–28.

[16] Blouin V, Brument N, Toublanc E, Raimbaud I, Moullier P, Salvetti A. Improving rAAV production and purification: towards the definition of a scaleable process. J Gene Med 2004;6(Suppl. 1):S223–8.

[17] Urabe M, Ding C, Kotin RM. Insect cells as a factory to produce adeno-associated virus type 2 vectors. Hum Gene Ther 2002;13(16):1935–43.

[18] McCarty DM, Young Jr SM, Samulski RJ. Integration of adeno-associated virus (AAV) and recombinant AAV vectors. Annu Rev Genet 2004;38:819–45.

[19] Thomas CE, Storm TA, Huang Z, Kay MA. Rapid uncoating of vector genomes is the key to efficient liver transduction with pseudotyped adeno-associated virus vectors. J Virol 2004;78(6):3110–22.

[20] McCarty DM. Self-complementary AAV vectors; advances and applications. Mol Ther J Am Soc Gene Ther 2008;16(10):1648–56.

[21] Nakai H, Storm TA, Kay MA. Recruitment of single-stranded recombinant adeno-associated virus vector genomes and intermolecular recombination are responsible for stable transduction of liver in vivo. J Virol 2000;74(20):9451–63.

[22] McCarty DM, Monahan PE, Samulski RJ. Self-complementary recombinant adeno-associated virus (scAAV) vectors promote efficient transduction independently of DNA synthesis. Gene Ther 2001;8(16):1248–54.

[23] Grieger JC, Samulski RJ. Packaging capacity of adeno-associated virus serotypes: impact of larger genomes on infectivity and postentry steps. J Virol 2005;79(15):9933–44.

[24] Allocca M, Doria M, Petrillo M, Colella P, Garcia-Hoyos M, Gibbs D, et al. Serotype-dependent packaging of large genes in adeno-associated viral vectors results in effective gene delivery in mice. J Clin Invest 2008;118(5):1955–64.

[25] Boutin S, Monteilhet V, Veron P, Leborgne C, Benveniste O, Montus MF, et al. Prevalence of serum IgG and neutralizing factors against adeno-associated virus (AAV) types 1, 2, 5, 6, 8, and 9 in the healthy population: implications for gene therapy using AAV vectors. Hum Gene Ther 2010;21(6):704–12.

[26] Mingozzi F, High KA. Immune responses to AAV vectors: overcoming barriers to successful gene therapy. Blood 2013;122(1):23–36.

[27] Chen J, Wu Q, Yang P, Hsu HC, Mountz JD. Determination of specific CD4 and CD8 T cell epitopes after AAV2- and AAV8-hF.IX gene therapy. Mol Ther J Am Soc Gene Ther 2006;13(2):260–9.

[28] Madsen D, Cantwell ER, O'Brien T, Johnson PA, Mahon BP. Adeno-associated virus serotype 2 induces cell-mediated immune responses directed against multiple epitopes of the capsid protein VP1. J Gen Virol 2009;90(Pt 11):2622–33.

[29] Mingozzi F, Maus MV, Hui DJ, Sabatino DE, Murphy SL, Rasko JE, et al. CD8(+) T-cell responses to adeno-associated virus capsid in humans. Nat Med 2007;13(4):419–22.

[30] Wu TL, Li H, Faust SM, Chi E, Zhou S, Wright F, et al. $CD8^+$ T cell recognition of epitopes within the capsid of adeno-associated virus 8-based gene transfer vectors depends on vectors' genome. Mol Ther J Am Soc Gene Ther 2014;22(1):42–51.

[31] Manno CS, Pierce GF, Arruda VR, Glader B, Ragni M, Rasko JJ, et al. Successful transduction of liver in hemophilia by AAV-Factor IX and limitations imposed by the host immune response. Nat Med 2006;12(3):342–7.

[32] Zhong L, Li B, Jayandharan G, Mah CS, Govindasamy L, Agbandje-McKenna M, et al. Tyrosine-phosphorylation of AAV2 vectors and its consequences on viral intracellular trafficking and transgene expression. Virology 2008;381(2):194–202.

[33] Zhong L, Li B, Mah CS, Govindasamy L, Agbandje-McKenna M, Cooper M, et al. Next generation of adeno-associated virus 2 vectors: point mutations in tyrosines lead to high-efficiency transduction at lower doses. Proc Natl Acad Sci USA 2008;105(22):7827–32.

[34] Lochrie MA, Tatsuno GP, Christie B, McDonnell JW, Zhou S, Surosky R, et al. Mutations on the external surfaces of adeno-associated virus type 2 capsids that affect transduction and neutralization. J Virol 2006;80(2):821–34.

[35] Mingozzi F, Anguela XM, Pavani G, Chen Y, Davidson RJ, Hui DJ, et al. Overcoming preexisting humoral immunity to AAV using capsid decoys. Sci Transl Med 2013;5(194):194ra92.

[36] Lisowski L, Dane AP, Chu K, Zhang Y, Cunningham SC, Wilson EM, et al. Selection and evaluation of clinically relevant AAV variants in a xenograft liver model. Nature 2014;506(7488):382–6.

[37] Davidoff AM, Gray JT, Ng CY, Zhang Y, Zhou J, Spence Y, et al. Comparison of the ability of adeno-associated viral vectors pseudotyped with serotype 2, 5, and 8 capsid proteins to mediate efficient transduction of the liver in murine and nonhuman primate models. Mol Ther J Am Soc Gene Ther 2005;11(6):875–88.

[38] Alemany R, Suzuki K, Curiel DT. Blood clearance rates of adenovirus type 5 in mice. J Gen Virol 2000;81(Pt 11):2605–9.

[39] Liu Q, Muruve DA. Molecular basis of the inflammatory response to adenovirus vectors. Gene Ther 2003;10(11):935–40.

[40] Wold W, Horwitz M. Adenoviruses. Fields Virol 2007;2:2395–436.

[41] Liu H, Jin L, Koh SB, Atanasov I, Schein S, Wu L, et al. Atomic structure of human adenovirus by cryo-EM reveals interactions among protein networks. Science 2010;329(5995):1038–43.

[42] Reddy VS, Natchiar SK, Stewart PL, Nemerow GR. Crystal structure of human adenovirus at 3.5 A resolution. Science 2010;329(5995):1071–5.

[43] Mizuta K, Matsuzaki Y, Hongo S, Ohmi A, Okamoto M, Nishimura H, et al. Stability of the seven hexon hypervariable region sequences of adenovirus types 1-6 isolated in Yamagata, Japan between 1988 and 2007. Virus Res 2009;140(1–2):32–9.

[44] Rux JJ, Burnett RM. Type-specific epitope locations revealed by X-ray crystallographic study of adenovirus type 5 hexon. Mol Ther J Am Soc Gene Ther 2000;1(1):18–30.

[45] Xu Z, Tian J, Smith JS, Byrnes AP. Clearance of adenovirus by Kupffer cells is mediated by scavenger receptors, natural antibodies, and complement. J Virol 2008;82(23):11705–13.

[46] Khare R, May SM, Vetrini F, Weaver EA, Palmer D, Rosewell A, et al. Generation of a Kupffer cell-evading adenovirus for systemic and liver-directed gene transfer. Mol Ther J Am Soc Gene Ther 2011;19(7):1254–62.

[47] Khare R, Reddy VS, Nemerow GR, Barry MA. Identification of adenovirus serotype 5 hexon regions that interact with scavenger receptors. J Virol 2012;86(4):2293–301.

[48] Sharma A, Li X, Bangari DS, Mittal SK. Adenovirus receptors and their implications in gene delivery. Virus Res 2009;143(2):184–94.

[49] Carlisle RC, Di Y, Cerny AM, Sonnen AF, Sim RB, Green NK, et al. Human erythrocytes bind and inactivate type 5 adenovirus by presenting Coxsackie virus-adenovirus receptor and complement receptor 1. Blood 2009;113(9):1909–18.

[50] Wolff G, Worgall S, van Rooijen N, Song WR, Harvey BG, Crystal RG. Enhancement of in vivo adenovirus-mediated gene transfer and expression by prior depletion of tissue macrophages in the target organ. J Virol 1997;71(1):624–9.

[51] Cotter MJ, Zaiss AK, Muruve DA. Neutrophils interact with adenovirus vectors via Fc receptors and complement receptor 1. J Virol 2005;79(23):14622–31.

[52] Othman M, Labelle A, Mazzetti I, Elbatarny HS, Lillicrap D. Adenovirus-induced thrombocytopenia: the role of von Willebrand factor and P-selectin in mediating accelerated platelet clearance. Blood 2007;109(7):2832–9.

[53] Wickham TJ, Mathias P, Cheresh DA, Nemerow GR. Integrins alpha v beta 3 and alpha v beta 5 promote adenovirus internalization but not virus attachment. Cell 1993;73(2):309–19.

[54] Wickham TJ, Filardo EJ, Cheresh DA, Nemerow GR. Integrin alpha v beta 5 selectively promotes adenovirus mediated cell membrane permeabilization. J Cell Biol 1994;127(1):257–64.

[55] Nicklin SA, Wu E, Nemerow GR, Baker AH. The influence of adenovirus fiber structure and function on vector development for gene therapy. Mol Ther J Am Soc Gene Ther 2005;12(3):384–93.

[56] Duffy MR, Bradshaw AC, Parker AL, McVey JH, Baker AH. A cluster of basic amino acids in the factor X serine protease mediates surface attachment of adenovirus/FX complexes. J Virol 2011;85(20):10914–9.

[57] Xu Z, Qiu Q, Tian J, Smith JS, Conenello GM, Morita T, et al. Coagulation factor X shields adenovirus type 5 from attack by natural antibodies and complement. Nat Med 2013;19(4):452–7.

[58] Russell WC. Adenoviruses: update on structure and function. J Gen Virol 2009;90(Pt 1):1–20.

[59] Lichtenstein DL, Toth K, Doronin K, Tollefson AE, Wold WS. Functions and mechanisms of action of the adenovirus E3 proteins. Int Rev Immunol 2004;23(1–2):75–111.

[60] Punga T, Kamel W, Akusjärvi G. Old and new functions for the adenovirus virus-associated RNAs. Future Virol 2013;8(4):343–56.

[61] Andersson MG, Haasnoot PC, Xu N, Berenjian S, Berkhout B, Akusjarvi G. Suppression of RNA interference by adenovirus virus-associated RNA. J Virol 2005;79(15):9556–65.

[62] Lu S, Cullen BR. Adenovirus VA1 noncoding RNA can inhibit small interfering RNA and microRNA biogenesis. J Virol 2004;78(23):12868–76.

[63] Kamel W, Segerman B, Oberg D, Punga T, Akusjarvi G. The adenovirus VA RNA-derived miRNAs are not essential for lytic virus growth in tissue culture cells. Nucleic Acids Res 2013;41(9):4802–12.

[64] Aparicio O, Carnero E, Abad X, Razquin N, Guruceaga E, Segura V, et al. Adenovirus VA RNA-derived miRNAs target cellular genes involved in cell growth, gene expression and DNA repair. Nucleic Acids Res 2010;38(3):750–63.

[65] Narvaiza I, Aparicio O, Vera M, Razquin N, Bortolanza S, Prieto J, et al. Effect of adenovirus-mediated RNA interference on endogenous microRNAs in a mouse model of multidrug resistance protein 2 gene silencing. J Virol 2006;80(24):12236–47.

[66] Crowther C, Mowa MB, Ely A, Arbuthnot P. Inhibition of hepatitis B virus replication in vivo using helper-dependent adenovirus vectors to deliver antiviral RNAi expression cassettes. Antivir Ther 2013;1(1):1.

[67] Mowa MB, Crowther C, Ely A, Arbuthnot P. Efficient silencing of hepatitis B virus by helper-dependent adenovirus vector-mediated delivery of artificial antiviral primary micro RNAs. Micro RNA 2012;1:19–25.

[68] Mowa MB, Crowther C, Ely A, Arbuthnot P. Inhibition of hepatitis B virus replication by helper dependent adenoviral vectors expressing artificial anti-HBV pri-miRs from a liver-specific promoter. BioMed Res Int 2014;718743, http://dx.doi.org/10.1155/2014/718743.

[69] Crowther C, Ely A, Hornby J, Mufamadi MS, Salazar F, Marion P, et al. Efficient inhibition of hepatitis B virus replication in vivo using Peg-Modified adenovirus vectors. Hum Gene Ther 2008;19(11):1325–31.

[70] Uprichard SL, Boyd B, Althage A, Chisari FV. Clearance of hepatitis B virus from the liver of transgenic mice by short hairpin RNAs. Proc Natl Acad Sci USA 2005;102(3):773–8.

[71] Mowa MB, Crowther C, Arbuthnot P. Therapeutic potential of adenoviral vectors for delivery of expressed RNAi activators. Expert Opin Drug Deliv 2010;7(12):1373–85.

[72] He TC, Zhou S, da Costa LT, Yu J, Kinzler KW, Vogelstein B. A simplified system for generating recombinant adenoviruses. Proc Natl Acad Sci USA 1998;95(5):2509–14.

[73] Parks RJ, Chen L, Anton M, Sankar U, Rudnicki MA, Graham FL. A helper-dependent adenovirus vector system: removal of helper virus by Cre-mediated excision of the viral packaging signal. Proc Natl Acad Sci USA 1996;93(24):13565–70.

[74] Umana P, Gerdes CA, Stone D, Davis JR, Ward D, Castro MG, et al. Efficient FLPe recombinase enables scalable production of helper-dependent adenoviral vectors with negligible helper-virus contamination. Nat Biotechnol 2001;19(6):582–5.

[75] Palmer D, Ng P. Improved system for helper-dependent adenoviral vector production. Mol Ther J Am Soc Gene Ther 2003;8(5):846–52.

[76] St George JA. Gene therapy progress and prospects: adenoviral vectors. Gene Ther 2003; 10(14):1135–41.

[77] Kreppel F, Kochanek S. Modification of adenovirus gene transfer vectors with synthetic polymers: a scientific review and technical guide. Mol Ther J Am Soc Gene Ther 2008; 16(1):16–29.

[78] Prill JM, Espenlaub S, Samen U, Engler T, Schmidt E, Vetrini F, et al. Modifications of adenovirus hexon allow for either hepatocyte detargeting or targeting with potential evasion from Kupffer cells. Mol Ther J Am Soc Gene Ther 2011;19(1):83–92.

[79] Hofherr SE, Shashkova EV, Weaver EA, Khare R, Barry MA. Modification of adenoviral vectors with polyethylene glycol modulates in vivo tissue tropism and gene expression. Mol Ther J Am Soc Gene Ther 2008;16(7):1276–82.

[80] Eto Y, Gao JQ, Sekiguchi F, Kurachi S, Katayama K, Maeda M, et al. PEGylated adenovirus vectors containing RGD peptides on the tip of PEG show high transduction efficiency and antibody evasion ability. J Gene Med 2005;7(5):604–12.

[81] Guidotti LG, Chisari FV. Noncytolytic control of viral infections by the innate and adaptive immune response. Annu Rev Immunol 2001;19:65–91.

[82] Schnell MA, Zhang Y, Tazelaar J, Gao GP, Yu QC, Qian R, et al. Activation of innate immunity in nonhuman primates following intraportal administration of adenoviral vectors. Mol Ther J Am Soc Gene Ther 2001;3(5 Pt 1):708–22.

[83] Taniguchi M, Seino K, Nakayama T. The NKT cell system: bridging innate and acquired immunity. Nat Immunol 2003;4(12):1164–5.

[84] Kolb M, Inman M, Margetts PJ, Galt T, Gauldie J. Budesonide enhances repeated gene transfer and expression in the lung with adenoviral vectors. Am J Respir Crit care Med 2001;164(5):866–72.

[85] Chen Q, Wei H, Sun R, Zhang J, Tian Z. Therapeutic RNA silencing of Cys-X3-Cys chemokine ligand 1 gene prevents mice from adenovirus vector-induced acute liver injury. Hepatology 2008;47(2):648–58.

[86] Thrasher AJ. Progress in lentiviral vector technologies. Hum Gene Ther 2013;24(2):117–8.

[87] Naldini L, Blomer U, Gallay P, Ory D, Mulligan R, Gage FH, et al. In vivo gene delivery and stable transduction of nondividing cells by a lentiviral vector. Science 1996;272(5259): 263–7.

[88] Fassati A. HIV infection of non-dividing cells: a divisive problem. Retrovirology 2006;3:74.

[89] Fassati A. Multiple roles of the capsid protein in the early steps of HIV-1 infection. Virus Res 2012;170(1–2):15–24.

[90] Schambach A, Zychlinski D, Ehrnstroem B, Baum C. Biosafety features of lentiviral vectors. Hum Gene Ther 2013;24(2):132–42.

[91] Sarkis C, Philippe S, Mallet J, Serguera C. Non-integrating lentiviral vectors. Curr Gene Ther 2008;8(6):430–7.

[92] Romani B, Engelbrecht S, Glashoff RH. Functions of Tat: the versatile protein of human immunodeficiency virus type 1. J Gen Virol 2010;91(Pt 1):1–12.

[93] Brown PH, Tiley LS, Cullen BR. Efficient polyadenylation within the human immunodeficiency virus type 1 long terminal repeat requires flanking U3-specific sequences. J Virol 1991;65(6):3340–3.

[94] Hope T, Trono D. Structure, expression, and regulation of the HIV genome. HIV InSite Knowl Base 2000; http://hivinsite.ucsf.edu/InSite.

[95] Nabel G, Baltimore D. An inducible transcription factor activates expression of human immunodeficiency virus in T cells. Nature 1987;326(6114):711–3.

[96] Kao SY, Calman AF, Luciw PA, Peterlin BM. Anti-termination of transcription within the long terminal repeat of HIV-1 by tat gene product. Nature 1987;330(6147):489–93.

[97] Schwartz S, Felber BK, Benko DM, Fenyo EM, Pavlakis GN. Cloning and functional analysis of multiply spliced mRNA species of human immunodeficiency virus type 1. J Virol 1990;64(6):2519–29.

[98] Yang Q, Lucas A, Son S, Chang LJ. Overlapping enhancer/promoter and transcriptional termination signals in the lentiviral long terminal repeat. Retrovirology 2007;4:4.

[99] Zaiss AK, Son S, Chang LJ. RNA 3′ readthrough of oncoretrovirus and lentivirus: implications for vector safety and efficacy. J Virol 2002;76(14):7209–19.

[100] Hargrove PW, Kepes S, Hanawa H, Obenauer JC, Pei D, Cheng C, et al. Globin lentiviral vector insertions can perturb the expression of endogenous genes in beta-thalassemic hematopoietic cells. Mol Ther J Am Soc Gene Ther 2008;16(3):525–33.

[101] Tiscornia G, Singer O, Verma IM. Design and cloning of lentiviral vectors expressing small interfering RNAs. Nat Protoc 2006;1(1):234–40.

[102] Liu YP, Haasnoot J, ter Brake O, Berkhout B, Konstantinova P. Inhibition of HIV-1 by multiple siRNAs expressed from a single microRNA polycistron. Nucleic Acids Res 2008;36(9):2811–24.

[103] Ivacik D, Ely A, Ferry N, Arbuthnot P. Sustained inhibition of hepatitis B virus replication in vivo using RNAi-activating lentiviruses. Gene Ther 2014;22(2):163–71.

[104] Cavazza A, Moiani A, Mavilio F. Mechanisms of retroviral integration and mutagenesis. Hum gene Ther 2013;24(2):119–31.

[105] Schroder AR, Shinn P, Chen H, Berry C, Ecker JR, Bushman F. HIV-1 integration in the human genome favors active genes and local hotspots. Cell 2002;110(4):521–9.

[106] Schmidt M, Schwarzwaelder K, Bartholomae C, Zaoui K, Ball C, Pilz I, et al. High-resolution insertion-site analysis by linear amplification-mediated PCR (LAM-PCR). Nat Methods 2007;4(12):1051–7.

[107] Cattoglio C, Pellin D, Rizzi E, Maruggi G, Corti G, Miselli F, et al. High-definition mapping of retroviral integration sites identifies active regulatory elements in human multipotent hematopoietic progenitors. Blood 2010;116(25):5507–17.

[108] Wang GP, Ciuffi A, Leipzig J, Berry CC, Bushman FD. HIV integration site selection: analysis by massively parallel pyrosequencing reveals association with epigenetic modifications. Genome Res 2007;17(8):1186–94.

[109] Barr SD, Leipzig J, Shinn P, Ecker JR, Bushman FD. Integration targeting by avian sarcoma-leukosis virus and human immunodeficiency virus in the chicken genome. J Virol 2005;79(18):12035–44.

[110] Hacein-Bey-Abina S, Garrigue A, Wang GP, Soulier J, Lim A, Morillon E, et al. Insertional oncogenesis in 4 patients after retrovirus-mediated gene therapy of SCID-X1. J Clin Investigation 2008;118(9):3132–42.

[111] Hacein-Bey-Abina S, von Kalle C, Schmidt M, Le Deist F, Wulffraat N, McIntyre E, et al. A serious adverse event after successful gene therapy for X-linked severe combined immunodeficiency. N Engl J Med 2003;348(3):255–6.

[112] Hacein-Bey-Abina S, Von Kalle C, Schmidt M, McCormack MP, Wulffraat N, Leboulch P, et al. LMO2-associated clonal T cell proliferation in two patients after gene therapy for SCID-X1. Science 2003;302(5644):415–9.

[113] Heckl D, Schwarzer A, Haemmerle R, Steinemann D, Rudolph C, Skawran B, et al. Lentiviral vector induced insertional haploinsufficiency of Ebf1 causes murine leukemia. Mol Ther J Am Soc Gene Ther 2012;20(6):1187–95.

[114] Montini E, Cesana D, Schmidt M, Sanvito F, Bartholomae CC, Ranzani M, et al. The genotoxic potential of retroviral vectors is strongly modulated by vector design and integration site selection in a mouse model of HSC gene therapy. J Clin Invest 2009;119(4):964–75.

[115] Arumugam PI, Higashimoto T, Urbinati F, Modlich U, Nestheide S, Xia P, et al. Genotoxic potential of lineage-specific lentivirus vectors carrying the beta-globin locus control region. Mol Ther J Am Soc Gene Ther 2009;17(11):1929–37.

[116] Gaszner M, Felsenfeld G. Insulators: exploiting transcriptional and epigenetic mechanisms. Nat Rev Genet 2006;7(9):703–13.

[117] Flowers CC, Woffendin C, Petryniak J, Yang S, Nabel GJ. Inhibition of recombinant human immunodeficiency virus type 1 replication by a site-specific recombinase. J Virol 1997; 71(4):2685–92.

[118] Ventura A, Meissner A, Dillon CP, McManus M, Sharp PA, Van Parijs L, et al. Cre-lox-regulated conditional RNA interference from transgenes. Proc Natl Acad Sci USA 2004;101(28):10380–5.

[119] Tiscornia G, Tergaonkar V, Galimi F, Verma IM. CRE recombinase-inducible RNA interference mediated by lentiviral vectors. Proc Natl Acad Sci USA 2004;101(19):7347–51.

[120] Berkhout B. Toward a durable anti-HIV gene therapy based on RNA interference. Ann N.Y Acad Sci 2009;1175:3–14.

CHAPTER 5

Delivery of Antiviral Nucleic Acids with Nonviral Vectors

5.1 INTRODUCTION

Efficient delivery of negatively charged nucleic acids to target cells is crucial to the implementation of gene therapy for viral infections. In certain cases, short, chemically modified antiviral nucleic acids may reach their target sites without the need for a vector [1–4]. An impressive example of this has been the use of naked antisense RNA to inhibit the function of miR-122 for treatment of hepatitis C virus (HCV) infection (Chapter 7). The candidate drug, miravirsen, includes modifications comprising phosphorothioates and locked nucleotides that improve stability in vivo and binding to the target RNA [5]. After subcutaneous injection of the drug, it enters the circulation and is capable of reaching its site of action in hepatocytes. Inhibition of viral replication by miravirsen has been shown in HCV-infected individuals who had been enrolled for a Phase 2a clinical trial. Although similar success has also been achieved with the cholesterol-lowering modified antisense molecule, mipomersen [6], in most cases it is necessary to use a carrier to transport a therapeutic sequence to the intended site of action. Naked nucleic acids may also be delivered to cells by using electroporation and gene guns (reviewed in ref. [7]). These methods are suited to prophylaxis or treatment of viral infections that is based on immunostimulation (Chapter 11). However, the formulations are generally not appropriate for delivery of nucleic acids that directly target the viral genes. Developing clinically relevant vectors to achieve efficient delivery of antiviral nucleic acids to infected cells has been important, but it is also particularly challenging.

Viral vectors are capable of transducing various cells and achieving transient or durable expression of antiviral transgenes (Chapter 4). Using viruses as vectors has exploited a fundamental property of these infectious agents, which is that they have evolved sophisticated mechanisms to deliver RNA or DNA to cells to enable their existence as parasitic nucleic acids. Nevertheless, despite successful coupling of therapy to this natural phenomenon, there are limitations to the use of viral vectors for antiviral application. Nonviral vectors (NVVs) have features that address some of the limitations of viral vectors; therefore, they have become attractive for gene therapy. Particularly important properties are

Gene Therapy for Viral Infections. http://dx.doi.org/10.1016/B978-0-12-410518-8.00005-3

that NVVs are amenable to large-scale preparation, they are compatible with delivery of various nucleic acids, and immunogenicity is low (reviewed in ref. [7]). To enable their use in clinical settings, improving efficiency of nucleic acid delivery by NVVs has been a priority. Thus, research has been aimed at understanding the precise relationships between physicochemical properties of NVVs and the mechanisms by which the vectors transport their cargos to target cells. Use of NVVs for gene therapy is a highly active and large field of research, and it is the essentials of using NVVs for viral gene therapy that are the focus of this chapter. The reader is referred to excellent reviews for broader coverage of the topic [7,8].

5.2 CONSIDERATIONS FOR OPTIMIZING NVV-MEDIATED DELIVERY OF THERAPEUTIC NUCLEIC ACIDS TO VIRUS-INFECTED CELLS

After systemic administration, efficient transport of nucleic acids to intended target tissues by existing NVVs is often inadequate. Impediments to the delivery are varied and include physical barriers, instability of the formulations, sequestration, imprecise tissue targeting, and inefficient trafficking of the therapeutic within cells. The potential problems associated with nucleic acid delivery are conveniently understood by considering the stages of NVV-mediated delivery of nucleic acids to the intended sites of action (Figure 5.1) and factors that may influence each step:

1. Neutralization of the charge of nucleic acids by NVVs is important to facilitate transport of RNA and DNA across negatively charged cell membranes.
2. Candidate drugs that comprise nucleic acids are large, and compaction by NVVs is important to confer drug-like properties.
3. Because nucleases are present in abundance in the plasma, NVVs should provide protection against degradation before reaching the target tissue.
4. Components of the NVVs should not have direct toxic effects.
5. NVVs should be capable of traversing barriers, such as are constituted by the hepatic endothelial fenestrations, which may restrict their access to target tissues.
6. The particles should preferentially be internalized by cells of virus-infected tissues.
7. Because endocytosis is commonly the mechanism by which NVVs are taken up, uninterrupted intracellular passage from the endosome to the lysosome would result in degradation of the candidate therapeutic. Therefore, escape from endosomes before their maturation is important to improve therapeutic efficacy.

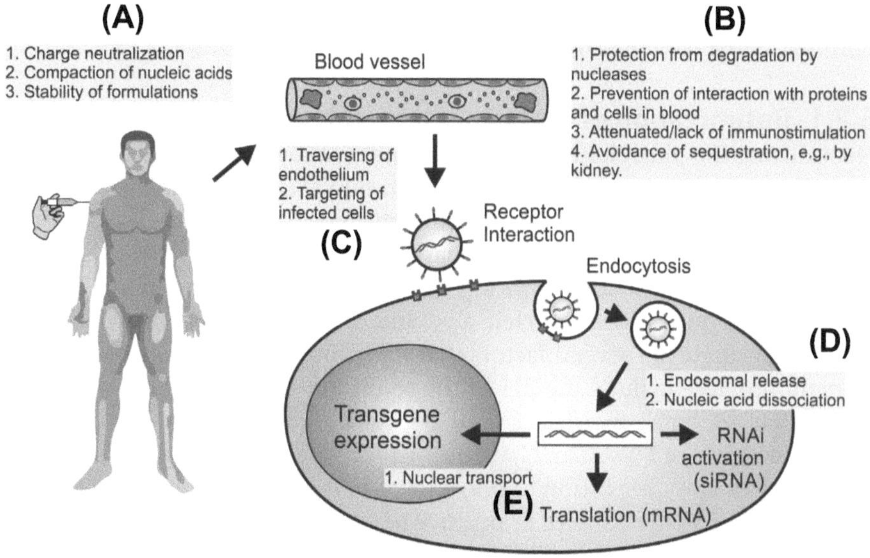

FIGURE 5.1 Delivering antiviral sequences using NVVs.
Several properties of NVVs are required for them to overcome the obstacles facing the carrying of antiviral sequences to infected cells. (A) The formulations should neutralize the negative charge and condense the therapeutic nucleic acids. To have utility in clinical settings, the NVVs should also be stable and amenable to long-term storage. (B) After entering the circulation, NVVs should provide protection against nucleases, have attenuated immunostimulatory effects, prevent interaction with blood components, and avoid sequestration. (C) Traversing endothelial barriers and interaction with specific cell types should be a feature to enable targeted therapy. (D) Escape from the maturing endosome to avoid degradation, then release of the nucleic acid load within the cytoplasm are important for efficacy. (E) In some cases, such as when expression cassettes comprising DNA are delivered, the vectors should also be capable of traversing the nuclear membrane. This is required to enable the therapeutic cargo to access the cellular gene expression machinery.

8. Within cells, the therapeutic nucleic acid should be dissociated from the carrier to allow action at the target.
9. In certain cases, such as with delivery of DNA containing a transgene or RNA molecules that are designed for transcriptional gene silencing, nuclear delivery of the cargo is required.
10. Elimination of NVVs by clearance in organs such as the lung, kidney, and liver diminishes therapeutic efficacy. Avoidance of this effect is important to ensure optimal pharmacokinetics of NVVs.
11. The vectors should not be immunogenic, which is important to limit toxicity and enable repeated administration of a therapeutic nucleic acid.
12. To allow long-term storage, vectors should be stable and resist degradation or aggregation.

Several methods have been used to overcome these potential hindrances and optimize delivery of nucleic acids using NVVs.

5.2.1 Optimization of Incorporation of Nucleic Acids into NVVs

Neutralizing the phosphate groups and condensation are important to confer drug-like properties of nucleic acids within NVVs. Delivery by the vectors to their targets is largely determined by the efficiency with which NVVs induce formation of uniformly small-sized particles that have some overall positive charge. Preparation of NVVs typically entails controlled formation of electrostatic interactions between nucleic acid and cationic vector components [8]. Predictably, there are several factors that determine the efficiency of complex formation. These include the pH of the solutions, the types of cationic molecules that are used, their sizes, and the ratio of the positive and negative charges within the NVV. The N/P ratio, which refers to the ratio of nitrogen residues of the binding cation to phosphates of the nucleic acids, is often used as an indicator of the balance of positive to negative charges. When short cationic oligomers are used in NVVs, nucleic acid molecules may be condensed individually to form the particles [9]. However, cationic polymers with multiple positive charges, which are more typically used, incorporate multiple DNA or RNA molecules into the NVVs during compaction [10]. Properties of the nucleic acids also affect formulation conditions. Short sequences, such as synthetic short interfering RNAs (siRNAs) and larger mRNAs, interact differently with cationic vector components to influence formation of the vector particles.

5.2.2 Stabilizing NVVs

NVVs that deliver nucleic acids efficiently to cells in culture often function poorly in vivo. Binding to plasma proteins; breakdown of vectors; aggregation; uptake by the reticuloendothelial system (RES); and sequestration in the liver, lungs, and kidneys may all contribute to poor efficacy of NVVs in vivo. A commonly used method of promoting vector stability and minimizing protein interaction has been through the decoration of NVVs with polyethylene glycol (PEG), an uncharged hydrophilic polymer. Coating with PEG provides a shield to the particles that serves as a steric impediment to interaction with blood components [11]. Prevention of opsonization and consumption by the RES contribute to improving stability of the particles [12]. The effect of PEG on NVVs varies and is dependent on factors such as the length of the polymer chain, its concentration within the NVV, and properties of the other components of the vector. Ideal conditions for shielding with PEG generally require empirical optimization. When using targeting ligands on NVVs, ensuring that the PEG does not impede binding of the targeting moiety to its cellular cognate is important.

5.2.3 Targeting of NVVs

Vectors with engineered tropism are useful to limit nonspecific delivery of nucleic acids [7,14,13]. To target an NVV, a ligand is typically incorporated on the vector's surface to enable specific interaction with a receptor on the targeted cell. Versatility of the approach is determined by the availability of suitable ligand–receptor pairs that may be exploited to confer tropism. To date, ligands that have been used include those for the asialoglycoprotein receptor on hepatocytes [15], the mannose receptor on macrophages [16], gp120, and CD4 on human immunodeficiency virus (HIV)-1-infected cells [17–19]. The asialoglycoprotein receptor, which is abundantly and exclusively expressed on hepatocytes, interacts specifically with carbohydrate moieties, particularly galactose and its derivatives. Thus, adding galactose moieties to the surface of NVVs may facilitate hepatotropism. Since first demonstrating this phenomenon 20 years ago [15], several different formulations using carbohydrate moieties attached to various NVVs have been used. Liver-targeting of siRNA has also been achieved with cholesterol conjugates, α-tocopherol conjugates [20], apolipoprotein A1 complexes [21], and in association with vitamin A-coupled liposomes [22]. Recently, adding *N*-acetyl galactosamine (NAG) to the surface of NVVs has been used successfully to advance siRNA-based therapy for hepatitis B virus (HBV) infection. Conjugation to NVVs of antibodies or aptamers with specific affinities to intended targets may also be used to direct the vectors to intended cell types (see Section 5.4.5).

5.2.4 Escape of Nucleic Acids into the Cytoplasm and Intracellular Trafficking

Mechanisms of cellular uptake of NVVs are influenced by the presence or absence of targeting molecules on the vectors. Without moieties that confer tropism, uptake is usually passive and results from interaction of positive charges on the NVVs with glycosaminoglycans (GAGs) such as heparin sulfate on the surface of cells [23,24]. Electrostatic interaction initiates uptake, which may occur by clathrin-dependent endocytosis or through macropinocytosis. When targeting molecules are included in an NVV, receptor-mediated endocytosis is typically responsible for the vector uptake (reviewed in ref. [13]). After internalization, location of the NVV complexes within the endosomal compartment has important implications for the ability of the therapeutic nucleic acids to be effective. In the maturing endosome, the pH decreases to 5–6 [25], which is lower than the normal cytosolic pH of 7.4. Degradation by lysosomal enzymes is triggered by the decreasing pH. Thus, disruption of the lysosomes' maturation and inhibition of the acidification within this subcellular compartment are important to enable therapeutic nucleic acids to be effective [8]. Several methods have been used to prevent lysosomal degradation. Buffering the lysosome with NVVs that have "proton sponge" potential leads to swelling and breakdown of the endosomes and then release of the nucleic acids into the cytoplasm. Degradation of maturing endosomes and lysosomes may also be affected by chloroquine or the use of membrane destabilizing peptides such as influenza HA-2 [26]. After release into the cytoplasm, cytoskeletal structures within the cytoplasmic environment may also place some limitation on the nucleic acids reaching their targets.

Restricted access of NVVs to the nucleus curbs delivery of nucleic acids to this subcellular compartment. With particles that have a size of 9–11 nm, passive diffusion across the nuclear pore complex (NPC) may occur [8,27]. However, most NVVs are considerably larger than the NPC, and nuclear delivery of DNA or RNA is only efficient when cell division is rapid. Some improvement in nuclear translocation may be achieved by including nuclear localization signals [8]. Nevertheless, overall nuclear transport with NVVs is low in quiescent cells. This is one of the main reasons why viral vectors are preferred to transduce cells with DNA that encodes expression cassettes, and NVVs are the vector of choice when cytoplasmic delivery of RNA is required.

5.2.5 Immunostimulation by NVVs

After systemic administration of NVVs, stimulation of the innate immune response may occur. Consequent release of proinflammatory cytokines, such as tumor necrosis factor-α, interferon-γ, and interleukin-6, may cause toxicity and an attenuation of efficacy of the antivirals [28,29]. Nucleic acids and the compounds within NVVs may be immunostimulatory [30]. Toll-like receptors (TLRs) located in endosomes, through their functioning as pattern recognition receptors, are important contributors to activation of the innate immune response (Chapter 1). In addition, cytoplasmic retinoic acid-induced gene-I-like proteins and nucleotide oligomerization domain-like receptors also contribute to the innate immunostimulation that may be effected by NVVs and their cargos. Certain features of nucleic acids are known to be immunostimulatory. These include unmethylated CpG islands on DNA, which activate TLR7 and TLR9, and duplex RNA that stimulates TLR3. Attenuation of immunostimulation may be achieved by avoiding potentially problematic sequences and through chemical modification of RNA. A widely used and effective modification of siRNAs has been inclusion of 2′-O-methyl residues. The altered nucleotide occurs naturally, is not toxic, attenuates immunostimulation, and improves stability of silencing sequences (Chapter 2). Modification of mRNA with pseudouridine residues also improves stability and diminishes immunostimulation by this exogenous nucleic acid when delivered with NVVs [31,32].

5.3 ANTIVIRAL NUCLEIC ACIDS THAT MAY BE DELIVERED WITH NVVs

The capabilities of NVVs have a bearing on selection of the most appropriate strategy for their use in antiviral gene therapy. Currently, using NVVs to transduce cells with DNA expression cassettes is limited. After systemic administration, efficiency of delivery of DNA with NVVs is generally modest and not as good as that achieved when using viral vectors. Therefore, DNA-containing NVVs are not suited to applications in which systemic administration is needed to transport the antiviral nucleic acid into a high proportion of virus-infected cells. However, use of NVVs in settings amenable to gene transfer topically or ex vivo is feasible. Accessibility of cervical epithelial cells infected with human papilloma virus (HPV) may be useful to mitigate problems associated with systemic administration. A recent study showed that a transfection reagent could be used to deliver DNA encoding transcription activator-like effector nucleases (TALENs) to target cells after topical administration in a murine model of HPV-induced cervical cancer [33]. Likewise, when the target cells are cultured ex vivo, efficient delivery and expression of the transgene is easier to achieve and problems associated with systemic administration of the antivirals are avoided. This has been applied to inactivate the HIV-1 co-receptor, CCR5, with sequences encoding zinc finger nucleases (ZFNs) that target the endogenous *ccr5* gene [34] (Chapter 8). Another example in which DNA delivery with NVVs may be useful for treating viral infections is in the application of DNA-based vaccination as a prophylactic or therapeutic immunotherapy (Chapter 11). Localized expression of the immunostimulatory proteins after subcutaneous, intradermal, or intramuscular administration of NVV formulations may activate humoral and cell-mediated immunity to a DNA-encoded viral antigen.

NVV-mediated delivery of DNA encoding transposons also has potential for antiviral gene therapy and is useful when sustained expression of a therapeutic sequence is desirable [35,36]. Transposons are capable of integrating into the genome of host cells. To achieve this, NVV-mediated delivery of two components is required: (1) a transposase-encoding sequence and (2) a donor DNA sequence containing the therapeutic gene. Transposase functions by recognizing terminal repeats in the transposon donor, excising the therapeutic sequence, and then inserting into the host genome. Advances in technology based on use of Sleeping Beauty [37] and piggyBac transposons [38] have been particularly exciting. Although concerns remain about possible genotoxic effects of integration of these sequences, transposons may have utility in applications in which viral gene therapy may be implemented by transduction of cells ex vivo [36].

The most popular use of NVVs is for delivering therapeutic RNA sequences. The advantage of using RNA, instead of DNA, in formulations of NVVs is that the therapeutic sequence need only be delivered to the cytoplasmic subcellular site of action. Transport of the nucleic acids across the nuclear membrane is not required. Introducing RNA into cells also avoids potentially genotoxic complications of recombination between transgenes and host DNA. The range of virus-disabling mechanisms that may be activated through use of NVVs to deliver RNA is broad. Synthetic siRNAs, virus-targeting aptamers, antisense, and ribozymes are all suited to delivery by these vectors. In addition, introducing mRNAs encoding engineered ZFNs, TALENs, and clustered regularly interspaced short palindromic repeats (CRISPR) and CRISPR-associated sequences adds gene editing technologies to the scope of NVVs for antiviral gene therapy (Chapter 3).

Therapeutic mRNAs are typically generated in vitro by using linearized DNA as a template for transcription by T7 RNA polymerase. Capping, polyadenylation, and incorporation of modified bases are important to prevent immunostimulation by exogenous RNA, improve translational efficiency, and increase stability of the antiviral nucleic acid [30,31]. A drawback of using therapeutic mRNA instead of transducing cells with expression cassettes is that the mRNA is not renewable. Repeated administration of the NVVs containing therapeutic sequences may be necessary to achieve adequate delivery to an entire population of virus-infected cells. Under such circumstances, eliciting of an immune response to the encoded therapeutic protein may compromise efficacy of repeat treatments. Activating RNA interference (RNAi) with synthetic sequences to silence pathology-causing genes, such as those expressed by viruses, has been a very active field of research. Many investigations have been performed to optimize formulations containing synthetic siRNAs, and various carrier molecules have been used in the development of antivirals. Use of siRNAs in NVVs is now being applied to the treatment of various viral infections and has advanced to clinical trial for therapy of HBV (http://www.arrowheadresearch.com/programs-overview).

5.4 CATEGORIES OF NVVs

NVVs that have been used for gene therapy are highly varied [7,30]. Advances in synthetic chemistry and the availability of new polymers, lipids, and nanoparticles are all contributing to improved methods of delivering nucleic acids for treatment of viral infections. There is no ideal vector formulation that is suited to all applications. Rather, NVVs are tailored to specific uses. Conveniently, components of lipoplex NVVs are amenable to a modular combination to allow for production of formulations with biological properties suited to specific applications. Many different cationic molecules, including multivalent compounds [39,40], with differing nucleic acid binding capacities and transfection efficiencies, have been utilized in NVVs. Incorporation of stabilizing, fusogenic, endosome-disrupting, and targeting constituents allows for semirational design of the formulations. Self-assembly of the NVVs may be achieved by noncovalent or covalent combination of the vector components, sonication, and extrusion to generate uniformly sized particles containing the nucleic acid cargo. Small sizes of the particle are generally preferred for use in vivo. This is important to improve biodistribution and avoid consumption by the RES. The range of different types of NVVs is vast and includes lipid-containing nucleic acid complexes (lipoplexes), polymer nucleic acid complexes (polyplexes), polymer-modified lipoplexes (lipopolyplexes), various conjugates, and other nanomaterials. The major categories of NVVs are discussed in the following subsections with a particular emphasis on their application to antiviral gene therapy.

5.4.1 Lipoplexes

NVVs derived from lipids are currently widely used and were also the first to have been described. In a study performed in 1980, Farley and colleagues demonstrated that it was possible to introduce viral DNA of SV40 into cultured kidney cells [41]. The investigation described the use of phosphatidylserine to form vesicles with an average diameter of 0.4 μm. Delivery of the DNA to cells with these particles was capable of initiating SV40 replication. An important additional observation was that the encapsulated DNA was resistant to degradation by DNase. This seminal finding led to the development of lipoplex NVVs, and the term "lipofection," which describes transfection by lipid carriers, was later coined [42].

Cationic lipids, which are capable of spontaneous formation of liposomes, have been central to the development of lipid-based NVVs. These positively charged lipids essentially comprise three components: a cationic head group that is responsible for DNA binding, a hydrophobic tail that is required for formation of the vesicular structure, and a linker that couples the cationic and hydrophobic components. Different cationic lipids have been used to formulate NVVs, and alterations of the properties of the vectors are based on manipulation of the three parts of the cationic lipid. Examples of such cationic lipids are *N*-[1-(2,3-dioleyloxy)propyl]-*N*,*N*,*N*-trimethylammonium chloride [42] and DC-Chol (3-[*N*-(*N*,*N*-dimethylaminoethane)-carbamoyl]cholesterol) [43], which include fatty acyl and cholesterol as their lipid components (Figure 5.2). Cationic lipids containing unsaturated acyl chains with a small head group adopt a conical structure, which may be used to influence shape and size of the lipoplexes [44]. Other cationic molecules can also be included to facilitate condensation of the nucleic acids and generation of uniformly small sized lipoplexes. An example is protamine, which is a naturally occurring nontoxic protein. In addition to the cationic lipids, lipoplexes typically also contain neutral helper lipids that serve to enhance stability and fusion of the lipoplex with the target cells. Commonly used helper and fusogenic lipids

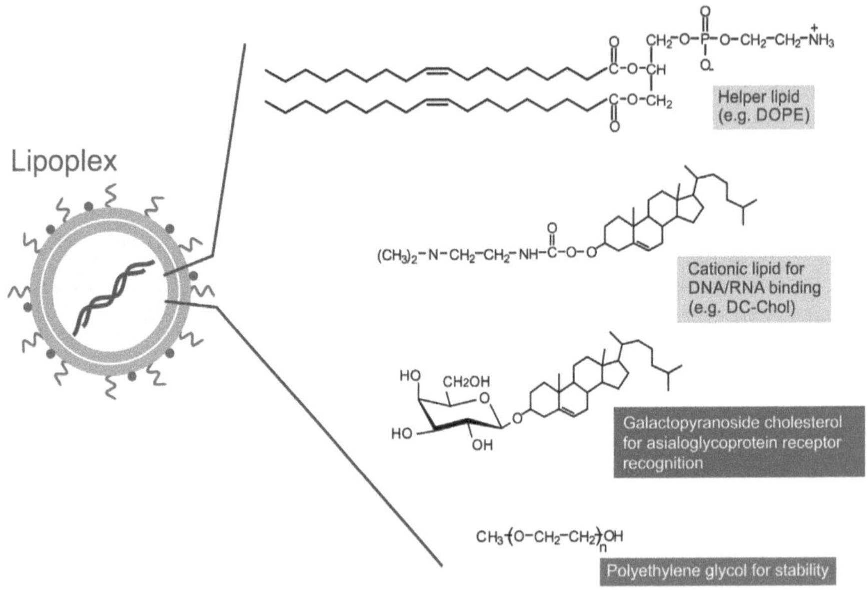

FIGURE 5.2 NVVs comprising lipid-based formulations.
Lipoplexes are typically engineered to generate vesicular nanoparticles. A helper fusogenic lipid (e.g., DOPE) forms the basis of the structure. To encapsulate the nucleic acid and condense the complex, a cationic lipid (e.g., DC-Chol) is required to bind the negatively charged DNA or RNA. Nucleic acids may also be bound to the surface of lipoplexes (not depicted here). Targeting molecules may be included to enable delivery of the therapeutic to infected cells. Galactose conjugates (e.g., galactopyranoside cholesterol) are commonly used in lipid-based and other NVVs to bind the asialoglycoprotein receptor of liver cells and confer hepatotropism on the vectors. To provide stability, limit interaction with circulating blood constituents, and attenuate immunostimulation, PEG may be added to the formulation. This is commonly achieved by using PEGylated DOPE to make up the lipoplexes.

are dioleoylphosphatidylethanolamine (DOPE) and cholesterol. To improve stability and limit interaction with host proteins and aggregation, PEG is often combined with the NVVs (see Section 5.2.2). This hydrophilic polymer may be conjugated to a helper lipid such as DOPE and included in the NVV during formulation of the lipoplexes [11]. PEG has also been used as a linker to couple targeting moieties to the vectors. When used in this way, the long chain of the polymer provides a spacer to enable optimal presentation of targeting molecule. A schematic of a typical lipoplex formulation is illustrated in Figure 5.2.

Stabilized nucleic acid-lipid particles (SNALPs) are a class of NVV that have been advanced for clinical application of RNAi-based viral gene silencing technology. Utility of the approach for treatment of various hepatic diseases has been demonstrated [45,46]. SNALPs, made up of cationic and fusogenic lipids, efficiently deliver nucleic acids to target cells and accomplish endosomal release of the siRNA cargos. Evidence indicates that the hepatotropism of SNALPs is conferred by a mechanism that involves apolipoprotein E (apoE)-dependent uptake of the ionizable lipid nanoparticles (LNPs) into hepatocytes [47]. Naturally, apoE is responsible for clearance of lipoproteins by the liver, and usurping of this pathway by SNALPs confers hepatotropism on the vectors. Interestingly, cationic particles that are made hepatotropic by addition of ligands for the asialoglycoprotein receptor are internalized by a different mechanism. SNALPs efficiently deliver siRNAs to liver cells in guinea pigs [48], nonhuman primates [49], and mice [45]. The first study using the vectors was aimed at developing therapy for HBV. Morrissey et al. used SNALPs that were decorated with PEG to generate neutral, hydrophilic particles with enhanced stability [45]. Anti-HBV siRNAs formulated within the SNALPs were injected intravenously into mice that replicated HBV after hydrodynamic injection of plasmid DNA. Serum markers of HBV replication were suppressed for up to 6 weeks when using a weekly dosing regimen of the siRNA-SNALPs (Chapter 6). Another application for use of SNALPs is in the treatment of Ebola hemorrhagic fever virus infection (Chapter 10). Efficacy was indeed observed after lethal exposure to Ebola virus to guinea pigs [48] and to nonhuman primates [49]. Refinements to the technology using ionizable LNPs as NVVs are ongoing [50]. Evidence shows a strong correlation between the pKa values of the cationic lipids and vector efficiency [50]. Optimal efficiency of the vectors was found when the pKa of the ionizable lipid was in the range of 6.2–6.5. SNALPs are currently also being used as vectors to deliver gene silencers used for treatment of hypercholesterolemia [51], hepatocellular carcinoma, and amyloidosis (reviewed in ref. [7]).

In another study using lipid-based vectors to deliver anti-HBV gene silencers, Hean et al. used cationic lipoplexes containing DOPE and a liver-targeting cholesterol galactoside component to formulate modified anti-HBV siRNAs [52]. HBV silencing was observed in HBV transgenic mice without evidence of toxicity or induction of an innate immune response. Other studies have also reported anti-HBV efficacy of siRNAs in lipoplex formulations containing a polyamine-conjugated cholesterol [53] or aminoxy cholesterol [52]. Aminoxy moieties were included to facilitate addition of PEG. The vectors were passively hepatotropic, and silencing of HBV replication was observed for a few weeks in transgenic mice [53].

An interesting development was the generation of lipidoid (lipid-like) molecules for use in NVVs. Candidate molecules that were capable of efficiently transfecting siRNAs were selected from a library of 1200 structurally diverse cationic compounds [54]. A follow-up study showed that formulations containing lipidoids were efficient vectors for systemic delivery of cholesterol-lowering siRNAs in vivo [55]. Although not yet applied to antiviral gene therapy, lipidoids may well find utility for this application.

5.4.2 Polymer Vectors

The condensation resulting from interaction of polymers with nucleic acids differs from that occurring when cationic lipids bind DNA or RNA. Typically, cationic polymers do not assemble into micelles. In addition, they contain multiple positive charges, which form multiple interactions between the polymer and nucleic acids. Commonly used polymers that have been used in the development of NVVs include polyethyline imine (PEI), poly L-lysine (PLL), and methacrylate- and carbohydrate-based polymers such as chitosan. Varied conformations of the polymers have been used, which include linear (e.g., PEI and PLL), branched (e.g., PEI), dendrimeric (e.g., polyamidoamine), and circular (e.g., cyclodextrin) structures (reviewed in refs [7,8]). PLL was one of the first polymers to be used to deliver DNA to cells [56,57]. Lysine is a basic amino acid; thus, the homopolymer is capable of binding acidic DNA or RNA through electrostatic interaction. Early studies showed that the polymer did indeed condense DNA, and coupling of a ligand for the asialoglycoprotein receptor conferred hepatotropism on the complexes [15,58]. However, complexes that include PLL as the main component have not been popular vectors for gene therapy, and the reason is that release of the loaded nucleic acids from the endosomes is not efficient. This is ascribed to poor proton binding by lysine at physiological pH [7,8]. Addition of other molecules, such as PEG [11], has been described, and in some cases delivery of nucleic acids with the PLL-based vectors has been improved.

With PEI, a nitrogen atom within the imine groups is located at every third position of the polymer, and these positive charges provide high nucleic acid binding capacity. Together with endosomolytic properties, PEI has become a popular choice of polymer for use in NVVs [7,8]. Structurally diverse types of PEI, which may have linear, branched, or dendrimeric shapes, have different biological properties and may be useful for particular applications to nucleic acid delivery. Linear PEIs with high molecular weight appear to have better ability to deliver nucleic acids, but they are associated with toxicity [59,60]. Conversely, low-molecular-weight PEIs are well tolerated, but delivery of RNA is less efficient [59,61]. To improve delivery and diminish toxicity, several variations on PEI-containing NVVs have been investigated [7]. Examples are the inclusion of PEG [62,63], alkylation [64,65], and cross-linking of the polymers with disulfide bridges to limit degradation [59].

Generating and screening libraries comprising derivatives of PEI and/or lipids is a promising development in gene therapy. To identify PEI-based NVVs that are capable of efficiently delivering siRNAs to endothelial cells, Dahlman and colleagues conjugated different PEIs with each of a large panel of epoxide-modified lipids [66]. Approximately 2000 different molecules, which constituted the library, were combined with PEG2000. The resulting nanoparticles were screened for the efficiency with which they delivered siRNAs to endothelial cells. Selected NVVs were functional in vivo without causing toxicity. They could be used to deliver silencing sequences to the normal pulmonary endothelium and to carry siRNAs to endothelial cells of primary and metastatic Lewis cancer cells in mice. Although not directly applicable to treatment of viral infections, library screening to identify vectors with defined properties will be useful for delivering antiviral sequences to infected cells of different tissues.

Chitosan, a natural nontoxic and biodegradable polymer, has also been used for delivery of nucleic acids (reviewed in ref. [67]). It is a linear polysaccharide that predominantly comprises D-glucosamine residues that are linked to each other through β-1,4-glycosidic bonds (Figure 5.3). Chitosan is derived from exoskeletons of crustaceans and is typically prepared by deproteinization, demineralization, and depigmentation of the shells. Strong alkaline treatment is also used to deacetylate *N*-acetyl-D-glucosamine residues that predominate in the chitin precursor. The amine groups of chitosan confer weakly basic properties that enable binding of nucleic acids through electrostatic interaction. In addition, the amine groups of the D-glucosamine provide for targeted addition of a range of moieties to chitosan. PEG has been coupled to these residues to enhance stability and limit unintended interaction with circulating proteins [68,69]. As with other NVVs, conjugation of galactose has been used

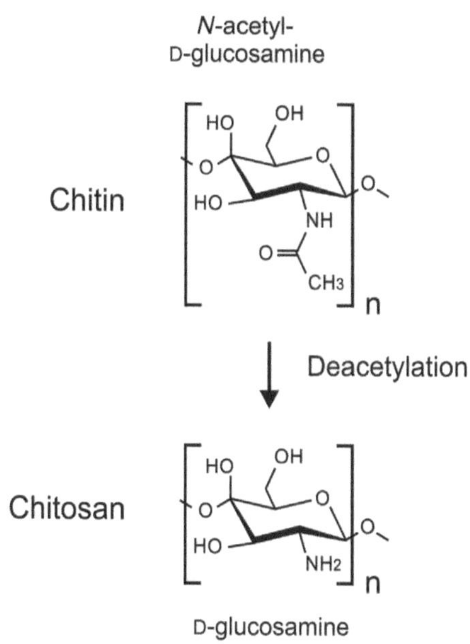

FIGURE 5.3 Chitosan preparation from chitin.
The *N*-acetyl-D-glucosamine units of chitin are linked to each other through β-1,4-glycosidic bonds in the polymer. Deacetylation of the monomers to form D-glucosamine residues converts chitin to chitosan.

to confer hepatotropism [70]. In one example of targeting of liver cells using this strategy, polyvinyl pyrrolidone was used as a spacer linker to couple the galactose moiety to the NVV [71]. Other ligands include mannose and folic acid moieties, which have been used to direct chitosan-based NVVs to antigen-presenting [72] and malignant [73] cells, respectively. Chitosan itself is not efficiently endosomolytic, and PEI has been conjugated to the polymer to improve entry into the cytoplasm [67].

One interesting example of use of chitosan for antiviral application is in vaccination against HBV [74]. Khatri et al. incorporated a DNA sequence encoding the HBV surface protein into a chitosan-based NVV. Intranasal administration of the vectors led to a strong humoral and cell-mediated immune response in mice, which was predicted to be protective. Moreover the nasal administration led to production of IgA in mucosal secretions, which was not detectable after intramuscular injection of naked DNA or alum-adsorbed HBV surface protein. Although chitosan formulations have useful features, they have not been widely used in protocols aimed at developing gene therapy for viral infections. Problems with stability and modest transfection efficiency of vectors containing the natural polymer have limited widespread use.

Selinsky and colleagues also used polymer-containing vectors to advance a gene-base vaccination strategy [75]. To prevent cytomegalovirus (CMV) infection, they incorporated sequences encoding the viral major surface glycoprotein B (gB) into NVVs. A nonionic poloxamer polymer and cationic surfactant were formulated together with plasmid DNA containing *gB*. Initial testing, which was performed by intramuscular injection of the formulations, showed induction of a humoral and cell-mediated immune response to the viral protein. The formulation has now been advanced to evaluation in clinical trials [76]. CMV infection is particularly serious in patients undergoing transplant, and the gene-based vaccine will be useful in this population.

5.4.3 Conjugates

Delivery platforms that are based on covalent attachment of therapeutic nucleic acids, particularly siRNAs, to a molecule that serves as a carrier have also shown promise for application to antiviral gene therapy. Coupling silencing and delivery components through conjugation may be used to generate formulations that conveniently comprise single molecules. Conjugates are typically smaller than lipoplex particles, and this facilitates their widespread distribution in vivo. In addition, because conjugates are modular, optimization of the function of individual components may be performed to improve efficacy and minimize toxicity.

The dynamic polyconjugate (DPC) platform, currently being developed by the Arrowhead Research Corporation, has impressive capabilities. The early

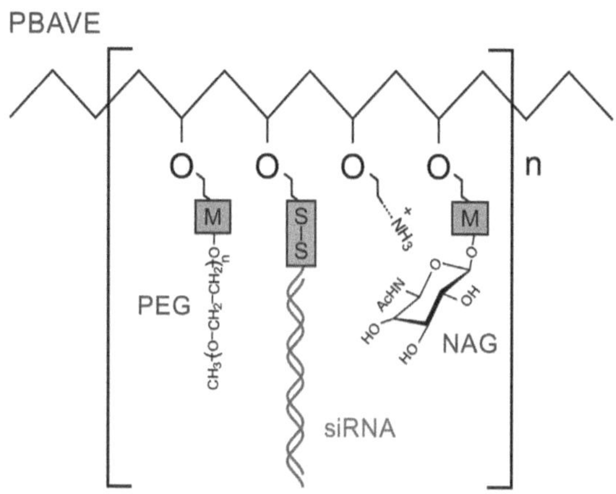

FIGURE 5.4 Representative use of conjugation to couple components to a backbone polymer and generate an NVV.
In this example, a DPC, based on use of PBAVE, is used to couple constituents that make up the antiviral NVV. Maleamate bonds (M) are acid labile and are broken in the maturing endosome to confer proton sponge capabilities on the remnant polymer. This facilitates release into the reducing cytoplasm, in which the disulfide bonds (S–S) are broken to release siRNAs to effect target gene silencing. *Adapted from refs [7,44].*

versions of the platform technology were based on use of an amphipathic polymer, polybutylamino vinyl ether (PBAVE) [77] (Figure 5.4). This backbone was used to couple shielding PEG molecules and hepatocyte-targeting NAG residues. Conjugation of carboxy dimethylmaleic anhydride derivatives of PEG and NAG to PBAVE generated a compound with acid-labile maleamate bonds. Disulfide bonds were used to conjugate the siRNA to the PBAVE polymer. The ingenious design was devised to ensure that the complex is stable in the circulation. However, after receptor-mediated endocytosis, the maleamate bonds between the PEG and NAG moieties and the PBAVE polymer are broken within the acidic milieu of the maturing endosome [78]. After release of the PEG and NAG components, the endosomolytic properties of the remnant are activated. After release of the siRNA-bound PBAVE into the reducing cytoplasmic environment, the disulfide bonds are then broken and the siRNAs are detached from the PBAVE. The RNAi activators are then free to effect target gene silencing. Several modifications of the technology have recently been reported, and use of other targeting moieties to deliver siRNAs to a range of tissues is being investigated. An interesting variation on the DPC methodology has been to separate the siRNA and PBAVE polymer parts [79]. Thus, siRNAs conjugated to cholesterol and the endosome-destabilizing with liver targeting components may be separately administered. Both are delivered to hepatocytes, where they function in concert to achieve markedly improved silencing efficacy. This modified DPC-based technology has been developed further by the Arrowhead Research Corporation for use in treatment of HBV [80]. The candidate drug has a melittin-like protein substituted for the PBAVE polymer, and two cholesterol-modified anti-HBV siRNAs are included to silence viral gene expression. The formulation is currently in clinical trial for treatment of HBV infection (http://www.arrowheadresearch.com).

Conjugates developed by Alnylam Pharmaceuticals are also undergoing preclinical and clinical trials for treatment of various liver diseases (http://www.alnylam.com/) [7,44]. The vectors comprise a chemically modified siRNA, which is conjugated through the 3′ end of the sense strand. Synthetic silencing molecules are coupled to a triantennary molecule that displays NAG residues and confers liver specificity on the conjugate. The trivalent conjugate was found to be an optimal configuration that has particularly high affinity for the asialoglycoprotein receptor. With application of the proprietary enhanced stabilizing chemistry, vectors could be generated that were efficiently hepatotropic and achieved good silencing efficacy after subcutaneous administration of the conjugate. Treatment of HBV infection, amyloidosis, hypercholesterolemia, and porphyria are diseases for which the technology is being developed. Assessment of efficacy against HBV has been measured in chronically infected chimpanzees. Although the results have not yet been published, significant inhibition of markers of viral replication was observed. Moreover serum activity of alanine transaminase in the chimpanzee with the highest viral titer decreased after the treatment. An application for an investigational new drug is planned for 2015.

5.4.4 Cell-Penetrating Peptides

Since their discovery approximately 25 years ago, cell-penetrating peptides (CPPs) have emerged as an interesting class of molecules that have usefulness for delivery of nucleic acids amongst other candidate drugs (reviewed in ref. [81]). CPPs typically comprise 5–30 amino acids and have been isolated from diverse sources, such as natural viral proteins, signal peptides, antimicrobial peptides, and artificial peptides derived from engineered libraries. CPPs have varied chemical and structural features and are categorized as being cationic, amphipathic, and hydrophobic. One of the first CPPs to have been described is a cationic arginine-rich peptide from the Tat protein of HIV-1. It was discovered after observing that Tat is rapidly internalized by cells [82,83]. An essential feature that appears to be important to use of CPPs in vectors is that they are capable of efficiently binding to cell membranes. Given that the turnover of cell membranes is a continuous process that involves rapid internalization and replacement [84], it is not surprising that peptides that bind to cell surfaces may be used to carry loaded drugs into cells. In the case of cationic CPPs, binding occurs to negatively charged GAGs, such as heparin sulfate [85]. The electrostatic interaction leads to clustering of the GAGs on the cell surface, which

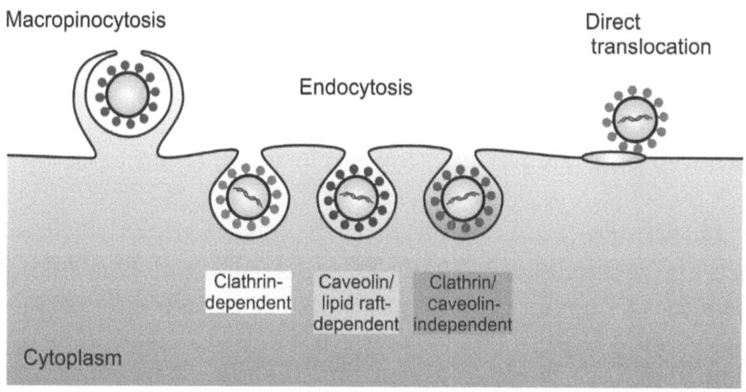

FIGURE 5.5 Mechanisms of cellular internalization of CPPs.
Different processes mediate uptake of particles that are decorated with CPPs. The particular uptake mechanism is dependent on the type of CPP, its concentration, and the therapeutic cargo, amongst other factors. Uptake of CPP-containing particles by macropinocytosis, clathrin-mediated endocytosis, caveolin/lipid raft-mediated endocytosis and clathrin/caveolin-independent endocytosis, direct translocation, and pore formation in the cell membrane have all been described.

in turn leads to direct translocation or endocytosis of the CPP. However, there is variation in the ways in which CPPs are internalized by cells (Figure 5.5). The particular mechanisms of uptake are influenced by the structural features, chemical properties, and concentration of the CPPs as well as the nature of the cargo [86–89]. CPPs mediate cellular uptake by direct translocation, pore formation, macropinocytosis, clathrin-mediated endocytosis, caveolin/lipid raft-mediated endocytosis, and clathrin/caveolin-independent endocytosis. When concentrations of CPPs are high, direct translocation into cells seems to be favored, whereas when the CPPs are present at lower concentration, endocytic mechanisms may predominate. The CPPs with their cargos are thought to escape the endosome through direct interaction with the membrane of the subcellular organelle or by proton sponge-mediated lysis [90]. However, some CPP conjugates may remain in the endosome for extended periods with reduced bioavailability.

To develop their use as vectors, CPPs have been coupled to therapeutic molecules by covalent or noncovalent binding. Oligonucleotides may be delivered in this way; thus, CPPs have been used to deliver synthetic siRNAs (reviewed in ref. [91]). Covalent attachment of oligonucleotides to CPPs may be through formation of disulfide bonds, which are broken in the reducing cytoplasmic environment. However noncovalent coupling of CPPs to their cargoes is more widely used. The advantage of using noncovalent attachment is that simpler protocols may be used to formulate the complexes. In addition to facilitating cellular internalization, interaction between the nucleic acids and CPPs may also serve to prevent degradation of the bound nucleic acid by circulating nucleases. Versatility of CPPs has been shown by demonstrating their compatibility with LNPs. By adding a cationic CPP derivative to LNPs, the properties of nucleic acid delivery were significantly enhanced [92]. To date, CPPs have not been particularly popular for the delivery of antiviral nucleic acids, but the approach seems feasible and likely to generate useful results.

5.4.5 Aptamers

Aptamers are single-stranded DNA or RNA molecules with high affinity binding capabilities for specific compounds. They are a versatile group of molecules with an extensive range of applications. In addition to generating aptamers that bind to and act directly on viral proteins [93], these oligonucleotides may be used as targeting molecules to deliver coupled antiviral nucleic acids (reviewed in refs [94,95]). Availability of aptamers with specificity for molecules on cells' surfaces is potentially very useful for regulating the tropism of NVVs. To isolate aptamers, Systematic Evolution of Ligands by EXponential enrichment (SELEX) is used [96–98]. The procedure essentially involves immobilization of the target molecule and then selection of single-stranded DNA or RNA from a library of the nucleic acids. More complex selection procedures using whole cells as the bait may also be used [99]. Iterative cycles of enrichment and amplification are used to purify aptamers with intended properties. During the SELEX procedure, steps are included to ensure that aptamers with optimal affinity are selected and the nucleic acids with nonspecific target binding properties are excluded.

Targeting aptamers have been used successfully to deliver HIV-1-silencing siRNAs to cells infected with the virus (Chapter 8). Aptamers that bind to CD4 [17] and the HIV-1-encoded gp120 glycoprotein [60,61], which is expressed on the surface of infected cells, have been used for target specificity. Two configurations of the coupled aptamer-siRNA complexes have been used. The aptamer may be contiguous and covalently bound to one of the strands of the siRNA. Alternatively, a bridging oligonucleotide may be used to stick the aptamer to the gene silencer [18,19,95]. After entering the cytoplasm, processing by Dicer separates the siRNA from the aptamer, and the RNAi activator is freed to effect silencing of viral gene expression. Uptake of the aptamer-siRNA complexes is thought to be by receptor-mediated endocytosis [94]. The mechanism by which endosomal escape occurs is yet to be elucidated. Although aptamers have only been used to target HIV-1 to date, it is likely that the technology will be applicable to treating other virus infections.

5.5 CONCLUSIONS

Although antiviral gene therapy has enormous potential, the technology has largely been hamstrung by limitations of methods available to deliver nucleic acids to virus-infected cells. Although viral vectors have shown good efficacy, they are not ideally suited to widespread use in clinical settings. Toxicity, immunostimulation, high costs of preparation, and the potential for causing mutation are some of the limitations of this class of vector (Chapter 4). NVVs do not have many of these drawbacks, and developing the vectors for viral gene therapy seems key to progress in the field. Particularly important are that they are (1) amenable to large-scale preparation that is needed for use in humans; (2) modular in composition, which enables convenient alteration of biological properties; (3) amenable to formulation with various different sized nucleic acids; and (4) usually modestly immunostimulatory.

Many different NVVs have been described, and several promising types of synthetic vectors are now available. These include lipoplexes, polymer complexes, conjugates to polymers, CPPs, and aptamers. Importantly, the categories of NVVs are not mutually exclusive in their formulation. Modularity provides versatility that enables the generation of vectors that incorporate features of the different NVV classes. An example is the coupling of cationic lipids with PEI to generate vectors with efficient and specific endothelial tropism [66]. As insights into the relationship between the physicochemical properties of NVVs and efficiency of delivering the nucleic acid payloads improves, the way is being paved for better design of vectors that may be used for viral gene therapy. A disadvantage of NVVs, when compared with their viral counterparts, is that the delivery of DNA-containing expression cassettes to target nuclei is generally poor after systemic administration. When used for delivering DNA cassettes, NVVs are mainly used ex vivo or to cause prophylactic or therapeutic immunostimulatory effects (Chapter 11). Nevertheless, the limitation on DNA delivery efficiency after systemic administration of NVVs is not necessarily a handicap. Current approaches to antiviral gene therapy are steering away from using DNA as the active therapeutic. Using RNA obviates the risk of mutagenesis and makes delivery of the therapeutic easier to achieve. Candidate RNA antivirals, with the exception of the small group that is being developed as transcriptional gene silencers, have their effects in the cytoplasm. Because transport across the nuclear membrane is not necessary, the requirements of RNA-containing vectors are not as stringent and make NVVs well suited to RNA-based antiviral gene therapy. Progress with synthetic RNA chemistry and the ability to attenuate the immunostimulatory properties of these nucleic acids has complemented development of NVVs for therapeutic application. Initiation of clinical trials to treat HBV infection with liver-targeting conjugates is an important development. Conclusions drawn from this trial, as well as outcomes from research using other protocols, will be valuable to inform broader implementation of treatment regimens that rely on use of NVVs to treat viral infections.

REFERENCES

[1] Monaghan M, Pandit A. RNA interference therapy via functionalized scaffolds. Adv Drug Deliv Rev 2011;63(4–5):197–208.

[2] Behlke MA. Progress towards in vivo use of siRNAs. Mol Ther J Am Soc Gene Ther 2006; 13(4):644–70.

[3] Behlke MA. Chemical modification of siRNAs for in vivo use. Oligonucleotides 2008; 18(4):305–19.

[4] Rettig GR, Behlke MA. Progress toward in vivo use of siRNAs-II. Mol Ther J Am Soc Gene Ther 2012;20(3):483–512.

[5] Janssen HL, Reesink HW, Lawitz EJ, Zeuzem S, Rodriguez-Torres M, Patel K, et al. Treatment of HCV infection by targeting microRNA. N Engl J Med 2013;368(18):1685–94.

[6] Raal FJ, Santos RD, Blom DJ, Marais AD, Charng MJ, Cromwell WC, et al. Mipomersen, an apolipoprotein B synthesis inhibitor, for lowering of LDL cholesterol concentrations in patients with homozygous familial hypercholesterolaemia: a randomised, double-blind, placebo-controlled trial. Lancet 2010;375(9719):998–1006.

[7] Yin H, Kanasty RL, Eltoukhy AA, Vegas AJ, Dorkin JR, Anderson DG. Non-viral vectors for gene-based therapy. Nat Rev Genet 2014;15(8):541–55.

[8] Mintzer MA, Simanek EE. Nonviral vectors for gene delivery. Chem Rev 2009;109(2):259–302.

[9] Blessing T, Remy J-S, Behr J-P. Template oligomerization of DNA-bound cations produces calibrated nanometric particles. J Am Chem Soc 1998;120(33):8519–20.

[10] Tang MX, Szoka FC. The influence of polymer structure on the interactions of cationic polymers with DNA and morphology of the resulting complexes. Gene Ther 1997;4(8):823–32.

[11] Bazile D, Prud'homme C, Bassoullet MT, Marlard M, Spenlehauer G, Veillard M. Stealth Me.PEG-PLA nanoparticles avoid uptake by the mononuclear phagocytes system. J Pharm Sci 1995;84(4):493–8.

[12] Alexis F, Pridgen E, Molnar LK, Farokhzad OC. Factors affecting the clearance and biodistribution of polymeric nanoparticles. Mol Pharm 2008;5(4):505–15.

[13] Molas M, Gomez-Valades AG, Vidal-Alabro A, Miguel-Turu M, Bermudez J, Bartrons R, et al. Receptor-mediated gene transfer vectors: progress towards genetic pharmaceuticals. Curr Gene Ther 2003;3(5):468–85.

[14] Miller N, Vile R. Targeted vectors for gene therapy. FASEB J 1995;9(2):190–9.

[15] Wu GY, Wu CH. Receptor-mediated gene delivery and expression in vivo. J Biol Chem 1988;263(29):14621–4.

[16] Ezekowitz R, Sastry K, Bailly P, Warner A. Molecular characterization of the human macrophage mannose receptor: demonstration of multiple carbohydrate recognition-like domains and phagocytosis of yeasts in Cos-1 cells. J Exp Med 1990;172(6):1785–94.

[17] Wheeler LA, Trifonova R, Vrbanac V, Basar E, McKernan S, Xu Z, et al. Inhibition of HIV transmission in human cervicovaginal explants and humanized mice using CD4 aptamer-siRNA chimeras. J Clin Invest 2011;121(6):2401–12.

[18] Zhou J, Li H, Li S, Zaia J, Rossi JJ. Novel dual inhibitory function aptamer-siRNA delivery system for HIV-1 therapy. Mol Ther J Am Soc Gene Ther 2008;16(8):1481–9.

[19] Zhou J, Swiderski P, Li H, Zhang J, Neff CP, Akkina R, et al. Selection, characterization and application of new RNA HIV gp 120 aptamers for facile delivery of Dicer substrate siRNAs into HIV infected cells. Nucleic Acids Res 2009;37(9):3094–109.

[20] Nishina K, Unno T, Uno Y, Kubodera T, Kanouchi T, Mizusawa H, et al. Efficient in vivo delivery of siRNA to the liver by conjugation of alpha-tocopherol. Mol Ther J Am Soc Gene Ther 2008;16(4):734–40.

[21] Kim SI, Shin D, Choi TH, Lee JC, Cheon GJ, Kim KY, et al. Systemic and specific delivery of small interfering RNAs to the liver mediated by apolipoprotein A-I. Mol Ther J Am Soc Gene Ther 2007;15(6):1145–52.

[22] Sato Y, Murase K, Kato J, Kobune M, Sato T, Kawano Y, et al. Resolution of liver cirrhosis using vitamin A-coupled liposomes to deliver siRNA against a collagen-specific chaperone. Nat Biotechnol 2008;26(4):431–42.

[23] Mislick KA, Baldeschwieler JD. Evidence for the role of proteoglycans in cation-mediated gene transfer. Proc Natl Acad Sci USA 1996;93(22):12349–54.

[24] Mounkes LC, Zhong W, Cipres-Palacin G, Heath TD, Debs RJ. Proteoglycans mediate cationic liposome-DNA complex-based gene delivery in vitro and in vivo. J Biol Chem 1998;273(40):26164–70.

[25] Ohkuma S, Poole B. Fluorescence probe measurement of the intralysosomal pH in living cells and the perturbation of pH by various agents. Proc Natl Acad Sci USA 1978;75(7):3327–31.

[26] Wharton SA, Martin SR, Ruigrok RW, Skehel JJ, Wiley DC. Membrane fusion by peptide analogues of influenza virus haemagglutinin. J Gen Virol 1988;69(Pt 8):1847–57.

[27] Bonner WM. Protein migration into nuclei. I. Frog oocyte nuclei in vivo accumulate micro-injected histones, allow entry to small proteins, and exclude large proteins. J Cell Biol 1975;64(2):421–30.

[28] Sellins K, Fradkin L, Liggitt D, Dow S. Type I interferons potently suppress gene expression following gene delivery using liposome(–)DNA complexes. Mol Ther J Am Soc Gene Ther 2005;12(3):451–9.

[29] Li SD, Huang L. Gene therapy progress and prospects: non-viral gene therapy by systemic delivery. Gene Ther 2006;13(18):1313–9.

[30] Zhang Y, Satterlee A, Huang L. In vivo gene delivery by nonviral vectors: overcoming hurdles? Mol Ther J Am Soc Gene Ther 2012;20(7):1298–304.

[31] Anderson BR, Muramatsu H, Jha BK, Silverman RH, Weissman D, Kariko K. Nucleoside modifications in RNA limit activation of 2′-5′-oligoadenylate synthetase and increase resistance to cleavage by RNase L. Nucleic Acids Res 2011;39(21):9329–38.

[32] Kariko K, Muramatsu H, Keller JM, Weissman D. Increased erythropoiesis in mice injected with submicrogram quantities of pseudouridine-containing mRNA encoding erythropoietin. Mol Ther J Am Soc Gene Ther 2012;20(5):948–53.

[33] Hu Z, Ding W, Zhu D, Yu L, Jiang X, Wang X, et al. TALEN-mediated targeting of HPV oncogenes ameliorates HPV-related cervical malignancy. J Clin Invest 2015;125(1):425–36.

[34] Holt N, Wang J, Kim K, Friedman G, Wang X, Taupin V, et al. Human hematopoietic stem/progenitor cells modified by zinc-finger nucleases targeted to CCR5 control HIV-1 in vivo. Nat Biotechnol 2010;28(8):839–47.

[35] VandenDriessche T, Ivics Z, Izsvak Z, Chuah MK. Emerging potential of transposons for gene therapy and generation of induced pluripotent stem cells. Blood 2009;114(8):1461–8.

[36] Di Matteo M, Belay E, Chuah MK, Vandendriessche T. Recent developments in transposon-mediated gene therapy. Expert Opin Biol Ther 2012;12(7):841–58.

[37] Mates L, Chuah MK, Belay E, Jerchow B, Manoj N, Acosta-Sanchez A, et al. Molecular evolution of a novel hyperactive Sleeping Beauty transposase enables robust stable gene transfer in vertebrates. Nat Genet 2009;41(6):753–61.

[38] Di Matteo M, Samara-Kuko E, Ward NJ, Waddingon SN, McVey JH, Chuah MK, et al. Hyperactive PiggyBac transposons for sustained and robust liver-targeted gene therapy. Mol Ther J Am Soc Gene Ther 2014;22(9):1614–24.

[39] Ahmad A, Evans HM, Ewert K, George CX, Samuel CE, Safinya CR. New multivalent cationic lipids reveal bell curve for transfection efficiency versus membrane charge density: lipid-DNA complexes for gene delivery. J Gene Med 2005;7(6):739–48.

[40] Farago O, Ewert K, Ahmad A, Evans HM, Gronbech-Jensen N, Safinya CR. Transitions between distinct compaction regimes in complexes of multivalent cationic lipids and DNA. Biophys J 2008;95(2):836–46.

[41] Fraley R, Subramani S, Berg P, Papahadjopoulos D. Introduction of liposome-encapsulated SV40 DNA into cells. J Biol Chem 1980;255(21):10431–5.

[42] Felgner PL, Gadek TR, Holm M, Roman R, Chan HW, Wenz M, et al. Lipofection: a highly efficient, lipid-mediated DNA-transfection procedure. Proc Natl Acad Sci USA 1987;84(21):7413–7.

[43] Li S, Gao X, Son K, Sorgi F, Hofland H, Huang L. DC-Chol lipid system in gene transfer. J Control Release 1996;39(2):373–81.

[44] Kanasty R, Dorkin JR, Vegas A, Anderson D. Delivery materials for siRNA therapeutics. Nat Mater 2013;12(11):967–77.

[45] Morrissey DV, Lockridge JA, Shaw L, Blanchard K, Jensen K, Breen W, et al. Potent and persistent in vivo anti-HBV activity of chemically modified siRNAs. Nat Biotechnol 2005;23(8):1002–7.

[46] Zimmermann TS, Lee AC, Akinc A, Bramlage B, Bumcrot D, Fedoruk MN, et al. RNAi-mediated gene silencing in non-human primates. Nature 2006;441(7089):111–4.

[47] Akinc A, Querbes W, De S, Qin J, Frank-Kamenetsk M, Jayaprakash KN, et al. Targeted delivery of RNAi therapeutics with endogenous and exogenous ligand-based mechanisms. Mol Ther J Am Soc Gene Ther 2010;18(7):1357–64.

[48] Geisbert TW, Hensley LE, Kagan E, Yu EZ, Geisbert JB, Daddario-DiCaprio K, et al. Postexposure protection of guinea pigs against a lethal Ebola virus challenge is conferred by RNA interference. J Infect Dis 2006;193(12):1650–7.

[49] Geisbert TW, Lee AC, Robbins M, Geisbert JB, Honko AN, Sood V, et al. Postexposure protection of non-human primates against a lethal Ebola virus challenge with RNA interference: a proof-of-concept study. Lancet 2010;375(9729):1896–905.

[50] Jayaraman M, Ansell SM, Mui BL, Tam YK, Chen J, Du X, et al. Maximizing the potency of siRNA lipid nanoparticles for hepatic gene silencing in vivo. Angew Chem 2012;51(34):8529–33.

[51] Fitzgerald K, Frank-Kamenetsky M, Shulga-Morskaya S, Liebow A, Bettencourt BR, Sutherland JE, et al. Effect of an RNA interference drug on the synthesis of proprotein convertase subtilisin/kexin type 9 (PCSK9) and the concentration of serum LDL cholesterol in healthy volunteers: a randomised, single-blind, placebo-controlled, phase 1 trial. Lancet 2014;383(9911):60–8.

[52] Hean J, Crowther C, Ely A, Ul Islam R, Barichievy S, Bloom K, et al. Inhibition of hepatitis B virus replication in vivo using lipoplexes containing altritol-modified antiviral siRNAs. Artif DNA PNA XNA 2010;1(1):17–26.

[53] Carmona S, Jorgensen MR, Kolli S, Crowther C, Salazar FH, Marion PL, et al. Controlling HBV replication in vivo by intravenous administration of triggered PEGylated siRNA-nanoparticles. Mol Pharm 2009;6(3):706–17.

[54] Akinc A, Zumbuehl A, Goldberg M, Leshchiner ES, Busini V, Hossain N, et al. A combinatorial library of lipid-like materials for delivery of RNAi therapeutics. Nat Biotechnol 2008;26(5):561–9.

[55] Frank-Kamenetsky M, Grefhorst A, Anderson NN, Racie TS, Bramlage B, Akinc A, et al. Therapeutic RNAi targeting PCSK9 acutely lowers plasma cholesterol in rodents and LDL cholesterol in nonhuman primates. Proc Natl Acad Sci USA 2008;105(33):11915–20.

[56] Laemmli UK. Characterization of DNA condensates induced by poly(ethylene oxide) and polylysine. Proc Natl Acad Sci USA 1975;72(11):4288–92.

[57] Olins DE, Olins AL, Von Hippel PH. Model nucleoprotein complexes: studies on the interaction of cationic homopolypeptides with DNA. J Mol Biol 1967;24(2):157–76.

[58] Wu GY, Wu CH. Receptor-mediated in vitro gene transformation by a soluble DNA carrier system. J Biol Chem 1987;262(10):4429–32.

[59] Breunig M, Lungwitz U, Liebl R, Goepferich A. Breaking up the correlation between efficacy and toxicity for nonviral gene delivery. Proc Natl Acad Sci USA 2007;104(36):14454–9.

[60] Godbey WT, Wu KK, Mikos AG. Size matters: molecular weight affects the efficiency of poly(ethylenimine) as a gene delivery vehicle. J Biomed Mater Res 1999;45(3):268–75.

[61] Grayson AC, Doody AM, Putnam D. Biophysical and structural characterization of polyethylenimine-mediated siRNA delivery in vitro. Pharm Res 2006;23(8):1868–76.

[62] Nguyen HK, Lemieux P, Vinogradov SV, Gebhart CL, Guerin N, Paradis G, et al. Evaluation of polyether-polyethyleneimine graft copolymers as gene transfer agents. Gene Ther 2000;7(2):126–38.

[63] Petersen H, Fechner PM, Martin AL, Kunath K, Stolnik S, Roberts CJ, et al. Polyethylenimine-graft-poly(ethylene glycol) copolymers: influence of copolymer block structure on DNA complexation and biological activities as gene delivery system. Bioconjug Chem 2002;13(4):845–54.

[64] Fortune JA, Novobrantseva TI, Klibanov AM. Highly effective gene transfection in vivo by alkylated polyethylenimine. J Drug Deliv 2011;2011:204058.

[65] Thomas M, Klibanov AM. Enhancing polyethylenimine's delivery of plasmid DNA into mammalian cells. Proc Natl Acad Sci USA 2002;99(23):14640–5.

[66] Dahlman JE, Barnes C, Khan OF, Thiriot A, Jhunjunwala S, Shaw TE, et al. In vivo endothelial siRNA delivery using polymeric nanoparticles with low molecular weight. Nat Nanotechnol 2014;9(8):648–55.

[67] Shi Q, Tiera MJ, Zhang X, Dai K, Benderdour M, Fernandes JC. Chitosan-DNA/siRNA nanoparticles for gene therapy. Non-Viral Gene Ther 2011;19:455–80.

[68] Ouchi T, Nishizawa H, Ohya Y. Aggregation phenomenon of PEG-grafted chitosan in aqueous solution. Polymer 1998;39(21):5171–5.

[69] Saito H, Wu X, Harris JM, Hoffman AS. Graft copolymers of poly (ethylene glycol)(PEG) and chitosan. Macromol Rapid Commun 1997;18(7):547–50.

[70] Gao S, Chen J, Xu X, Ding Z, Yang YH, Hua Z, et al. Galactosylated low molecular weight chitosan as DNA carrier for hepatocyte-targeting. Int J Pharm 2003;255(1–2):57–68.

[71] Park IK, Ihm JE, Park YH, Choi YJ, Kim SI, Kim WJ, et al. Galactosylated chitosan (GC)-graft-poly(vinyl pyrrolidone) (PVP) as hepatocyte-targeting DNA carrier. Preparation and physicochemical characterization of GC-graft-PVP/DNA complex (1). J Control Release 2003;86(2–3):349–59.

[72] Kim TH, Nah JW, Cho MH, Park TG, Cho CS. Receptor-mediated gene delivery into antigen presenting cells using mannosylated chitosan/DNA nanoparticles. J Nanosci Nanotechnol 2006;6(9–10):2796–803.

[73] Lee D, Lockey R, Mohapatra S. Folate receptor-mediated cancer cell specific gene delivery using folic acid-conjugated oligochitosans. J Nanosci Nanotechnol 2006;6(9–10):2860–6.

[74] Khatri K, Goyal AK, Gupta PN, Mishra N, Vyas SP. Plasmid DNA loaded chitosan nanoparticles for nasal mucosal immunization against hepatitis B. Int J Pharm 2008;354(1–2):235–41.

[75] Selinsky C, Luke C, Wloch M, Geall A, Hermanson G, Kaslow D, et al. Research paper a DNA-based vaccine for the prevention of human cytomegalovirus-associated diseases. Hum Vaccin 2005;1(1):16–23.

[76] Smith LR, Wloch MK, Chaplin JA, Gerber M, Rolland AP. Clinical development of a cytomegalovirus DNA vaccine: from product concept to pivotal phase 3 trial. Vaccines 2013;1(4):398–414.

[77] Rozema DB, Lewis DL, Wakefield DH, Wong SC, Klein JJ, Roesch PL, et al. Dynamic poly-Conjugates for targeted in vivo delivery of siRNA to hepatocytes. Proc Natl Acad Sci USA 2007;104(32):12982–7.

[78] Rozema DB, Ekena K, Lewis DL, Loomis AG, Wolff JA. Endosomolysis by masking of a membrane-active agent (EMMA) for cytoplasmic release of macromolecules. Bioconjug Chem 2003;14(1):51–7.

[79] Wong SC, Klein JJ, Hamilton HL, Chu Q, Frey CL, Trubetskoy VS, et al. Co-injection of a targeted, reversibly masked endosomolytic polymer dramatically improves the efficacy of cholesterol-conjugated small interfering RNAs in vivo. Nucleic Acid Ther 2012;22(6):380–90.

[80] Wooddell CI, Rozema DB, Hossbach M, John M, Hamilton HL, Chu Q, et al. Hepatocyte-targeted RNAi therapeutics for the treatment of chronic hepatitis B virus infection. Mol Ther J Am Soc Gene Ther 2013;21(5):973–85.

[81] Milletti F. Cell-penetrating peptides: classes, origin, and current landscape. Drug Discov Today 2012;17(15–16):850–60.

[82] Green M, Ishino M, Loewenstein PM. Mutational analysis of HIV-1 Tat minimal domain peptides: identification of trans-dominant mutants that suppress HIV-LTR-driven gene expression. Cell 1989;58(1):215–23.

[83] Green M, Loewenstein PM. Autonomous functional domains of chemically synthesized human immunodeficiency virus Tat trans-activator protein. Cell 1988;55(6):1179–88.

[84] Steinman RM, Mellman IS, Muller WA, Cohn ZA. Endocytosis and the recycling of plasma membrane. J Cell Biol 1983;96(1):1–27.

[85] Ziegler A, Seelig J. Binding and clustering of glycosaminoglycans: a common property of mono- and multivalent cell-penetrating compounds. Biophys J 2008;94(6):2142–9.

[86] Lundin P, Johansson H, Guterstam P, Holm T, Hansen M, Langel U, et al. Distinct uptake routes of cell-penetrating peptide conjugates. Bioconjug Chem 2008;19(12):2535–42.

[87] Madani F, Lindberg S, Langel U, Futaki S, Graslund A. Mechanisms of cellular uptake of cell-penetrating peptides. J Biophys 2011;2011:414729.

[88] Mayor S, Pagano RE. Pathways of clathrin-independent endocytosis. Nat Rev Mol Cell Biol 2007;8(8):603–12.

[89] Koren E, Torchilin VP. Cell-penetrating peptides: breaking through to the other side. Trends Mol Med 2012;18(7):385–93.

[90] El-Sayed A, Futaki S, Harashima H. Delivery of macromolecules using arginine-rich cell-penetrating peptides: ways to overcome endosomal entrapment. AAPS J 2009;11(1):13–22.

[91] Margus H, Padari K, Pooga M. Cell-penetrating peptides as versatile vehicles for oligonucleotide delivery. Mol Ther J Am Soc Gene Ther 2012;20(3):525–33.

[92] Asai T, Tsuzuku T, Takahashi S, Okamoto A, Dewa T, Nango M, et al. Cell-penetrating peptide-conjugated lipid nanoparticles for siRNA delivery. Biochem Biophys Res Commun 2014;444(4):599–604.

[93] Held DM, Kissel JD, Patterson JT, Nickens DG, Burke DH. HIV-1 inactivation by nucleic acid aptamers. Front Biosci 2006;11:89–112.

[94] Zhou J, Rossi JJ. Therapeutic potential of aptamer-siRNA conjugates for treatment of HIV-1. BioDrugs Clin Immunother Biopharm Gene Ther 2012;26(6):393–400.

[95] Zhou J, Neff CP, Swiderski P, Li H, Smith DD, Aboellail T, et al. Functional in vivo delivery of multiplexed anti-HIV-1 siRNAs via a chemically synthesized aptamer with a sticky bridge. Mol Ther J Am Soc Gene Ther 2013;21(1):192–200.

[96] Gold L, Brown D, He Y-Y, Shtatland T, Singer BS, Wu Y-Y. From oligonucleotide shapes to genomic SELEX: novel biological regulatory loops. Proc Natl Acad Sci USA 1997;94(1):59–64.

[97] Gold L, Polisky B, Uhlenbeck O, Yarus M. Diversity of oligonucleotide functions. Annu Rev Biochem 1995;64(1):763–97.

[98] Guo KT, Ziemer G, Paul A, Wendel HP. Cell-SELEX: novel perspectives of aptamer-based therapeutics. Int J Mol Sci 2008;9(4):668–78.

[99] Ohuchi S. Cell-SELEX technology. BioRes Open Access 2012;1(6):265–72.

CHAPTER 6

Gene Therapy for Chronic Hepatitis B Virus Infection

6.1 INTRODUCTION

During the early part of the twentieth century, significant outbreaks of hepatitis occurred after vaccination campaigns [1]. At the time, addition of human serum to vaccines was a common practice. These epidemics were likely to have resulted from hepatitis B virus (HBV) infection and were described many years before the virus was discovered. A timeline from this point, which summarizes important events in HBV research, is depicted in Figure 6.1. Transmission of the infectious agent after administration of human blood or serum suggested that the pathogen was distinct from the type A hepatitis, which typically caused infection after ingestion of contaminated food and water. The most notable outbreak of hepatitis was that which occurred in 1941 among many US military personnel after a yellow fever vaccination program. Approximately 50,000 cases were reported, and as many as 330,000 people may have been infected by the contaminated vaccines. A study performed in 1985 on military veterans revealed that HBV infection was responsible for the hepatitis outbreak [2]. Interestingly, although many people were infected, very few became chronic carriers of the virus. This observation is in agreement with the now well-established epidemiological feature of HBV infection, which is that persistence rarely occurs after transmission of the virus to adults. Rather, chronicity arises when infection occurs during early life.

Discovery of the Australia antigen (Au-Ag), which is now known as the HBV surface antigen (HBsAg), was reported in 1965 [3,4] and was a seminal development in HBV research. The breakthrough with finding the Au-Ag was made by Baruch Blumberg, and the importance of the discovery was recognized by awarding him the Nobel Prize for Physiology or Medicine in 1976. Interestingly, Blumberg's discovery of HBV had an element of serendipity. Before finding the virus, his work was aimed at identifying genetic markers of disease predisposition and involved characterization of protein heterogeneity. To achieve this, immunological studies were performed on patients who had received several donations of blood or blood products. During this investigation, Harvey Alter, Blumberg's colleague, discovered a new antigen that was

Gene Therapy for Viral Infections. http://dx.doi.org/10.1016/B978-0-12-410518-8.00006-5

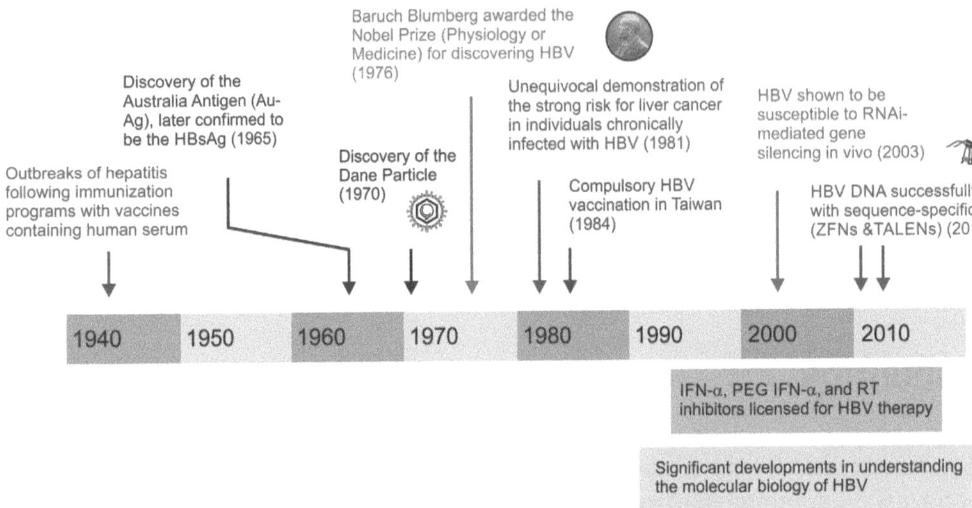

FIGURE 6.1 Timeline of the major milestones in HBV research.

also associated with hepatitis. The protein was commonly found in samples from Australian aboriginals and was therefore called the Au-Ag. Shortly thereafter, serum hepatitis (SH) antigen was described, but it soon became clear that Au-Ag and SH were the same markers [5]. Identifying the antigen linked to hepatitis was a discovery that enabled identification of people who were potentially infectious. It was then also established that individuals could be carriers of the virus without manifesting symptoms of hepatitis.

Improvement in immunoassays used to detect the Au-Ag [6] and identification of subtypes of the antigen [7] provided a basis for making significant advances in understanding the biology of HBV. The filamentous and spherical particles that are present in abundance in the blood of HBV-infected individuals were found to be noninfectious and could be used as vaccines [8–10]. This was particularly important in the campaign to prevent the cirrhosis and liver cancer that frequently complicate chronic HBV infection [11]. The larger infectious Dane particles with more typical virus-like structure were discovered in 1970 [12], and the viral genome was shown to comprise partly double-stranded DNA. This feature proved to be technically useful because it enabled convenient labeling of core particle-associated DNA by the viral polymerase [13]. Subsequent analysis by electron microscopy revealed that the viral DNA was circular in structure [14]. The sequence of the genome of HBV was described in three main publications appearing during 1979 [15–17], and verification that the DNA supported viral replication was provided after injection of chimpanzees with cloned HBV sequences [18]. Unraveling the DNA replication mechanism [19] and dependence of HBV on reverse transcription led to the use of nucleoside and nucleotide analogs for therapy (recently reviewed in ref. [20]). Discovery of the HBV e antigen (HBeAg) was another clinically useful development [21]. The presence of this viral protein correlates well with viral replication rate and individuals' infectiousness. Highly sensitive detection of circulating HBV DNA using the polymerase chain reaction enabled more detailed characterization of the infection. Surprisingly, the mechanism by which HBV enters hepatocytes remained elusive for many years. Evidence that the sodium-dependent taurocholate cotransporting polypeptide (NTCP) serves as a receptor for the virus was a recent discovery [22]. Demonstrating that HBV replication could be very effectively silenced in vivo using RNA interference (RNAi) [23] was an important stimulus for investigating gene therapy as a mode of treating infection with the virus.

6.1.1 HBV Epidemiology

According to the World Health Organization, approximately two billion people have been infected with HBV [24]. In most cases, the infection is self-limiting and individuals become immune to HBV after its elimination (reviewed in refs [25–27]). However, when the virus is acquired at a young age, HBV often persists and leads to the chronic carrier state. There are currently approximately 350 million chronic carriers in the world, and it is these individuals who are at risk for complicating cirrhosis and liver cancer (reviewed in ref. [28]). The mechanism of hepatocarcinogenesis induced by HBV infection is currently incompletely understood. Roles for viral integrants, transcriptional activation of growth regulatory genes by viral proteins, and disruption of cell signaling pathways are implicated [28]. HBV is highly infectious, approximately 100 times more infectious than human immunodeficiency virus (HIV)-1, and is spread parenterally. Among children, the virus may be transmitted perinatally from a highly replicative carrier mother to her child. Horizontal spread among children may occur, but the mechanisms of this process are poorly understood. In adults, self-limiting acute infection typically develops after sexual contact with an HBV-infected individual, use of contaminated needles among intravenous drug users, unsterile body piercing, tattooing practices, or sharing of razors and toothbrushes. The major geographical regions where HBV infection is endemic are east and southeast Asia, the western Pacific islands, and sub-Saharan Africa (Figure 6.2). Interestingly, childhood transmission differs in Asian and in African settings. Asian women of childbearing age are more commonly highly replicative carriers than are their African counterparts [6,27,29]. As a result, perinatal mother-to-child transmission of HBV is more common in Asia than it is in Africa. Infection with the virus among African children usually occurs after 6 months of age and occurs by horizontal spread [29].

Early approaches to HBV vaccination entailed administration of plasma-derived preparations. These virion-free HBsAg extracts from HBV-infected individuals achieved good protective immunity in chimpanzees [9] and later in humans [8,10].

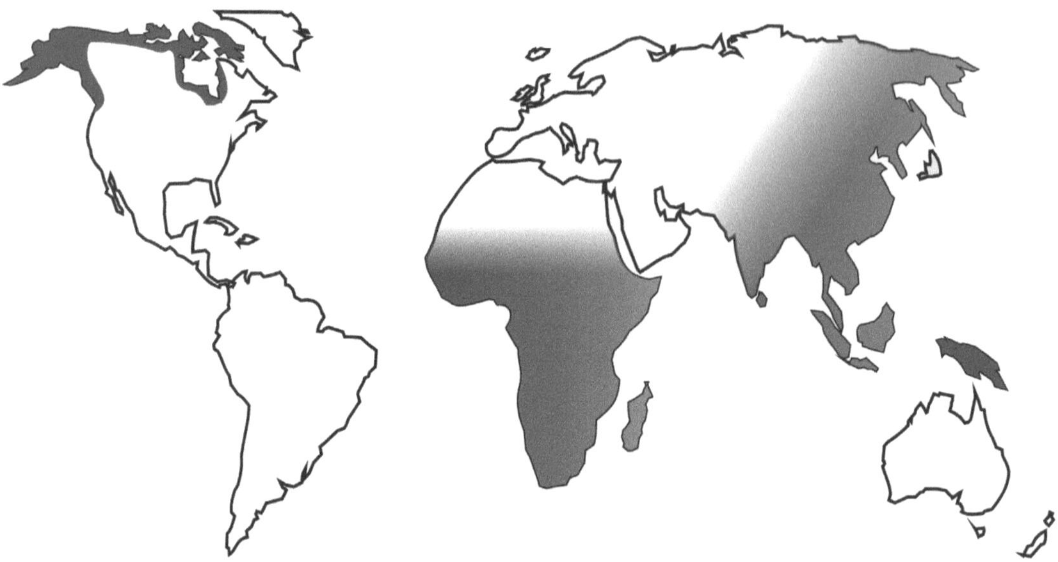

FIGURE 6.2 **Geographical distribution of areas where chronic HBV infection is highly endemic.**
Darkly shaded regions of the global map indicate where chronic HBV infection occurs with an approximate prevalence of more than 8% of the general population. Most HBV carriers are located in sub-Saharan Africa, the western Pacific islands, and East and Southeast Asia. Although there is a high prevalence in northern North America, the numbers of chronic carriers in this region are low.

Nevertheless, worries about the use of blood-derived products spurred investigation of alternative vaccination approaches, and generation of recombinant HBsAg was a logical way of addressing these concerns. The lack of appropriate eukaryotic posttranslational modification of HBsAg expressed in bacteria meant that these organisms were not suitable for vaccine production. Although cultured mammalian cells may be used to make HBsAg, which is similar to that produced during a natural HBV infection, yields from cultured cells are inadequate for use in large-scale vaccination programs. These factors led to the development of yeast strains that could generate large amounts of immunogenic HBsAg [30,31]. Although the yeast-derived antigen is not identical to the naturally occurring HBsAg, excellent anti-HBV immunity is achievable, and the protein is now the standard vaccine that is widely used in immunization programs throughout the world. In addition to the use of recombinant proteins, gene-based vaccination is being developed for prophylactic and therapeutic use against HBV (Chapter 11).

Although vaccination is part of the Expanded Program of Immunization of many countries, some endemic areas still do not have compulsory HBV vaccination programs. As a result, the complications of HBV infection are likely to continue to be a significant global public health problem for many years. Availability of effective ways of eliminating the virus remains a priority, and detailed understanding of the molecular biology of HBV is fundamental to developing new approaches to HBV therapy.

6.1.2 HBV Molecular Biology

HBV is the prototype member of the *Hepadnavirus* family. There are eight genotypes (A–H) of the virus, which have intergroup genome sequence divergence of 8% and particular geographical distributions [32]. The intact virion, or Dane particle, has an outer endoplasmic reticulum-derived lipid envelope with embedded small, middle, and large surface proteins (reviewed in ref. [33]). The viral capsid, containing the HBV partly double-stranded relaxed circular DNA (rcDNA) genome, is located within the membranous envelope. rcDNA comprises a minus strand of ~3.2 kb that encompasses the complete HBV genome and a plus strand of variable shorter length.

Interestingly, the mechanism by which HBV gains access to hepatocytes has been the subject of many years of frustrating investigation. Although the pre-S1 domain of HBV has long been known to be essential for the entry of HBV into hepatocytes [34], the cellular cognate has remained elusive. Only recently has the NTCP been identified as the receptor that mediates HBV infection of liver cells [22]. The elegant strategy to identify the receptor entailed modification of pre-S1 peptides with photoactivatable residues and near-zero distance photo-crosslinking followed by HBV infection of primary hepatocytes of the Asian tree shrew (*Tupaia belangeri*). Tandem affinity purification combined with mass spectrometry and deep sequencing led to identifying the NTCP as an HBV receptor. This bile transporting protein is expressed in the liver, but it is absent in undifferentiated hepatocyte lines and disappears from primary hepatocytes after a few days in culture. Importantly, expression of NTCP in cells that do not normally produce the protein rendered them susceptible to HBV infection.

After infection of hepatocytes, the virion envelope is removed and the core particle is transported to the nucleus. rcDNA is released from the capsid and then "repaired" to form covalently closed circular DNA (cccDNA) (Figure 6.3). cccDNA contains promoters with binding sites for several hepatocyte transcription factors (reviewed in ref. [35]). The replication intermediate serves as template for synthesis of pregenomic RNA (pgRNA) and mRNA encoding the HBV proteins (Figure 6.4). cccDNA exists as a stable minichromosome and its persistence within hepatocyte nuclei has been a challenge to eradicating HBV from chronic carriers. Therefore, new gene therapy technologies that use sequence-specific DNA binding proteins to disable cccDNA have been of considerable interest [36]. Likewise, because viral RNAs are essential for HBV replication, the therapeutic potential of inactivating these sequences has been vigorously explored.

There are four promoters and two enhancer elements within the cccDNA: basic core promoter and enhancer II, enhancer I, preS1, pre-S2, and X promoters (reviewed in ref. [35]; Figure 6.4). These transcription regulatory elements are responsible for directing cellular RNA polymerase II (Pol II) to initiate synthesis of *pregenomic/preC*, *preS2*, *preS1*, and *X* mRNAs. There is one transcription

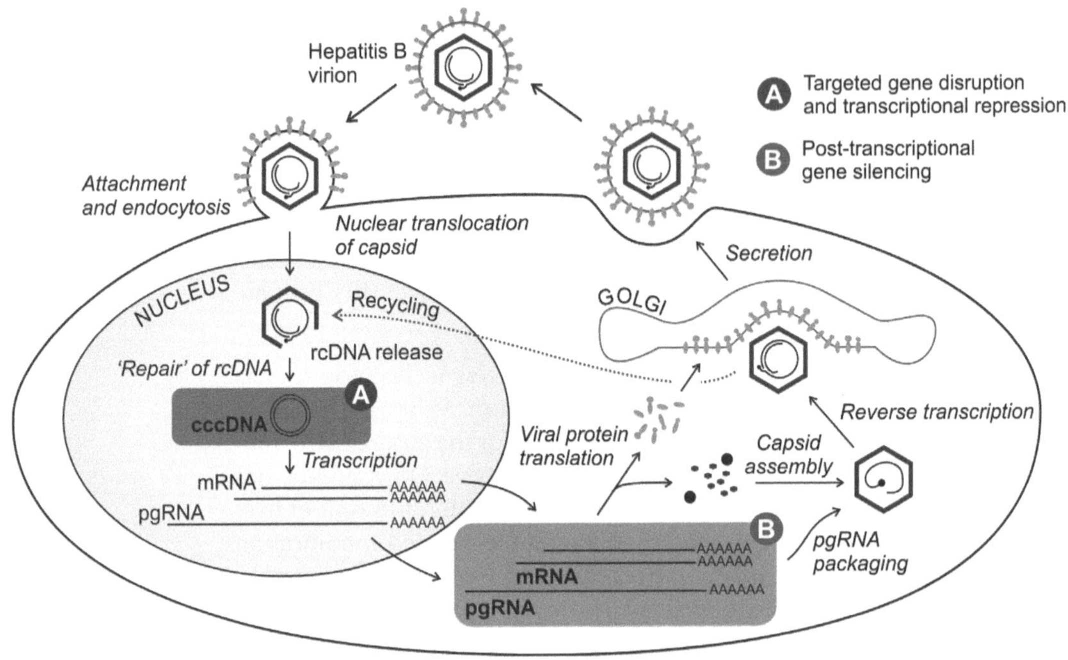

FIGURE 6.3 HBV replication cycle.

Stages involved in the replication of HBV are schematically illustrated. After attachment and endocytosis, nuclear release then repair of rcDNA leads to the formation of the cccDNA viral transcription template. The shaded rectangular shapes depicted in the nucleus and cytoplasm indicate the potential viral targets for sequence-specific nucleases (A) and RNAi activators (B). Together with viral polymerase, pgRNA is packaged into capsid particles that comprise core protein. Reverse transcription of pgRNA to form rcDNA occurs within the capsid, and recycling of newly synthesized rcDNA provides an additional source of cccDNA. Nascent virions are secreted via the endoplasmic reticulum and Golgi apparatus. The HBeAg, which is also secreted via the Golgi, and the X antigen are not depicted here. cccDNA, covalently closed circular DNA; rcDNA relaxed circular DNA; pgRNA, pregenomic RNA; ER, endoplasmic reticulum.

termination signal; therefore, all of the viral mRNAs have common 3′ sequences. Splicing of the HBV transcripts does occur [8], but the mechanisms of its regulation and functional significance are incompletely defined. Reverse transcription of spliced pgRNA gives rise to defective HBV DNA (dDNA). Although this dDNA has been identified in HBV carriers, it is not associated with particular disease states and its abundance is related to integrity of the viral polymerase (Pol) [39]. Evidence also indicates that HBV evolved mechanisms of suppressing mRNA splicing at cryptic sites to facilitate viral replication [40].

In a remarkably compact arrangement of its genome, the four HBV open reading frames (ORFs) overlap with each other on the cccDNA (Figure 6.3) [33]. This economical use of its genome is possible by making use of different codon triplets encoded by the cccDNA. The largest ORF generates the multifunctional Pol protein. It is required for pgRNA encapsidation [41], and it possesses reverse transcriptase (RT), primase, DNA-dependent DNA polymerase,

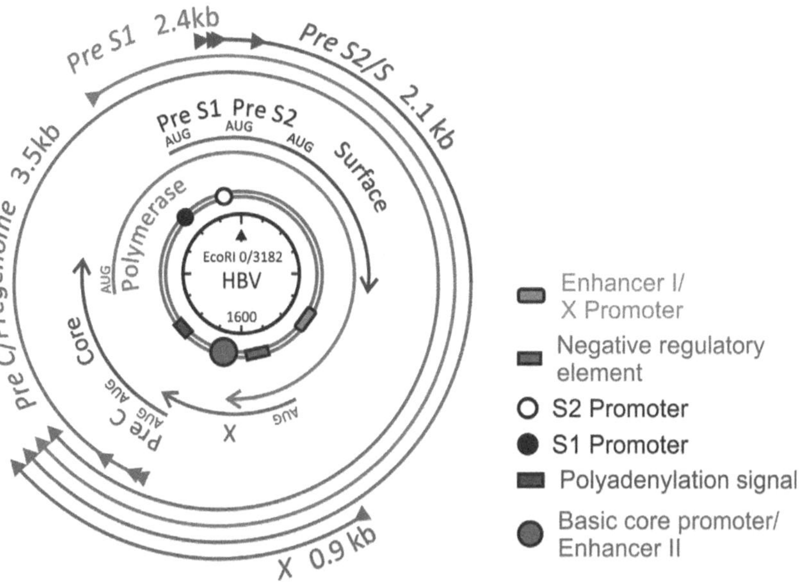

FIGURE 6.4 Organization of HBV genome.
HBV cccDNA, indicated schematically at the center of the figure, is derived from virion rcDNA. Nucleotide coordinates are calculated from the unique *Eco*RI site that occurs in the HBV DNA (top center). Approximate location of promoters, enhancers, and *cis* regulatory elements are indicated as circular and rectangular symbols on the cccDNA template. The four ORFs encoding Core, Pol, Surface, and X proteins are shown as arrows immediately surrounding the genome. The four major viral transcripts, which are initiated at different promoters and terminate at a single polyadenylation site, are indicated as outermost arrows.

and RNAse H activities [19,42]. Pol comprises 833 amino acids, and its ORF overlaps in its entirety with sequences encoding the Surface proteins. There are three in-frame translation initiation codons of the Surface region. Translation from the most 5′ preS1 AUG gives rise to the large Surface protein. When initiated from the preS2 start codon, the middle Surface protein is translated and the third AUG of the ORF gives rise to the small Surface protein. Therefore, all three Surface proteins have similar C-terminal ends and differ according to the presence or absence of additional N-terminal sequences. The *core* ORF has two initiation codons. A signal located at the N-terminal region directs the synthesized protein to the endoplasmic reticulum where it is processed then secreted as the HBeAg. The presence of HBeAg in clinical settings is conveniently used to assess replication rate of the virus. The HBeAg-positive carrier state, such as often occurs in Asia, has a high replication rate whereas the HBeAg-negative carrier state, such as is typical in sub-Saharan Africa, usually has a low rate of HBV replication [6,27]. Translation that is initiated from the core AUG gives rise to the HBV core protein comprising 183 amino acids [45]. Usually,

120 core protein dimers assemble to form an HBV capsid with T4 icosahedral triangulation number. Smaller sized capsids with T3 structure and made up of 90 core protein dimers occur less frequently. Structural assembly and immunodominant loop domains are located within the N-terminal 140 amino acids. The C-terminal region comprises a linker, arginine-rich nucleic acid binding domains, and a putative nuclear localization sequence.

Formation of virions commences in the cytoplasm with incorporation of Pol and bound HBV pregenomic RNA into capsid particles (Figure 6.3) [19,42–44]. The highly structured stem-loop epsilon packaging signal is required for this step and is important for the priming of reverse transcription of pregenomic RNA. Specificity of the incorporation of the HBV RNA is likely to involve a chaperone protein, which may be a serine arginine protein kinase [46]. The unusual primase function of Pol involves covalent attachment of the first deoxynucleotide to a tyrosine residue of the enzyme. RT activity of Pol within the capsid forms the HBV rcDNA. This complicated process involves two template strand switches, RNase H degradation of pregenomic sequences, and RNA- and DNA-dependent polymerization [19,33]. Only mature capsids containing a complete minus strand and a partial plus DNA strand are enveloped [47]. HBV budding occurs via the endoplasmic reticulum and involves two endocyte-specific proteins: gamma2-adaptin and Nedd4 ubiquitin ligase [48,49]. In addition, HBV release from the infected hepatocyte involves participation of cellular machinery concerned with generation of multivesicular bodies before the intact virions are secreted [50]. Some rcDNA-containing capsids may be retained within hepatocytes and be transported back to the nucleus where they serve as a further source of cccDNA (Figure 6.4).

6.1.3 Models Used to Test Candidate HBV Therapies

Availability of models that accurately simulate all stages of HBV infection and replication is crucially important for advancement of more effective therapies. Gaining a detailed understanding of the molecular biology of HBV replication has provided useful insights required for generating surrogate models, but convenient methods of reproducing all stages of HBV infection are still not readily available (summarized in Table 6.1). A major factor that has impeded progress has been the slowness of clarifying the mechanism of HBV entry into hepatocytes. With description of the NTCP as an HBV receptor and demonstration that liver-derived cells transduced with sequences encoding this protein are susceptible to HBV infection [22], improved models may soon become more accessible. In addition, targeted inhibition of NTCP provides a new approach to identifying anti-HBV agents [51]. Cultured cell types that are infectable with hepatitis B virions have previously been scarce. Primary human hepatocytes (reviewed in ref. [52]) and the HepaRG cell line [53,54] may be infected with HBV, but they have not been widely used in the development of gene therapy

Table 6.1 Cell Culture and Animal Models of HBV Replication

	Advantages	Disadvantages
Cell Culture Models		
Transient transfection of cultured cells	Very convenient and rapid. cccDNA formation may be simulated.	Replication stages of HBV life cycle are simulated, but infection of cultured hepatocytes does not occur. Amount of HBV DNA that is transfected may not reflect typical dose of viral DNA introduced into a cell during natural HBV infection. Effects of candidate therapeutics on transfected plasmid DNA needs to be discriminated from inhibitory effects on cccDNA.
Stable transfection of cultured cells	Continuous production of viral particles that mimics production of HBV during chronic infection. HBV replication may be induced, such as is the case with HepAD38 cells. cccDNA formation occurs in human lines (e.g., HepG2.2.15 cells)	Infection of cells does not occur. Replication may vastly exceed that occurring during natural HBV infection.
HepaRG cell infection	Early stages of infection may be reproduced using virions to infect cells in culture. cccDNA formation occurs.	Carefully controlled procedures for cell differentiation are required to enable infection.
Infection of NTCP-expressing cells	Early stages of infection may be reproduced using virions to infect cells in culture. cccDNA formation occurs.	Although promising, the model has not yet been extensively studied.
Murine Models		
Hepatocyte transfection using hydrodynamic tail vein injection	Very convenient and rapid.	cccDNA formation does not occur. However, generation of cccDNA-like DNA has recently been described when using plasmids containing LoxP sites in combination with Cre recombinase [73]. Infection of cells does not occur. Replication rate may not accurately reflect replication in infected hepatocytes. The injection procedure is hepatotoxic itself.
HBV transgenic mice	Convenient simulation of continuous HBV replication in vivo.	cccDNA formation does not occur. Infection of cells does not occur.
Xenografted uPA SCID mice	All stages of HBV replication, including infection, are reproduced in these animals.	Specialized facilities are required to house and treat the animals. uPA SCID mice are immunocompromised, which may confound interpretation of efficacy of drug candidates.

Continued...

Table 6.1 Cell Culture and Animal Models of HBV Replication *Continued*

	Advantages	Disadvantages
Other Animal Models		
DHBV	Convenient animal model that has similarity to HBV infection.	DHBV has significant differences to HBV. Extrapolation from DHBV model to human condition may not be accurate.
WHV	Convenient animal model that has similarity to HBV infection. As with HBV, hepatocarcinogenesis occurs commonly following infection.	Specialized and costly housing is required for maintaining the animals. WHV has significant differences from HBV. The mechanism of hepatocarcinogenesis differs between humans and woodchucks. Extrapolation from the woodchuck model to human condition may not be accurate.
Tree shrew (*T. belangeri*)	Animals are infectable with HBV.	Specialized housing is required for maintaining the animals.
Chimpanzee	Chimpanzees may be infected with HBV, and these animals simulate the human infection.	Ethical considerations limit use of chimpanzees. Specialized and costly housing facilities are required.

for HBV. Plasmids that encode reporter constructs with fused HBV sequences or replication-competent greater-than-genome-length HBV mRNA (Figure 6.5) have successfully been used to evaluate the efficacy of gene therapy strategies in transient cell transfection experiments [55,56]. Cells that have replication-competent HBV DNA stably integrated, such as the HepG2.2.15 line [57], have also been used to assess antivirals. Although the complete natural HBV life cycle is not simulated in these models, they provide convenient means of measuring the value of drug candidates. HepAD38 cells also stably produce HBV and have the useful advantage of inducible replication after withdrawal of tetracycline from the culture medium [58]. The recent rapid advances in the generation of induced pluripotent stem cells (iPSCs) and their differentiation into hepatocytes and liver organoids [59,60] may soon find application in the simulation of natural HBV infection. Moreover, conferring resistance to hepatocytes by genetic modification of induced pluripotent stem cell-derived liver cells and patient infusion may have therapeutic utility.

As with cell culture models of HBV, there is a dearth of small animals that are infectable with HBV. Because the mechanism by which HBV infects hepatocytes in vivo is only now being established, genetic manipulation of animals to make them susceptible to HBV infection has been slow. Research on hepatitis C virus (HCV) drug development has been faced with similar challenges,

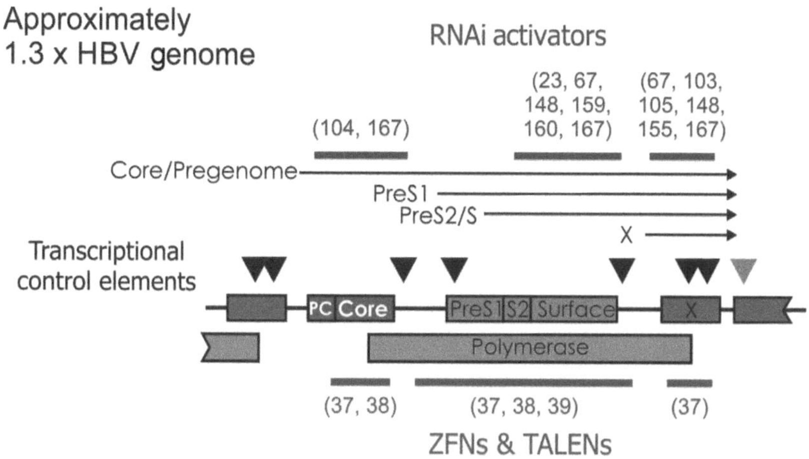

FIGURE 6.5 Schematic of typical greater-than-genome length replication-competent HBV sequences, with sites successfully targeted using RNAi activators, TALENs, and ZFPs indicated. The generic example of a 1.3 × HBV genome, comprising ~4.2 kb, shows sites targeted by RNAi activators, ZFNs, and TALENs. The overlapping *surface*, *core*, *polymerase*, and *HBx* viral ORFs are indicated by labeled rectangles. The four arrows above the viral DNA represent the HBV transcripts that are initiated from different viral promoter elements but have a common 3′ end (cf. Figure 6.4). Essential *cis* elements controlling transcription are indicated by black and gray arrowheads. Regions of the viral genome that have commonly and successfully been targeted by anti-HBV RNAi activators and sequence-specific nucleases are indicated by bold lines above and below the HBV DNA. Some of the relevant numbered references cited in the text that describe each of the anti-HBV sequences are given in parentheses.

but recent development of HCV infection-susceptible mice will go some way to addressing the problem [61]. In addition to their use for measuring efficacy of antivirals, animal models are equally important for gauging toxicity of drug candidates.

A convenient method of simulating HBV replication in vivo makes use of hydrodynamic injection [62]. The procedure depends on intrahepatic hydrostatic backpressure to transfect up to 40% of hepatocytes in vivo. This is achieved after rapid tail vein injection of a large volume, typically 10% of the mouse body weight, of a DNA-containing saline solution. When used to deliver plasmids encoding HBV pgRNA sequences, transient viral replication occurs in the mice [63]. Effects of gene therapies on markers of HBV proliferation may then be conveniently assessed. A drawback is that the procedure itself causes a transient hepatitis, which may complicate analysis of the effectiveness of antiviral agents. Nevertheless the convenience of the method has led to its widespread use for short-term in vivo studies. HBV transgenic mice also have been used in studies aimed at developing new antivirals [64]. A greater-than-genome-length HBV sequence, which is integrated into the genomes of these animals, sustains constitutive viral replication. Although hepatocyte infection does not occur, long-term replication of HBV in the transgenic animals approximates some attributes of the chronic HBV infection of humans [67,68].

An important limitation of transgenic and hydrodynamic injection models is that HBV cccDNA is not formed in mouse hepatocytes [69]. Therefore, determining elimination of cccDNA in these animals, an important requirement of HBV drug development, is not possible. Use of severe combined immunodeficient (SCID) mice with xenografted human hepatocytes allows for this [70]. In these mice, the hepatotoxic urokinase plasminogen activator gene selectively eliminates murine hepatocytes [71]. Engraftment of the SCID animals with human hepatocytes leads to the formation of chimeric mice with humanized livers. Infection with HBV then imitates all stages of the natural viral replication [72]. Although the model has the significant advantage of producing the cccDNA intermediate, procedures required to generate the chimeric animals are challenging. As a result, humanized SCID mice have not been widely used to assess therapeutic potential of HBV gene therapies. Another shortcoming of the model is that the immunocompromised state of the xenografted SCID mice eliminates the role of host immune responses to HBV during a therapeutic response.

A recent innovative simulation of cccDNA formation in mice entailed incorporation of LoxP sites into a chimeric intron of HBV sequences within precursor plasmid DNA [73]. Co-administration of Cre recombinase-expressing sequences during hydrodynamic injection of mice resulted in excision of the plasmid backbone and formation of recombinant circular DNA of 3.3 kb. The LoxP sequences remaining in the cccDNA mimic did not disrupt HBV replication because they were located within an engineered intron and were spliced from viral mRNA after transcription. Assessment revealed that the recombinant cccDNA with concomitant viral replication persisted in immunocompetent mice for up to 9 weeks.

Other hepadnaviruses, such as duck hepatitis B virus (DHBV) [74,75] and woodchuck hepatitis virus (WHV) [76], simulate the human condition of HBV infection and may be useful to assess antiviral efficacy. However, nonhuman hepadnaviruses are genetically and immunologically divergent from HBV, and reliable inferences to the human condition may not be possible. Asian tree shrews (*T. belangeri*) may be infected with HBV and provide a useful model [77]. However, these animals have not yet been widely used for the analysis of efficacy of HBV gene therapy candidates. Chimpanzees are infectable by HBV [78–80]; therefore, they could be used to test antiviral drug efficacy. Limitations pertaining to ethical issues, cost, and availability of suitable housing restrict the use of chimpanzees for preclinical assessment of HBV gene therapy.

6.1.4 Current and New HBV Therapies

Therapy for HBV infection should ideally achieve sustained suppression of viral replication and reversal of liver inflammation to prevent complications of cirrhosis and liver cancer [20,81]. This is assessed in a clinical setting by measuring circulating viral antigens (HBsAg and HBeAg), HBV DNA, serum alanine transaminase (ALT), and histological assessment of liver biopsies. Currently available therapies are nucleoside/nucleotide inhibitors of HBV RT and interferon (IFN-α). IFN preparations, which have immunostimulatory and antiviral actions, are administered by injection for a finite period. RT inhibitors are generally well tolerated but may need to be given to patients for many years to achieve continued suppression of HBV replication [82]. However, without eliminating cccDNA, there is a risk for rebound after treatment withdrawal. Therefore, improved HBV treatments should ideally have a sustained effect that is associated with elimination of cccDNA.

IFN-α was the first antiviral shown to have efficacy against HBV [83,84] and was approved for use as an anti-HBV agent in 1992. Unmodified and polyethylene glycol (PEG)-conjugated IFNs are currently licensed for therapy of HBV [20,85]. The PEG formulation is preferred because it requires less frequent weekly administration and is associated with a better antiviral response. Although the antiviral efficacy of IFN-α is variable and may be influenced by many factors such as the HBV genotype, treatment with IFN-α has been associated with the best curative efficacy of all anti-HBV agents [86]. However, IFN-α treatment is not commonly used in HBV regimens [87]. Administration of the drug is often complicated by debilitating side effects, and its administration is contraindicated in patients with decompensated cirrhosis. Expense, injection rather than oral route of administration, and lack of availability in resource-poor settings are further disadvantages of IFN-α therapy. Recently, IFN-α has been shown to inhibit miR-122 function through a mechanism that involves sequestration by mRNA encoded by *NT5C3*, an IFN-stimulated gene [88]. Unlike with HCV, HBV replication is inhibited by miR-122. Thus, the effect of IFN-α on miR-122 may negatively affect the antiviral action against HBV.

Nucleoside and nucleotide analogs are widely used for HBV therapy (reviewed in refs [20,81,85,89–91]). After intracellular phosphorylation, they mimic naturally occurring nucleotides and competitively inhibit HBV Pol during reverse transcription of viral pregenomic RNA. The resulting premature termination of DNA synthesis and inhibition of priming of HBV DNA may cause a profound decrease in HBV replication. There are five currently licensed anti-HBV nucleoside/nucleotide analogs: lamivudine, adefovir, entecovir, tenofovir, and telbivudine. In some cases, emergence of viral resistance is a problem [90]. Resistance to lamivudine, which is widely used for HBV treatment, is well documented and occurs commonly [92,93]. However, recently developed nucleoside and nucleotide antivirals, such as entecavir and tenofovir, have a high genetic barrier to resistance and are less prone to selection of HBV escape mutants. Although these drugs have excellent short-term safety profiles, their possible toxicity after use for several years is yet to be excluded.

New distinct classes of small-molecule drug candidates are emerging as promising anti-HBV agents. Examples include the nucleocapsid destabilizing heteroaryldihydropyrimidines [94,95] and phenylpropenamides [96]. Results, to date, have shown that the agents are effective against the virus in vivo and in cell culture-based assays. An endoplasmic reticulum glycosidase inhibitor impeded HBV particle assembly and reduced WHV replication in vivo [97].

6.2 GENE THERAPY FOR HBV INFECTION

To date, most approaches to gene therapy for HBV infection have centered on developing methods of disabling gene expression of the virus. In general, procedures have involved nucleic acid hybridization to HBV RNA sequences and construction of proteins that bind and edit viral DNA. Design of the antivirals has been based on simple rules that are easily applied in molecular biology laboratories, which in turn has facilitated research on the topic. Early methods entailed use of antisense sequences and ribozymes. With the discovery of RNAi and demonstration that this powerful and specific gene silencing pathway may be harnessed with exogenous sequences, RNAi-based HBV gene silencing has become more popular. Interesting alternative methods, such as use of RNase P to degrade artificial HBV RNA-containing tRNA intermediates [98], have also been used.

6.2.1 General Considerations for Engineering and Delivery of Artificial HBV-Targeting RNAi Activators

Although algorithms are available to aid in the selection of HBV silencers [99,100], empirical assessment is essential for optimization. As with many antiviral gene silencers, RNAi activators targeting HBV should be potent (effective at low concentrations), silence viral targets specifically without induction of an immune response, have a durable effect, cause minimal if any disruption of the endogenous micro RNA (miR) pathway, prevent emergence of viral escape mutants, and be amenable to dose regulation.

The importance of identifying potent anti-HBV sequences is that unintended disruption of the endogenous regulatory pathways and interaction with cellular mRNA is diminished when therapeutic RNAi activators are used at low concentrations. Two mechanisms may cause silencing of cellular mRNA by exogenous RNAi activators: nonspecific interaction of the guide strand seed region with a cellular target or unintended incorporation of the passenger strand into the RNA-induced silencing complex. Judicious target sequence selection and chemical modifications of synthetic short interfering RNAs (siRNAs) may be used to minimize these effects. Interestingly, it has been reported that expressed gene silencers display lower off-target silencing when compared with synthetic siRNAs [101], but the mechanisms underlying this observation are not clear.

Achieving reliably efficient and safe hepatotropic delivery of potentially therapeutic RNAi activators remains difficult and is an important priority for successful HBV gene therapy. Therefore, carriers of anti-HBV sequences are required to take antiviral sequences to their nuclear or cytoplasmic sites of action within hepatocytes after systemic administration. Ideally, only a single vector administration should need to be given to effect long-term HBV silencing. Endothelial cells within the liver form a fenestrated barrier between the blood and hepatocytes that prevents particles that have a diameter greater than 100 nm from coming into contact with hepatocytes. Thus, to be effective, viral and nonviral vectors (NVVs) should have a uniform size that is smaller than 100 nm (reviewed in ref. [25]). In general, NVVs have been used to deliver synthetic siRNAs and recombinant viral vectors utilized for delivery of antiviral expression cassettes. As synthetic formulations, NVVs are amenable to large-scale preparation, which is useful for clinical application (Chapter 5). Currently available NVVs are inadequate for in vivo hepatotropic delivery of anti-HBV DNA cassettes to the hepatocyte nucleus. However, NVVs are capable of efficient delivery of smaller synthetic siRNAs to their cytoplasmic site of action.

6.3 DEVELOPING RNAi ACTIVATORS AS TREATMENT OF HBV INFECTION

Various expressed and synthetic RNAi activators, delivered to liver cells using viral and NVVs, have been used successfully to counter HBV replication. Expressed sequences include short hairpin RNAs (shRNAs), long hairpin RNAs (lhRNAs), and single and multimeric artificial pri-miRs. Various chemical modifications have been incorporated into synthetic siRNAs to enhance their efficacy. The results from specific investigations performed in cell culture and murine models are summarized below and in Figure 6.5.

6.3.1 Expressed RNAi Activators That Silence HBV

Interestingly, the first demonstration that RNAi activation has antiviral therapeutic potential came from a study that used shRNA expression cassettes to knock down HBV replication in mice [23]. This observation was supported by subsequent studies that also used U6 Pol III promoters to generate HBV-silencing shRNAs. An investigation aimed at identifying suitable target sites in the *X* region reported on the screening of efficacy of a panel of U6 Pol III shRNA expression cassettes [64,66,102]. Highly effective silencing (90–95% knockdown) was reported with some of the expression cassettes. As expected, there was a correlation between processing efficiency and bias in favor of selecting the intended guide strand [64]. Efficacy of shRNA expression cassettes targeting other viral sequences [66,103–105] was confirmed in other investigations (Figure 6.5). Secreted markers of HBV replication were reported to be diminished in all of these studies. Improved efficacy of RNAi activators, when used in conjunction with the licensed HBV drug lamivudine, was shown in one study [106] but has not been a particularly active line of investigation. U6 cassettes that target contiguous *X* sequences with lhRNAs were also tested for antiviral efficacy [107]. These lhRNAs had a duplex stem of ~60 bp and were designed to augment silencing efficacy by generating three virus-targeting siRNAs. Although good inhibition of HBV replication was achieved, the effect was mostly as a result of silencing by siRNAs generated from the stem base of the lhRNA. Dicer processing of the lhRNAs was not equal across the span of the double-stranded region and diminished toward the loop side part of the lhRNA. As an alternative to the powerful U6 transcription regulator, the H1 and $tRNA^{Lys3}$ Pol III promoters were used to transcribe *X*-targeting anti-HBV shRNAs [108]. Successful inhibition of viral replication in cultured cells and in vivo was achieved.

Identifying factors that influence the efficacy of expressed HBV gene silencers has also been the subject of investigation. Target accessibility and thermodynamic stability profiles of the duplex region of miR mimics have been reported to be important [109]. However, a separate study reported that the HBV epsilon packaging signal was a good RNAi target sequence [110]. This highly structured sequence with duplex stem and loop regions is bound by cellular proteins and viral Pol [43], but it remains susceptible to RNAi-based silencing [110]. Although *surface*, *precore*/*core*, and *X* sequences have been targeted most (Figure 6.5), results suggest that no particular sequence of the viral genome is especially sensitive to silencing by expression cassettes. Rather, it is the sequence-specific properties of the RNAi activators that influence their effectiveness.

The significant finding that overproduction of anti-HBV shRNAs from U6 Pol III promoters is potentially severely toxic was a caveat to researchers working in the field of therapeutic RNAi [111]. Since this observation, a priority has been to express RNAi-activating sequences in a regulated manner and at the lowest possible dose to be effective. Studies have attempted to simulate natural miR biogenesis by producing artificial Pol II pri-miR anti-HBV sequences [109,112–114]. Diminished disruption of the endogenous RNAi pathway, improved silencing potency, and compatibility with liver-specific Pol II promoters have been demonstrated. Natural occurrence of pri-miR sequences as polycistrons has also been exploited to generate multimeric anti-HBV cassettes [115]. Simultaneous targeting of viral sequences was developed to improve antiviral efficacy and limit emergence of viral escape mutants. With the artificial pri-miR-31-derived anti-HBV trimers, position- and sequence-specific effects influenced antiviral efficacy and processing of individual antiviral sequences [115]. However, function of each monomer could be restored by changing the order of the monomers in the cassette or by substituting the pri-miR31 backbone with a pri-miR30a scaffold. This useful modular feature should be adaptable to other gene silencing applications that require concurrent targeting of multiple sites. Although the exact number of simultaneously acting HBV-silencing RNAi activators required to prevent escape is not established, it is likely that a trimeric cassette will be adequate. For the considerably larger and more flexible HIV-1 genome, four silencing components of an artificial policistronic pri-miR cassette were predicted to be sufficient to prevent emergence of viral escape [116,117].

6.3.1.1 Delivering Anti-HBV RNAi-Activating Expression Cassettes

Adeno-associated virus (AAV) and adenovirus (Ad) vectors have been used successfully to deliver anti-HBV expression cassettes to the liver in vivo (reviewed in refs [25,118]). Exploiting the natural hepatotropism of AAV serotype 8 (AAV-8) was used to develop liver-specific vectors [119], which have been used to deliver HBV gene silencing expression cassettes [120,121]. Silencing of HBV replication in transgenic mice was sustained over a period of 22 weeks. Thereafter, tapered inhibition of HBV replication correlated with animals developing a vector-neutralizing immune response. A second administration of vectors containing serotype 9 capsid effected silencing of HBV replication. A recent study confirmed that anti-HBV AAVs were also capable of effectively countering pathology arising in the livers of transgenic mice [122].

HBV infection itself has recently been shown to influence the efficiency of hepatocyte infection by AAVs [123]. This was observed after administering AAVs to cells in culture and after injection of the vectors into HBV transgenic mice. However, the entry of AAVs into HBV-positive and HBV-negative cells was indistinguishable, and it appeared that efficiency of delivery of the vector DNA to the nucleus was the important difference. More detailed analysis showed that HBV X protein (HBx) function and activity of the phosphatidylinositol-3-kinase pathway were required for augmented transgene expression. This interesting observation suggests that AAV-mediated gene transduction could be exploited for enhanced HBV gene therapy.

The highly hepatotropic properties of Ads after systemic administration are useful for application to RNAi-based HBV therapy. Intravenous administration of first-generation Ads that express anti-HBV shRNAs from Pol III promoters inhibited viral replication in HBV transgenic mice [64,66]. PEGylation of these first-generation vectors attenuated problematic vector immunostimulation and improved HBV knockdown after repeat administration of the Ads [65]. Another study aimed to avoid host immunostimulation by using helper-dependent (HD) Ads [124]. Modest inhibition of HBV replication in transgenic mice was observed, but this is likely to be a result of low silencing efficacy of the transduced expression cassette. A more recent study has shown that HD Ads containing anti-HBx-targeting U6 shRNA cassettes were indeed capable of HBV replication in vivo [125]. Mono- and trimeric artificial anti-HBV expressed from a Pol II promoter and delivered by HD Ads also inhibited HBV replication in transgenic mice [126]. Polymer modification of HD Ad vectors and preadministration of immunosuppressive glucocorticoids may further attenuate immunostimulation to make these vectors safer for therapeutic application.

Recombinant lentiviral vectors are well characterized and have been particularly popular for gene therapy of inherited diseases and HIV-1 infection [127]. The vectors have also been used to disable HBV replication [128]. Administration to neonatal HBV transgenic mice achieved sustained inhibition of viral replication in the animals. Although the stable expression of gene silencing sequences from integrated proviral sequences is useful, modest efficiency of hepatocyte transduction after systemic administration to adult animals has limited their development for HBV therapy. However, an application for lentiviral vectors may be found with ex vivo delivery. Such an approach would entail transduction of hepatocytes in culture followed by their hepatic reinfusion into patients. Ex vivo engraftment of hepatocytes has been demonstrated in nonhuman primates [129,130] and in therapy of hyperbilirubinemic Gunn rats [131,132]. The impressive recent progress that has been made in propagating hepatocyte-like cells from iPSCs [133,134] may make an ex vivo approach to RNAi-based HBV therapy feasible in the near future.

In one of the first clinical trials reporting on the utility of RNAi against HBV, a formulation termed NUC B1000 was evaluated for its anti-HBV activity in individuals with chronic HBV infection [135,136]. A plasmid encoding four different HBV-targeting shRNAs was complexed to cholesteryl spermidine nanoparticles and then administered to patients. After NUC B1000 administration (5 mg of a single DNA dose), patients displayed side effects such as pharyngitis, chills, muscle pain, and fever after ~4–7 h, which were attenuated after administration of an antipyretic. Induction in interleukin (IL)-8 was observed in one participant and increased serum concentrations of IFN-γ and IL-10 were detected in two patients after NUC B1000 administration. Importantly, concentrations of HBsAg and HBV DNA were unchanged and suggested that an increase in the dose may be required to diminish these markers of viral replication. More recently, a candidate treatment regimen for HBV, based on the use of a synthetic RNAi activator, has reached phase II of clinical trials (see below, section 6.3.2).

6.3.2 Design and Delivery of Synthetic siRNAs against HBV

Several early studies reported successful inhibition of HBV replication with single [137–142] or multiple [143,144] HBV-targeting siRNAs. As with expressed RNAi activators, sites encompassing most regions of HBV have shown susceptibility to RNAi-based knockdown with synthetic RNAi activators (Figure 6.5).

Various chemical modifications have been made to anti-HBV synthetic sequences to improve siRNA stability, diminish immunostimulation and non-specific translational suppression. In one of the first studies using chemically modified anti-HBV siRNAs, a mixture of alterations to natural nucleotides was incorporated into *surface* and *X*-targeting sequences [145]. These included phosphorothioate linkages and inverted abasic residues at the 3′ and 5′ ends of the sense strand. Some residues also had 2′-F, 2′O-methyl and deoxyribose modifications. Interestingly, when administered as naked molecules, these anti-HBV siRNAs were capable of inhibiting HBV replication in vivo. However, the high dose that was necessary to cause this effect was impractical for therapy. Therefore, use was made of stable nucleic acid lipid particle (SNALP) formulations to deliver the gene silencers [146]. These lipoplexes assemble into uniform small-sized particles that are well suited to the delivery of siRNAs to the liver (Chapter 5). Indeed, SNALPs and their derivatives have also been used successfully for hepatotropic delivery of siRNAs targeting Ebola virus [147] (Chapter 10). When used to deliver the chemically modified HBV inhibitors, markers of HBV replication were suppressed in mice over a 6-week period [146]. The advantages of chemical modification for immunoattenuation of siRNAs were also demonstrated in this study.

Other chemical modifications of anti-HBV siRNAs entailed the use of altritol six-carbon sugars in place of the naturally occurring five-carbon ribose in *X*-targeting silencers [148–150]. The additional hydroxyl group at the 3′ position of the modified siRNAs is directed to the minor groove of the duplex to improve helix stability and target silencing. Compared with natural siRNAs, anti-HBV efficacy and specificity were improved with altritol-containing siRNAs [150]. Other recently published work improved on anti-HBV efficacy of natural siRNAs by using guanidinopropyl modifications [151,152]. These moieties achieve some charge neutralization of siRNAs, which is useful to improve the delivery of nucleic acid-based gene silencers. The modification also attenuates off-target interaction of the seed with mRNA sequences. Anti-HBV efficacy of siRNAs in lipoplex formulations was also characterized in HBV transgenic mice [142]. Complexes included a polyamine-conjugated cholesterol [153] or aminoxy cholesterol lipid to facilitate postcoupling of "stealth" PEG moieties [142]. Passive hepatotropism was achieved with these vectors, and silencing of HBV replication was observed over a period of a few weeks.

Activation of the innate immune response by HBV-targeting siRNAs has been used in an interesting approach to augmenting antiviral efficacy [154–157]. In one of the first studies reporting on this effect, 5′ triphosphates were incorporated into siRNAs targeting the *HBx* sequence [154]. In cultured cells, these 5′ triphosphorylated RNAi activators, termed 3p-HBx-siRNAs, achieved viral gene silencing that was associated with induction of a strong type I IFN response. Using a murine hydrodynamic injection model, stronger HBV silencing was also observed in mice receiving 3p-HBx-siRNAs in lipoplex formulations. These results were subsequently corroborated in a similar study using 3p-siRNAs that targeted the highly conserved regions of the *surface*, *polymerase*, and *X* ORFs [155,156]. The bifunctional 3p-siRNAs were specific to their viral targets and strongly induced *retinoic acid-induced gene I* with resultant activation of an IFN response. 3p-siRNAs also exerted more effective and durable silencing of viral replication in cultured cells and in HBV transgenic mice. The combination of effects of these siRNAs—gene silencing with reversal of HBV-induced immune tolerance—is significant. If found to have no toxicity, then the use of 3p-siRNAs for treating HBV infection will be of considerable interest.

Using an alternative approach to generate siRNA-like molecules, hydrolysis of in vitro-generated long duplex RNA was performed [158,159]. Recombinant Dicer or *Escherichia coli* RNase III was used to form these so-called endoribonuclease-prepared siRNAs (esiRNAs). (Note that these esiRNAs are distinct from the endogenous siRNAs, which are also termed esiRNAs, which have been described in *Drosophila melanogaster* [160] and oocytes of mice [161].) The exogenous esiRNAs were effective against HBV in cell culture and in vivo [162]. A combination of gene silencing sequences may improve efficacy and diminish the possibility of viral escape. However, a pool of esiRNAs or siRNAs [143,144] may have the undesirable consequence of increasing off-target silencing.

In a recent study using synthetic anti-HBV siRNAs, dynamic polyconjugates (DPCs) were used for hepatotropic delivery of the siRNAs in vivo [163] (also see Chapter 6). This DPC complex was targeted to liver cells using *N*-acetylgalactosamine-conjugated melittin-like peptide (NAG-MLP) as a ligand. Cholesterol-conjugated siRNAs could be conveniently administered together with the DPCs without the need for more complicated methods for siRNA incorporation into the complexes [164]. General utility of the vectors was assessed by measuring decreases in circulating coagulation factor VII (F7), expressed from an endogenous hepatic gene, in mice and nonhuman primates (cynomolgus monkeys) [163]. F7 concentrations in the plasma were reduced by ~30–60% after co-injection with 1 mg/kg NAG-MLP and 2 mg/kg cholesterol-siRNA, whereas co-injection with 3 mg/kg NAG-MLP and 2 mg/kg cholesterol-siRNA resulted in ~97–99% knockdown. These effects were observed without evidence of toxicity. In HBV transgenic mice, HBsAg secretion was suppressed by ~80% for 4 weeks after co-injecting the formulation containing an anti-HBV siRNA. Viral particle equivalents were also decreased by ~90% for 4 weeks after single administration. The DPC and cholesterol-siRNA combination, even at high dose, appeared to be safe because serum activities of liver-derived ALT, aspartate transaminase, and alkaline phosphatase were not elevated. Efficiency and safety of co-delivery of NAG-MLP together with two cholesterol-siRNAs have led to the development of the technology to a stage II clinical trial that is being conducted by the Arrowhead Research Corporation (http://www.arrowheadresearch.com/programs-overview). The candidate drug, called ARC-520, is currently at the most advanced stage of any gene therapy for HBV infection.

Using siRNA conjugates to treat HBV infection is also at an advanced stage of development by Alnylam Pharmaceuticals (http://www.alnylam.com/product-pipeline/hepatitis-b-virus-infection/). The formulation comprises a triantennary linker attached to *N*-acetylgalactosamine that is conjugated to a chemically modified siRNA (Chapter 5). Investigations performed in chimpanzees show good efficacy of the candidate liver-targeting drug, but outcomes from the complete study have not yet been published.

6.4 ALTERNATIVE RNA-BASED METHODS OF SILENCING HBV REPLICATION

RNA-based alternatives to RNAi activators have also been used to inhibit HBV replication. Antisense sequences and ribozymes were among the first HBV gene expression inhibitors that were tested [165]. Investigations with antisense sequences that inhibit HBV replication have been ongoing, although reports in the literature of their efficacy have recently been infrequent [166,167]. In addition to targeting HBV sequences, antisense molecules have been used against the abhydrolase domain, a putative HBV host factor [169]. Ribozymes are RNA sequences that act enzymatically without the requirement for proteins. They frequently act in *cis* to mediate self-cleavage [170]. The hammerhead ribozyme is such an endoribonuclease and has been developed to act in *trans* to effect cleavage of diverse target sequences. It comprises a small catalytic core (helix II) and 5′ and 3′ double-stranded regions (helix I and helix III, respectively) (Figure 6.6). Hammerhead ribozymes may be designed to target foreign sequences in *trans*

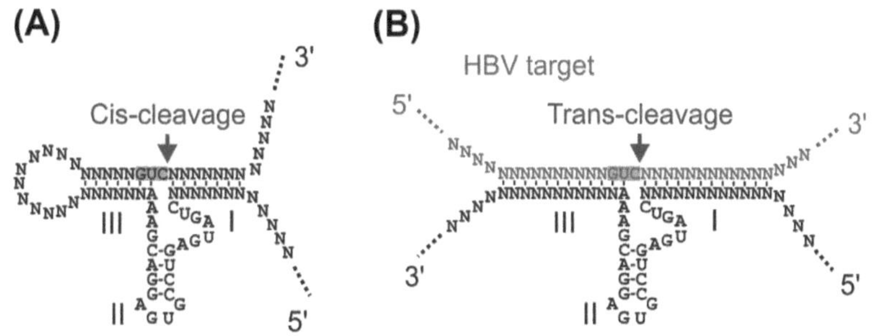

FIGURE 6.6 ***Cis*- and *trans*-cleavage by hammerhead ribozymes.**
(A) Scheme of *cis*-cleavage reaction showing the *cis*-acting ribozyme sequence and its *cis*-recognition target with a 5′ GUC 3′ cleavage triplet (boxed). (B) Ribozyme *trans*-cleavage of HBV RNA. The lengths of hybridizing helices I and III domains are typically longer for the *trans*-cleavage reaction than for *cis*-cleavage.

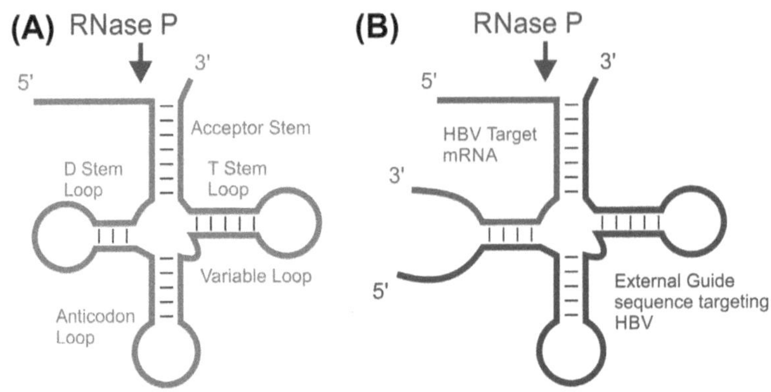

FIGURE 6.7 **Targeting HBV using an EGS to mediate RNase P degradation.**
(A) Natural maturation of tRNA involves endonucleolytic cleavage of the tRNA precursor transcript by the RNase P ribonucleoprotein. (B) EGS may be engineered to generate structures with similarity to natural tRNA RNase P substrates. Cleavage of the target HBV mRNA results in destabilization and inhibition of viral gene expression.

by generating RNA molecules with complementary sequences in the helix I and helix III regions that flank the catalytic core within helix II. Both expressed and synthetic *trans*-acting ribozymes have been used to disable viral genes, and as with siRNAs, synthetic ribozymes targeting HBV are amenable to chemical modification to confer specific biological properties [171]. The only sequence constraint for targeting RNA with *trans*-acting ribozymes is the presence of a cleavable 5′ NUH 3′ motif (where H represents C, U or A) and the most common natural cleavage triplet is 5′ GUC 3′ [172]. A limitation of the artificial ribozymes is their inefficient intracellular catalytic activity, which has resulted in modest inhibition of HBV replication [168,173–177]. Multimeric ribozymes that initially cleave in *cis* to release many *trans*-cleaving units have been used in an attempt to augment ribozyme efficacy [178]. However, improvements using this approach were minimal, and inhibition of HBV replication was inadequate for therapeutic use. The superior efficacy of RNAi-based HBV gene silencing has led to greater efforts being focused on RNAi-mediated gene silencing, and there has been a decline in research on ribozyme-mediated inhibition of HBV gene expression.

Recently, activation of endogenous RNase P ribonucleoprotein enzyme was used to inactivate HBV replication [98] (Figure 6.7). The approach to countering HBV gene expression entailed use of an external guide sequence (EGS) to direct RNase P to cleave HBV RNA. The EGS, when hybridized to HBV target mRNA, generates a secondary RNA structure that mimics that of a tRNA precursor. Therefore, the EGS enabled creation of an artificial RNase P substrate to cause cleavage of the viral RNA. When tested in cell culture and in vivo, good suppression of viral replication was observed when using an attenuated *Salmonella* bacterial strain as vector.

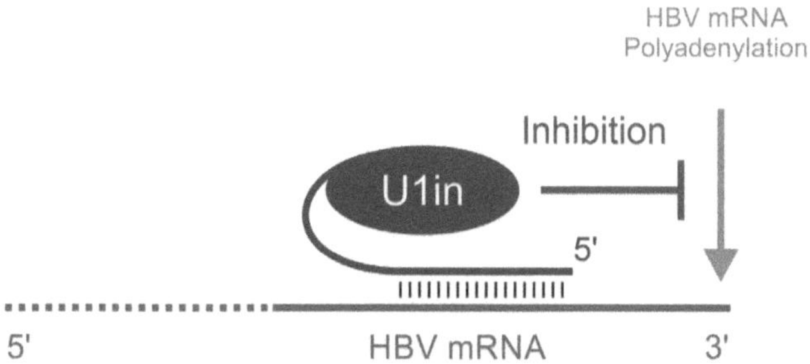

FIGURE 6.8 Targeting HBV using U1in to inhibit HBV mRNA polyadenylation.
An engineered U1in sequence is designed to hybridize to a complementary HBV sequence and mediate inhibition of viral mRNA polyadenylation.

Using another alternative approach, Blazquez and colleagues showed that U1 small nuclear RNA ribonucleoprotein (snRNP)-based inhibition (U1in) could be used effectively to augment RNAi-mediated inhibition of HBV replication [179,180] (Figure 6.8). U1 snRNP plays a role in mRNA splicing and may bind to mRNA to inhibit polyadenylation with a resultant decrease of target gene expression. Exogenous U1in sequences are designed to hybridize to the 3′ ends of target mRNA. Using this approach, impressive inhibition of markers of HBV replication were demonstrated in cultured cells and in vivo [180].

6.5 GENE EDITING FOR THE INACTIVATION OF VIRAL cccDNA

Stability of cccDNA has been the major impediment to successful elimination of HBV. Although currently available therapies are capable of suppressing the virus, withdrawal of nucleoside and nucleotide analogs that act post-transcriptionally is associated with reactivation of replication. Because RNAi-based gene silencing also occurs post-transcriptionally, cccDNA is similarly unlikely to be affected by exogenous RNAi activators. This impression was confirmed in a study that analyzed the effects of anti-HBV RNAi activators on cccDNA in cultured hepatocyte-derived cells [181]. Formation of new cccDNA molecules was inhibited by gene silencing, but concentrations of established cccDNA were unaffected by the HBV-targeting shRNAs. Thus, RNAi-based gene therapy for HBV may suffer from similar limitations to those of existing licensed HBV therapies. Therefore, recent developments in engineering of proteins that are capable of sequence-specific DNA binding have generated enthusiasm [182] (Chapter 3). Specific sequence recognition provides the means for targeted disabling of HBV DNA. Early investigations used zinc finger proteins (ZFPs) to inhibit HBV cccDNA function [183]. More recently, transcription activator-like effector (TALE) derivatives have been used [36,37]. Engineered homing endonucleases (meganucleases) that target HBV have also been investigated.

Their use is the subject of a recent patent submission (WO/2010/136,981 and US.2012/0,171,191A1), but published results on the subject are not yet available. Clustered regularly interspaced short palindromic repeats (CRISPR) with CRISPR-associated (Cas) proteins have interesting potential for therapeutic DNA manipulation [184,185], and their efficacy against HBV has recently been demonstrated [186].

Utility of ZFPs targeting a hepadnavirus was initially investigated by Zimmerman and colleagues [183]. In this study, polydactyl (multifingered) ZFPs were engineered to inhibit DHBV gene expression by binding to 9- or 18-bp sequences within the enhancer region of DHBV cccDNA. Analysis using electrophoretic mobility shift assay and surface plasmon resonance identified two ZFPs that bound target sequences with high affinity. Eukaryotic expression cassettes that produced the selected ZFPs were then engineered for assessment of intracellular efficacy. Transfection of avian hepatoma cells reduced DHBV pgRNA expression by 41.6%. There was a concomitant decrease in the production of DHBV surface protein, core protein, and viral particle equivalents. These ZFPs are likely to have had their effects by competitively inhibiting binding of transcription factors to *cis* elements of the DHBV enhancer within the viral cccDNA. Because ZFP expression was transient, this effect is unlikely to be sufficiently sustained to be therapeutically beneficial. Nevertheless, demonstrating an effect on hepadnaviral cccDNA was significant and provided the basis for investigating utility of functionally enhanced designer DNA binding proteins that target HBV cccDNA.

Coupling nucleases to DNA binding proteins has now been widely used to introduce double-stranded breaks (DSBs) at specific target sites. Repairing of DSBs typically occurs by nonhomologous end joining (NHEJ) or homology-directed repair (reviewed in refs. [187,188]; Chapter 3). Insertions, deletions, and substitutions at sites of DSBs may frequently occur as a result of the error-prone nature of NHEJ. Therefore, by specifically introducing DSBs into cccDNA, designer nucleases may be used to cause disabling mutations of this viral replication intermediate. Because HBV has a highly compact genome arrangement with overlapping ORFs, introduction of mutations is likely to render the cccDNA defective (Figures 6.4 and 6.5).

Zinc finger nucleases (ZFNs) and TALE nucleases (TALENs) are typically generated by fusing a FokI nuclease domain to the C-terminal region of ZFP or TALE sequences that have targeted DNA binding properties [189,190]. To cleave strands of a DNA duplex, interaction of two FokI monomers is required at the intended cleavage site [191,192]. Therefore, ZFNs and TALENs are designed to act in pairs that comprise so-called left and right subunits. Thus, each of the monomers making up the ZFN or TALEN binds to sequences flanking the target in a head-to-head arrangement. A spacer between the DNA cognates is important to accommodate each of the FokI nucleases and facilitate target cleavage. Requirements for binding of both subunits at the target limit nonspecific effects, which may be further diminished by the use of nucleases that are functional only after formation of obligate heterodimers [193].

To assess the antiviral efficacy of ZFNs, Cradick and colleagues engineered sequence-specific nucleases that targeted viral sequences within the *core* ORF and overlapping polyadenylation signal [38] (Figure 6.5). Co-transfection of cells with ZFN-expression plasmids together with an HBV replication-competent plasmid caused mutation at the viral target. Inhibition of HBV replication was verified by a decrease in pgRNA synthesis. Detailed analysis of mutagenic effects demonstrated that ZFN cleavage resulted in frame shift mutations that would lead to core protein truncation and impaired viral replication. Significantly, specific mutation of the all-important cccDNA target was not verified in this study. However, specific mutation of HBV sequences was an important development.

Recently, substituting ZFPs with the plant bacterial TALEs has been used successfully to generate a new class of designer nucleases. The advantage of TALEs is that a single monomer within the TALE DNA binding domain confers individual nucleotide specificity with little effect of neighboring monomers [194,195]. With ZFNs, target specificity and cleavage efficiency are influenced by context-dependent positioning of each of the finger motifs. Efficiency of DNA binding by ZFPs is difficult to predict, and identifying proteins with intended target binding properties entails complicated selection procedures [196]. TALE design strategies are more robust, and concatamerization of monomers to form sequence-specific DNA binding domains can reliably be achieved [197]. DNA binding subunits are typically designed to target sequences of 19 bp, which confers good sequence specificity [198]. TALENs have been used to target multiple endogenous genes [199], and their potential for use in anti-HBV strategies has recently emerged [36,37].

Four HBV-specific TALENs were initially engineered to target multifunctional conserved sites within the *surface* (S TALEN), *core* (C TALEN), and *polymerase* (P1 and P2 TALENs) ORFs [36] (Figure 6.5). Co-transfection of liver-derived Huh7 cells showed that the TALEN targeting the *surface* ORF inhibited HBsAg secretion by 80%. Efficacy was confirmed by more stringent studies performed on the HepG2.2.15 cell line. These cells stably replicate HBV, and the S-TALEN-encoding sequences achieved inhibition of HBsAg secretion by ~60%. A T7E1 endonuclease assay showed that targeted disruption of 31–35% of putative cccDNA occurred after transfection of S-TALEN-expressing plasmids. The C-TALEN caused 12% targeted disruption of cccDNA and, as expected, did not inhibit HBsAg secretion. Efficacy in vivo of the TALENs was confirmed after hydrodynamic tail vein injection of mice with plasmids that included the TALEN-expressing cassettes and HBV replication-competent elements. HBsAg concentrations in serum samples from mice treated with the S-TALEN were 95% lower than those of control animals, whereas intrahepatic HBV c antigen expression was also markedly inhibited by the C-TALEN. The number of circulating viral particle equivalents was significantly decreased for both the groups of animals. The T7E1 endonuclease assay also revealed efficient targeted mutation for both TALENs: 58–68% targeted mutation for the S-TALEN and 62–87% targeted mutation for the C-TALEN. Importantly, mice do not form cccDNA during HBV replication [67]. Therefore, mutations were induced in the HBV replication-competent target plasmid that was injected during the hydrodynamic injection and not in cccDNA.

Cytotoxicity resulting from off-target effects of TALENs is an important consideration [200–203]. To address whether this is potentially problematic for HBV-targeting TALENs, genome-wide analysis was performed to identify homologous target sites within the mouse (*Mus musculus*) and human genomes [36]. Sites that could potentially be cleaved by the HBV-targeting TALENs were not found, which is in agreement with the lack of cytotoxicity that was observed after TALEN-expressing plasmid transfection of cultured liver cells and hepatocytes in vivo.

The therapeutic potential of anti-HBV TALENs was recently corroborated in a study using TALENs with targets in the *core* ORF and the RNaseH region of the *polymerase* ORF [37]. Significant decreases in HBeAg, HBsAg, and pgRNA concentrations were observed in Huh7 cells after their transfection with TALEN-expressing plasmids. The antiviral effect was observed for viral isolates of genotype B, C, and D, confirming that TALENs targeting conserved HBV sequences are effective against several genotypes. Efficacy in vivo was confirmed when using a murine hydrodynamic injection model. This study also showed synergistic antiviral synergy when combining TALENs with IFN-α in Huh7 cell cultures. Improving efficacy of TALENs when using them in combination with licensed drugs is an interesting approach that has also been used to augment RNAi-based inhibition of HBV replication [106].

One member of the panel of TALENs studied by Bloom and colleagues, P1, had a silencing effect on HBV replication without causing target DNA cleavage [36]. Moreover, individual left and right TALEN subunits of P1 were independently capable of causing inhibition of HBV replication. The observation suggests that endonuclease function is not required for the inhibitory effect on HBV proliferation. P1 nuclease has cognates that overlap the HBV enhancer I region of the viral genome. A mechanism involving transient competitive inhibition of transcription factors required for enhancer I activity is likely to be occurring, which is reminiscent of ZFP-mediated inhibition of DHBV replication [183]. The HBV enhancer I sequence regulates viral gene transcription and protein expression through binding of hepatocyte transcription factors (reviewed by [35,204,205]).

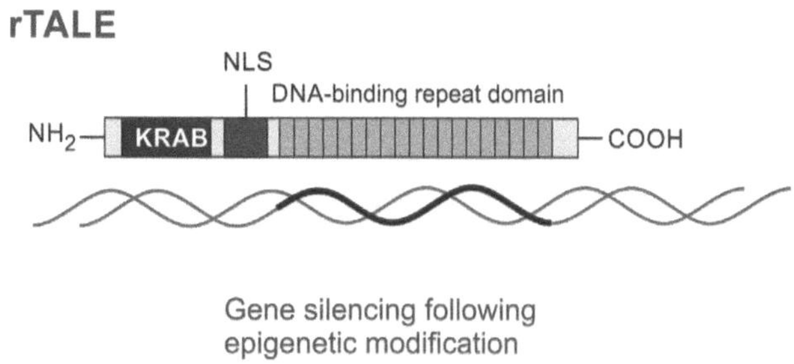

FIGURE 6.9 Disabling HBV cccDNA using a sequence-specific rTALE.
An engineered sequence-specific DNA binding domain is coupled to a repressor domain, such as the KRAB. rTALEs effect gene silencing by targeted introduction of epigenetic changes, which inhibit viral gene expression over a long period.

The P1 left TALEN may inhibit binding of retinoic acid response element and regulatory factor X1 to prevent the cooperative binding of hepatocyte nuclear factor-4 and retinoid X receptor-α/peroxisome proliferator-activated receptor heterodimers. The P1 right TALEN subunit binds within a region of the essential LSR element of HBV enhancer I. The viral sequence binds several hepatocyte transcription factors to activate *X* gene expression [206] and, in conjunction with enhancer II, may facilitate surface protein expression [207]. Repression of transcription by HBV-targeting sequence-specific DNA binding proteins is interesting and potentially a valuable property for therapy. Because HBV DNA integration into the host genome is a frequent occurrence, target cleavage by TALENs may induce chromosomal breaks and potentially carcinogenic effects. Therefore, coupling repressor elements, such as the Krüppel-associated box (KRAB) [208,209], to TALEs to form repressor TALEs (rTALEs) is of interest (Figure 6.9). Using rTALEs to achieve long-term silencing of HBV sequences without target cleavage may overcome unintended genotoxicity.

Eight guide RNAs targeting HBV sequences were assessed for their ability to cause CRISPR/Cas9-mediated inactivation of viral replication [186]. The two most effective guides were identified after transfection of cultured Huh7 cells and measurement of viral antigens. Thereafter, evaluation was performed in mice that were subjected to hydrodynamic injection. Markers of HBV replication could be inhibited in the animals, which is encouraging for the potential therapeutic utility of this gene editing approach. However, definitive proof of efficacy against the viral cccDNA intermediate remains to be established.

6.6 CONCLUSIONS

Chronic HBV infection remains an important global health issue. The many individuals who carry the virus, along with the high risk and severity of complicating cirrhosis and liver cancer, make HBV persistence a serious problem.

Although the available vaccines effectively prevent infection with the virus, they are of little use to people who are already chronically infected with HBV. New-generation HBV vaccines, particularly those based on gene transfer, may have prophylactic and therapeutic potential and are discussed in Chapter 11. Limitations of currently available therapies and attenuated host immune responses to the virus mean that the infection is rarely eliminated. Progress in gene therapy has shown that the approach offers potential to devise strategies based on rational design to eliminate the infection.

A large body of work has shown that harnessing RNAi can be used successfully to inhibit HBV replication. Very impressive silencing of HBV genes has been demonstrated. In most cases, HBV gene silencing does not seem to have been complicated by toxic side effects. The bifunctional 3p-siRNAs, which stimulate an innate immune response to HBV and have a direct silencing effect on the virus, offer interesting potential therapeutic benefits [154–156]. Nevertheless, control of innate immunostimulation by 3p-siRNAs, and whether this approach meets stringent safety requirements for therapeutic application, needs to be ascertained. Although not a popular line of investigation, disabling of HBV host dependency factors may be feasible for therapy. Defining host targets that may be safely inactivated to disrupt HBV replication without causing unintended side effects has been difficult. Possible candidates are genes encoding the recently discovered NTCP receptor for HBV [22] and the human abhydrolase domain-containing 2 [169]. Whether RNAi could be used to eliminate HBV from infected individuals still remains to be established. The post-transcriptional silencing mechanism suggests that RNAi activators would have little effect on cccDNA already existing within hepatocytes, which has been shown to be the case in cultured cells [181]. Nevertheless developments with the assessment of RNAi-based therapy in clinical trials and advanced preclinical trials are exciting, and the outcome of these investigations is awaited with interest.

Availability of engineered sequence-specific DNA binding proteins has been an important development in the field of HBV gene therapy. Using nuclease derivatives to introduce mutations into cccDNA provides a tool that has considerable therapeutic potential. To date, ZFNs and TALENs have been used successfully to introduce mutations into HBV genes [36–38]. Evidence indicates that TALENs have superior specificity, efficacy, and are easier to generate [198]. Use of CRISPR/Cas nucleases against HBV has recently been reported [186], and if good target specificity can be demonstrated [210], then the technology will have exciting therapeutic potential. Apart from off-target cleavage by all classes of designer nucleases, action at sites of HBV integration in host DNA could be problematic. Resulting DSBs may predispose to chromosomal breaks, which could be carcinogenic. Introducing epigenetic changes at the viral targets by coupling transcriptional repressors, such as a KRAB domain to generate HBV-specific rTALEs, could overcome this problem. The approach should bring about more durable viral gene silencing than is caused by simple competitive inhibition by ZFPs or TALEs for transcription factor binding to viral *cis* elements.

Advances in gene therapy for HBV infection, as with gene therapy for most diseases, have been hampered by difficulties associated with achieving efficient delivery of therapeutic sequences. Despite the enhanced susceptibility of hepatocytes to gene transduction by AAVs [123], generation of vectors that are safe, economical to produce, and efficiently hepatotropic after repeated systemic administration has been demanding. Although clinical trials using gene therapy for HBV treatment are in progress, global use of the approach is not imminent. Nevertheless, the power of rational design of methods to disable the virus together with developments in related fields will no doubt facilitate widespread HBV treatment based on the use of nucleic acids.

REFERENCES

[1] Gerlich WH. Medical virology of hepatitis B: how it began and where we are now. Virol J 2013;10:239.

[2] Seeff LB, Beebe GW, Hoofnagle JH, Norman JE, Buskell-Bales Z, Waggoner JG, et al. A serologic follow-up of the 1942 epidemic of post-vaccination hepatitis in the United States Army. N Engl J Med 1987;316(16):965–70.

[3] Blumberg BS. Hepatitis B: the hunt for a killer virus. Princeton, NJ, USA: Princeton University Press; 2002.

[4] Blumberg BS, Alter HJ. A "new" antigen in leukemia sera. JAMA 1965;191(7):541–6.

[5] Prince AM. Relation of Australia and SH antigens. Lancet 1968;2(7565):462–3.

[6] Ling CM, Overby LR. Prevalence of hepatitis B virus antigen as revealed by direct radioimmune assay with 125 I-antibody. J Immunol 1972;109(4):834–41.

[7] Le Bouvier GL, McCollum RW, Hierholzer Jr WJ, Irwin GR, Krugman S, Giles JP. Subtypes of Australia antigen and hepatitis-B virus. JAMA 1972;222(8):928–30.

[8] Maupas P, Goudeau A, Coursaget P, Drucker J, Bagros P. Immunisation against hepatitis B in man. Lancet 1976;1(7974):1367–70.

[9] Purcell RH, Gerin JL. Hepatitis B subunit vaccine: a preliminary report of safety and efficacy tests in chimpanzees. Am J Med Sci 1975;270(2):395–9.

[10] Szmuness W, Stevens CE, Harley EJ, Zang EA, Oleszko WR, William DC, et al. Hepatitis B vaccine: demonstration of efficacy in a controlled clinical trial in a high-risk population in the United States. N Engl J Med 1980;303(15):833–41.

[11] Beasley RP, Hwang LY, Lin CC, Chien CS. Hepatocellular carcinoma and hepatitis B virus. A prospective study of 22,707 men in Taiwan. Lancet 1981;2(8256):1129–33.

[12] Dane DS, Cameron CH, Briggs M. Virus-like particles in serum of patients with Australia-antigen-associated hepatitis. Lancet 1970;1(7649):695–8.

[13] Kaplan PM, Greenman RL, Gerin JL, Purcell RH, Robinson WS. DNA polymerase associated with human hepatitis B antigen. J Virol 1973;12(5):995–1005.

[14] Robinson WS, Clayton DA, Greenman RL. DNA of a human hepatitis B virus candidate. J Virol 1974;14(2):384–91.

[15] Charnay P, Pourcel C, Louise A, Fritsch A, Tiollais P. Cloning in *Escherichia coli* and physical structure of hepatitis B virion DNA. Proc Natl Acad Sci USA 1979;76(5):2222–6.

[16] Pasek M, Goto T, Gilbert W, Zink P, Schaller H, MacKay P, et al. Hepatitis B virus genes and their expression in *E. coli*. Nature 1979;282(5739):575–9.

[17] Valenzuela P, Gray P, Quiroga M, Zaldivar J, Goodman HM, Rutter WJ. Nucleotide sequence of the gene coding for the major protein of hepatitis B virus surface antigen. Nature 1979;280(5725):815–9.

[18] Will H, Cattaneo R, Koch HG, Darai G, Schaller H, Schellekens H, et al. Cloned HBV DNA causes hepatitis in chimpanzees. Nature 1982;299(5885):740–2.

[19] Seeger C, Ganem D, Varmus HE. Biochemical and genetic evidence for the hepatitis B virus replication strategy. Science 1986;232(4749):477–84.

[20] Halegoua-De Marzio D, Hann H-W. Then and now: the progress in hepatitis B treatment over the past 20 years. World J Gastroenterol 2014;20(2):401–13.

[21] Magnius LO, Espmark A. A new antigen complex co-occurring with Australia antigen. Acta Pathologica Microbiol Scand Sect B Microbiol Immunol 1972;80(2):335–7.

[22] Yan H, Zhong G, Xu G, He W, Jing Z, Gao Z, et al. Sodium taurocholate cotransporting polypeptide is a functional receptor for human hepatitis B and D virus. eLife 2012;1:e00049.

[23] McCaffrey AP, Nakai H, Pandey K, Huang Z, Salazar FH, Xu H, et al. Inhibition of hepatitis B virus in mice by RNA interference. Nat Biotechnol 2003;21(6):639–44.

[24] WHO. Hepatitis B virus Fact sheet No. 204 (Revised 2008). 2000. http://www.who.int/mediacentre/factsheets/fs204/en/index.html.

[25] Arbuthnot P, Ely A, Weinberg MS. Hepatic delivery of RNA interference activators for therapeutic application. Curr Gene Ther 2009;9(2):91–103.

[26] Block TM, Guo H, Guo J-T. Molecular virology of hepatitis B virus for clinicians. Clin Liver Dis 2007;11(4):685–706.

[27] Beasley RP, Hwang LY. Hepatocellular carcinoma and hepatitis B virus. Semin Liver Dis 1984;4(2):113–21.

[28] Arbuthnot P, Kew M. Hepatitis B virus and hepatocellular carcinoma. Int J Exp Pathol 2001;82(2):77–100.

[29] Botha JF, Ritchie MJ, Dusheiko GM, Mouton HW, Kew MC. Hepatitis B virus carrier state in black children in Ovamboland: role of perinatal and horizontal infection. Lancet 1984;1(8388):1210–2.

[30] Harford N, Cabezon T, Crabeel M, Simoen E, Rutgers A, De Wilde M. Expression of hepatitis B surface antigen in yeast. Dev Biol Stand 1983;54:125–30.

[31] Valenzuela P, Medina A, Rutter WJ, Ammerer G, Hall BD. Synthesis and assembly of hepatitis B virus surface antigen particles in yeast. Nature 1982;298(5872):347–50.

[32] Kramvis A, Kew M, Francois G. Hepatitis B virus genotypes. Vaccine 2005;23(19):2409–23.

[33] Huovila AP, Eder AM, Fuller SD. Hepatitis B surface antigen assembles in a post-ER, pre-Golgi compartment. J Cell Biol 1992;118(6):1305–20.

[34] Le Seyec J, Chouteau P, Cannie I, Guguen-Guillouzo C, Gripon P. Infection process of the hepatitis B virus depends on the presence of a defined sequence in the pre-S1 domain. J Virol 1999;73(3):2052–7.

[35] Moolla N, Kew M, Arbuthnot P. Regulatory elements of hepatitis B virus transcription. J Viral Hepat 2002;9(5):323–31.

[36] Bloom K, Ely A, Mussolino C, Cathomen T, Arbuthnot P. Inactivation of hepatitis B virus replication in cultured cells and in vivo with engineered transcription activator-like effector nucleases. Mol Ther J Am Soc Gene Ther 2013;21(10):1889–97.

[37] Chen J, Zhang W, Lin J, Wang F, Wu M, Chen C, et al. An efficient antiviral strategy for targeting hepatitis B virus genome using transcription activator-like effector nucleases. Mol Ther J Am Soc Gene Ther 2014;22(2):303–11.

[38] Cradick TJ, Keck K, Bradshaw S, Jamieson AC, McCaffrey AP. Zinc-finger nucleases as a novel therapeutic strategy for targeting hepatitis B virus DNAs. Mol Ther 2010;18(5):947–54.

[39] Preiss S, Littlejohn M, Angus P, Thompson A, Desmond P, Lewin SR, et al. Defective hepatitis B virus DNA is not associated with disease status but is reduced by polymerase mutations associated with drug resistance. Hepatology 2008;48(3):741–9.

[40] Huang C, Xie MH, Liu W, Yang B, Yang F, Huang J, et al. A structured RNA in hepatitis B virus post-transcriptional regulatory element represses alternative splicing in a sequence-independent and position-dependent manner. FEBS J 2011;278(9):1533–46.

[41] Bartenschlager R, Junker-Niepmann M, Schaller H. The P gene product of hepatitis B virus is required as a structural component for genomic RNA encapsidation. J Virol 1990;64(11):5324–32.

[42] Pollack JR, Ganem D. Site-specific RNA binding by a hepatitis B virus reverse transcriptase initiates two distinct reactions: RNA packaging and DNA synthesis. J Virol 1994;68(9):5579–87.

[43] Wang G-H, Seeger C. The reverse transcriptase of hepatitis B virus acts as a protein primer for viral DNA synthesis. Cell 1992;71(4):663–70.

[44] Tavis JE, Perri S, Ganem D. Hepadnavirus reverse transcription initiates within the stem-loop of the RNA packaging signal and employs a novel strand transfer. J Virol 1994;68(6):3536–43.

[45] Steven AC, Conway JF, Cheng N, Watts NR, Belnap DM, Harris A, et al. Structure, assembly, and antigenicity of hepatitis B virus capsid proteins. Adv Virus Res 2005;64:125–64.

[46] Chen C, Wang JC, Zlotnick A. A kinase chaperones hepatitis B virus capsid assembly and captures capsid dynamics in vitro. PLoS Pathog 2011;7(11):e1002388.

[47] Gerelsaikhan T, Tavis JE, Bruss V. Hepatitis B virus nucleocapsid envelopment does not occur without genomic DNA synthesis. J Virol 1996;70(7):4269–74.

[48] Hartmann-Stuhler C, Prange R. Hepatitis B virus large envelope protein interacts with gamma2-adaptin, a clathrin adaptor-related protein. J Virol 2001;75(11):5343–51.

[49] Rost M, Mann S, Lambert C, Doring T, Thome N, Prange R. Gamma-adaptin, a novel ubiquitin-interacting adaptor, and Nedd4 ubiquitin ligase control hepatitis B virus maturation. J Biol Chem 2006;281(39):29297–308.

[50] Lambert C, Doring T, Prange R. Hepatitis B virus maturation is sensitive to functional inhibition of ESCRT-III, Vps4, and gamma 2-adaptin. J Virol 2007;81(17):9050–60.

[51] Watashi K, Sluder A, Daito T, Matsunaga S, Ryo A, Nagamori S, et al. Cyclosporin A and its analogs inhibit hepatitis B virus entry into cultured hepatocytes through targeting a membrane transporter NTCP. Hepatology 2014;59(5):1726–37.

[52] Glebe D, Urban S. Viral and cellular determinants involved in hepadnaviral entry. World J Gastroenterol 2007;13(1):22–38.

[53] Gripon P, Rumin S, Urban S, Le Seyec J, Glaise D, Cannie I, et al. Infection of a human hepatoma cell line by hepatitis B virus. Proc Natl Acad Sci USA 2002;99(24):15655–60.

[54] Guillouzo A, Corlu A, Aninat C, Glaise D, Morel F, Guguen-Guillouzo C. The human hepatoma HepaRG cells: a highly differentiated model for studies of liver metabolism and toxicity of xenobiotics. Chem Biol Interact 2007;168(1):66–73.

[55] Sells MA, Chen ML, Acs G. Production of hepatitis B virus particles in Hep G2 cells transfected with cloned hepatitis B virus DNA. Proc Natl Acad Sci USA 1987;84(4):1005–9.

[56] Sells MA, Zelent AZ, Shvartsman M, Acs G. Replicative intermediates of hepatitis B virus in HepG2 cells that produce infectious virions. J Virol 1988;62(8):2836–44.

[57] Hirschman SZ, Price P, Garfinkel E, Christman J, Acs G. Expression of cloned hepatitis B virus DNA in human cell cultures. Proc Natl Acad Sci USA 1980;77(9):5507–11.

[58] Ladner SK, Otto MJ, Barker CS, Zaifert K, Wang GH, Guo JT, et al. Inducible expression of human hepatitis B virus (HBV) in stably transfected hepatoblastoma cells: a novel system for screening potential inhibitors of HBV replication. Antimicrob Agents Chemother 1997;41(8):1715–20.

[59] Huch M, Dorrell C, Boj SF, van Es JH, Li VS, van de Wetering M, et al. In vitro expansion of single $Lgr5^+$ liver stem cells induced by Wnt-driven regeneration. Nature 2013;494(7436):247–50.

[60] Takebe T, Sekine K, Enomura M, Koike H, Kimura M, Ogaeri T, et al. Vascularized and functional human liver from an iPSC-derived organ bud transplant. Nature 2013;499(7459):481–4.

[61] Dorner M, Horwitz JA, Robbins JB, Barry WT, Feng Q, Mu K, et al. A genetically humanized mouse model for hepatitis C virus infection. Nature 2011;474(7350):208–11.

[62] Liu F, Song Y, Liu D. Hydrodynamics-based transfection in animals by systemic administration of plasmid DNA. Gene Ther 1999;6(7):1258–66.

[63] Yang PL, Althage A, Chung J, Chisari FV. Hydrodynamic injection of viral DNA: a mouse model of acute hepatitis B virus infection. Proc Natl Acad Sci USA 2002;99(21):13825–30.

[64] Carmona S, Ely A, Crowther C, Moolla N, Salazar FH, Marion PL, et al. Effective inhibition of HBV replication in vivo by anti-HBx short hairpin RNAs. Mol Ther 2006;13(2):411–21.

[65] Crowther C, Ely A, Hornby J, Mufamadi MS, Salazar F, Marion P, et al. Efficient inhibition of hepatitis B virus replication in vivo, using polyethylene glycol-modified adenovirus vectors. Hum Gene Ther 2008;19(11):1325–31.

[66] Uprichard SL, Boyd B, Althage A, Chisari FV. Clearance of hepatitis B virus from the liver of transgenic mice by short hairpin RNAs. Proc Natl Acad Sci USA 2005;102(3):773–8.

[67] Guidotti LG, Matzke B, Schaller H, Chisari FV. High-level hepatitis B virus replication in transgenic mice. J Virol 1995;69(10):6158–69.

[68] Marion P, Salazar F, Liittschwager K, Bordier BB, Seeger C, Winters MA, et al. A transgenic mouse lineage useful for testing antivirals targeting hepatitis B virus. In: Raymond FS, Jean-Pierre S, Rice CM, editors. Frontiers in viral hepatitis. Amsterdam: Elsevier; 2003: 97–210.

[69] Raney AK, Eggers CM, Kline EF, Guidotti LG, Pontoglio M, Yaniv M, et al. Nuclear covalently closed circular viral genomic DNA in the liver of hepatocyte nuclear factor 1 alpha-null hepatitis B virus transgenic mice. J Virol 2001;75(6):2900–11.

[70] Meuleman P, Libbrecht L, De Vos R, de Hemptinne B, Gevaert K, Vandekerckhove J, et al. Morphological and biochemical characterization of a human liver in a uPA-SCID mouse chimera. Hepatology 2005;41(4):847–56.

[71] Heckel JL, Sandgren EP, Degen JL, Palmiter RD, Brinster RL. Neonatal bleeding in transgenic mice expressing urokinase-type plasminogen activator. Cell 1990;62(3):447–56.

[72] Meuleman P, Leroux-Roels G. The human liver-uPA-SCID mouse: a model for the evaluation of antiviral compounds against HBV and HCV. Antivir Res 2008;80(3):231–8.

[73] Qi Z, Li G, Hu H, Yang C, Zhang X, Leng Q, et al. Recombinant covalently closed circular hepatitis B virus DNA induces prolonged viral persistence in immunocompetent mice. J Virol 2014;88(14):8045–56.

[74] Mason WS, Seal G, Summers J. Virus of Pekin ducks with structural and biological relatedness to human hepatitis B virus. J Virol 1980;36(3):829–36.

[75] Tennant BC, Gerin JL. The woodchuck model of hepatitis B virus infection. ILAR J 2001;42(2):89–102.

[76] Summers J, Smolec JM, Snyder R. A virus similar to human hepatitis B virus associated with hepatitis and hepatoma in woodchucks. Proc Natl Acad Sci USA 1978;75(9):4533–7.

[77] von Weizsäcker F, Köck J, MacNelly S, Ren S, Blum HE, Nassal M. The tupaia model for the study of hepatitis B virus. Hepatitis B and D protocols. Springer; 2004. p. 153–61.

[78] Barker LF, Chisari FV, McGrath PP, Dalgard DW, Kirschstein RL, Almeida JD, et al. Transmission of type B viral hepatitis to chimpanzees. J Infect Dis 1973;127(6):648–62.

[79] Bertoni R, Sette A, Sidney J, Guidotti LG, Shapiro M, Purcell R, et al. Human class I supertypes and CTL repertoires extend to chimpanzees. J Immunol 1998;161(8):4447–55.

[80] Guidotti LG, Rochford R, Chung J, Shapiro M, Purcell R, Chisari FV. Viral clearance without destruction of infected cells during acute HBV infection. Science 1999;284(5415):825–9.

[81] Sohn JA, Litwin S, Seeger C. Mechanism for CCC DNA synthesis in hepadnaviruses. PloS One 2009;4(11):e8093.

[82] Lee KW, Yun T, Song EK, Na II, Shin H, Bang SM, et al. A pilot study of bortezomib in Korean patients with relapsed or refractory myeloma. J Korean Med Sci 2005;20(4):598–602.

[83] Greenberg HB, Pollard RB, Lutwick LI, Gregory PB, Robinson WS, Merigan TC. Effect of human leukocyte interferon on hepatitis B virus infection in patients with chronic active hepatitis. N Engl J Med 1976;295(10):517–22.

[84] Weimar W, Heijtink RA, Schalm SW, van Blankenstein M, Schellekens H, Masurel N, et al. Fibroblast interferon in HBsAg-positive chronic active hepatitis. Lancet 1977;2(8051):1282.

[85] Loomba R, Liang TJ. Treatment of chronic hepatitis B. Antivir Ther 2007;12(Suppl. 3):H33–41.

[86] Buster EH, Hansen BE, Lau GK, Piratvisuth T, Zeuzem S, Steyerberg EW, et al. Factors that predict response of patients with hepatitis B e antigen-positive chronic hepatitis B to peginterferon-alfa. Gastroenterology 2009;137(6):2002–9.

[87] Zoulim F, Perrillo R, Hepatitis B. reflections on the current approach to antiviral therapy. J Hepatol 2008;48(Suppl. 1):S2–19.

[88] Hao J, Jin W, Li X, Wang S, Zhang X, Fan H, et al. Inhibition of alpha interferon (IFN-alpha)-induced microRNA-122 negatively affects the anti-hepatitis B virus efficiency of IFN-alpha. J Virol 2013;87(1):137–47.

[89] Ferir G, Kaptein S, Neyts J, De Clercq E. Antiviral treatment of chronic hepatitis B virus infections: the past, the present and the future. Rev Med Virol 2008;18(1):19–34.

[90] Keeffe EB, Dieterich DT, Pawlotsky JM, Benhamou Y. Chronic hepatitis B: preventing, detecting, and managing viral resistance. Clin Gastroenterol Hepatol 2008;6(3):268–74.

[91] Arora G, Keeffe EB. Chronic hepatitis B with advanced fibrosis or cirrhosis: impact of antiviral therapy. Rev Gastroenterol Disord 2007;7(2):63–73.

[92] Delaney WE 4th, Yang H, Westland CE, Das K, Arnold E, Gibbs CS, et al. The hepatitis B virus polymerase mutation rtV173L is selected during lamivudine therapy and enhances viral replication in vitro. J Virol 2003;77(21):11833–41.

[93] Fu L, Cheng YC. Role of additional mutations outside the YMDD motif of hepatitis B virus polymerase in L(-)SddC (3TC) resistance. Biochem Pharmacol 1998;55(10):1567–72.

[94] Stray SJ, Bourne CR, Punna S, Lewis WG, Finn MG, Zlotnick A. A heteroaryldihydropyrimidine activates and can misdirect hepatitis B virus capsid assembly. Proc Natl Acad Sci USA 2005;102(23):8138–43.

[95] Weber O, Schlemmer KH, Hartmann E, Hagelschuer I, Paessens A, Graef E, et al. Inhibition of human hepatitis B virus (HBV) by a novel non-nucleosidic compound in a transgenic mouse model. Antivir Res 2002;54(2):69–78.

[96] King RW, Ladner SK, Miller TJ, Zaifert K, Perni RB, Conway SC, et al. Inhibition of human hepatitis B virus replication by AT-61, a phenylpropenamide derivative, alone and in combination with (-)beta-L-2′,3′-dideoxy-3′-thiacytidine. Antimicrob Agents Chemother 1998;42(12):3179–86.

[97] Block TM, Lu X, Mehta AS, Blumberg BS, Tennant B, Ebling M, et al. Treatment of chronic hepadnavirus infection in a woodchuck animal model with an inhibitor of protein folding and trafficking. Nat Med 1998;4(5):610–4.

[98] Xia C, Chen YC, Gong H, Zeng W, Vu GP, Trang P, et al. Inhibition of hepatitis B virus gene expression and replication by ribonuclease P. Mol Ther J Am Soc Gene Ther 2013;21(5):995–1003.

[99] Trejo-Avila L, Elizondo-Gonzalez R, Trujillo-Murillo Kdel C, Zapata-Benavides P, Rodriguez-Padilla C, Rivas-Estilla AM. Antiviral therapy: inhibition of hepatitis C virus expression by RNA interference directed against the NS5B region of the viral genome. Ann Hepatol 2007;6(3):174–80.

[100] Schopman NC, Liu YP, Konstantinova P, ter Brake O, Berkhout B. Optimization of shRNA inhibitors by variation of the terminal loop sequence. Antivir Res 2010;86(2):204–11.

[101] Klinghoffer RA, Magnus J, Schelter J, Mehaffey M, Coleman C, Cleary MA. Reduced seed region-based off-target activity with lentivirus-mediated RNAi. RNA 2010;16(5):879–84.

[102] Kim YH, Lee JH, Paik NW, Rho HM. RNAi-based knockdown of HBx mRNA in HBx-transformed and HBV-producing human liver cells. DNA Cell Biol 2006;25(7):412–7.

[103] Shlomai A, Shaul Y. Inhibition of hepatitis B virus expression and replication by RNA interference. Hepatology 2003;37(4):764–70.

[104] Ren X, Luo G, Xie Z, Zhou L, Kong X, Xu A. Inhibition of multiple gene expression and virus replication of HBV by stable RNA interference in 2.2.15 cells. J Hepatol 2006;44(4):663–70.

[105] Ren XR, Zhou LJ, Luo GB, Lin B, Xu A. Inhibition of hepatitis B virus replication in 2.2.15 cells by expressed shRNA. J Viral Hepat 2005;12(3):236–42.

[106] Chen Y, Du D, Wu J, Chan CP, Tan Y, Kung HF, et al. Inhibition of hepatitis B virus replication by stably expressed shRNA. Biochem Biophys Res Commun 2003;311(2):398–404.

[107] Weinberg MS, Ely A, Barichievy S, Crowther C, Mufamadi S, Carmona S, et al. Specific inhibition of HBV replication in vitro and in vivo with expressed long hairpin RNA. Mol Ther J Am Soc Gene Ther 2007;15(3):534–41.

[108] Dyer V, Ely A, Bloom K, Weinberg M, Arbuthnot P. tRNA Lys3 promoter cassettes that efficiently express RNAi-activating antihepatitis B virus short hairpin RNAs. Biochem Biophys Res Commun 2010;398(4):640–6.

[109] Keck K, Volper EM, Spengler RM, Long DD, Chan CY, Ding Y, et al. Rational design leads to more potent RNA interference against hepatitis B virus: factors effecting silencing efficiency. Mol Ther 2009;17(3):538–47.

[110] Sun D, Rosler C, Kidd-Ljunggren K, Nassal M. Quantitative assessment of the antiviral potencies of 21 shRNA vectors targeting conserved, including structured, hepatitis B virus sites. J Hepatol 2010;52(6):817–26.

[111] Grimm D, Streetz KL, Jopling CL, Storm TA, Pandey K, Davis CR, et al. Fatality in mice due to oversaturation of cellular microRNA/short hairpin RNA pathways. Nature 2006;441(7092):537–41.

[112] Ely A, Naidoo T, Mufamadi S, Crowther C, Arbuthnot P. Expressed anti-HBV primary microRNA shuttles inhibit viral replication efficiently in vitro and in vivo. Mol Ther 2008;16(6):1105–12.

[113] Ely A, Arbuthnot P. Silencing hepatitis B virus replication with antiviral pri-miR shuttles generated from liver-specific Pol II promoter. In: Catalano AJ, editor. Gene silencing: theory, techniques and applications. Nova Publishers; 2010.

[114] Giering JC, Grimm D, Storm TA, Kay MA. Expression of shRNA from a tissue-specific pol II promoter is an effective and safe RNAi therapeutic. Mol Ther J Am Soc Gene Ther 2008;16(9):1630–6.

[115] Ely A, Naidoo T, Arbuthnot P. Efficient silencing of gene expression with modular trimeric Pol II expression cassettes comprising microRNA shuttles. Nucleic Acids Res 2009;37(13):e91.

[116] ter Brake O, t Hooft K, Liu YP, Centlivre M, von Eije KJ, Berkhout B. Lentiviral vector design for multiple shRNA expression and durable HIV-1 inhibition. Mol Ther J Am Soc Gene Ther 2008;16(3):557–64.

[117] Berkhout B. Toward a durable anti-HIV gene therapy based on RNA interference. Ann NY Acad Sci 2009;1175:3–14.

[118] Mowa MB, Crowther C, Arbuthnot P. Therapeutic potential of adenoviral vectors for delivery of expressed RNAi activators. Expert Opin Drug Deliv 2010;7(12):1373–85.

[119] Grimm D, Pandey K, Nakai H, Storm TA, Kay MA. Liver transduction with recombinant adeno-associated virus is primarily restricted by capsid serotype not vector genotype. J Virol 2006;80(1):426–39.

[120] Chen CC, Ko TM, Ma HI, Wu HL, Xiao X, Li J, et al. Long-term inhibition of hepatitis B virus in transgenic mice by double-stranded adeno-associated virus 8-delivered short hairpin RNA. Gene Ther 2007;14(1):11–9.

[121] Chen CC, Sun CP, Ma HI, Fang CC, Wu PY, Xiao X, et al. Comparative study of anti-hepatitis B virus RNA interference by double-stranded adeno-associated virus serotypes 7, 8, and 9. Mol Ther J Am Soc Gene Ther 2009;17(2):352–9.

[122] Chen CC, Chang CM, Sun CP, Yu CP, Wu PY, Jeng KS, et al. Use of RNA interference to modulate liver adenoma development in a murine model transgenic for hepatitis B virus. Gene Ther 2012;19(1):25–33.

[123] Hosel M, Lucifora J, Michler T, Holz G, Gruffaz M, Stahnke S, et al. Hepatitis B virus infection enhances susceptibility towards adeno-associated viral vector transduction in vitro and in vivo. Hepatology 2014;59(6):2110–20.

[124] Rauschhuber C, Xu H, Salazar FH, Marion PL, Ehrhardt A. Exploring gene-deleted adenoviral vectors for delivery of short hairpin RNAs and reduction of hepatitis B virus infection in mice. J Gene Med 2008;10(8):878–89.

[125] Crowther C, Mowa MB, Ely A, Arbuthnot PB. Inhibition of hepatitis B virus replication in vivo using helper-dependent adenovirus vectors to deliver antiviral RNAi expression cassettes. Antivir Ther 2014;19:363–73.

[126] Mowa MB, Crowther C, Ely A, Arbuthnot P. Efficient silencing of hepatitis B virus by helper-dependent adenovirus vector-mediated delivery of artificial antiviral primary micro RNAs. Micro RNA 2012;1:19–25.

[127] Liu SL, Lerman MI, Miller AD. Putative phosphatidylinositol 3-kinase (PI3K) binding motifs in ovine betaretrovirus Env proteins are not essential for rodent fibroblast transformation and PI3K/Akt activation. J Virol 2003;77(14):7924–35.

[128] Ivacik D, Ely A, Ferry N, Arbuthnot P. Sustained inhibition of hepatitis B virus replication in vivo using RNAi-activating lentiviruses. Gene Ther 2015;22(2):163–71.

[129] Birraux J, Wildhaber BE, Jond C, Belli DC, Menzel O. Ex vivo hepatocyte gene therapy: increased biosafety protocol for transduction in suspension with lentiviral vectors and immediate transplantation (SLIT). Transplantation 2010;89(1):127–129.

[130] Menzel O, Birraux J, Wildhaber BE, Jond C, Lasne F, Habre W, et al. Biosafety in ex vivo gene therapy and conditional ablation of lentivirally transduced hepatocytes in nonhuman primates. Mol Ther 2009;17(10):1754–60.

[131] Birraux J, Menzel O, Wildhaber B, Jond C, Nguyen TH, Chardot C. A step toward liver gene therapy: efficient correction of the genetic defect of hepatocytes isolated from a patient with Crigler-Najjar syndrome type 1 with lentiviral vectors. Transplantation 2009;87(7):1006–12.

[132] Nguyen TH, Birraux J, Wildhaber B, Myara A, Trivin F, Le Coultre C, et al. Ex vivo lentivirus transduction and immediate transplantation of uncultured hepatocytes for treating hyperbilirubinemic Gunn rat. Transplantation 2006;82(6):794–803.

[133] Wright M, Tanner JA, Miller AD. Quantitative single-step purification of dinucleoside polyphosphates. Anal Biochem 2003;316(1):135–8.

[134] Miller AD. The problem with cationic liposome/micelle-based non-viral vector systems for gene therapy. Curr Med Chem 2003;10(14):1195–211.

[135] Miller AD. Identification of Hyal2 as the cell-surface receptor for jaagsiekte sheep retrovirus and ovine nasal adenocarcinoma virus. Curr Top Microbiol Immunol 2003;275:179–99.

[136] Gish RG, Satishchandran C, Young M, Pachuk C. RNA interference and its potential applications to chronic HBV treatment: results of a Phase I safety and tolerability study. Antivir Ther 2011;16(4):547–54.

[137] Giladi H, Ketzinel-Gilad M, Rivkin L, Felig Y, Nussbaum O, Galun E. Small interfering RNA inhibits hepatitis B virus replication in mice. Mol Ther 2003;8(5):769–76.

[138] Hamasaki K, Nakao K, Matsumoto K, Ichikawa T, Ishikawa H, Eguchi K. Short interfering RNA-directed inhibition of hepatitis B virus replication. FEBS Lett 2003;543(1–3):51–4.

[139] Klein C, Bock CT, Wedemeyer H, Wustefeld T, Locarnini S, Dienes HP, et al. Inhibition of hepatitis B virus replication in vivo by nucleoside analogues and siRNA. Gastroenterology 2003;125(1):9–18.

[140] Konishi M, Wu CH, Wu GY. Inhibition of HBV replication by siRNA in a stable HBV-producing cell line. Hepatology 2003;38(4):842–50.

[141] Ying C, De Clercq E, Neyts J. Selective inhibition of hepatitis B virus replication by RNA interference. Biochem Biophys Res Commun 2003;309(2):482–4.

[142] Carmona S, Jorgensen MR, Kolli S, Crowther C, Salazar FH, Marion PL, et al. Controlling HBV replication in vivo by intravenous administration of triggered PEGylated siRNA-nanoparticles. Mol Pharm 2009;6(3):706–17.

[143] Chen Z, Xu ZF, Ye JJ, Yao HP, Zheng S, Ding JY. Combination of small interfering RNAs mediates greater inhibition of human hepatitis B virus replication and antigen expression. J Zhejiang Univ Sci B 2005;6(4):236–41.

[144] Wu KL, Zhang X, Zhang J, Yang Y, Mu YX, Liu M, et al. Inhibition of hepatitis B virus gene expression by single and dual small interfering RNA treatment. Virus Res 2005;112 (1–2):100–7.

[145] Morrissey DV, Blanchard K, Shaw L, Jensen K, Lockridge JA, Dickinson B, et al. Activity of stabilized short interfering RNA in a mouse model of hepatitis B virus replication. Hepatology 2005;41(6):1349–56.

[146] Morrissey DV, Lockridge JA, Shaw L, Blanchard K, Jensen K, Breen W, et al. Potent and persistent in vivo anti-HBV activity of chemically modified siRNAs. Nat Biotechnol 2005;23(8):1002–7.

[147] Liu SL, Duh FM, Lerman MI, Miller AD. Role of virus receptor Hyal2 in oncogenic transformation of rodent fibroblasts by sheep betaretrovirus env proteins. J Virol 2003;77(5): 2850–8.

[148] Fisher M, Abramov M, Van Aerschot A, Rozenski J, Dixit V, Juliano RL, et al. Biological effects of hexitol and altritol-modified siRNAs targeting B-Raf. Eur J Pharmacol 2009;606(1–3):38–44.

[149] Fisher M, Abramov M, Van Aerschot A, Xu D, Juliano RL, Herdewijn P. Inhibition of MDR1 expression with altritol-modified siRNAs. Nucleic Acids Res 2007;35(4):1064–74.

[150] Hean J, Crowther C, Ely A, Ul Islam R, Barichievy S, Bloom K, et al. Inhibition of hepatitis B virus replication in vivo using lipoplexes containing altritol-modified antiviral siRNAs. Artif DNA, PNA XNA 2010;1(1):17–26.

[151] Brzezinska J, D'Onofrio J, Buff MC, Hean J, Ely A, Marimani M, et al. Synthesis of 2-O-guanidinopropyl-modified nucleoside phosphoramidites and their incorporation into siRNAs targeting hepatitis B virus. Bioorg Med Chem 2011;20:1594–606.

[152] Marimani MD, Ely A, Buff MC, Bernhardt S, Engels JW, Arbuthnot P. Inhibition of hepatitis B virus replication in cultured cells and in vivo using 2'-*O*-guanidinopropyl modified siRNAs. Bioorg Med Chem 2013;21(20):6145–55.

[153] Islam RU, Hean J, van Otterlo WA, de Koning CB, Arbuthnot P. Efficient nucleic acid transduction with lipoplexes containing novel piperazine- and polyamine-conjugated cholesterol derivatives. Bioorg Med Chem Lett 2009;19(1):100–3.

[154] Han Q, Zhang C, Zhang J, Tian Z. Reversal of hepatitis B virus-induced immune tolerance by an immunostimulatory 3p-HBx-siRNAs in a retinoic acid inducible gene I-dependent manner. Hepatology 2011;54(4):1179–89.

[155] Chen X, Qian Y, Yan F, Tu J, Yang X, Xing Y, et al. 5'-triphosphate-siRNA activates RIG-I-dependent type I interferon production and enhances inhibition of hepatitis B virus replication in HepG2.2.15 cells. Eur J Pharmacol 2013;721(1–3):86–95.

[156] Ebert G, Poeck H, Lucifora J, Baschuk N, Esser K, Esposito I, et al. 5' Triphosphorylated small interfering RNAs control replication of hepatitis B virus and induce an interferon response in human liver cells and mice. Gastroenterology 2011;141(2):696–706. e1–3.

[157] Han Q, Zhang C, Zhang J, Tian Z. Involvement of activation of PKR in HBx-siRNA-mediated innate immune effects on HBV inhibition. PloS One 2011;6(12):e27931.

[158] Myers JW, Jones JT, Meyer T, Ferrell Jr JE. Recombinant Dicer efficiently converts large dsRNAs into siRNAs suitable for gene silencing. Nat Biotechnol 2003;21(3):324–8.

[159] Yang D, Buchholz F, Huang Z, Goga A, Chen CY, Brodsky FM, et al. Short RNA duplexes produced by hydrolysis with *Escherichia coli* RNase III mediate effective RNA interference in mammalian cells. Proc Natl Acad Sci USA 2002;99(15):9942–7.

[160] Ghildiyal M, Seitz H, Horwich MD, Li C, Du T, Lee S, et al. Endogenous siRNAs derived from transposons and mRNAs in Drosophila somatic cells. Science 2008;320(5879):1077–81.

[161] Tam OH, Aravin AA, Stein P, Girard A, Murchison EP, Cheloufi S, et al. Pseudogene-derived small interfering RNAs regulate gene expression in mouse oocytes. Nature 2008;453(7194):534–8.

[162] Xuan B, Qian Z, Hong J, Huang W. EsiRNAs inhibit hepatitis B virus replication in mice model more efficiently than synthesized siRNAs. Virus Res 2006;118(1–2):150–5.

[163] Wooddell CI, Rozema DB, Hossbach M, John M, Hamilton HL, Chu Q, et al. Hepatocyte-targeted RNAi therapeutics for the treatment of chronic hepatitis B virus infection. Mol Ther J Am Soc Gene Ther 2013;21(5):973–85.

[164] Wong SC, Klein JJ, Hamilton HL, Chu Q, Frey CL, Trubetskoy VS, et al. Co-injection of a targeted, reversibly masked endosomolytic polymer dramatically improves the efficacy of cholesterol-conjugated small interfering RNAs in vivo. Nucleic Acid Ther 2012;22(6):380–90.

[165] Offensperger WB, Blum HE, Gerok W. Therapy of hepadnavirus infection using antisense oligonucleotides. Intervirology 1995;38(1–2):113–9.

[166] Offensperger WB, Offensperger S, Blum HE. Antisense therapy of hepatitis B virus infection. Mol Biotechnol 1998;9(2):161–70.

[167] zu Putlitz J, Wieland S, Blum HE, Wands JR. Antisense RNA complementary to hepatitis B virus specifically inhibits viral replication. Gastroenterology 1998;115(3):702–13.

[168] Beck J, Nassal M. Efficient hammerhead ribozyme-mediated cleavage of the structured hepatitis B virus encapsidation signal in vitro and in cell extracts, but not in intact cells. Nucleic Acids Res 1995;23(24):4954–62.

[169] Ding X, Yang J, Wang S. Antisense oligonucleotides targeting abhydrolase domain containing 2 block human hepatitis B virus propagation. Oligonucleotides 2011;21(2):77–84.

[170] Thomson JB, Tuschl T, Eckstein F. The hammerhead ribozyme. In: Eckstein F, Lilley DMJ, editors. Catalytic RNA. Berlin, Germany: Springer-Verlag; 1997. p. 173–96.

[171] Tan TM, Zhou L, Houssais S, Seet BL, Jaenicke S, Peter F, et al. Intracellular inhibition of hepatitis B virus S gene expression by chimeric DNA-RNA phosphorothioate minimized ribozyme. Antisense Nucleic Acid Drug Dev 2002;12(4):257–64.

[172] Usman N, Stuchcomb DT. Design, synthesis and function of therapeutic hammerhead ribozymes. In: Eckstein F, Lilley DMJ, editors. Catalytic RNA. Berlin, Germany: Springer-Verlag; 1997. p. 243–64.

[173] Feng Y, Kong YY, Wang Y, Qi GR. Inhibition of hepatitis B virus by hammerhead ribozyme targeted to the poly(A) signal sequence in cultured cells. Biol Chem 2001;382(4):655–60.

[174] Kim YK, Junn E, Park I, Lee Y, Kang C, Ahn JK. Repression of hepatitis B virus X gene expression by hammerhead ribozymes. Biochem Biophys Res Commun 1999;257(3):759–65.

[175] Passman M, Weinberg M, Kew M, Arbuthnot P. In situ demonstration of inhibitory effects of hammerhead ribozymes that are targeted to the hepatitis Bx sequence in cultured cells. Biochem Biophys Res Commun 2000;268(3):728–33.

[176] von Weizsacker F, Blum HE, Wands JR. Cleavage of hepatitis B virus RNA by three ribozymes transcribed from a single DNA template. Biochem Biophys Res Commun 1992;189(2):743–8.

[177] Weinberg M, Passman M, Kew M, Arbuthnot P. Hammerhead ribozyme-mediated inhibition of hepatitis B virus X gene expression in cultured cells. J Hepatol 2000;33(1):142–51.

[178] Weinberg MS, Ely A, Passman M, Mufamadi SM, Arbuthnot P. Effective anti-hepatitis B virus hammerhead ribozymes derived from multimeric precursors. Oligonucleotides 2007; 17(1):104–12.

[179] Blazquez L, Fortes P. U1 snRNP control of 3 -end processing and the therapeutic application of U1 inhibition combined with RNA interference. Curr Mol Med 2013.

[180] Blazquez L, Gonzalez-Rojas SJ, Abad A, Razquin N, Abad X, Fortes P. Increased in vivo inhibition of gene expression by combining RNA interference and U1 inhibition. Nucleic Acids Res 2012;40(1):e8, http://dx.doi.org/10.1093/nar/gkr956.

[181] Starkey JL, Chiari EF, Isom HC. Hepatitis B virus (HBV)-specific short hairpin RNA is capable of reducing the formation of HBV covalently closed circular (CCC) DNA but has no effect on established CCC DNA in vitro. J Gen Virol 2009;90(Pt 1):115–26.

[182] Schiffer JT, Aubert M, Weber ND, Mintzer E, Stone D, Jerome KR. Targeted DNA mutagenesis for the cure of chronic viral infections. J Virol 2012;86(17):8920–36.

[183] Zimmerman KA, Fischer KP, Joyce MA, Tyrrell DL. Zinc finger proteins designed to specifically target duck hepatitis B virus covalently closed circular DNA inhibit viral transcription in tissue culture. J Virol 2008;82(16):8013–21.

[184] Cong L, Ran FA, Cox D, Lin S, Barretto R, Habib N, et al. Multiplex genome engineering using CRISPR/Cas systems. Science 2013;339(6121):819–23.

[185] Hwang WY, Fu Y, Reyon D, Maeder ML, Tsai SQ, Sander JD, et al. Efficient genome editing in zebrafish using a CRISPR-Cas system. Nat Biotechnol 2013;31(3):227–9.

[186] Lin SR, Yang HC, Kuo YT, Liu CJ, Yang TY, Sung KC, et al. The CRISPR/Cas9 system facilitates clearance of the intrahepatic HBV templates in vivo. Mol Ther Nucleic Acids 2014;3:e186.

[187] Bernstein KA, Rothstein R. At loose ends: resecting a double-strand break. Cell 2009; 137(5):807–10.

[188] Lieber MR. The mechanism of double-strand DNA break repair by the nonhomologous DNA end-joining pathway. Annu Rev Biochem 2010;79:181–211.

[189] Kim YG, Cha J, Chandrasegaran S. Hybrid restriction enzymes: zinc finger fusions to Fok I cleavage domain. Proc Natl Acad Sci USA 1996;93(3):1156–60.

[190] Christian M, Cermak T, Doyle EL, Schmidt C, Zhang F, Hummel A, et al. Targeting DNA double-strand breaks with TAL effector nucleases. Genetics 2010;186(2):757–61.

[191] Bitinaite J, Wah DA, Aggarwal AK, Schildkraut I. FokI dimerization is required for DNA cleavage. Proc Natl Acad Sci USA 1998;95(18):10570–5.

[192] Wah DA, Hirsch JA, Dorner LF, Schildkraut I, Aggarwal AK. Structure of the multimodular endonuclease FokI bound to DNA. Nature 1997;388(6637):97–100.

[193] Doyon Y, Vo TD, Mendel MC, Greenberg SG, Wang J, Xia DF, et al. Enhancing zinc-finger-nuclease activity with improved obligate heterodimeric architectures. Nat Methods 2011;8(1): 74–9.

[194] Boch J, Scholze H, Schornack S, Landgraf A, Hahn S, Kay S, et al. Breaking the code of DNA binding specificity of TAL-type III effectors. Science 2009;326(5959):1509–12.

[195] Moscou MJ, Bogdanove AJ. A simple cipher governs DNA recognition by TAL effectors. Science 2009;326(5959):1501.

[196] Maeder ML, Thibodeau-Beganny S, Sander JD, Voytas DF, Joung JK. Oligomerized pool engineering (OPEN): an "open-source" protocol for making customized zinc-finger arrays. Nat Protoc 2009;4(10):1471–501.

[197] Morbitzer R, Elsaesser J, Hausner J, Lahaye T. Assembly of custom TALE-type DNA binding domains by modular cloning. Nucleic Acids Res 2011;39(13):5790–9.

[198] Mussolino C, Morbitzer R, Lutge F, Dannemann N, Lahaye T, Cathomen T. A novel TALE nuclease scaffold enables high genome editing activity in combination with low toxicity. Nucleic Acids Res 2011;39(21):9283–93.

[199] Hockemeyer D, Wang H, Kiani S, Lai CS, Gao Q, Cassady JP, et al. Genetic engineering of human pluripotent cells using TALE nucleases. Nat Biotechnol 2011;29(8):731–4.

[200] Li T, Huang S, Zhao X, Wright DA, Carpenter S, Spalding MH, et al. Modularly assembled designer TAL effector nucleases for targeted gene knockout and gene replacement in eukaryotes. Nucleic Acids Res 2011;39(14):6315–25.

[201] Ding Q, Lee YK, Schaefer EA, Peters DT, Veres A, Kim K, et al. A TALEN genome-editing system for generating human stem cell-based disease models. Cell Stem Cell 2013;12(2): 238–51.

[202] Roos WP, Kaina B. DNA damage-induced cell death by apoptosis. Trends Mol Med 2006; 12(9):440–50.

[203] Bree RT, Neary C, Samali A, Lowndes NF. The switch from survival responses to apoptosis after chromosomal breaks. DNA Repair (Amst) 2004;3(8–9):989–95.

[204] Quasdorff M, Protzer U. Control of hepatitis B virus at the level of transcription. J Viral Hepat 2010;17(8):527–36.

[205] Bar-Yishay I, Shaul Y, Shlomai A. Hepatocyte metabolic signalling pathways and regulation of hepatitis B virus expression. Liver Int 2011;31(3):282–90.

[206] Fukai K, Takada S, Yokosuka O, Saisho H, Omata M, Koike K. Characterization of a specific region in the hepatitis B virus enhancer I for the efficient expression of X gene in the hepatic cell. Virology 1997;236(2):279–87.

[207] Su H, Yee JK. Regulation of hepatitis B virus gene expression by its two enhancers. Proc Natl Acad Sci USA 1992;89(7):2708–12.

[208] Friedmann T, Roblin R. Gene therapy for human genetic disease? Science 1972; 175(4025):949–55.

[209] Munyon W, Kraiselburd E, Davis D, Mann J. Transfer of thymidine kinase to thymidine kinaseless L cells by infection with ultraviolet-irradiated herpes simplex virus. J Virol 1971; 7(6):813–20.

[210] Fu Y, Foden JA, Khayter C, Maeder ML, Reyon D, Joung JK, et al. High-frequency off-target mutagenesis induced by CRISPR-Cas nucleases in human cells. Nat Biotechnol 2013.

CHAPTER 7

Gene Therapy for Hepatitis C Virus Infection

7.1 DISCOVERY OF HEPATITIS C VIRUS

The search for hepatitis C virus (HCV) was prompted by studies published in the 1970s that reported on patients who developed hepatitis after blood transfusion (Figure 7.1). These cases of apparent viral infection, not caused by hepatitis A virus or hepatitis B virus, were termed non-A, non-B hepatitis (NANBH) [1–3]. To identify HCV, detailed characterization of human serum samples was performed on experimentally infected chimpanzees [4]. Isolation and sequence analysis of the HCV genome were then reported in 1989 [5]. The methodology entailed preparation of a cDNA library from RNA isolated from experimentally inoculated chimpanzees and then screening of cDNA-encoded proteins with patients' sera to identify the cDNA that was specific to NANBH infections. At the time, this was an impressive feat and was the first identification of a viral pathogen using applied molecular biology techniques.

This early characterization of HCV led rapidly to more detailed understanding of the virus. HCV was found to be an RNA virus with a positive single-stranded genome, which belongs to the family of *Flaviviridae*. The *Hepacivirus* genus, for which HCV is a member, is one of four genera within the *Flaviviridae* family. Other members of the *Hepacivirus* genus are canine hepacivirus [6], the first non-primate hepacivirus, bat hepacivirus [7], rodent hepacivirus, and GB virus B (GBV-B). GBV-B was originally isolated in 1967 from a surgeon, with the initials of G.B., who had acute unexplained hepatitis [8]. When passaged through tamarins and New World monkeys, serum from this patient caused hepatitis. However, despite this apparently significant observation, GBV-B has not again been isolated from patients with hepatitis.

7.2 EPIDEMIOLOGY AND CLINICAL SIGNIFICANCE OF HCV INFECTION

The global prevalence of people with anti-HCV increased to more than 185 million by 2005, which represents 2.8% of the world's human population [9]. The infection is most common in inhabitants of North African, Middle Eastern, and Central and East Asian countries, where it may occur in up to 10% of the

Gene Therapy for Viral Infections. http://dx.doi.org/10.1016/B978-0-12-410518-8.00007-7

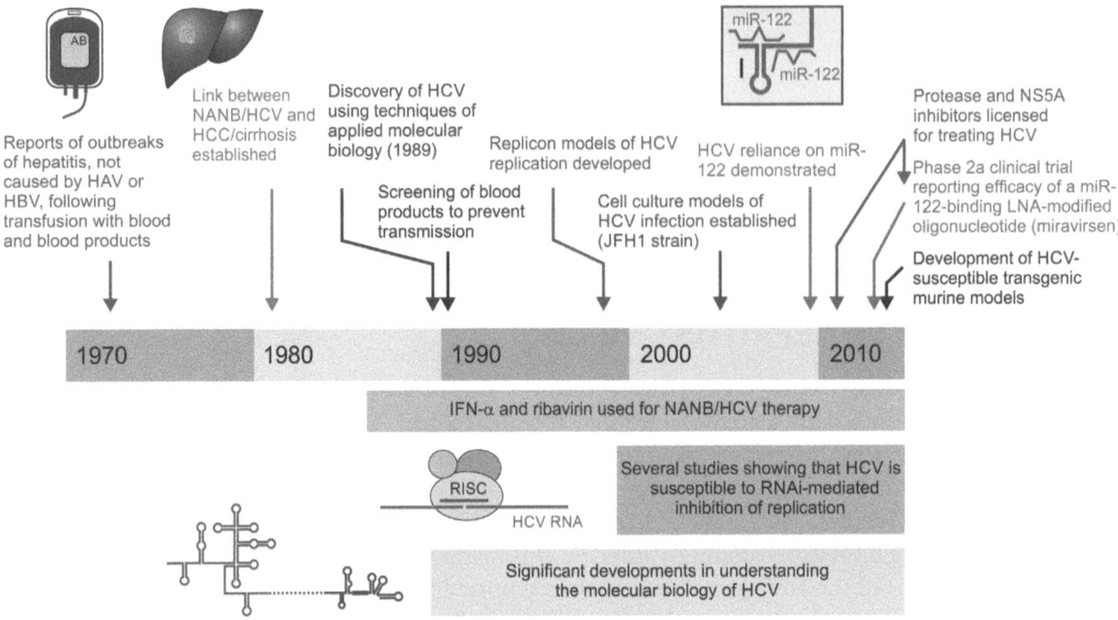

FIGURE 7.1 Timeline of the major milestones in HCV research.

general populations (Figure 7.2). The infection follows an insidious clinical course, and approximately half of the people who are infected with HCV are not aware that they carry the virus [10]. The virus is eliminated in ~20% of cases and persists in the remaining 80%. It is estimated that cirrhosis will develop 20–30 years after HCV infection in 25–30% of cases [11,12]. Life-threatening end-stage liver disease, including hepatocellular carcinoma, occurs in more than 25% of these cirrhotic patients. In the absence of serious complications, symptoms of HCV infection are usually nonspecific and include fatigue, arthralgia, abdominal pain, and loss of appetite. Coupled with the lack of a convenient immunoassay to identify active viral replication, the vague features make it difficult to identify and monitor HCV-infected patients. In the United States, HCV is now a more significant underlying factor contributing to mortality than is infection caused by human immunodeficiency virus (HIV)-1 [13]. A particularly important causative factor is that HCV infection is diagnosed approximately half as often as is HIV-1 infection in fatal cases ascribed to either of the infections [11,14].

Infection with HCV occurs as a result of parenteral transmission. Transfusion of HCV-infected blood was a major source of infection, but with the widespread availability of tests to screen for markers of the virus in donated blood, transfusion-related transmission of HCV has been minimized [11]. In addition to screening of blood-derived products, diminished HCV transmission may

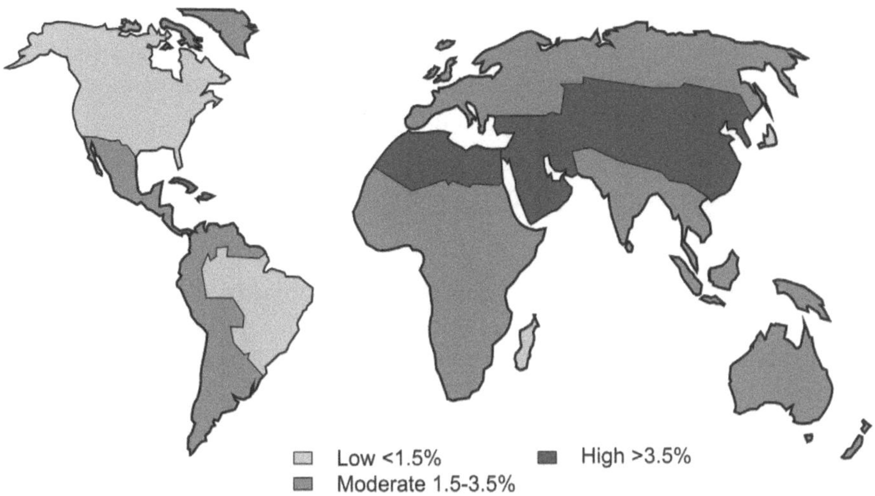

FIGURE 7.2 Geographical distribution of areas where chronic HCV infection is most prevalent. Shaded regions of the global map indicate prevalence of HCV infection throughout the world. Most HCV-infected individuals are located in North Africa, the Middle East, and central and East Asian countries. *Adapted from Mohd Hanafiah et al. [9].*

also be a result of behavioral changes that have been aimed at limiting HIV-1 spread. The overall incidence of HCV infections in the United States has now considerably diminished [11,15]. However, transmission of the virus among intravenous drug users (IDUs) is assuming greater importance, particularly in developed countries [11,16]. Identifying these individuals is understandably difficult and complicates management of their HCV infections. Managing HCV transmission is also made difficult by confounding infection-predisposing variables. The role of HCV transmission through sexual contact is particularly relevant. It appears that sexual transmission of HCV is very rare in otherwise healthy individuals [17,18]. However, because having multiple sex partners may be more common in IDUs who do not acknowledge their drug practices, greater importance of sexual transmission may incorrectly be ascribed as the source of HCV transmission [11]. Understanding the role of sexual contact in the spread of the virus is further confused by factors that do indeed increase the risk of HCV infection by this route. When there is co-infection with HIV-1, risk for acquiring HCV infection in women and men is increased [17,18]. In addition, other factors that compromise genital mucosal integrity, such as herpes virus and syphilis infection, increase HCV transmission through sexual contact.

Extrahepatic manifestations of HCV infection are a frequent occurrence and are found in 40–70% of cases [19,20]. Signs and symptoms may result from the production of autoimmune antibodies and include cryoglobulinemia, autoimmune thyroiditis, diabetes mellitus, lymphoproliferative disorders, renal disease, and various dermatological conditions. HCV itself may also infect nonhepatic cells such as peripheral blood leukocytes [21] and cells found in the central nervous system [22]. The full clinical significance of infection of extrahepatic tissues by HCV is not yet fully clear.

7.3 HCV GENOME AND VIRAL REPLICATION

As a member of the *Hepacivirus* genus, HCV has a single-stranded RNA genome comprising ~9.6 kb of positive/sense polarity (Figure 7.3). The genome is highly plastic and sequence diversity is a prominent feature of HCV isolates. The seven main genotypes, numbered in the order of their discovery, have sequence divergence of up to 35% [23,24]. There are also multiple HCV subtypes, named with lower case letters, which have a lower sequence diversity of ~20%. Recombinant forms of the HCV genome have also been described [25,26], which further add to the high degree of variability of the HCV genome. With the exception of the conserved 5′ nontranslated region (NTR), *core*, and envelope (E) regions, sequence differences between isolates are distributed throughout the genome [27]. The genotypes have

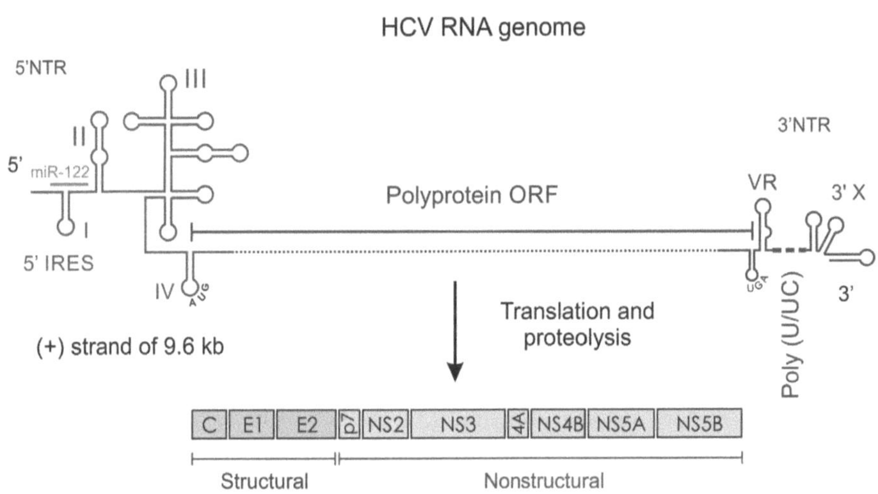

FIGURE 7.3 Organization of HCV genome.

The HCV positive strand RNA with polyprotein-encoding sequence and the 5′ and 3′ NTRs are schematically shown. The 10 structural and nonstructural proteins derived from processing of the precursor protein are indicated. The 5′ NTR comprises four helical domains (I, II, III, and IV), which are important for the initiation of HCV RNA translation. miR-122 binds to HCV RNA at helical domain I. The 3′ NTR also comprises several loops, which are required for efficient viral translation. These loops are the variable region (VR), poly (U/C) sequence, and 3′ X. C, Core; E1, envelope-1; E2, envelope-2; NS2, nonstructural protein-2; NS3, nonstructural protein-3; NS4A, nonstructural protein 4A; NS4B, nonstructural protein 4B; NS5A, nonstructural protein 5A; NS5B, nonstructural protein 5B.

distinct geographical distributions; for example, genotype 1 is common in North America, Europe, and Japan, whereas genotypes 4 and 5 are found mainly in Africa. Clinical progression and response to therapy are also reported to vary according to genotype. Subgenotype 1b HCV is associated with a particularly aggressive clinical course and is more frequently found in patients with cirrhosis and decompensated liver disease [28–30]. Outcomes from HCV treatment may be influenced by the viral genotype [31]. Before the recent licensing of HCV protease inhibitors, when the combination of polyethylene glycol (PEG)-interferon (IFN)-α and ribavirin were the standard of care (Figure 7.1), cure from infection with genotypes 1 and 4 was achieved in less than 50% of cases [32–34]. A sustained virological response occurs in 70–80% of treated cases infected with genotypes 2 and 3, whereas elimination of genotypes 5 and 6 is achieved with intermediate success. Infection with all HCV genotypes should improve with recently licensed and candidate drugs in clinical trials [31,35].

The uncapped HCV genome of 9.6 kb has 5′ and 3′ NTRs and encodes a single large polyprotein of approximately 3000 amino acids [36] (reviewed in ref. [37]; Figure 7.3). Initiation of translation is guided by the 330-nt internal ribosomal entry site (IRES) that is located within the 5′ NTR. The IRES is highly conserved and has a complex structure with several stem, loop, and bulge regions that fold into four helical domains (I, II, III, and IV) in an arrangement that is critical for translation initiation [38,39]. The IRES binds to the 40S human ribosomal subunit in the absence of eukaryotic translation initiation factors. Assembly of the complete ribosome is later dependent on maintenance of the HCV IRES structure. The conserved 3′ NTR also plays a role in regulating HCV translation. As with the 5′ NTR, the 3′ NTR interacts with the 40S ribosomal subunit, and there is some overlap in the RNA-binding sites to the ribosome [40]. Stimulation of translation occurs by a mechanism that occurs subsequent to the initial round of translation. It is thought that interaction of the 3′ NTR with the 40S ribosomal subunit retains viral RNA on the ribosomes. This explains why the later rounds of viral RNA translation, rather than the earlier rounds, are stimulated by effects of the 3′ NTR in *cis*.

Demonstration that HCV replication is dependent on microRNA (miR)-122 was a particularly interesting finding [41]. This effect is mediated by binding of miR-122 to two conserved sequences at the 5′ end of HCV RNA, termed site 1 (S1) and site 2 (S2) (Figure 7.4). Because miRs typically bind to 3′ untranslated regions of genes to cause translational suppression (Chapter 2), the stimulatory effect miR-122 on HCV replication is particularly curious. As with RNAi-mediated gene silencing, the mechanism of HCV replication stimulation involves recruitment of RNA-induced silencing complex (RISC) components and requires interaction of the miR-122 seed with the HCV RNA

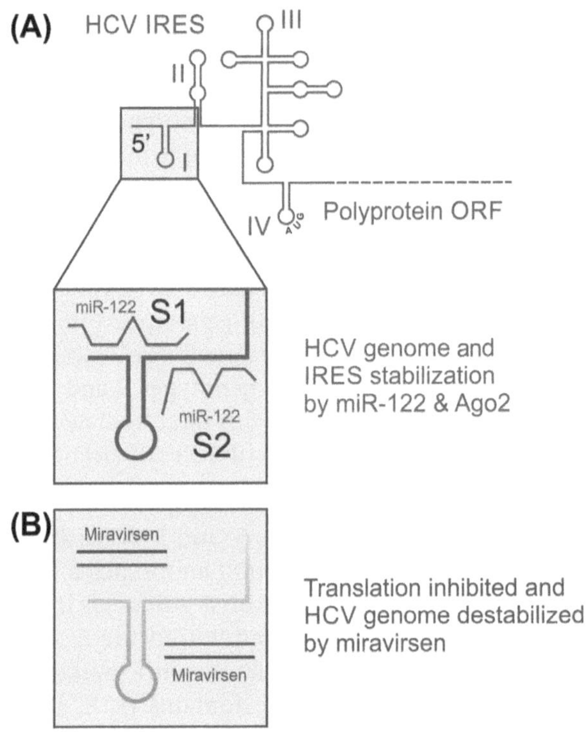

FIGURE 7.4 Binding sites for miR-122 within the 5′ NTR and mechanism of action of miravirsen (SPC3649).

(A) There are two binding sites for miR-122, S1 and S2, within the helix I domain of the 5′ NTR of the HCV genome. Interaction of miR-122 with the 5′ NTR results in Ago2-mediated stabilization of HCV RNA and enhanced translation of the viral polyprotein. (B) Miravirsen, an miR-122-complementary oligonucleotide, binds and disables this important HCV dependency factor.

cognate [42]. The effect is at least partially mediated by stimulation of HCV RNA translation [43,44] and may involve enhancement of the viral IRES function [45]. In addition, miR-122 may increase the stability of HCV RNA [44,46]. The requirement of HCV for miR-122 prompted investigation of the therapeutic effect of inhibition of function of this host dependency factor. Miravirsen (SPC3649), which is a modified oligonucleotide that binds to and inhibits miR-122 function, has thus been developed as a candidate HCV therapeutic [47].

The large precursor polyprotein of HCV is processed by proteolytic cleavage to form 10 proteins that have structural and nonstructural functions [37,48] (Table 7.1 and Figure 7.3). The three structural proteins—core, E1, and E2—are derived from the N-terminal region. They perform functions required for the formation of the capsid (core) and envelope (E1 and E2) of the virion. The seven nonstructural proteins are essential for the membrane-associated

Table 7.1 HCV-Encoded Structural and Nonstructural Proteins

Protein	Function	Type
C	Capsid formation	Structural
E1	Envelope	
E2	Envelope	
P7	Viroporin	Nonstructural
NS2	Cysteine protease and assembly	
NS3	Serine protease, helicase, NTPase	
NS4A	Cofactor of NS3	
NS4B	Membranous web formation	
NS5A	RNA binding and RNA replication	
NS5B	RNA-dependent RNA polymerase (RdRp)	

cytoplasmic replication of HCV. These proteins include p7, NS2, NS3, NS4A, NS4B, NS5A, and NS5B (reviewed in ref. [49]). Maturation of the HCV proteins occurs through a series of protein cleavage reactions that involve cellular proteases during formation of the structural proteins and viral proteases for maturation of the nonstructural proteins. Initially, the polyprotein is targeted to the endoplasmic reticulum by an internal signal peptide that is located between the core protein and E1 sequence [50] (Figure 7.5). Signal peptidase releases the core protein with coupled signal peptide, and subsequent action by an intramembrane-cleaving signal peptide protease frees the mature core protein. The core protein is then released and interacts with lipid droplets (LDs), where cholesterol and triglycerides are stored within the hepatocyte. The cellular signal peptidase also cleaves the polyprotein to release E1, E2, and p7. The dimeric NS2-NS3 cysteine protease is formed by a mechanism that entails interaction of two NS2-NS3 molecules and self-cleavage at the NS2/NS3 junction [51]. The NS3 serine protease is stabilized and directed to the ER by the NS4A co-factor protein [52]. This NS3–NS4 complex then causes cleavage at the junctions of the NS3, NS4A, NS5A, and NS5B proteins to release the remaining nonstructural HCV proteins.

Interestingly, proteolytic action of the NS3–NS4A complex is also responsible for inhibiting the host innate immune response to HCV. As with most viral infections, pathogen-associated molecular patterns activate pattern recognition receptors (PRRs), such as the retinoic acid inducible gene-I and Toll-like receptor-3, to cause downstream antiviral defenses (Chapter 1). NS3–NS4A acts on the PRR adaptor proteins, mitochondrial antiviral signaling protein [53], and Toll or interleukin domain-containing adaptor-inducing IFN-β [54]. Proteolysis of these mediators of innate immunostimulation counters cellular responses to HCV infection.

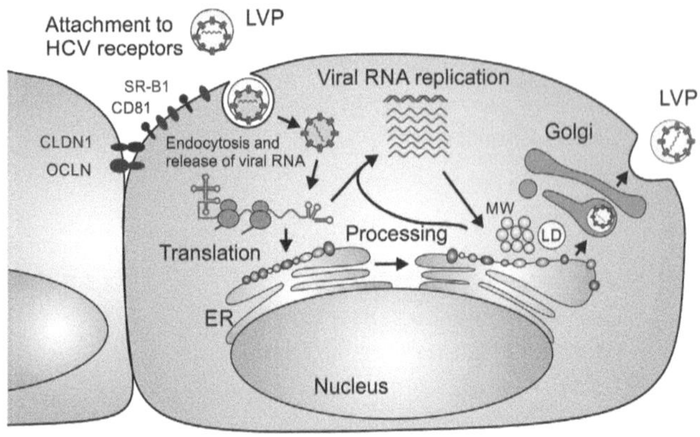

FIGURE 7.5 **HCV replication cycle.**
Stages involved in the replication of HCV are schematically illustrated. After attachment to receptors, which include SR-B1, CD81, CLDN1, and OCLN, the LVP is internalized by endocytosis and HCV RNA is released. Translation results in formation of the polyprotein precursor, which is processed to give rise to the individual structural and nonstructural proteins. Viral RNA replication occurs through a mechanism that involves nonstructural proteins and use of the genomic RNA to form the minus (–) strand RNA, which in turn serves as a template for synthesis of the plus (+) strand genomic RNA. Assembly of the HCV LVPs involves participation of LDs, the membranous web, and secretion via the endoplasmic reticulum and Golgi. SR-B1, scavenger receptor class B type I; ER, endoplasmic reticulum; MW, membranous web.

HCV particle formation occurs on specialized areas of the ER, termed the membranous web [55,56], and is coupled to secretion of lipoproteins from the hepatocyte (Figure 7.5). Formation of the aggregates of spherules that comprise the membranous web is induced by the viral NS4B and NS5A proteins. The process of HCV particle formation has been divided into three stages: (1) initial interaction of core protein with LDs at the cytosolic side of the ER; (2) transport of other components of the HCV replication complexes (RCs) to the LDs and assembly of viral capsids; and then (3) maturation and secretion of the viral particles via the ER and Golgi, which involves incorporation of E1 and E2 proteins and envelopment within a lipid bilayer [57].

Interaction of the core protein with LDs is essential for replication of HCV. There are two domains of the mature HCV core protein, termed D1 and D2, and it is the D2 region that binds to the LDs [58,59]. This domain comprises two amphipathic α-helices and an intervening hydrophobic loop structure. Correct folding of the helices requires a hydrophobic environment, which is likely to occur through interaction with the phospholipid bilayer of the LDs. Mutations within the D2 domain that inhibit LD binding disrupt the release of HCV from hepatocytes [60,61]. Conversely, stimulation of HCV replication may occur as a result of a D2 mutation that enhances LD interaction [62]. Numerous residues of the core protein are required for the secretion of HCV [63].

However, compensatory mutations in p7 and NS2 may rescue the core-related defect, indicating that these nonstructural proteins also have a role in HC virion formation.

The NS5A protein plays a crucial role in facilitating the replication of HCV RNA and trafficking of the RC to the LDs and assembly of the virion (reviewed in ref. [57]). This viral protein has been successfully targeted with development by Gilead Sciences of the recently licensed drug, Ledipasvir (http://www.gilead.com/research/pipeline). There are three domains of the NS5A protein: DI, DII, and DIII. DI and parts of DII are needed for HCV RNA replication [64], DI facilitates trafficking of the RC, whereas DIII is required for specific interaction with the core protein of the LD [65]. Mutational analysis of NS5A DIII sequences revealed that a single serine residue is required for early virus replication, and phosphorylation of this amino acid by casein kinase II is essential for facilitating HCV replication [66]. However, in an overall hyperphosphorylated state of NS5A, p58, is associated with inhibition of viral replication [67]. The phosphorylation status of NS5A is also influenced by lipid kinase phosphatidylinositol 4-kinase III-α (PI4KIIIα), which is needed for HCV replication [68]. This host enzyme catalyzes the increased synthesis of phosphatidylinositol 4-phosphate that occurs in HCV-infected cells. Interaction of PI4KIIIα with a conserved sequence comprising seven amino acids in the carboxy-terminal region of DI of NS5A augments enzyme activity of the kinase. PI4KIIIα does not have a direct kinase action on NS5A, but hyperphosphorylation of NS5A results from mutation-induced disruption of binding between NS5A and PI4KIIIα. Therefore, PI4KIIIα interacts with DI of NS5A to result in an overall hypophosphorylated state that facilitates HCV RNA replication.

After the initiation phase of HCV assembly, a complex process that involves participation of host factors and HCV nonstructural proteins is required for virion maturation and secretion from the infected hepatocyte [69]. Core protein causes clustering of LDs at the periphery of the hepatocyte nuclei through a mechanism that requires microtubules and involves displacement of adipocyte differentiation-related protein from the LD surface [70]. Interaction with tubulin and actin proteins to form microtubular tracks for the RCs may also play a role [71]. p7, NS2, and NS3A are also required for the assembly of HCV particles. The small hydrophobic p7 viroporin protein, which comprises 63 amino acids, is essential for the propagation of viral particles in vivo [72] and in culture models [69,73]. p7 functions by oligomerization to form cation-selective pores that reduce the acidity of intracellular organelles [74]. Interestingly, HCV generation may be enhanced by some p7 mutations [75], and there is variability in the efficiency of viral particle formation that results from chimeric HCV genomes that contain p7 sequences from different strains [73]. NS2 plays a role in late stages of HCV particle production that is independent of its role as a protease [76]. NS2 interacts with the E2 glycoprotein and enables HCV particle production through a mechanism that appears to entail facilitation of the association of E2 glycoprotein with the core particles in the nascent virions.

The NS5B protein functions as an RNA-dependent RNA polymerase (RdRp) and is central to the replication of the HCV RNA genome. The enzyme is highly prone to error, which results in a high degree of genome variability [77]. NS5B has canonical features of nucleic acid polymerases, but it also demonstrates unique characteristics, such as the extensive interactions between the fingers and thumb domains of the polymerase that encircle the enzyme's active site [78]. HCV RNA replication occurs through the production of a negative-stranded RNA sequence using the positive-stranded RNA genome as a template. In turn, the positive-stranded genome is produced from the negative-stranded template. Initiation of HCV negative and positive strand synthesis is likely to rely on sequences present in the genomic 3′ NTR and 5′ NTR, respectively. Evidence indicates that the positive strand is present in 20- to 100-fold excess over the negative strand. The lower concentration of the negative strand led to investigating the utility of targeting this essential viral replication intermediate to treat the viral infection [79,80].

The HCV RdRp is capable of initiating RNA synthesis de novo, but it requires accessory functions of NS3A [81] and NS5A [64] during viral RNA synthesis. The role of NS3A in HCV RNA replication is independent of its protease function. NS3A contains a DExH/D helicase/NTPase domain [81,82], which may be important for separation of newly synthesized HCV RNA from its template, displacement of RNA-binding proteins, and resolution of secondary structures [49]. Interaction between NS5B and NS5A bound to Cyclophylin A (CypA), a host factor, induces conformational changes in the HCV RNA replication machinery that facilitates NS5B-mediated function [83]. A requirement for CypA is interesting because antagonists of CypA, such as cyclosporin A [83] and the nonimmunosuppressive analog DEBIO-025 [84], have antiviral efficacy.

The E1 and E2 envelope proteins each have glycosylated N-terminal ectodomains and hydrophobic C-terminal anchors [85]. The membrane-bound C-terminal region is important for the association of E1 and E2 to form the envelope glycoprotein heterodimer. The LDs to which RNA-containing HCV capsids are associated are very-low-density lipoprotein (VLDL) precursors, and release of HCV particles is coupled to the VLDL secretion pathway [86]. During the process, the apolipoproteins ApoB, ApoC, and ApoE are acquired by the particles before secretion via the Golgi [87–89]. The complex formed between lipoprotein and viral elements led to use of the term lipoviroparticles (LVPs) to describe circulating HCV particles, which also have a lower buoyant density than their intracellular precursors. LVPs have varied shape and their structure differs according to the source from which they are isolated [90]. The physical association of VLDL with HCV particles is likely to provide protection for the virus against the host's neutralizing immune response [49,91].

7.3.1 HCV Entry into Hepatocytes

HCV has a narrow species tropism and naturally only infects humans and chimpanzees [92]. The entry of HCV into hepatocytes involves LVP interaction with multiple host factors, some of which have species-specific properties to confer the narrow host tropism of HCV (reviewed in ref. [49]). Initial studies indicated that HCV interaction with the LDL receptor is partially responsible for virion uptake [93]. Binding of E2 to glycosaminoglycans on hepatocyte cell membranes was subsequently also shown to be important for HCV uptake [94]. The mechanism involves interaction of positively charged residues on the hypervariable region of E2 with heparin sulfate proteoglycans. After these early events, E1–E2 virion glycoproteins bind to the human scavenger receptor class B type I (SCARB1) [95] and CD81 [96], among others (Figure 7.5). HCV appears to enter the hepatocyte via the tight junctions, where interactions with occludin (OCLN) [97] and claudin-1 (CLDN1) [98] are important. The essential minimal factors that are required for HCV infection of murine hepatocytes are human CD81 and OCLN [99]. An important property of HCV is that the virus may be spread from one cell to another by direct contact without traversing extracellular compartments [100,101].

7.4 MODELS OF HCV INFECTION

Several different models have been developed to simulate parts or all of the HCV replication cycle (Table 7.2). The simplest and most convenient methods for assessing HCV gene silencing have involved coupling of viral sequences to reporter genes, such as *Renilla* luciferase. Reporter gene activity is then used as the indicator of efficacy of anti-HCV drug candidates. This approach provides useful information, but it has significant limitations [80]. The plasmid-encoded reporter mRNA and HCV mRNA sequences (1) do not localize to the same intracellular sites as do natural HCV elements, (2) interact differently with viral and cellular proteins, and (3) have different secondary and tertiary structures.

Early studies that aimed to reproduce parts of the HCV replication cycle in cultured cells used the subgenomic replicon systems (reviewed in refs [102,103]; Figure 7.6). Replicons lack sequences encoding the structural elements that are required to propagate HCV. Development of the replicon system was prompted by the observation that other sense strand viruses do not require structural proteins for replication of genomic RNA sequences. The first HCV replicons were derived from HCV genotype 1 [104]. Structural genes encoding C, E1, and E2 were removed. The neomycin phosphotransferase (*neo*) gene, which confers resistance to G418, was fused in frame to the open reading frame encoding the

Table 7.2 Models Commonly Used to Assess the Efficacy of HCV-Targeting Gene Therapy

	Model	Advantages	Disadvantages
Cell culture	Reporter gene fusion	Convenient Rapid	Replication of virus is not simulated. Target structures and accessibility differ from those of natural HCV during infection.
	Replicon	Convenient Rapid Good simulation of functions of nonstructural HCV proteins	Complete life cycle of virus not replicated. Inherent disadvantages of cell culture models (e.g., lack of immune response, delivery efficacy not assessed).
	Infection of cultured cells	All stages of life cycle replicated Convenient	As above, disadvantages that are characteristic of cell culture models (e.g., lack of immune response, delivery efficacy not assessed).
In vivo	Hydrodynamic transfection	Convenient Rapid	Replication of virus is not simulated. Target structures and accessibility differ from those of natural HCV during infection. Liver pathogenesis not reproduced.
	uPA/SCID mouse human hepatocyte xenograft	Simulates all stages of HCV replication in vivo Human hepatocytes infected	Technically demanding. Incomplete host immunity to HCV. Liver pathogenesis not reproduced.
	HCV infection-susceptible transgenic mice	Convenient All stages of HCV infection replicated HCV-specific immune responses intact Potentially a good challenge model	Differences between natural and murine HCV infection needs to be established.
	Chimpanzee	Complete mimicking of HCV replication Intact immune response similar to that of humans	Ethical concerns. Expensive. Specialized facilities required.

N-terminal 12 amino acids of C. To enable expression of nonstructural HCV proteins, an encephalomyelitis virus IRES was inserted downstream of *neo*. Propagation of the HCV-derived sequences entails transcription of replicon RNA in vitro, then transfection of Huh7 cells, and selection with G418. Several

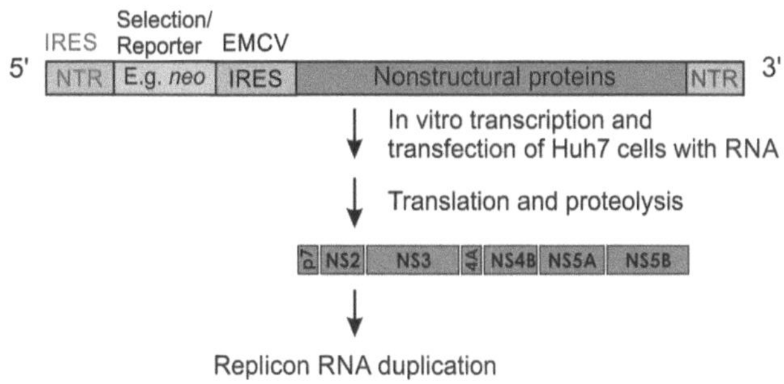

FIGURE 7.6 Essential features of typical HCV replicons.

Structural protein-encoding sequences of the HCV genome are removed and replaced with a selection or reporter gene. The 5′ NTR and 3′ NTR are retained together with the sequences encoding the nonstructural proteins. To enable translation of the nonstructural proteins, an additional IRES is included, such as that which is derived from the encephalomyocarditis virus. Use of a DNA template to transcribe these sequences in vitro, followed by transfection of Huh7 cells, leads to duplication of the replicon RNA. In the case of using the *neo* gene, antibiotic selection enables enrichment with replicon-containing cells. EMCV, encephalomyocarditis virus; *neo*, neomycin resistance gene.

variations on the replicon system have been described, and adaptive mutants have been isolated that are capable of more efficient replication [105]. Replicons derived from different HCV genotypes have been propagated, and reporter genes have been incorporated to facilitate measurement of their proliferation.

Recapitulating all stages of the HCV life cycle in cultured cells was achieved in 2005 after the development of methods of propagating HCV in tissue culture [106,107]. Sequences of the JFH1 strain of HCV genotype 2a, cloned from an individual with fulminant hepatitis, were shown to be capable of supporting replication of the virus in cultured human hepatoma cells (Huh7). Although convenient for assessing the efficacy of anti-HCV drugs, the model of course has limitations because it does not recapitulate all features of HCV infection in vivo.

Studies on chimpanzees were crucial to the identification of HCV and thereafter provided useful insights into the mechanisms of immune responses and disease progression during HCV infection. These animals, unlike other primates, provide a model that closely simulates all stages of natural HCV infection that occur in humans. However, there are significant ethical and cost-related considerations that limit the use of these animals. As a result, concerted efforts have been devoted to developing small animal models of HCV infection [92]. As with HBV, immunocompromised uPA/severe combined immunodeficient (SCID) mice (Chapter 6) have been xenografting with human hepatocytes and then successfully infected with HCV [108]. The $Fah^{-}/Rag2^{-/-}$ $IL2rg^{-/-}$ (FRG) mice, which are similarly immunocompromised and tolerate xenografting with HCV infection-susceptible human hepatocytes, have also been used to model HCV infection in vivo [109]. Human hepatocyte infection, virus replication, and innate immunity are faithfully reproduced in these animals and have proved useful to model HCV infection. However, the mice do not provide information on the role of the adaptive immune response during a therapeutic response. An added complication of the models is that the procedures for population of mouse livers with human hepatocytes and infection with HCV are technically demanding. There is an associated high mortality and variability of infection kinetics in inexperienced hands.

Insights into the mechanisms of HCV entry into hepatocytes have recently provided the means to achieve genetic modification required to render mice susceptible to HCV infection. In one of the first studies, recombinant adenoviral vectors were used to deliver sequences encoding human CD81, SCARB1, CLDN1, and/or OCLN to mice [110]. The results demonstrated that HCV did indeed infect mice that received the human orthologous genes, although only *CD*81 and *OCLN* were necessary for the infection. To overcome the problems of innate immunostimulation that accompanies adenoviral vector administration, transgenic mice that express the human *CD*81, *SCARB*1, *CLDN*1, and *OCLN* genes in a liver-specific manner were propagated [111]. As with treatment of mice with recombinant adenoviruses, analysis showed that human CD81 and OCLN were the essential factors required for HCV infection of murine hepatocytes in vivo. It is likely that murine orthologs of the other cellular uptake factors were capable of supporting HCV entry into hepatocytes. In the transgenic animals, replication was significantly enhanced in immunocompromised mice. Viral particles from mice were also capable of infecting naïve Huh-7.5 cells, confirming that all stages of the HCV life cycle could be recapitulated in the animal model. Generation of these transgenic mice represents an important advance in the development of models that may be used for HCV drug development. However, persistence of replication is dependent on compromised innate immunity. Availability of mice that are immunocompetent, yet still susceptible to chronic HCV infection, would be a further improvement.

7.5 CURRENT AND NEW HCV TREATMENT

Ever improving insights into HCV molecular biology have led to the development of numerous promising anti-HCV drug candidates [49]. For several years, the standard of care has been a combination of ribavirin and PEG-IFN-α, but this is changing with the approval of newer direct acting antivirals (DAAs). Efficacy of therapy has varied and has been influenced by viral genotypes and host-encoded factors. For example, HCV genotype 1 is most resistant to treatment [112], and polymorphisms of the interleukin-28 gene influence the response to therapy [113]. The objective of achieving IFN-free treatment regimens, which should be better tolerated than current therapies, is now feasible. In 2011, the HCV NS3 protease inhibitors, telepravir and bocepravir, were licensed for treatment of the infection in patients infected with the genotype 1 strain of the virus. Newer NS3 antagonists (e.g., Neceprevir) and DAAs that inhibit functioning of HCV polymerase and NS5A are now at advanced stages of clinical development. The NS5A antagonists (e.g., Ledipasvir and Daclatasvir) generally demonstrate very high potency and low-to-medium barriers to resistance. NS5B viral polymerase inhibitors include nucleoside/nucleotide analogs (e.g., Sofosbuvir) and non-nucleoside (e.g., ABT-333) analogs. Although the potency of these drugs is high, they are usually effective against specific HCV genotypes or subgenotypes and resistance to antiviral effects may emerge. Host-targeting agents (HTAs) have also reached phases II and phase III of the clinical trial. Drugs in this class have highly varied mechanisms of action and impose a high barrier to HCV resistance. HTAs at an advanced stage of development include the cyclophylin inhibitors (e.g., Alisporivir) [114]; the miR-122 antagonist (Miravirsen or SPC3649) [47]; and viral entry inhibitor, ITX-5061 [115]. The expanded repertoire of anti-HCV agents augurs well for the future of HCV treatment. Recent licensing to Gilead of a combination drug comprising Lepidasvir with Sofosbuvir is a particularly important development for attaining the goal of IFN-free treatment of HCV infection.

7.6 COUNTERING HCV WITH NUCLEIC ACIDS

Because HCV is an RNA virus that replicates in the cytoplasm, nucleic acid-based approaches to target the viral genetic material directly have been widely used. These antiviral strategies have largely entailed the use of RNAi activators, aptamers, antisense sequences, and ribozymes. Although utility has been demonstrated, the error-prone nature of the viral NS5B RdRp results in the introduction of variability in viral targets to diminish antiviral efficacy of therapeutic nucleic acids. Therefore, highly conserved regions of the HVC genome, such as the 5′ NTR and 3′ NTR, have been favored as targets of hybridizing therapeutic sequences. Combinatorial approaches that target multiple HCV sequences and silencing of host factors have been used to limit emergence of resistant strains. An example is disabling of miR-122, a host factor required by HCV, to inhibit viral replication and provide a barrier to resistance. Gene editing as a means of disabling HCV host factors has also recently been used. Two studies have reported on the use of transcription activator-like effector nucleases to inactivate expression of ApoB100 [116] and diacylglycerol acyltransferase-1 [117] to render target cells resistant to HCV infection. However, the approach has not been widely used, and the feasibility for treatment of HCV infection is yet to be established.

7.6.1 Targeting HCV with RNAi Activators

Many studies have demonstrated that HCV is susceptible to RNAi-mediated gene inactivation (Table 7.3). Both synthetic and expressed RNAi activators have been used against the virus, and early studies targeting the 5′ NTR demonstrated that both types of RNAi activators are capable of disabling HCV replication [118,119]. Results showed a significant inhibition of markers of proliferation by synthetic and expressed sequences. Further support for the suitability of targeting the 5′ NTR was provided by more recent studies [120,121]. Prabhu and colleagues initially used a screening assay that measured inhibition by short interfering RNAs (siRNAs) of reporter gene expression when fused to IRES sequences from each of six HCV genotypes [122]. The siRNAs that targeted loop II, but not loop III, of the IRES were very effective. Moreover, the loop II-targeting siRNAs effectively inhibited core protein expression from full-length infectious clones of the viral genotype 1. Chandra et al. reported on the use of combinations of 5′ NTR-targeting siRNAs that were incorporated into liposomal nanoparticles [121]. Delivery of the siRNAs to liver cells

Table 7.3 Selected Examples of RNAi-Based Anti-HCV Strategies

Antiviral agent with reference	Viral target	Current stage of antiviral application	Efficacy
miR122 antagonist (miravirsen) [47,156].	*5′ NTR*	Phase IIa clinical trial	Very good. Antagonism of a viral host factor provides a high barrier to resistance.
Combinatorial synthetic [121] and expressed [120] antiviral RNAi activators.	*5′ NTR*	Cell culture and preclinical assessment in a xenograft murine model	Very good. Effective against different viral genotypes and provides a high barrier to resistance.
Expressed RNAi activators [123].	*3′ NTR*	Cell culture replicon	Good efficacy.
Expressed shRNAs in lentiviral vectors [125].	*NS3* and *NS5B*	Cell culture replicon	Good efficacy that was better than 5′ *NTR* targeting.
siRNAs [130].	*NS5B*	Cell culture replicon	Good efficacy that was attenuated after development of mutations at the replicon's target site. Better efficacy with administration of two siRNAs.
Expressed shRNAs with defined guide strand bias delivered with recombinant adeno-associated viruses (AAVs) [80]	HCV minus strand	Cell culture infection	Better effect on plus strand is favored.
Anti-HCV mimics of the miR-17-92 cluster delivered with recombinant AAVs [135]	*5′ NTR*, *core*, and *NS5B*	Cell culture and murine reporter assay	Good efficacy with minimal toxicity.
sshRNAs that are Dicer-independent [143,144]	*5′ NTR*	uPA/SCID humanized mice infected with HCV	>99% reduction in viral load with a high barrier to viral resistance when used in combination.

was demonstrated, and some of the siRNAs were capable of highly effective silencing of replicons and HCV replication after infection of cultured cells. The siRNA nanosomes did not cause liver injury in mice, and they were effective against the virus when tested in a murine xenograft model. Differences in the susceptibility of the 3′ NTR to RNAi-mediated silencing have been reported. Korf and colleagues showed replicon silencing with RNAi activators that targeted this region [123], but poor efficacy against the target was reported in two other studies [79,80]. Subtle variations in the structures of the RNAi activators and replicons used to assess efficacy are thought to be responsible for the differences.

Dose-dependent inhibition of reproduction of the HCV replicon was also demonstrated in one of the early studies aimed at using RNAi to inhibit HCV gene expression [124]. This report provided an early indication of the importance of HCV escape from hybridizing nucleic acids. It was observed that HCV replicon variants with target sites that differed by as few as 3 nt evaded silencing, and only when guide and HCV target sequences were completely complementary was silencing effective. In addition to the 5′ NTR, other regions of the HCV genome have been targeted with RNAi activators. Increased efficacy against HCV replicons was shown with siRNAs and short hairpin RNAs (shRNAs) against NS3 and NS5B, which had superior silencing when compared to 5′ NTR-targeting RNAi activators [125]. RNAi-activating viral gene silencers against *core*, *E1*, *E2*, *NS3*, *NS4B*, and *NS5B* sequences have also achieved impressive knockdown that ranged from 70% to 93% [126–128]. In an investigation aimed at targeting the viral *NS5B* gene, and one of the first studies performed to test efficacy in vivo, mice were subjected to hydrodynamic injection [129]. Co-administration of an HCV target fused to the Firefly luciferase gene and anti-HCV shRNA expression cassettes achieved significant target knockdown. Although targeting viral protein-encoding sequences may be useful, their mutability may compromise therapeutic efficacy. Concerns about the potential problems of accumulation of HCV mutations that evade silencing effects were vindicated by Wilson and Richardson [130]. In their study, repeated administration of *NS5A*-targeting siRNAs to replicon-expressing cells eventually resulted in selection of escape mutants. Point mutations were observed at the target site, but replicons were sensitive to silencing effects of siRNAs directed to a different genomic location, which supported the notion that simultaneously using multiple gene silencers to target different sites is useful to prevent emergence of escape mutants.

Targeting the minus strand of HCV is an interesting strategy to augment the efficacy of gene silencing efficacy. Virus replication entails formation of a minus strand RNA intermediate, which serves as the template for synthesis of the plus strand by NS5B. The minus strand is typically present in cells in 20- to 100-fold lower abundance than the plus strand. Because guide strands of synthetic or expressed RNAi activators may comprise plus or minus strand-targeting sequences, most studies reporting on RNAi efficacy have not distinguished silencing effects of the plus or minus viral RNA strand. Extensive analysis of a panel of siRNAs that had predicted plus or minus guide strand-targeting bias was initially performed [79]. Results from this analysis demonstrated that the negative strand was resistant to RNAi-based silencing. In another study, highly effective HCV-targeting shRNAs were used to generate asymmetric RNAi activators that were specifically active against either of the viral RNA strands [80]. Analyses confirmed that the minus strand of HCV is not accessible to the RNAi machinery, but plus strand sequences in the 3′ end of the 5′ NTR, *core*, and *NS5B* were good targets for silencing. The reasons for resistance of the minus strand to RNAi are not entirely clear. Association of the viral RNA minus strand with viral proteins within the membranous web and existence as part of a duplex RNA may be protective against exogenous RNAi activators. Different susceptibilities of RNAs that serve as protein synthesis templates or viral nucleic replication intermediates accord with the observations on rotavirus RNA [131]. The rotavirus nucleic acid replication template within the viroplasm was resistant to RNAi-based silencing, but the protein-encoding sequence could be silenced using RNAi.

7.6.1.1 Combinatorial RNAi to Limit Emergence of Escape Mutants

Mathematical modeling has indicated that use of at least four HIV-1-targeting RNAi activators is required to prevent viral escape by selected mutant strains of the virus [132]. Given that HCV is also prone to mutation and has a comparably sized genome, it is likely that HIV-1 and HCV require a similar number of RNAi activators to prevent escape mutation. The earliest attempts to generate multiple HCV gene silencing sequences from a single expression cassette entailed the use of U6 Pol III long hairpin RNA (lhRNA) expression cassettes [133]. G:U wobbles were encoded in the duplex region of the lhRNAs, and these were included to attenuate the innate immune response to the long duplex RNAs. Effective silencing of an HCV replicon was achieved in cultured cells, and emergence of escape mutants was prevented. Despite these encouraging results, use of lhRNA expression cassettes to counter HCV replication has not been widely used. Variability in the processing specificity and efficiency of siRNA generation from lhRNAs is concerning [134]. Consequently, the design of HCV-targeting cassettes has been modified to include monomeric components that each generate HCV silencers. Yang et al. developed an artificial polycistronic primary miR expression cassette that was derived from the miR-17-92 cluster [135]. Using this approach, five independent HCV-targeting guides were generated, three of which targeted the 5′ NTR and one to each of the *core* and *NS5B* sequences. Each of the artificial miRs was active against the virus. Recombinant adeno-associated viruses (AAVs) containing these polycistronic cassettes demonstrated highly effective silencing without evidence of toxicity when tested against HCV in cultured cells and in a murine reporter gene assay. These observations are in keeping with other previous reports that showed improved safety and effective inhibition of HBV [136] and HIV-1 [137] replication when using artificial primary miR cassettes instead of Pol III shRNA expression cassettes.

The importance of transcription regulation of HCV-targeting shRNAs was verified in a study performed on nonhuman primates [138]. Three U6 Pol III shRNA expression cassettes were incorporated into recombinant AAVs. These vectors were then administered to cynomolgus monkeys. Dose-dependent toxicity was observed, which correlated with previously reported disruption of the endogenous miR pathway [139,140]. Importantly, a second generation of the vectors was propagated in which the transcription of the shRNAs was attenuated. After administration of AAVs containing the modified cassettes to the nonhuman primates, hepatotoxicity was eliminated, but shRNA expression was sustained for 50 days.

HCV IRES-targeting synthetic shRNAs (sshRNAs) have also shown promise for therapeutic use [141,142] and may be multimerized [143,144]. These sshRNAs, which function similarly to endogenous miR-451, are Dicer-independent gene silencers and have gene inhibitory effects after Ago2-mediated processing and target cleavage [145,146]. An advantage of this mechanism is that off-target silencing caused by passenger strand incorporation into RISC is minimized. As synthetic oligonucleotides, they are amenable to chemical modification, and inclusion of 2′-O-Me residues into the sshRNAs improved stability and diminished immunostimulatory properties [145]. Initial investigations showed good target inhibition when lipid nanoparticle formulations containing sshRNAs were administered to mice that expressed a reporter-HCV target fusion. Subsequent studies using uPA/SCID mice with human hepatocyte xenografts revealed that the sshRNA lipid nanoparticle formulations were tolerated well and achieved good target silencing [143]. However, administration of individual sshRNAs led to mutations in the target sites. When a combination of two sshRNAs was formulated and administered to HCV-infected mice, viral escape took longer to develop. The observation that viral escape is not completely suppressed by two sshRNAs is not unexpected and is in agreement with the estimated requirement for a combination of four independent HCV gene silencers to provide an adequate genetic barrier to the virus (see above, section 7.6.1.1) [132]. Although shRNAs transcribed in vitro [146] and endonuclease-prepared siRNAs [147] may also be used in a combinatorial approach, these HCV-targeting sequences have not gained widespread favor. Off-target effects caused by innate immunostimulation and unintended silencing of cellular genes may be problematic.

The promising results that have recently been obtained from assessment of nonviral siRNA conjugates to silence HBV replication [148] may well be applicable to advancing RNAi-based therapy for HCV infection. The vectors were designed to be hepatotropic and escape endosomal degradation efficiently (Chapters 4 and 6). The candidate drugs are currently undergoing tests in clinical trial (http://www.arrowheadresearch.com/programs-overview), and outcomes will be valuable to advancing therapeutic silencing of HCV.

7.6.1.2 Silencing HCV Host Factors

Silencing of individual genes in genome-wide analyses has been useful to identify host factors that are required by HCV to infect and replicate in hepatocytes. The first such study detected 26 human host factors needed by HCV [149]. Requirement for miR-122 [41] was verified by the demonstration that the RNAi pathway itself is necessary for HCV replication [149]. Subsequent genome-wide silencing studies used more detailed analyses to identify host factors that are needed for early or late stages of HCV replication [150,151]. Because host genes are not prone to mutation, and bypassing of host factors by viruses is improbable, silencing of host factors may be useful to inhibit viral replication and prevent emergence of escape mutants. Apart from miR-122, other cellular factors have been successfully silenced and counter HCV replication. These include entry factors such as SR-BI, CD81, CLDN1, and OCLN [152–154]. In addition, knockdown of proteasome alpha-subunit 7 and Hu antigen R, which are cellular co-factors required for later stages of HCV replication, neutralizes the virus [123]. Although the results are encouraging, host factors may perform essential cellular functions and exclusion of toxicity will be important before the approach can be exploited for therapeutic use.

7.6.2 Strategies Using Antisense Technologies against HCV

7.6.2.1 Inhibiting HCV Replication by Targeting miR-122 with Antisense Sequences

miR-122 binds to two sites, S1 and S2, in the 5′ NTR of the HCV IRES to facilitate translation of the viral polyprotein (see above, section 7.3, and Figure 7.4) [41]. Interaction of miR-122 with the HCV IRES involves recruitment of Ago proteins to stimulate HCV polyprotein translation through enhancement of IRES function [43,44] and stabilization of the viral RNA [44–46]. Dependence of HCV on this unusual mechanism of miR action led rapidly to investigating the use of antagonists of miR-122 to counter HCV replication. Miravirsen (SPC3649) is a modified oligonucleotide that is complementary to miR-122 [155]. To stabilize interaction with the hepatocyte miR cognate, locked nucleotides were incorporated into the sequence to generate a locked nucleic acid (LNA) oligonucleotide. Initial work revealed that the naked LNA-modified oligonucleotides, formulated in saline, were effectively taken up by hepatocytes after systemic administration. A long-lasting but reversible inhibition of miR-122 function was achieved after intravenous systemic administration of doses of 10 mg/kg of the LNA-modified oligonucleotide to nonhuman primates. A subsequent investigation showed that administration of the drug candidate to chimpanzees chronically infected with HCV achieved durable inhibition of viral replication [156]. The animals tolerated the drug administration well, and there was no evidence of viral escape from the effects of SPC3649. Follow-up investigations in chimpanzees and early human trials confirmed that the drug was safe [157]. The recently reported phase IIa clinical trial in humans, supported by Santaris Pharma, assessed the efficacy of miravirsen in 36 patients infected with HCV genotype 1 [47]. A dose-dependent decrease in serum HCV RNA was demonstrated. The highest dose of 7 mg/kg achieved a 3.0 log10 decrease in IU of viral RNA per milliliter of serum. In four patients who received the highest dose, HCV RNA was undetectable in their serum samples during the 14-week follow-up period. As with the studies performed on chimpanzees, no adverse effects were observed, and viral escape from inhibition by miravirsen was not detected. Antiviral efficacy and imperceptible side effects are particularly encouraging.

7.6.2.2 Countering HCV Replication with Other Antisense Approaches

In addition to targeting miR-122, antisense oligonucleotides that hybridize to HCV RNA sequences have been used to cause inhibition of viral replication [158,159]. As with many potentially therapeutic oligonucleotides, chemical modifications have been introduced to improve specificity, stability, and efficacy against the targets. In addition to synthetic antisense sequences, expressed antisense molecules have also been tested for efficacy against HCV [160]. Again, the 5′ NTR has been a popular target site, and good antiviral efficacy has been reported by several groups. One HCV-targeting antisense oligonucleotide, a 20-mer phosphorothioate oligodeoxynucleotide targeting the 5′ NTR and called ISIS 14,803, was tested in early clinical trials. Although initial studies were promising, further analysis showed modest efficacy of this candidate drug [158].

7.6.3 HCV-Targeting Cleaving and *trans*-Splicing Ribozymes

Before discovery of the RNAi pathway, HCV-targeting ribozymes were widely used in the drive to develop HCV gene therapy. Both *trans* cleaving and splicing ribozymes have been investigated. The cleaving group includes derivatives of hairpin, hammerhead, hepatitis delta virus (HDV), and RNase P ribozymes. Essentially, cleaving ribozymes are engineered to enable the cutting of HCV RNA. The principle underlying the approach entails artificial separation of enzymatic and substrate components of the natural ribozymes. By designing RNA sequences to form active ribozymes when bound to sites within the HCV target, it is possible to engineer enzymatic RNAs that target an intended viral sequence. In the case of RNase P, which is naturally responsible for cleavage of tRNA precursors, an external guide sequence (EGS) is designed to hybridize to HCV sequences and induce formation of a structure that simulates that of the natural tRNA precursor. RNase P processing of the EGS and HCV RNA complex results in cleavage of the HCV target.

Artificial hairpin ribozymes that target the 5′ NTR and 3′ NTR of HCV have been shown to disable HCV targets when using a reporter gene assay [161,162]. However, the effect was not sufficient to be of use for clinical application. HDV derivatives have also been designed to target HCV [163,164]. In one of the studies the specific on/off system described by Bergeron and colleagues [165,166] was used to engineer 5′ NTR-targeting HDV sequences [163] (Figure 7.7(A)). The advantage of this configuration is that specificity of target RNA cleavage is improved. A blocker domain present on the HDV ribozyme prevents cleavage activity in the absence of the target. When exposed to the HCV sequence, the biosensor domain of the ribozyme hybridizes to the viral target to release the blocker from inhibiting cleavage in *trans*. Efficacy against the target was good when tested in vitro, but it was disappointingly modest in cultured cells. In a notable development of ribozyme drugs, chemically modified synthetic

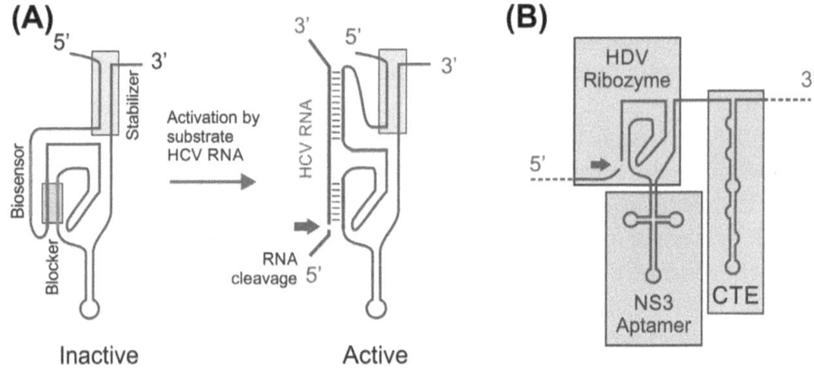

FIGURE 7.7 Targeting HCV using HDV ribozyme-containing sequences to generate (A) ribozymes or (B) aptamers that target HCV.

(A) The on/off system improves specificity by taking advantage of the requirement for a particular configuration of the ribozyme to be active. In the inactive configuration, a blocker domain prevents cleavage activity of the HDV ribozyme in the absence of the target. When exposed to the HCV sequence, the biosensor domain of the ribozyme hybridizes to the viral target to release the blocker from inhibiting cleavage in *trans*. A stabilizer sequence is also required to maintain the HDV ribozyme structure.

(B) Expression and release of NS3-binding aptamers from a Pol II expression cassette. The transcribed RNA includes an HDV ribozyme, the NS3-binding aptamer together with a minimal mutant of CTE. This design enables processing in *cis* and cytoplasmic location of the active released NS3-binding aptamer. NS3, nonstructural protein-3; CTE, constitutive transport element.

trans-acting hammerhead ribozymes were successfully used to inhibit HCV replication [167–169]. The candidate drug, called Hepatozyme™, showed good stability and promising dose-dependent efficacy, which was enhanced by IFN-α co-administration. However, concerns about toxicity led to the premature termination of the drug's development. DNAzymes are an attractive alternative to ribozymes because of their improved stability and easier synthesis. Again, some efficacy has been observed with these HCV-cleaving nucleic acids [170–173]. However, the inhibitory efficacy was inadequate to advance the lead DNAzymes to preclinical and clinical assessment.

Trans-splicing ribozymes also have potential applications against HCV [174]. These enzymatic RNA molecules function by specifically cleaving a target sequence, and then splicing RNA downstream of the target. The property is useful because a therapeutic sequence may be ligated to a specific target site. To generate HCV-splicing type I introns, Ryu and colleagues engineered HCV-cleaving ribozymes that recognized the viral IRES [175]. Ligation of diphtheria toxin A (DTA)-encoding sequences downstream of the targeted site enabled expression of this toxic protein and resultant programmed cell death in cells expressing HCV. The advantage of the approach is that viral target cleavage and splicing of a therapeutic gene are selectively and simultaneously achieved. However, DTA may not be the ideal therapeutic gene, and use of other artificial donors, such as IFN-encoding sequences, may be more therapeutically applicable [174,175].

7.6.4 Aptamers against HCV

Aptamers have useful properties that may be exploited in the development of HCV-disabling agents. Nucleic acids with good target specificity and affinity may be conveniently prepared using the systematic evolution of ligands by exponential enrichment (SELEX) procedure (Chapter 6). Several different HCV targets, which include NS3 and NS5B, Core, and E1/E2 proteins and RNA functional domains, such as the viral IRES, have been used to select potentially therapeutic aptamers.

Early investigations targeting NS3 described the use of SELEX to generate aptamers that interacted with this HCV nonstructural protein [176]. High-affinity binding, with a constant of 10 nM, and noncompetitive inhibition of protease activity by 90% were demonstrated in vitro. Subsequent studies by the same group involved the incorporation of the most effective NS3-binding aptamers into Pol II expression cassettes [177]. The ingenious system included an HDV ribozyme element together with a minimal mutant of constitutive transport element (CTE) (Figure 7.7(B)). This design enabled processing in *cis* by the ribozyme and cytoplasmic location of the released NS3-binding aptamer. The HDV ribozyme-aptamer-CTE (HAC) elements were also arranged in tandem, and augmented efficacy against NS3 protease was observed in tetrameric cassettes. The same group of researchers also targeted the helicase domain of NS3, and strong inhibition of strand separation of a partially duplexed DNA substrate was observed in vitro [178]. High-affinity competitive inhibition of HCV RNA binding to NS3 was reported in another study [179]. In this report, analysis of the minimal binding domains revealed that the secondary structure at the 3′ end of the HCV RNA negative strand and the aptamer domains were similar. Potential usefulness as a therapeutic against HCV was suggested by attaining partial inhibition of intracellular HCV RNA synthesis from an HCV replicon. Although these results are encouraging, similarities between helicases of HCV and cellular origins may be problematic, causing off-target effects of helicase-binding aptamers [174].

A DNA aptamer selected for interaction with NS5B of HCV genotype 3a was capable of high-affinity binding and efficient inhibition of polymerase activity of the viral protein [180]. However, the effect was specific to genotype, and the aptamer was not active against NS5B from genotype 1 isolates. Another study also reported on the potential utility of DNA aptamers that bind to NS5B [181]. Efficient inhibition of HCV RNA template binding and blocking of initiation and elongation of RNA synthesis was achieved in vitro. Moreover, replication of the JFH1 HCV isolate in cultured cells was inhibited by the NS5B-binding DNA aptamer. More recently, RNA aptamers, which included nucleotides with 2′-fluoro moieties, were used successfully to inhibit NS5B function [182]. The aptamers competitively inhibited HCV RNA synthesis and attenuated viral replication in cultured cells. These effects were observed without evidence of off-target effects or the appearance of escape mutants. Safe and efficient delivery to hepatocytes was achieved by conjugation to cholesterol or liver-targeting galactose-PEG molecules. The same study reported that intracellular expression of a 2′-hydroxy-aptamer from a Pol III cassette had a similar efficacy against HCV replication. Disruption of formation of infectious HCV particles was recently achieved with aptamers directed to the viral Core protein [183]. The aptamers appeared to have their effects by interfering with association of the core protein with NS5A and intracellular LDs to prevent viral RNA incorporation into nascent virions.

In an interesting approach to using aptamers to block viral replication, RNA sequences were selected for their ability to bind to helical domains III and IV of the viral IRES [184]. Investigations using surface plasmon resonance showed good target binding, with dissociation constants in the low-nanomolar range for most of the aptamers. Good inhibitory efficacy against IRES-mediated translation was observed in vitro and in cells. Detailed sequence analysis revealed that the aptamers' consensus sequence was complementary to the apical loop sequence of domain III. Thus, binding to the loop III region was dependent on sequence and structure of the target.

A potential advantage of using aptamers for viral therapies is that inhibitory effects may result from interactions in an extracellular environment and do not necessarily need to take place within cells. Such an effect was demonstrated with DNA aptamers that target the E1/E2 structural proteins of HCV [185]. Interaction of the viral particles with their hepatocyte receptors was inhibited by the DNA aptamers, with consequent reduction in viral replication.

7.7 CONCLUSIONS

Discovery of HCV in 1989 was a landmark in modern medicine. Identification of the viral sequences using techniques of molecular biology was the first application of such an approach to detect and characterize a viral pathogen. Since the discovery of the virus, progress in understanding the fundamentals of HCV replication has been impressive. Studies have provided useful insights into an interesting virus and have enabled rapid development of strategies to improve management of the HCV infection. Mechanisms of viral reproduction, detailed structural and functional analysis of individual viral proteins, the error-prone nature of RNA replication, and selection for escape mutants have been particularly important. Persistence of HCV is frequently associated with hepatocellular carcinoma and cirrhosis. Because vaccinations against HCV are currently not available, treatment to eliminate the infection is an important priority to prevent life-threatening complications of infection. Understanding the molecular biology of HCV has facilitated advancement of many new antiviral drugs. Some have recently been licensed, and many are at advanced stages of preclinical and clinical trials. The range of antivirals is broad and includes DAAs and HTAs [49]. Nucleic acid-based HCV-disabling agents have also been developed and may have effects as DAAs or HTAs.

The finding that HCV replication is dependent on miR-122 was an interesting and important observation [41]. Stimulation of viral gene expression, as opposed to the more common natural inhibitory effects of miRs, entails facilitation of the function of the viral IRES and enhanced translation of the viral polyprotein. Dependence of HCV on miR-122 led rapidly to development of miravirsen, an LNA oligonucleotide that hybridizes and disables the function of miR-122. Good efficacy against HCV in preclinical assessments [156] was followed by successful clinical trial studies of the candidate drug [47]. Because miR-122 has important functions in hepatocytes, disruption of the function of this gene regulator may have been expected to result in unintended consequences. However, neither adverse effects nor HCV escape from the drug were observed in patients receiving miravirsen. Collectively, these findings suggest that the potential utility of this nucleic acid-based drug is good.

As a virus that replicates its RNA genome in the cytoplasm, HCV is well suited to gene therapy-based strategies that specifically cleave the viral target RNA. Early studies entailed the use of HCV-specific ribozymes, but utilizing these enzymatic RNAs has recently been superseded by more effective strategies that activate RNAi to target HCV RNA. Many different methods have been used to activate RNAi to inhibit HCV replication. Expressed and synthetic RNAi activators, in various configurations, have been used successfully by many researchers. The favored viral targets have been the conserved 5′ NTR, which contains the essential viral IRES. Silencing that should be adequate for therapeutic use has been achieved when tested in various models of replication of the virus. As with gene therapy strategies for most diseases, overcoming challenges of delivering therapeutic sequences remains important. However, the experience of administering miravirsen as a subcutaneous injection of naked nucleic acid is particularly interesting. Most approaches aimed at advancing RNAi activators to treat HCV infection have required viral or non-viral vectors to deliver the gene silencing sequences.

The error-prone nature of NS5B enables the virus to mutate rapidly and evolve strains that are capable of evading inhibitory effects of RNAi activators. Verification that this is indeed the case was shown during early [124] and more recent [143] studies on HCV-targeting RNAi activators. To overcome this problem, combinatorial RNAi approaches have been used that simultaneously target multiple sites within conserved regions of the viral genome [135,143]. On the basis of the analysis of preventing HIV-1 escape from RNAi [132], it appears that a minimum of four HCV-targeting RNAi activators should provide a complete barrier to resistance. Targeting of host factors using gene silencing technology may also be used to prevent viral escape. The rationale for this approach is that avoiding dependence on host factors would require a major viral genetic alteration, which would constitute a formidable barrier to survival of the virus. Use of miravirsen to disable miR-122 is an example of success of the approach.

Anticipating the role of gene therapy for future management of HCV infection is difficult. Harnessing RNAi shows promise as a therapeutic approach, but other nucleic acids, such as aptamers, may come to the fore. The achievement with miravirsen suggests that nucleic acids will play an important role in future treatment of the viral infection. Utility of gene therapy for HCV infection will also be dependent on advances in related fields. Small-molecule drugs have the considerable advantage of oral administration and compatibility with large-scale synthesis needed for widespread clinical use. Outcomes from treatment with the recently licensed and highly effective antivirals could dramatically change priorities of HCV therapy. Nevertheless, the nimble way in which adaptations can be made to antiviral gene therapies is advantageous for this class of HCV-targeting drugs. Combination with other drug classes and tailoring of treatment according to specific characteristics of individuals' infections may be a feature of new regimens of HCV therapy.

REFERENCES

[1] Feinstone SM, Kapikian AZ, Purcell RH, Alter HJ, Holland PV. Transfusion-associated hepatitis not due to viral hepatitis type A or B. N Engl J Med 1975;292(15):767–70.

[2] Norkrans G, Frosner G, Hermodsson S, Nenonen N, Iwarson S. The epidemiological pattern of hepatitis A, B, and non-A, non-B in Sweden. Scand J Gastroenterology 1978;13(7):873–7.

[3] Rakela J, Redeker AG. Chronic liver disease after acute non-A, non-B viral hepatitis. Gastroenterology 1979;77(6):1200–2.

[4] Bradley DW, Maynard JE, Popper H, Cook EH, Ebert JW, McCaustland KA, et al. Posttransfusion non-A, non-B hepatitis: physicochemical properties of two distinct agents. J Infect Dis 1983;148(2):254–65.

[5] Choo QL, Kuo G, Weiner AJ, Overby LR, Bradley DW, Houghton M. Isolation of a cDNA clone derived from a blood-borne non-A, non-B viral hepatitis genome. Science 1989;244(4902):359–62.

[6] Kapoor A, Simmonds P, Gerold G, Qaisar N, Jain K, Henriquez JA, et al. Characterization of a canine homolog of hepatitis C virus. Proc Natl Acad Sci USA 2011;108(28):11608–13.

[7] Quan PL, Firth C, Conte JM, Williams SH, Zambrana-Torrelio CM, Anthony SJ, et al. Bats are a major natural reservoir for hepaciviruses and pegiviruses. Proc Natl Acad Sci USA 2013;110(20):8194–9.

[8] Deinhardt F, Holmes AW, Capps RB, Popper H. Studies on the transmission of human viral hepatitis to marmoset monkeys. I. Transmission of disease, serial passages, and description of liver lesions. J Exp Med 1967;125(4):673–88.

[9] Mohd Hanafiah K, Groeger J, Flaxman AD, Wiersma ST. Global epidemiology of hepatitis C virus infection: new estimates of age-specific antibody to HCV seroprevalence. Hepatology 2013;57(4):1333–42.

[10] Denniston MM, Klevens RM, McQuillan GM, Jiles RB. Awareness of infection, knowledge of hepatitis C, and medical follow-up among individuals testing positive for hepatitis C: National Health and Nutrition Examination Survey 2001–2008. Hepatology 2012;55(6):1652–61.

[11] Klevens RM, Hu DJ, Jiles R, Holmberg SD. Evolving epidemiology of hepatitis C virus in the United States. Clin Infect Dis 2012;55(Suppl. 1):S3–9.

[12] Seeff LB. Natural history of chronic hepatitis C. Hepatology 2002;36(5 Suppl. 1):S35–46.

[13] Ly KN, Xing J, Klevens RM, Jiles RB, Ward JW, Holmberg SD. The increasing burden of mortality from viral hepatitis in the United States between 1999 and 2007. Ann Intern Med 2012;156(4):271–8.

[14] Campsmith ML, Rhodes PH, Hall HI, Green TA. Undiagnosed HIV prevalence among adults and adolescents in the United States at the end of 2006. J Acquir Immune Defic Syndr 2010;53(5):619–24.

[15] Klevens RM, Miller J, Vonderwahl C, Speers S, Alelis K, Sweet K, et al. Population-based surveillance for hepatitis C virus, United States, 2006–2007. Emerg Infect Dis 2009;15(9):1499–502.

[16] Cornberg M, Razavi HA, Alberti A, Bernasconi E, Buti M, Cooper C, et al. A systematic review of hepatitis C virus epidemiology in Europe, Canada and Israel. Liver Int 2011;31(Suppl. 2):30–60.

[17] Terrault NA. Sexual activity as a risk factor for hepatitis C. Hepatology 2002;36(5 Suppl. 1):S99–105.

[18] Tohme RA, Holmberg SD. Is sexual contact a major mode of hepatitis C virus transmission? Hepatology 2010;52(4):1497–505.

[19] Cacoub P, Renou C, Rosenthal E, Cohen P, Loury I, Loustaud-Ratti V, et al. Extrahepatic manifestations associated with hepatitis C virus infection. A prospective multicenter study of 321 patients. The GERMIVIC. Groupe d'Etude et de Recherche en Medecine Interne et Maladies Infectieuses sur le Virus de l'Hepatite C. Medicine 2000;79(1):47–56.

[20] Galossi A, Guarisco R, Bellis L, Puoti C. Extrahepatic manifestations of chronic HCV infection. J Gastrointest Liver Dis 2007;16(1):65.

[21] Muller HM, Pfaff E, Goeser T, Kallinowski B, Solbach C, Theilmann L. Peripheral blood leukocytes serve as a possible extrahepatic site for hepatitis C virus replication. J Gen Virol 1993;74(Pt 4):669–76.

[22] Winston A, Garvey L, Scotney E, Yerrakalva D, Allsop JM, Thomson EC, et al. Does acute hepatitis C infection affect the central nervous system in HIV-1 infected individuals? J Viral Hepat 2010;17(6):419–26.

[23] Okamoto H, Kurai K, Okada S, Yamamoto K, Lizuka H, Tanaka T, et al. Full-length sequence of a hepatitis C virus genome having poor homology to reported isolates: comparative study of four distinct genotypes. Virology 1992;188(1):331–41.

[24] Nakano T, Lau GM, Lau GM, Sugiyama M, Mizokami M. An updated analysis of hepatitis C virus genotypes and subtypes based on the complete coding region. Liver Int 2012;32(2):339–45.

[25] Kalinina O, Norder H, Magnius LO. Full-length open reading frame of a recombinant hepatitis C virus strain from St Petersburg: proposed mechanism for its formation. J Gen Virol 2004;85(Pt 7):1853–7.

[26] Noppornpanth S, Lien TX, Poovorawan Y, Smits SL, Osterhaus AD, Haagmans BL. Identification of a naturally occurring recombinant genotype 2/6 hepatitis C virus. J Virol 2006;80(15):7569–77.

[27] Zein NN. Clinical significance of hepatitis C virus genotypes. Clin Microbiol Rev 2000;13(2):223–35.

[28] Nousbaum JB, Pol S, Nalpas B, Landais P, Berthelot P, Brechot C. Hepatitis C virus type 1b (II) infection in France and Italy. Collaborative study group. Ann Intern Med 1995;122(3):161–8.

[29] Silini E, Bono F, Cividini A, Cerino A, Bruno S, Rossi S, et al. Differential distribution of hepatitis C virus genotypes in patients with and without liver function abnormalities. Hepatology 1995;21(2):285–90.

[30] Zein NN, Rakela J, Krawitt EL, Reddy KR, Tominaga T, Persing DH. Hepatitis C virus genotypes in the United States: epidemiology, pathogenicity, and response to interferon therapy. Collaborative study group. Ann Intern Med 1996;125(8):634–9.

[31] Liang TJ, Ghany MG. Current and future therapies for hepatitis C virus infection. N Engl J Med 2013;368(20):1907–17.

[32] Zhu Y, Chen S. Antiviral treatment of hepatitis C virus infection and factors affecting efficacy. World J Gastroenterology: WJG 2013;19(47):8963–73.

[33] Ghany MG, Strader DB, Thomas DL, Seeff LB. American Association for the Study of Liver D. Diagnosis, management, and treatment of hepatitis C: an update. Hepatology 2009;49(4):1335–74.

[34] Manns MP, Wedemeyer H, Cornberg M. Treating viral hepatitis C: efficacy, side effects, and complications. Gut 2006;55(9):1350–9.

[35] Gane EJ, Stedman CA, Hyland RH, Ding X, Svarovskaia E, Symonds WT, et al. Nucleotide polymerase inhibitor sofosbuvir plus ribavirin for hepatitis C. N Engl J Med 2013;368(1):34–44.

[36] Grakoui A, Wychowski C, Lin C, Feinstone SM, Rice CM. Expression and identification of hepatitis C virus polyprotein cleavage products. J Virol 1993;67(3):1385–95.

[37] Bartenschlager R, Frese M, Pietschmann T. Novel insights into hepatitis C virus replication and persistence. Adv Virus Res 2004;63:71–180.

[38] Otto GA, Puglisi JD. The pathway of HCV IRES-mediated translation initiation. Cell 2004;119(3):369–80.

[39] Pestova TV, Kolupaeva VG, Lomakin IB, Pilipenko EV, Shatsky IN, Agol VI, et al. Molecular mechanisms of translation initiation in eukaryotes. Proc Natl Acad Sci USA 2001;98(13):7029–36.

[40] Bai Y, Zhou K, Doudna JA. Hepatitis C virus 3′UTR regulates viral translation through direct interactions with the host translation machinery. Nucleic Acids Res 2013;41(16):7861–74.

[41] Jopling CL, Yi M, Lancaster AM, Lemon SM, Sarnow P. Modulation of hepatitis C virus RNA abundance by a liver-specific MicroRNA. Science 2005;309(5740):1577–81.

[42] Shimakami T, Yamane D, Welsch C, Hensley L, Jangra RK, Lemon SM. Base pairing between hepatitis C virus RNA and microRNA 122 3′ of its seed sequence is essential for genome stabilization and production of infectious virus. J Virol 2012;86(13):7372–83.

[43] Henke JI, Goergen D, Zheng J, Song Y, Schuttler CG, Fehr C, et al. microRNA-122 stimulates translation of hepatitis C virus RNA. EMBO J 2008;27(24):3300–10.

[44] Conrad KD, Giering F, Erfurth C, Neumann A, Fehr C, Meister G, et al. MicroRNA-122 dependent binding of Ago2 protein to hepatitis C virus RNA is associated with enhanced RNA stability and translation stimulation. PloS One 2013;8(2):e56272.

[45] Jangra RK, Yi M, Lemon SM. Regulation of hepatitis C virus translation and infectious virus production by the microRNA miR-122. J Virol 2010;84(13):6615–25.

[46] Norman KL, Sarnow P. Modulation of hepatitis C virus RNA abundance and the isoprenoid biosynthesis pathway by microRNA miR-122 involves distinct mechanisms. J Virol 2010;84(1):666–70.

[47] Janssen HL, Reesink HW, Lawitz EJ, Zeuzem S, Rodriguez-Torres M, Patel K, et al. Treatment of HCV infection by targeting micro RNA. N Engl J Med 2013;368(18):1685–94.

[48] Penin F, Dubuisson J, Rey FA, Moradpour D, Pawlotsky JM. Structural biology of hepatitis C virus. Hepatology 2004;39(1):5–19.

[49] Scheel TK, Rice CM. Understanding the hepatitis C virus life cycle paves the way for highly effective therapies. Nat Med 2013;19(7):837–49.

[50] McLauchlan J, Lemberg MK, Hope G, Martoglio B. Intramembrane proteolysis promotes trafficking of hepatitis C virus core protein to lipid droplets. EMBO J 2002;21(15):3980–8.

[51] Lorenz IC, Marcotrigiano J, Dentzer TG, Rice CM. Structure of the catalytic domain of the hepatitis C virus NS2-3 protease. Nature 2006;442(7104):831–5.

[52] Wolk B, Sansonno D, Krausslich HG, Dammacco F, Rice CM, Blum HE, et al. Subcellular localization, stability, and trans-cleavage competence of the hepatitis C virus NS3-NS4A complex expressed in tetracycline-regulated cell lines. J Virol 2000;74(5):2293–304.

[53] Li XD, Sun L, Seth RB, Pineda G, Chen ZJ. Hepatitis C virus protease NS3/4A cleaves mitochondrial antiviral signaling protein off the mitochondria to evade innate immunity. Proc Natl Acad Sci USA 2005;102(49):17717–22.

[54] Gitlin L, Stone JK, Andino R. Poliovirus escape from RNA interference: short interfering RNA-target recognition and implications for therapeutic approaches. J Virol 2005;79(2):1027–35.

[55] Egger D, Wolk B, Gosert R, Bianchi L, Blum HE, Moradpour D, et al. Expression of hepatitis C virus proteins induces distinct membrane alterations including a candidate viral replication complex. J Virol 2002;76(12):5974–84.

[56] Gosert R, Egger D, Lohmann V, Bartenschlager R, Blum HE, Bienz K, et al. Identification of the hepatitis C virus RNA replication complex in Huh-7 cells harboring subgenomic replicons. J Virol 2003;77(9):5487–92.

[57] Jones DM, McLauchlan J. Hepatitis C virus: assembly and release of virus particles. J Biol Chem 2010;285(30):22733–9.

[58] Hope RG, Murphy DJ, McLauchlan J. The domains required to direct core proteins of hepatitis C virus and GB virus-B to lipid droplets share common features with plant oleosin proteins. J Biol Chem 2002;277(6):4261–70.

[59] Boulant S, Montserret R, Hope RG, Ratinier M, Targett-Adams P, Lavergne JP, et al. Structural determinants that target the hepatitis C virus core protein to lipid droplets. J Biol Chem 2006;281(31):22236–47.

[60] Boulant S, Targett-Adams P, McLauchlan J. Disrupting the association of hepatitis C virus core protein with lipid droplets correlates with a loss in production of infectious virus. J Gen Virology 2007;88(Pt 8):2204–13.

[61] Miyanari Y, Atsuzawa K, Usuda N, Watashi K, Hishiki T, Zayas M, et al. The lipid droplet is an important organelle for hepatitis C virus production. Nat Cell Biol 2007;9(9):1089–97.

[62] Shavinskaya A, Boulant S, Penin F, McLauchlan J, Bartenschlager R. The lipid droplet binding domain of hepatitis C virus core protein is a major determinant for efficient virus assembly. J Biol Chem 2007;282(51):37158–69.

[63] Murray CL, Jones CT, Tassello J, Rice CM. Alanine scanning of the hepatitis C virus core protein reveals numerous residues essential for production of infectious virus. J Virol 2007;81(19):10220–31.

[64] Tellinghuisen TL, Foss KL, Treadaway JC, Rice CM. Identification of residues required for RNA replication in domains II and III of the hepatitis C virus NS5A protein. J Virol 2008;82(3):1073–83.

[65] Appel N, Zayas M, Miller S, Krijnse-Locker J, Schaller T, Friebe P, et al. Essential role of domain III of nonstructural protein 5A for hepatitis C virus infectious particle assembly. PLoS Pathog 2008;4(3):e1000035.

[66] Tellinghuisen TL, Foss KL, Treadaway J. Regulation of hepatitis C virion production via phosphorylation of the NS5A protein. PLoS Pathog 2008;4(3):e1000032.

[67] Neddermann P, Quintavalle M, Di Pietro C, Clementi A, Cerretani M, Altamura S, et al. Reduction of hepatitis C virus NS5A hyperphosphorylation by selective inhibition of cellular kinases activates viral RNA replication in cell culture. J Virol 2004;78(23):13306–14.

[68] Reiss S, Harak C, Romero-Brey I, Radujkovic D, Klein R, Ruggieri A, et al. The lipid kinase phosphatidylinositol-4 kinase III alpha regulates the phosphorylation status of hepatitis C virus NS5A. PLoS Pathog 2013;9(5):e1003359.

[69] Jones CT, Murray CL, Eastman DK, Tassello J, Rice CM. Hepatitis C virus p7 and NS2 proteins are essential for production of infectious virus. J Virol 2007;81(16):8374–83.

[70] Boulant S, Douglas MW, Moody L, Budkowska A, Targett-Adams P, McLauchlan J. Hepatitis C virus core protein induces lipid droplet redistribution in a microtubule- and dynein-dependent manner. Traffic 2008;9(8):1268–82.

[71] Lai CK, Jeng KS, Machida K, Lai MM. Association of hepatitis C virus replication complexes with microtubules and actin filaments is dependent on the interaction of NS3 and NS5A. J Virol 2008;82(17):8838–48.

[72] Sakai A, Claire MS, Faulk K, Govindarajan S, Emerson SU, Purcell RH, et al. The p7 polypeptide of hepatitis C virus is critical for infectivity and contains functionally important genotype-specific sequences. Proc Natl Acad Sci USA 2003;100(20):11646–51.

[73] Steinmann E, Penin F, Kallis S, Patel AH, Bartenschlager R, Pietschmann T. Hepatitis C virus p7 protein is crucial for assembly and release of infectious virions. PLoS Pathog 2007;3(7):e103.

[74] Wozniak AL, Griffin S, Rowlands D, Harris M, Yi M, Lemon SM, et al. Intracellular proton conductance of the hepatitis C virus p7 protein and its contribution to infectious virus production. PLoS Pathog 2010;6(9):e1001087.

[75] Russell RS, Meunier JC, Takikawa S, Faulk K, Engle RE, Bukh J, et al. Advantages of a single-cycle production assay to study cell culture-adaptive mutations of hepatitis C virus. Proc Natl Acad Sci USA 2008;105(11):4370–5.

[76] Yi M, Ma Y, Yates J, Lemon SM. Trans-complementation of an NS2 defect in a late step in hepatitis C virus (HCV) particle assembly and maturation. PLoS Pathog 2009;5(5):e1000403.

[77] Martell M, Esteban JI, Quer J, Genesca J, Weiner A, Esteban R, et al. Hepatitis C virus (HCV) circulates as a population of different but closely related genomes: quasispecies nature of HCV genome distribution. J Virol 1992;66(5):3225–9.

[78] Lesburg CA, Cable MB, Ferrari E, Hong Z, Mannarino AF, Weber PC. Crystal structure of the RNA-dependent RNA polymerase from hepatitis C virus reveals a fully encircled active site. Nat Struct Biol 1999;6(10):937–43.

[79] Smith RM, Smolic R, Volarevic M, Wu GY. Positional effects and strand preference of RNA interference against hepatitis C virus target sequences. J Viral Hepat 2007;14(3):194–212.

[80] Lisowski L, Elazar M, Chu K, Glenn JS, Kay MA. The anti-genomic (negative) strand of hepatitis C virus is not targetable by shRNA. Nucleic Acids Res 2013;41(6):3688–98.

[81] Raney KD, Sharma SD, Moustafa IM, Cameron CE. Hepatitis C virus non-structural protein 3 (HCV NS3): a multifunctional antiviral target. J Biol Chem 2010;285(30):22725–31.

[82] Suzich JA, Tamura JK, Palmer-Hill F, Warrener P, Grakoui A, Rice CM, et al. Hepatitis C virus NS3 protein polynucleotide-stimulated nucleoside triphosphatase and comparison with the related pestivirus and flavivirus enzymes. J Virol 1993;67(10):6152–8.

[83] Watashi K, Hijikata M, Hosaka M, Yamaji M, Shimotohno K. Cyclosporin A suppresses replication of hepatitis C virus genome in cultured hepatocytes. Hepatology 2003;38(5):1282–8.

[84] Paeshuyse J, Kaul A, De Clercq E, Rosenwirth B, Dumont JM, Scalfaro P, et al. The non-immunosuppressive cyclosporin DEBIO-025 is a potent inhibitor of hepatitis C virus replication in vitro. Hepatology 2006;43(4):761–70.

[85] Lavie M, Goffard A, Dubuisson J. Assembly of a functional HCV glycoprotein heterodimer. Curr Issues Mol Biol 2007;9(2):71–86.

[86] Huang H, Sun F, Owen DM, Li W, Chen Y, Gale Jr M, et al. Hepatitis C virus production by human hepatocytes dependent on assembly and secretion of very low-density lipoproteins. Proc Natl Acad Sci USA 2007;104(14):5848–53.

[87] Coller KE, Heaton NS, Berger KL, Cooper JD, Saunders JL, Randall G. Molecular determinants and dynamics of hepatitis C virus secretion. PLoS Pathog 2012;8(1):e1002466.

[88] Chang KS, Jiang J, Cai Z, Luo G. Human apolipoprotein e is required for infectivity and production of hepatitis C virus in cell culture. J Virol 2007;81(24):13783–93.

[89] Gastaminza P, Cheng G, Wieland S, Zhong J, Liao W, Chisari FV. Cellular determinants of hepatitis C virus assembly, maturation, degradation, and secretion. J Virol 2008;82(5):2120–9.

[90] Lindenbach BD, Meuleman P, Ploss A, Vanwolleghem T, Syder AJ, McKeating JA, et al. Cell culture-grown hepatitis C virus is infectious in vivo and can be recultured in vitro. Proc Natl Acad Sci USA 2006;103(10):3805–9.

[91] Bartenschlager R, Penin F, Lohmann V, Andre P. Assembly of infectious hepatitis C virus particles. Trends Microbiol 2011;19(2):95–103.

[92] Bukh J. Animal models for the study of hepatitis C virus infection and related liver disease. Gastroenterology 2012;142(6):1279–87. e3.

[93] Agnello V, Abel G, Elfahal M, Knight GB, Zhang QX. Hepatitis C virus and other flaviviridae viruses enter cells via low density lipoprotein receptor. Proc Natl Acad Sci USA 1999;96(22):12766–71.

[94] Barth H, Schafer C, Adah MI, Zhang F, Linhardt RJ, Toyoda H, et al. Cellular binding of hepatitis C virus envelope glycoprotein E2 requires cell surface heparan sulfate. J Biol Chem 2003;278(42):41003–12.

[95] Scarselli E, Ansuini H, Cerino R, Roccasecca RM, Acali S, Filocamo G, et al. The human scavenger receptor class B type I is a novel candidate receptor for the hepatitis C virus. EMBO J 2002;21(19):5017–25.

[96] Pileri P, Uematsu Y, Campagnoli S, Galli G, Falugi F, Petracca R, et al. Binding of hepatitis C virus to CD81. Science 1998;282(5390):938–41.

[97] Liu S, Yang W, Shen L, Turner JR, Coyne CB, Wang T. Tight junction proteins claudin-1 and occludin control hepatitis C virus entry and are downregulated during infection to prevent superinfection. J Virol 2009;83(4):2011–4.

[98] Evans MJ, von Hahn T, Tscherne DM, Syder AJ, Panis M, Wolk B, et al. Claudin-1 is a hepatitis C virus co-receptor required for a late step in entry. Nature 2007;446(7137):801–5.

[99] Ploss A, Evans MJ, Gaysinskaya VA, Panis M, You H, de Jong YP, et al. Human occludin is a hepatitis C virus entry factor required for infection of mouse cells. Nature 2009;457(7231):882–6.

[100] Brimacombe CL, Grove J, Meredith LW, Hu K, Syder AJ, Flores MV, et al. Neutralizing antibody-resistant hepatitis C virus cell-to-cell transmission. J Virol 2011;85(1):596–605.

[101] Timpe JM, Stamataki Z, Jennings A, Hu K, Farquhar MJ, Harris HJ, et al. Hepatitis C virus cell-cell transmission in hepatoma cells in the presence of neutralizing antibodies. Hepatology 2008;47(1):17–24.

[102] Bartenschlager R, Lohmann V. Novel cell culture systems for the hepatitis C virus. Antivir Res 2001;52(1):1–17.

[103] Kato N, Shimotohno K. Systems to culture hepatitis C virus. Curr Top Microbiol Immunol 2000;242:261–78.

[104] Lohmann V, Korner F, Koch J, Herian U, Theilmann L, Bartenschlager R. Replication of subgenomic hepatitis C virus RNAs in a hepatoma cell line. Science 1999;285(5424):110–3.

[105] Blight KJ, Norgard EA. HCV replicon systems. In: Tan SL, editor. Hepatitis C viruses: genomes and molecular biology. Norfolk, UK; 2006:311–52.

[106] Wakita T, Pietschmann T, Kato T, Date T, Miyamoto M, Zhao Z, et al. Production of infectious hepatitis C virus in tissue culture from a cloned viral genome. Nat Med 2005;11(7):791–6.

[107] Zhong J, Gastaminza P, Cheng G, Kapadia S, Kato T, Burton DR, et al. Robust hepatitis C virus infection in vitro. Proc Natl Acad Sci USA 2005;102(26):9294–9.

[108] Mercer DF, Schiller DE, Elliott JF, Douglas DN, Hao C, Rinfret A, et al. Hepatitis C virus replication in mice with chimeric human livers. Nat Med 2001;7(8):927–33.

[109] Bissig KD, Wieland SF, Tran P, Isogawa M, Le TT, Chisari FV, et al. Human liver chimeric mice provide a model for hepatitis B and C virus infection and treatment. J Clin Invest 2010;120(3):924–30.

[110] Dorner M, Horwitz JA, Robbins JB, Barry WT, Feng Q, Mu K, et al. A genetically humanized mouse model for hepatitis C virus infection. Nature 2011;474(7350):208–11.

[111] Dorner M, Horwitz JA, Donovan BM, Labitt RN, Budell WC, Friling T, et al. Completion of the entire hepatitis C virus life cycle in genetically humanized mice. Nature 2013;501(7466):237–41.

[112] Manns MP, McHutchison JG, Gordon SC, Rustgi VK, Shiffman M, Reindollar R, et al. Peginterferon alfa-2b plus ribavirin compared with interferon alfa-2b plus ribavirin for initial treatment of chronic hepatitis C: a randomised trial. Lancet 2001;358(9286):958–65.

[113] Thompson AJ, Muir AJ, Sulkowski MS, Ge D, Fellay J, Shianna KV, et al. Interleukin-28B polymorphism improves viral kinetics and is the strongest pretreatment predictor of sustained virologic response in genotype 1 hepatitis C virus. Gastroenterology 2010;139(1):120–9. e18.

[114] Gallay PA, Lin K. Profile of alisporivir and its potential in the treatment of hepatitis C. Drug Des Dev Ther 2013;7:105–15.

[115] Sulkowski MS, Kang M, Matining R, Wyles D, Johnson VA, Morse GD, et al. Safety and antiviral activity of the HCV entry inhibitor ITX5061 in treatment-naive HCV-infected adults: a randomized, double-blind, phase 1b study. J Infect Dis 2014;209(5):658–67.

[116] Ding Q, Lee YK, Schaefer EA, Peters DT, Veres A, Kim K, et al. A TALEN genome-editing system for generating human stem cell-based disease models. Cell Stem Cell 2013;12(2):238–51.

[117] Sung PS, Murayama A, Kang W, Kim MS, Yoon SK, Fukasawa M, et al. Hepatitis C virus entry is impaired by claudin-1 downregulation in diacylglycerol acyltransferase-1-deficient cells. J Virol 2014;88(16):9233–44.

[118] Yokota T, Sakamoto N, Enomoto N, Tanabe Y, Miyagishi M, Maekawa S, et al. Inhibition of intracellular hepatitis C virus replication by synthetic and vector-derived small interfering RNAs. EMBO Reports 2003;4(6):602–8.

[119] Seo MY, Abrignani S, Houghton M, Han JH. Small interfering RNA-mediated inhibition of hepatitis C virus replication in the human hepatoma cell line Huh-7. J Virol 2003;77(1):810–2.

[120] Prabhu R, Garry RF, Dash S. Small interfering RNA targeted to stem-loop II of the 5′ untranslated region effectively inhibits expression of six HCV genotypes. Virol J 2006;3:100.

[121] Chandra PK, Kundu AK, Hazari S, Chandra S, Bao L, Ooms T, et al. Inhibition of hepatitis C virus replication by intracellular delivery of multiple siRNAs by nanosomes. Mol Ther: J Am Soc Gene Ther 2012;20(9):1724–36.

[122] Prabhu R, Vittal P, Yin Q, Flemington E, Garry R, Robichaux WH, et al. Small interfering RNA effectively inhibits protein expression and negative strand RNA synthesis from a full-length hepatitis C virus clone. J Med Virol 2005;76(4):511–9.

[123] Korf M, Jarczak D, Beger C, Manns MP, Kruger M. Inhibition of hepatitis C virus translation and subgenomic replication by siRNAs directed against highly conserved HCV sequence and cellular HCV cofactors. J Hepatol 2005;43(2):225–34.

[124] Randall G, Grakoui A, Rice CM. Clearance of replicating hepatitis C virus replicon RNAs in cell culture by small interfering RNAs. Proc Natl Acad Sci USA 2003;100(1):235–40.

[125] Takigawa Y, Nagano-Fujii M, Deng L, Hidajat R, Tanaka M, Mizuta H, et al. Suppression of hepatitis C virus replicon by RNA interference directed against the NS3 and NS5B regions of the viral genome. Microbiol Immunol 2004;48(8):591–8.

[126] Ali Ashfaq U, Ansar M, Sarwar MT, Javed T, Rehman S, Riazuddin S. Post-transcriptional inhibition of hepatitis C virus replication through small interference RNA. Virol J 2011;8:112.

[127] Ansar M, Ashfaq UA, Shahid I, Sarwar MT, Javed T, Rehman S, et al. Inhibition of full length hepatitis C virus particles of 1a genotype through small interference RNA. Virol J 2011;8:203.

[128] Kim M, Shin D, Kim SI, Park M. Inhibition of hepatitis C virus gene expression by small interfering RNAs using a tri-cistronic full-length viral replicon and a transient mouse model. Virus Res 2006;122(1–2):1–10.

[129] McCaffrey AP, Meuse L, Pham TT, Conklin DS, Hannon GJ, Kay MA. RNA interference in adult mice. Nature 2002;418(6893):38–9.

[130] Wilson JA, Richardson CD. Hepatitis C virus replicons escape RNA interference induced by a short interfering RNA directed against the NS5b coding region. J Virol 2005;79(11):7050–8.

[131] Silvestri LS, Taraporewala ZF, Patton JT. Rotavirus replication: plus-sense templates for double-stranded RNA synthesis are made in viroplasms. J Virol 2004;78(14):7763–74.

[132] Leonard JN, Schaffer DV. Computational design of antiviral RNA interference strategies that resist human immunodeficiency virus escape. J Virol 2005;79(3):1645–54.

[133] Watanabe T, Sudoh M, Miyagishi M, Akashi H, Arai M, Inoue K, et al. Intracellular-diced dsRNA has enhanced efficacy for silencing HCV RNA and overcomes variation in the viral genotype. Gene Ther 2006;13(11):883–92.

[134] Weinberg MS, Ely A, Barichievy S, Crowther C, Mufamadi S, Carmona S, et al. Specific inhibition of HBV replication in vitro and in vivo with expressed long hairpin RNA. Molecular therapy. J Am Soc Gene Ther 2007;15(3):534–41.

[135] Yang X, Haurigot V, Zhou S, Luo G, Couto LB. Inhibition of hepatitis C virus replication using adeno-associated virus vector delivery of an exogenous anti-hepatitis C virus microRNA cluster. Hepatology 2010;52(6):1877–87.

[136] Ely A, Naidoo T, Arbuthnot P. Efficient silencing of gene expression with modular trimeric pol II expression cassettes comprising microRNA shuttles. Nucleic Acids Res 2009;37(13):e91.

[137] Aagaard LA, Zhang J, von Eije KJ, Li H, Saetrom P, Amarzguioui M, et al. Engineering and optimization of the miR-106b cluster for ectopic expression of multiplexed anti-HIV RNAs. Gene Ther 2008;15(23):1536–49.

[138] Suhy DA, Kao SC, Mao T, Whiteley L, Denise H, Souberbielle B, et al. Safe, long-term hepatic expression of anti-HCV shRNA in a nonhuman primate model. Mol Ther: J Am Soc Gene Ther 2012;20(9):1737–49.

[139] Grimm D, Streetz KL, Jopling CL, Storm TA, Pandey K, Davis CR, et al. Fatality in mice due to oversaturation of cellular microRNA/short hairpin RNA pathways. Nature 2006;441(7092):537–41.

[140] Grimm D, Wang L, Lee JS, Schurmann N, Gu S, Borner K, et al. Argonaute proteins are key determinants of RNAi efficacy, toxicity, and persistence in the adult mouse liver. J Clin Invest 2010;120(9):3106–19.

[141] Ge Q, Ilves H, Dallas A, Kumar P, Shorenstein J, Kazakov SA, et al. Minimal-length short hairpin RNAs: the relationship of structure and RNAi activity. RNA 2010;16(1):106–17.

[142] Ilves H, Kaspar RL, Wang Q, Seyhan AA, Vlassov AV, Contag CH, et al. Inhibition of hepatitis C IRES-mediated gene expression by small hairpin RNAs in human hepatocytes and mice. Ann N Y Acad Sci 2006;1082:52–5.

[143] Dallas A, Ilves H, Ma H, Chin DJ, Maclachlan I, Klumpp K, et al. Inhibition of hepatitis C virus in chimeric mice by short synthetic hairpin RNAs: sequence analysis of surviving virus shows added selective pressure of combination therapy. J Virol 2014;88(9):4647–56.

[144] Ma H, Dallas A, Ilves H, Shorenstein J, MacLachlan I, Klumpp K, et al. Formulated minimal-length synthetic small hairpin RNAs are potent inhibitors of hepatitis C virus in mice with humanized livers. Gastroenterology 2014;146(1):63–6. e5.

[145] Ge Q, Dallas A, Ilves H, Shorenstein J, Behlke MA, Johnston BH. Effects of chemical modification on the potency, serum stability, and immunostimulatory properties of short shRNAs. RNA 2010;16(1):118–30.

[146] Wang Q, Contag CH, Ilves H, Johnston BH, Kaspar RL. Small hairpin RNAs efficiently inhibit hepatitis C IRES-mediated gene expression in human tissue culture cells and a mouse model. Mol Ther: J Am Soc Gene Ther 2005;12(3):562–8.

[147] Kronke J, Kittler R, Buchholz F, Windisch MP, Pietschmann T, Bartenschlager R, et al. Alternative approaches for efficient inhibition of hepatitis C virus RNA replication by small interfering RNAs. J Virol 2004;78(7):3436–46.

[148] Wooddell CI, Rozema DB, Hossbach M, John M, Hamilton HL, Chu Q, et al. Hepatocyte-targeted RNAi therapeutics for the treatment of chronic hepatitis B virus infection. Mol Ther: J Am Soc Gene Ther 2013;21(5):973–85.

[149] Randall G, Panis M, Cooper JD, Tellinghuisen TL, Sukhodolets KE, Pfeffer S, et al. Cellular cofactors affecting hepatitis C virus infection and replication. Proc Natl Acad Sci USA 2007;104(31):12884–9.

[150] Li Q, Brass AL, Ng A, Hu Z, Xavier RJ, Liang TJ, et al. A genome-wide genetic screen for host factors required for hepatitis C virus propagation. Proc Natl Acad Sci USA 2009;106(38):16410–5.

[151] Tai AW, Benita Y, Peng LF, Kim SS, Sakamoto N, Xavier RJ, et al. A functional genomic screen identifies cellular cofactors of hepatitis C virus replication. Cell Host Microbe 2009;5(3):298–307.

[152] Jahan S, Khaliq S, Samreen B, Ijaz B, Khan M, Ahmad W, et al. Effect of combined siRNA of HCV E2 gene and HCV receptors against HCV. Virol J 2011;8:295.

[153] Jahan S, Samreen B, Khaliq S, Ijaz B, Khan M, Siddique MH, et al. HCV entry receptors as potential targets for siRNA-based inhibition of HCV. Genet Vaccines Ther 2011;9:15.

[154] Zeisel MB, Koutsoudakis G, Schnober EK, Haberstroh A, Blum HE, Cosset FL, et al. Scavenger receptor class B type I is a key host factor for hepatitis C virus infection required for an entry step closely linked to CD81. Hepatology 2007;46(6):1722–31.

[155] Elmen J, Lindow M, Schutz S, Lawrence M, Petri A, Obad S, et al. LNA-mediated microRNA silencing in non-human primates. Nature 2008;452(7189):896–9.

[156] Lanford RE, Hildebrandt-Eriksen ES, Petri A, Persson R, Lindow M, Munk ME, et al. Therapeutic silencing of microRNA-122 in primates with chronic hepatitis C virus infection. Science 2010;327(5962):198–201.

[157] Warwick Z, Churchill D. Nucleoside analogues in 2008. J HIV Ther 2008;13(1):3.

[158] McHutchison JG, Patel K, Pockros P, Nyberg L, Pianko S, Yu RZ, et al. A phase I trial of an antisense inhibitor of hepatitis C virus (ISIS 14803), administered to chronic hepatitis C patients. J Hepatol 2006;44(1):88–96.

[159] McCaffrey AP, Meuse L, Karimi M, Contag CH, Kay MA. A potent and specific morpholino antisense inhibitor of hepatitis C translation in mice. Hepatology 2003;38(2):503–8.

[160] Gonzalez-Carmona MA, Vogt A, Heinicke T, Quasdorff M, Hoffmann P, Yildiz Y, et al. Inhibition of hepatitis C virus gene expression by adenoviral vectors encoding antisense RNA in vitro and in vivo. J Hepatol 2011;55(1):19–28.

[161] Jarczak D, Korf M, Beger C, Manns MP, Kruger M. Hairpin ribozymes in combination with siRNAs against highly conserved hepatitis C virus sequence inhibit RNA replication and protein translation from hepatitis C virus subgenomic replicons. FEBS J 2005;272(22):5910–22.

[162] Welch PJ, Tritz R, Yei S, Leavitt M, Yu M, Barber J. A potential therapeutic application of hairpin ribozymes: in vitro and in vivo studies of gene therapy for hepatitis C virus infection. Gene Ther 1996;3(11):994–1001.

[163] Levesque MV, Levesque D, Briere FP, Perreault JP. Investigating a new generation of ribozymes in order to target HCV. PloS One 2010;5(3):e9627.

[164] Yu YC, Mao Q, Gu CH, Li QF, Wang YM. Activity of HDV ribozymes to trans-cleave HCV RNA. World J Gastroenterol 2002;8(4):694–8.

[165] Bergeron LJ, Perreault JP. Target-dependent on/off switch increases ribozyme fidelity. Nucleic Acids Res 2005;33(4):1240–8.

[166] Bergeron LJ, Reymond C, Perreault JP. Functional characterization of the SOFA delta ribozyme. RNA 2005;11(12):1858–68.

[167] Lee PA, Blatt LM, Blanchard KS, Bouhana KS, Pavco PA, Bellon L, et al. Pharmacokinetics and tissue distribution of a ribozyme directed against hepatitis C virus RNA following subcutaneous or intravenous administration in mice. Hepatology 2000;32(3):640–6.

[168] Macejak DG, Jensen KL, Jamison SF, Domenico K, Roberts EC, Chaudhary N, et al. Inhibition of hepatitis C virus (HCV)-RNA-dependent translation and replication of a chimeric HCV poliovirus using synthetic stabilized ribozymes. Hepatology 2000;31(3):769–76.

[169] Macejak DG, Jensen KL, Pavco PA, Phipps KM, Heinz BA, Colacino JM, et al. Enhanced antiviral effect in cell culture of type 1 interferon and ribozymes targeting HCV RNA. J Viral Hepat 2001;8(6):400–5.

[170] Kumar D, Chaudhury I, Kar P, Das RH. Site-specific cleavage of HCV genomic RNA and its cloned core and NS5B genes by DNAzyme. J Gastroenterol Hepatol 2009;24(5):872–8.

[171] Lee B, Kim KB, Oh S, Choi JS, Park JS, Min DH, et al. Suppression of hepatitis C virus genome replication in cells with RNA-cleaving DNA enzymes and short-hairpin RNA. Oligonucleotides 2010;20(6):285–96.

[172] Roy S, Gupta N, Subramanian N, Mondal T, Banerjea AC, Das S. Sequence-specific cleavage of hepatitis C virus RNA by DNAzymes: inhibition of viral RNA translation and replication. J Gen Virol 2008;89(Pt 7):1579–86.

[173] Trepanier J, Tanner JE, Momparler RL, Le ON, Alvarez F, Alfieri C. Cleavage of intracellular hepatitis C RNA in the virus core protein coding region by deoxyribozymes. J Viral Hepat 2006;13(2):131–8.

[174] Lee CH, Kim JH, Lee SW. Prospects for nucleic acid-based therapeutics against hepatitis C virus. World J Gastroenterol 2013;19(47):8949–62.

[175] Ryu KJ, Kim JH, Lee SW. Ribozyme-mediated selective induction of new gene activity in hepatitis C virus internal ribosome entry site-expressing cells by targeted trans-splicing. Mol Ther: J Am Soc Gene Ther 2003;7(3):386–95.

[176] Fukuda K, Vishnuvardhan D, Sekiya S, Hwang J, Kakiuchi N, Taira K, et al. Isolation and characterization of RNA aptamers specific for the hepatitis C virus nonstructural protein 3 protease. Eur J Biochem/FEBS 2000;267(12):3685–94.

[177] Nishikawa F, Kakiuchi N, Funaji K, Fukuda K, Sekiya S, Nishikawa S. Inhibition of HCV NS3 protease by RNA aptamers in cells. Nucleic Acids Res 2003;31(7):1935–43.

[178] Nishikawa F, Funaji K, Fukuda K, Nishikawa S. In vitro selection of RNA aptamers against the HCV NS3 helicase domain. Oligonucleotides 2004;14(2):114–29.

[179] Hwang B, Cho JS, Yeo HJ, Kim JH, Chung KM, Han K, et al. Isolation of specific and high-affinity RNA aptamers against NS3 helicase domain of hepatitis C virus. RNA 2004;10(8):1277–90.

[180] Jones LA, Clancy LE, Rawlinson WD, White PA. High-affinity aptamers to subtype 3a hepatitis C virus polymerase display genotypic specificity. Antimicrob Agents Chemother 2006;50(9):3019–27.

[181] Bellecave P, Cazenave C, Rumi J, Staedel C, Cosnefroy O, Andreola ML, et al. Inhibition of hepatitis C virus (HCV) RNA polymerase by DNA aptamers: mechanism of inhibition of in vitro RNA synthesis and effect on HCV-infected cells. Antimicrob Agents Chemother 2008;52(6):2097–110.

[182] Lee CH, Lee YJ, Kim JH, Lim JH, Kim JH, Han W, et al. Inhibition of hepatitis C virus (HCV) replication by specific RNA aptamers against HCV NS5B RNA replicase. J Virol 2013;87(12):7064–74.

[183] Shi S, Yu X, Gao Y, Xue B, Wu X, Wang X, et al. Inhibition of hepatitis C virus production by aptamers against the core protein. J Virol 2014;88(4):1990–9.

[184] Kikuchi K, Umehara T, Fukuda K, Kuno A, Hasegawa T, Nishikawa S. A hepatitis C virus (HCV) internal ribosome entry site (IRES) domain III-IV-targeted aptamer inhibits translation by binding to an apical loop of domain IIId. Nucleic Acids Res 2005;33(2):683–92.

[185] Yang D, Meng X, Yu Q, Xu L, Long Y, Liu B, et al. Inhibition of hepatitis C virus infection by DNA aptamer against envelope protein. Antimicrob Agents Chemother 2013;57(10):4937–44.

CHAPTER 8

Gene Therapy for HIV-1 Infection

8.1 DISCOVERY AND EARLY MAJOR DEVELOPMENTS IN RESEARCH ON HIV

Human immunodeficiency virus (HIV)-1 belongs to the *Lentivirus* genus of the *Retroviridae* family. The original infections of humans with HIV-1 probably occurred ~100 years ago (reviewed in ref. [1]). Practices relating to killing and butchering apes for food in West and central Africa are likely to have led to zoonotic transmission of simian immunodeficiency virus (SIV). Detailed analysis indicates that it is the SIVcpz lentivirus, which naturally infects chimpanzees and has considerable sequence similarity to HIV-1, that gave rise to the human disease. It is interesting to note that HIV-2 has a different origin to that of HIV-1. The SIVsmm from the sooty mangabey monkey is the origin of HIV-2, and infection of humans with this virus is largely restricted to parts of West Africa. The clinical course and mode of transmission of HIV-2 differ considerably from those of HIV-1. HIV-2 loads in infected individuals tend to be lower and progression to acquired immune deficiency syndrome (AIDS) is less common.

Bushmeat hunting has been a long tradition in parts of West Africa; therefore, reasons for the relatively recent emergence of HIV infection have been puzzling. Colonial ways at the turn of the nineteenth to twentieth centuries are likely to have been important contributors to initiating the pandemic [2]. Factors such as development of transport networks, social disruption caused by forced labor, rapid urbanization, and arm-to-arm vaccination programs are likely to have been important. Although several transmissions of SIV from apes to humans occurred, it is infection by the main (M) group that is responsible for the current pandemic [1,3,4]. There are nine subtypes of this group, of which subtypes B and C are the most important. The subtype C virus is that which predominates in sub-Saharan Africa, where the infection is particularly common. Subtype B infection occurs more commonly in Europe and North and South America.

The first cases of AIDS caused by HIV-1 were described by Gottlieb and colleagues in 1981 [5]. The study reported on *Pneumocystis carinii* pneumonia and

Gene Therapy for Viral Infections. http://dx.doi.org/10.1016/B978-0-12-410518-8.00008-9

extensive mucosal candidiasis against a background of an immunocompromised state occurring in four men who had sex with men. Two years after this initial article, and the publication of other similar clinical studies, the virus was isolated from a lymph node of a patient with AIDS [6]. This breakthrough study was performed by a group of French clinicians and virologists working at the Pasteur Institute. The importance of the finding eventually led to the awarding of the Nobel Prize in Physiology or Medicine to Françoise Barré-Sinoussi and Luc Montagnier in 2008 (Figure 8.1).

8.2 EPIDEMIOLOGY OF HIV-1 INFECTION

Since the early 1980s, HIV-1 has spread widely, and the resultant global burden of disease imposed by the virus increased considerably. During the past 30 years, approximately 25 million people have died from AIDS-related illnesses. However, with the development of more effective antiretrovirals, implementation of treatment programs, and preventative measures, new infections and complications of the viral infection have started to decrease [3]. Total AIDS-related deaths in 2005 were 2.3 million and 1.6 million in 2012 [7]. The prevalence of HIV-1 infection is estimated to have increased from 31 million in 2002 to 35.3 million in 2012, and the upturn is largely as a result of improved treatment of HIV-1-infected individuals. However, regarding new cases of HIV-1 infection, the incidence has diminished from 3.3 million in 2002 to 2.3 million in 2012 [3,7]. Most people infected with HIV reside in sub-Saharan Africa (Figure 8.2). The resource-poor settings in many parts of this region make complications of the infection particularly grave.

HIV infection occurs as a result of parenteral contact with individuals carrying the virus. Sexual transmission, intravenous drug administration using shared needles, and mother-to-child transmission that may occur perinatally or during breast feeding are the most important modes of acquiring the infection. Several factors influence the risk of HIV sexual transmission (reviewed in ref. [3]). Increased circulating viral load [8], genital ulcers, pregnancy [9], and receptive anal intercourse [10] augment the rates of infection. Treatment with antiretrovirals [11], condom use, and male circumcision [12] diminish the risk of HIV transmission. Behavioral factors that are associated with a higher likelihood of transmission include having multiple sexual partners, excessive alcohol use, and violence of men against their female partners.

8.3 PROTEINS ENCODED BY HIV-1

The genome of HIV-1 encodes 15 proteins, which may be categorized as enzymatic, structural, regulatory, and accessory (Figure 8.3). Structural proteins are responsible for the formation of the viral capsid and envelope. Viral enzymes,

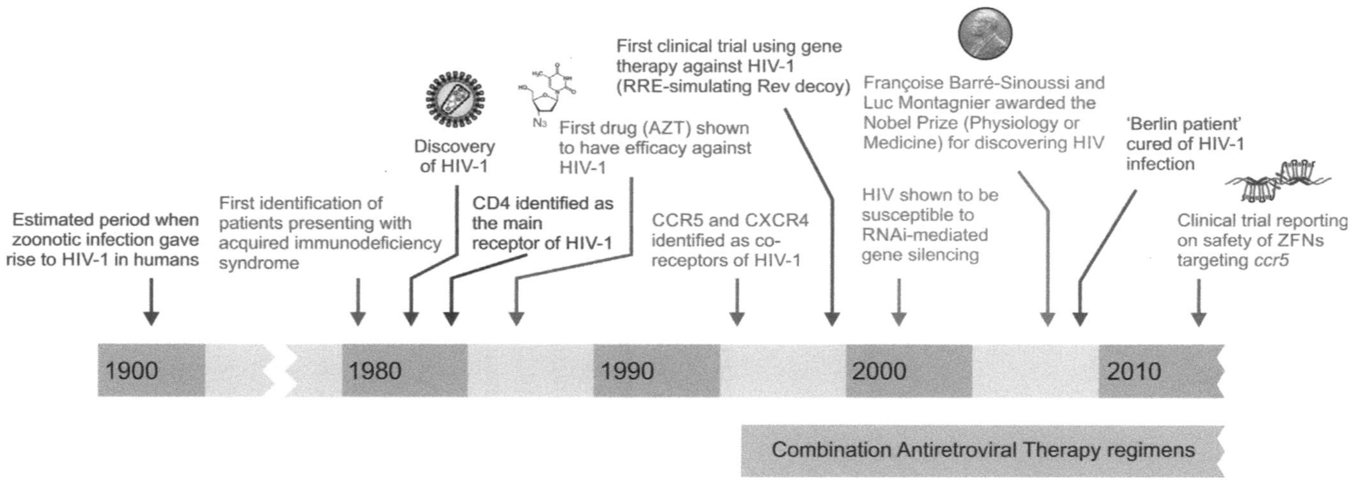

FIGURE 8.1 Timeline of the major milestones in research on HIV-1.

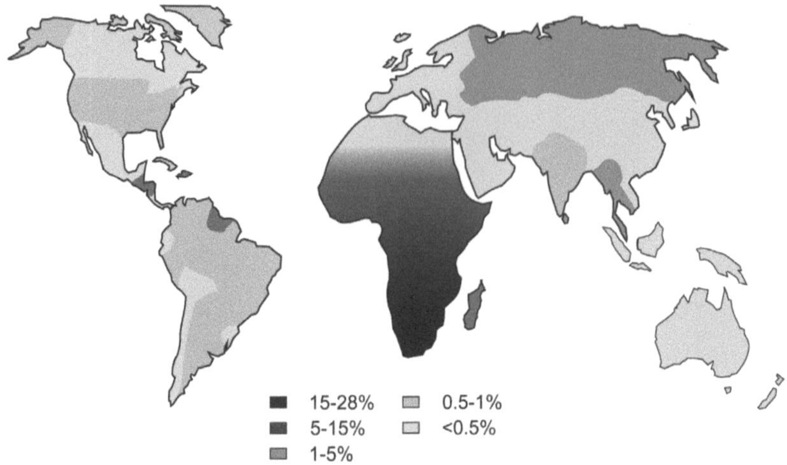

FIGURE 8.2 Approximate global prevalence of HIV-1 infection in adult populations.
Shading of the map indicates the proportion of the general adult populations who are infected with HIV-1. Most HIV-1-infected individuals are from sub-Saharan Africa, and the prevalence is particularly high in southern Africa.

reverse transcriptase (RT), protease (PR), and integrase (IN) are formed from the Gag-Pol precursor. The Gag-Pol polyprotein is formed after translational read through of the *gag* stop codon. This occurs in ~5% of events of ribosomes traversing the sequence and results from the effects of a pseudoknot in the mRNA, which is located immediately downstream of the stop codon [13]. Processing of the precursor is effected by PR, which also plays a central role in the maturation of other viral proteins. The importance of PR in HIV-1 replication led to development of drugs that function as PR inhibitors, which have been used successfully to counter replication of the virus. PR is itself part of the precursor Gag-Pol precursor. To perform its functions on viral proteins, PR is initially released from Gag-Pol by an autocleavage mechanism [14,15]. RT and IN are respectively responsible for the synthesis of complementary DNA from the viral RNA genome and integration of this DNA into the host genome to form the provirus (discussed below). As with PR inhibitors, the importance of RT and IN for HIV-1 replication also led to development of drugs that inactivate these enzymes.

Gag is encoded by *gag* sequences, and derivatives of this precursor contribute to formation of the capsid of immature noninfectious viral particles [16]. Maturation of viral particles is brought about by PR-mediated processing of Gag and leads to rearrangement of the viral capsid after formation of the matrix (MA, p17), capsid (CA, p24), nucleocapsid (NC, p7), and p6 proteins (Figures 8.3 and 8.4). PR also plays a crucial role in maturation of the envelope proteins [17]. The large gp160 precursor, encoded by *env*, is cleaved to form

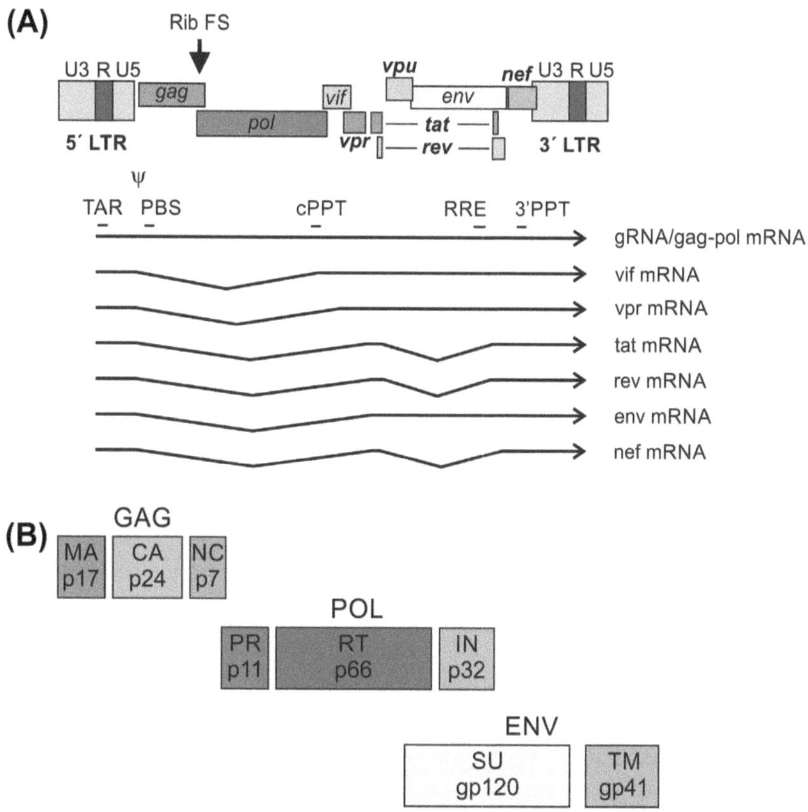

FIGURE 8.3 Organization of the provirus of HIV-1.

(A) Long terminal repeats (LTRs), comprising U3, R, and U5 sequences, flank the protein-coding regions and *cis*-elements required for control of viral replication. The U3 sequence in the 5′ LTR contains promoter elements that regulate initiation of transcription of viral RNAs. Termination of transcription occurs at the junction of the R and U5 sequences at the 3′ LTR. The *gag* open reading frame (ORF) encodes proteins responsible for packaging of the viral genome within the core particle, the *pol* ORF generates the viral enzymes, and *env* the envelope glycoproteins. The position of the ribosomal frame shift site (Rib FS), which leads to read through during translation and formation of the env-encoded proteins, is shown by the vertical arrow. Regulatory (Tat and Rev) and accessory (Vif, Vpu, Vpr, and Nef) proteins are formed from fully spliced viral mRNAs that predominate during early stages of the infection. Noncontiguous sequences encoding Tat and Rev proteins are reconstituted by splicing. Positions of the Tat response element (TAR), Rev response element (RRE), packaging sequence (ψ), primer binding site (PBS), central and 3′ polypurine tracts (cPPT and 3′PPT) are indicated on the unspliced genomic RNA. (B) The POL precursor protein is processed to form protease (PR), reverse transciptase (RT), and integrase (IN). PR cleaves the GAG protein to result in formation of matrix (MA), capsid (CA) and nucleocapsid (NC) proteins. Surface (SU, gp120) and transmembrane (TM, gp41) envelope glycoproteins are generated after processing of the ENV precursor.

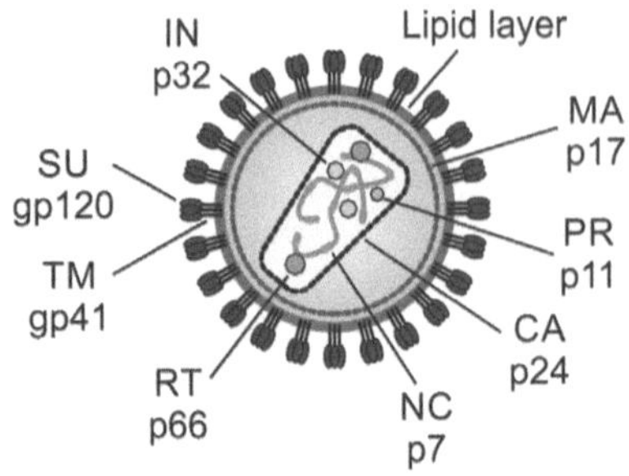

FIGURE 8.4 Schematic illustration of the structure of the HIV-1 particle.
The cone-shaped capsid, comprising p24, encapsidates the viral genome together with viral enzymes RT, PR, and IN as well as the nucleocapsid and accessory proteins. Two copies of the RNA genome, comprising strands of 9.7 kb with sense polarity, are encapsidated. A shell comprising the matrix protein surrounds the core particle, which in turn is enveloped by a lipid layer derived from the previously infected cell. Trimers of the envelope glycoproteins, gp41 and gp120, are embedded in the envelope layer.

transmembrane gp41 (TM) and surface gp120 (SU) glycoproteins. Trimers of each of these envelope proteins interact with each other, are embedded in the cell membrane, and are arranged on the surface of the virion after budding [18] (Figures 8.4 and 8.5).

Tat (*trans*-activating protein) and Rev (regulator of expression of virion proteins) are regulatory proteins that are required for transcription and nuclear export of unspliced viral RNA (discussed below). There are four accessory proteins, which were originally so termed because their functions were not essential for viral replication in vitro, but they were important for establishing infection under natural conditions [19]. The four accessory proteins are Vif (Viral infectivity factor), Vpr (Viral protein R), Vpu (Viral protein unique), and Nef (Negative factor). Vpx (Viral protein X) is found in HIV-2 instead of Vpr [20]. Vif is essential for overcoming the effects of apolipoprotein B mRNA-editing-catalytic peptide-like-3G (APOBEC-3G; Table 8.1). This host restriction factor deaminates C residues of the minus strand of DNA:RNA hybrids to result in G-to-A mutations in the sense strand [21]. By binding APOBEC-3G, Vif prevents encapsidation of the restriction factor and maintains integrity of the viral genome [22]. Vpu also counters the effects of a host restriction factor. Tetherin, produced by HIV-infected host cells, has an antiviral effect by inhibiting release of the virion from the cell surface [23]. Vpu inhibits the activity of tetherin by downregulating the host protein at the cell surface [24], causing degradation of tetherin [25], and by direct interaction within the cell membrane [26]. In the case of HIV-2, Vpx inhibits function of the sterile alpha motif histidine aspartic acid domain-containing protein-1 (SAMHD1) [27,28]. This host restriction factor hydrolyzes intracellular deoxynucleotide triphosphates to diminish their concentrations to a level below that which is required for activity of RT. This mechanism has recently been disputed, and the RNase function of SAMHD1

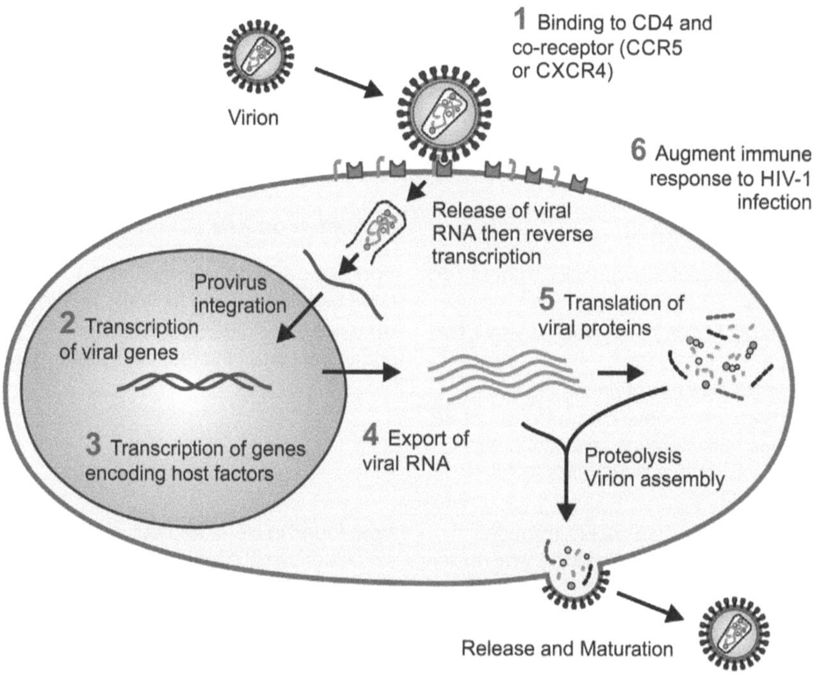

FIGURE 8.5 Schematic illustration of the replication cycle of HIV-1 with stages that have been targeted using gene therapy indicated.
Surface glycoproteins of the virion interact with CD4 and co-receptors (CCR5 and CXCR4) on the surface of the target cells. Fusion, entry of the capsid, and release of the viral genome molecules are followed by reverse transcription. The DNA copy is transported to the nucleus where integration into the host cell's genome occurs to form the provirus. Exported transcripts are translated to form viral enzymatic, structural, regulatory, and accessory proteins. Genomic RNA is selectively packaged during a budding process that initiates formation of the virions. Viral particles then undergo maturation, during which PR-mediated cleavage of GAG and POL precursors leads to reorganization of the core and viral genome. Stages of the viral replication cycle numbered 1–5 have been targets for inhibition using gene therapy. In addition, gene therapy is being developed to enhance the host's immune response to the viral infection (6) (see Table 8.2).

seems more important for the restriction of HIV-2 replication [29]. The effects of Nef are to elevate viral replication and pathogenesis in vivo. This regulatory protein disrupts T-cell receptor (TCR) function by lowering the receptor's signaling threshold, inhibition of cell death, and transcriptional activation to promote viral replication [30].

8.4 REPLICATION OF HIV-1

The virion of HIV-1 has a surrounding envelope, which is derived from the plasma membrane of the previously infected cell (Figure 8.4). Two copies of

Table 8.1 Host factors and restriction factors that facilitate and impede HIV replication

Host Factor	**Mechanism of Action**	
LEDGF/p75	Tethers viral DNA to host DNA as an initiating step of formation of the provirus [48].	
TREX1	Degrades viral excess cytoplasmic DNA to prevent the activation of an innate (interferon) immune response [46].	
Restriction Factor	**Mechanism of Action**	**HIV Effect on Restriction Factor**
Tetherin	Inhibits budding of viral particles from the cell membrane.	**Vpu** binds and causes degradation to downregulate tetherin function [24–26].
APOBEC3G/F	Cytosine deaminase that mutates viral DNA.	**Vif** binds to APOBEC3G to prevent encapsidation of the restriction factor [22].
TRIM5α	Disrupts the process of uncoating of viral capsid particles. Differences in the structure of TRIM5α from humans and nonhuman primates influence susceptibility to HIV-1 infection [42].	
SAMDH1	Degrades cellular deoxyribonucleotide triphosphates and RNA to inhibit viral nucleic acid synthesis.	**Vpx**, found in HIV-2 and not in HIV-2, inhibits enzymatic function [18,19].

Abbreviations: LEDGF/p75: Lens epithelium-derived growth factor/p75; TREX1: Three prime Repair EXonuclease1; Vpu: Viral protein unique; Vif: Viral infectivity factor; Vpx: Viral protein R; APOBEC-3G: Apolipoprotein B mRNA-editing-catalytic peptide-like-3G; TRIM5α: Tripartite motif five alpha; SAMHD1: Sterile alpha motif histidine aspartic acid domain-containing protein 1.

the RNA genome, together with RT, PR, and IN, are located within the capsid. Inclusion of two genomes within each virion provides a mechanism for enhancing genetic diversity and rescue of genetic integrity in the event of mutation of one of the molecules of the viral genome. The MA protein surrounds the capsid and is located on the inside of the virion's envelope. The viral envelope glycoproteins are gp120 and gp41 and are named according to their approximate molecular weights in kiloDaltons. These surface proteins are arranged in trimers and are heavily glycosylated [31]. This glycan shield provides protection to functional domains of the envelope proteins. The shield, together with the burying of the receptor binding sites within deep pockets of gp120, restricts access to the viral protein by bulky neutralizing antibodies [32,33]. Together with the variability of the viral envelope proteins, these factors have made development of a vaccine against HIV-1 infection very difficult (Chapter 11). The limited success of vaccine development has been one of the main disappointments in the field of HIV-1 research. gp120 is exposed for interaction with the main virion receptor (CD4) [34,35] and co-receptors (CC chemokine receptor type 5 [CCR5] and C-X-C chemokine receptor type 4 [CXCR4]) [36–38] located on the target cells. $CD4^+$ T lymphocytes are the main target of HIV-1 infection, but other cells that also express CD4, such as dendritic cells, monocytes, and macrophages, are also susceptible to the virus. In some cases,

HIV-1 may infect and cause pathology in cells that do not express CD4, such as renal epithelial cells and astrocytes.

After attachment of gp120 to CD4, the viral envelope fuses with the cell membrane and gp41 facilitates this process [39] (Figure 8.5). Thereafter, the viral capsid is released into the cytoplasm. The core particle is uncoated to release the viral genome before commencement of reverse transcription within the reverse transcription complex [40]. It is important to note that the tripartite motif-5α (TRIM5α) proteins, which function as innate host restriction factors, may disrupt this uncoating step of retroviral replication (reviewed in ref. [41]; Table 8.1). TRIM5α proteins recognize and interact with specific determinants of p24 and then form a surrounding lattice to cause premature dissociation of the capsid core. TRIM5α proteins of humans and nonhuman primates have subtle differences in their structures to affect the host range of HIV-1. Although effective against some retroviruses, human TRIM5α does not effectively block HIV-1 infection. However, TRIM5α of macaques and other simian species effectively restricts HIV-1 infection in these primates [42].

The genomic RNA of HIV-1 is flanked by repeat (R) sequences at both ends and unique 5′ (U5) and unique 3′ (U3) elements, which are respectively at the 5′ and 3′ ends of the genomic RNA (Figure 8.3). Reverse transcription of the RNA genome is primed by cellular $tRNA^{Lys3}$ (reviewed in ref. [43]). This tRNA is packaged into the capsid through interaction with Gag and Gag-Pol precursor proteins and pairs with 18 nt at the primer binding site (PBS). The PBS is located immediately downstream of the U5 sequence and the hybridizing $tRNA^{Lys3}$ provides a 3′ terminus for the synthesis of viral DNA by RT (reviewed in refs [44,45]). RT has RNA-dependent DNA polymerase, DNA-dependent DNA polymerase, and RNaseH functions, which are all essential for the synthesis of HIV duplex DNA. Another particularly important feature of RT is that the enzyme is prone to forming errors, which result in mutation during DNA synthesis and allows the virus to evolve rapidly. After pairing of the $tRNA^{Lys3}$, DNA synthesis occurs in a 5′–3′ direction to form a sequence that is complementary to the R sequence at the 5′ end of the viral genome. Reverse transcription then occurs through a series of two template strand switches that are coupled to RNaseH-mediated degradation of the viral RNA template. The first strand jump results from detachment of the newly synthesized strand of DNA and then pairing with the R sequence at the 3′ end of the genomic RNA. The entire viral genome is then copied to form complementary DNA of minus strand polarity. Hybridized genomic RNA is degraded by RNaseH, but the purine-rich RNA sequences comprising central and 3′ polypurine tracts (cPPT and 3′PPT, respectively) are relatively resistant to degradation. The PPT remnants of the viral genome then serve as primers to initiate synthesis of the plus strand of the DNA duplex. A second strand jump occurs when the extended 3′PPT is shifted to the 3′ end of the initially synthesized minus DNA strand, and the mechanism involves hybridization of complementary R sequences located at either end of the newly synthesized DNA. On completion of reverse transcription, two long terminal repeats (LTRs) comprising U5, R, and U3 sequences are formed at either end of the duplex. The three prime Repair EXonuclease1 (TREX1) host factor enables HIV-1 to bypass the innate immunostimulatory effects of cytosolic viral DNA [46]. TREX1 degrades excess cytosolic viral DNA to limit activation of the interferon response. When TREX1 expression is silenced, HIV DNA accumulates and causes innate immunostimulation.

After reverse transcription of the viral genome, a pre-integration complex (PIC) comprising viral and host cellular proteins is formed [47]. A feature of the HIV-1 PIC is that it is capable of traversing the nuclear pores without the need for degradation of the nuclear membrane, which is the case for gammaretroviruses. This property has been used successfully for the development of lentiviral vectors, which are capable of transducing nondividing cells (Chapter 4). Integration of viral DNA, generated after reverse transcription of the genomic RNA, initially entails processing of the 3′ ends of the viral DNA and then cleavage of the host DNA with strand transfer and repair at the junctions to form the provirus (reviewed in ref. [48]). IN binds to the LTRs of the viral DNA and causes removal of dinucleotides, comprising GT residues, from the 3′ ends of the viral DNA. Resulting CA overhangs at the 5′ ends are reactive intermediates of the integration mechanism. The strand transfer reaction then entails cleavage of host DNA to generate a staggered overhang. Ligation of the recessed viral 3′ hydroxy ends to the 5′ phosphate of the host's DNA then occurs. The unpaired nucleotides of the viral DNA are removed, and gaps within the host DNA are then repaired to result in the duplication of sequences of 5 bp on either side of the provirus. Actively transcribed regions of the host DNA, which may be in exons or introns, are preferred sites for integration [49,50]. The integration reaction may be constituted in vitro, but IN activity is significantly enhanced by cellular proteins. The lens epithelium-derived growth factor (LEDGF)/p75 protein of host cell origin was initially implicated in integration of lentiviruses because of the high affinity that this protein has for IN [51,52]. The LEDGF component of the name was derived from the identification of the protein's involvement in growth of lens epithelial cells [53], whereas the p75 part of the name comes from independent characterization of the protein as a transcriptional activator [54]. LEDGF/p75 serves a tethering function to connect the viral DNA to host DNA through interaction with IN [48]. LEDGF/p75 has an IN-binding domain that couples the viral enzyme and has an AT hook motif that interacts with AT-rich sequences of DNA. It is interesting to note, but perhaps not surprising, that HIV integration is favored at host sites where LEDGF/p75 plays a transcriptional modulatory role and at AT-rich sequences [55].

The provirus is crucially important to pathogenesis and management of HIV-1-related disease because infected individuals permanently harbor the DNA replication intermediate. A state of latency occurs when the viral genes are dormant and replication of HIV-1 is inactive. Under these conditions, HIV-1 is not susceptible to antiviral drugs and detection by the host's immune system is avoided. Elimination of latently infected cells and disruption of proviral sequences are very difficult and are the main reasons for the limitations of current therapies for HIV-1. Gene therapy, and in particular application of gene editing technologies, potentially provides the means for eliminating or disrupting integrated sequences of HIV-1.

The proviral sequence relies on host transcriptional machinery and *cis*-elements located in the LTRs to regulate viral gene expression. The viral transcripts undergo similar post-transcriptional modifications, particularly splicing, capping, and polyadenylation, to which the cellular Pol II transcripts are subjected. The U3 sequence at the 5′ LTR contains promoter elements that control initiation of transcription of viral mRNA (reviewed in refs [56,57]), whereas transcription termination is caused by U5 elements within the 3′ LTR [58] (Figure 8.3). In addition, HIV-1 encodes its own *trans*-activator of transcription, Tat. The Tat-responsive element (TAR), which comprises a stem–loop (SL) sequence at the 5′ end of the viral transcripts, is responsible for mediating the regulatory function of Tat on viral gene expression [59]. The enhancing effect on transcription is caused by concerted effects of Tat together with cellular proteins on the host's RNA Pol II. Regulation of expression of viral proteins is achieved through intricate mechanisms that modulate splicing of viral transcripts and by causing a reading frame shift during translation of the *gag/pol* sequences [44,60] (discussed above). The fully spliced transcripts that encode regulatory proteins, Tat, Rev, and Nef, are the first to accumulate. mRNAs encoding Env and accessory proteins, together with the full-length viral transcripts required for translation of Gag, Pol, and formation of the viral genome, are unspliced or singly spliced and accumulate later.

Export of unspliced mRNA from the nucleus is normally restricted. This mechanism is essential to prevent anomalous protein translation, but it would hamper transport of genomes of HIV-1 to the cytoplasm. The control is disrupted through specific and efficient interaction of the arginine-rich viral Rev protein with the Rev-responsive element (RRE) located within *env* (Figure 8.3). In conjunction with chromosome region maintenance-1 and RanGTP, Rev facilitates export of unspliced viral mRNA from the nucleus. Of course, fully spliced transcripts, which encode Tat, Rev, and Nef, are efficiently exported from the nucleus during the early stages of HIV-1 replication. This timing ensures that Tat and Rev are initially produced in adequate amounts to facilitate viral mRNA transcription and export of unspliced transcripts that occurs during later stages of replication.

Packaging of unspliced viral RNA is another important function that requires careful regulation. The sequences that are important for packaging are located close to the TAR element, which in turn is immediately downstream of the transcription initiation site [60] (Figure 8.3). The packaging element comprises four SL structures, of which SL3 appears to be the most important for efficient binding to the viral Gag protein and incorporation into the capsid [61–63]. A dimer initiation signal (DIS) is also located within the sequence comprising the TAR and SLs [64]. The DIS is made up of autocomplementary palindromic sequences, which enable formation of "kissing loops" between two viral sequences, and the packaging of two viral genomes.

8.4.1 MicroRNAs and HIV-1 Replication

In addition to proteins, HIV-1 also encodes microRNAs (miRs) that play a role in regulating viral replication and pathogenesis of disease (reviewed in refs [65,66]). Progress on the topic has been rapid, but understanding the complex and intricate interplay between the regulatory RNAs and target sequences of the virus and of the host cells is not yet complete. Early studies that aimed to identify miRs derived from HIV-1 were based on bioinformatic analysis to identify structural motifs that were characteristic of the hairpins found in precursors of mature miRs. Several different miR candidates have now been identified at different sites of the viral genome, including those sequences encoding Nef [67], Rev [68], TAR [69,70], and RT [71] sequences. An additional potential source of viral gene silencing sequences may be duplex RNA arising after transcription of viral sequences with antisense polarity. Such transcripts may be derived from promoter elements located downstream of randomly integrated provirus, and resultant formation of double-stranded RNA may serve as a template for the RNA interference (RNAi) processing machinery [72].

The hairpin structure of the TAR element of HIV-1 is similar to that of miR precursors (Chapter 2); therefore, it prompted interest in this sequence as a source of HIV-1-encoded miR. TAR-derived RNAs are processed by Dicer, and processing bias favors the 3′ side of the sequence of the hairpin, miR-TAR-3p, for formation of the mature guide [66,73]. It is interesting to note that miR-TAR causes an inhibitory effect on viral gene expression at the level of transcription. miR generated from RT-encoding sequences has an opposite effect and increases the transcription of viral mRNA [71]. Thus, the two virus-encoded miRs may act in concert to regulate whether the virus enters a state of latency or active replication. TAR-derived miR also interacts with the 3′ untranslated regions of several genes involved in modulating programmed cell death, such as *Caspase 8*, *Aiolos*, and *Ikaros* [70,74]. This mechanism is thought to modulate conditions within infected cells to enhance survival of the virus. The finding of TAR-derived miRs within exosomes suggests that an intercellular signaling function may also be at play to modulate pathogenesis of the viral infection [75].

Cellular miRs also play a role in regulating HIV-1 replication. The mechanisms are complex and may involve direct interaction with viral targets or modulation of expression of cellular sequences to influence HIV-1 replication. For example, miR-29a inhibits viral replication in $CD4^+$ cells through its binding to *nef* sequences in the 3′ LTR [76]. Expression of Cyclin T1, required for Tat-activated HIV-1 transcription and differentiation of monocytes to macrophages, is inhibited by cellular miR-198 to cause diminished replication of the virus [77]. High levels of expression of miR-28, miR-125b, miR-150, and miR-382 are enriched in resting $CD4^+$ cells and associated with a state of latency [78]. A recent interesting finding has been that concentrations of certain miRs in the serum of HIV-1-infected individuals correlate with counts of $CD4^+$ cells [79]. Particularly important were miR-33a and miR-29b, which were present in higher concentrations in elite controllers (see below, section 8.5.1). Moreover, when overexpressed in cultured cells, viral production was diminished by these miRs.

8.5 PATHOGENESIS OF AIDS

Depletion of $CD4^+$ cells and concomitant immune dysfunction are fundamental to the pathogenesis of AIDS in HIV-1-infected individuals [80]. In addition, chronic immune activation and inflammation are significant contributors to the disease state. The severity of immunodeficiency and inflammation are the main determinants of the clinical course that HIV-1-infected individuals follow.

Mathematical modeling, based on detailed analysis of evolution and phylogenetic characteristics of *env* genes from many subjects, revealed that transmission of a founder virus is the initiating event of an acute infection [81]. *env* from founder viruses exhibit characteristic phenotypic features, such as dependence on CCR5 as a co-receptor and enhanced resistance to broadly neutralizing antibodies (bNAbs), which distinguish the sequences from those recovered from chronically infected individuals. After establishment of the infection, rapid proliferation of the virus is accompanied by a spike in viral loads and intense production of proinflammatory cytokines [82]. Because viral load is a determinant of transmission of the infection, acutely infected individuals are themselves highly infectious. Induction of a powerful innate and adaptive immune response causes circulating viral particle equivalents to diminish to a set point level. The viral load at the set point is a determinant of the speed with which an untreated individual will progress to full-blown AIDS [83]. The set point is typically established 2–3 months after initial exposure to the virus, and the concentration of the serum viral RNA is determined by the effectiveness of the immune response to HIV-1. The immunological pressure that is brought to bear on the virus has the effect of selecting mutants that are capable of escaping from the neutralizing effects.

HIV-1 is cytopathic to the CD4+ cells that it infects, which rapidly leads to depletion of this population of T cells and attendant immunocompromised state. Therefore, measuring numbers of circulating CD4+ cells has been a valuable marker for the monitoring of HIV-1-infected individuals. Until recently, the mechanisms by which HIV-1 causes a cytopathic effect have been incompletely understood. Death of CD4+ cells resulting from direct toxicity, syncytium formation, and consequences of inflammation play a role, and induction of apoptosis is central to loss of CD4+ cells [84–86]. A recent interesting finding is that activation of caspase-1, rather than caspase-3, is the main mechanism causing death of CD4+ cells [87]. Caspase-1-initiated cell death is triggered by abortive viral replication and is an intensely inflammatory form of programmed cell death. The process is accompanied by production and release of proinflammatory cytokines. Significantly, the effect of apoptosis caused by caspase-1 activation couples the inflammatory state and depletion of CD4+ cells, which are hallmarks of HIV-1 infection.

Profound depletion of CD4+ cells in the gastrointestinal tract is a highly significant contributor to the pathogenesis of AIDS [88]. Unlike the case with circulating CD4+ cells, CD4+ depletion in the gastrointestinal tract does not recover after initiation of combination antiretroviral therapy (cART) [89]. Gut-associated lymphoid tissue is a reservoir of HIV-1 and plays an important role in the persistence of the virus in HIV-1-infected individuals. The compromised immunity in the gut disrupts the effectiveness of the barrier to intestinal microbes. Greater permeability results in microbial products, especially lipopolysaccharide (LPS), reaching the portal then systemic circulations [3,88]. Activation of Toll-like receptor (TLR)-4 by LPS, direct interaction of HIV-1 with TLR7 and TLR8, co-infection with other viruses, and disruption of T helper-17 and regulatory T cells all contribute to the immunostimulatory and inflammatory effects of infection with HIV-1.

8.5.1 Factors Affecting Viral Pathogenesis

The varied response of individuals to infection with HIV-1 has been a subject of intense interest. Some people infected with the virus follow a mild clinical course that is characterized by low rates of viral replication and slow progression to AIDS [86]. So-called long-term nonprogressors (LTNPs), who comprise 2–15% of the population of HIV-1-infected individuals, maintain normal CD4+ counts with moderate-to-low viral loads in the absence of cART over a period of ~8 years. Elite controllers are an even rarer group of HIV-1-infected individuals, and these people typically have very low or even undetectable viral loads for a prolonged period. The precise definitions of LTNPs and elite controllers have varied in distinct settings. Differences in assays, such as sensitivities of viral load determinations, have complicated comparative studies [83]. Nevertheless, the characteristics of these groups of individuals have generated considerable interest. Because they demonstrate a "functional cure," which is control of HIV-1 in the absence of antiretroviral therapy, simulating the condition using gene therapy may be valuable for therapeutic application. Control of HIV-1 replication in LTNPs and elite controllers may occur as a result of

- Effective immune-mediated suppression of HIV-1;
- Resistance of CD4$^+$ cells to infection with the virus;
- Predominance of replication-defective proviral sequences; and
- Attenuated inflammation in infected individuals, which in turn would decrease the availability of HIV-1-susceptible CD4$^+$ cells.

Several studies have been performed to gain an understanding of the precise mechanisms of control of HIV-1 in LTNPs and elite controllers. Mutations in *ccr5* are not a common feature of LTNPs and elite controllers [80–83], and evidence indicates that CD4$^+$ cells from these individuals are typically susceptible to infection with the virus [84]. In addition, proviral sequences detected in CD4$^+$ cells from LTNPs and elite controllers are competent to replicate [85]. However, the number of copies of proviral sequences in CD4$^+$ cells is lower in people who demonstrate a slow progression to AIDS.

The mechanisms by which hosts achieve immunological control of the virus are complex and involve interplay between several factors [88]. A comprehensive analysis of host genome factors in HIV-1-infected individuals demonstrated that certain types of human leukocyte antigens (HLAs) are associated with protection or poorer prognosis [80]. However, there is considerable variation in the linkage of alleles with the elite controller state. In individuals with the HLA B*27 allele, a vigorous virus-controlling immune response to epitopes of Gag and p24 of HIV-1 subtype B is mounted [89]. The presence of HLA B*57 and HLA B*13 is associated with more effective control of the virus [80,90,91]. Linkage between alleles of the major histocompatability complex class locus and elite control of the HIV-1 infection has also been described. This association is likely to relate to an influence on the function of CD8-mediated responses to HIV-1 infection. Indeed, it has been shown that CD8$^+$ cells from some elite controllers have an enhanced functional efficacy in response to exposure to Gag. However, this property of CD8$^+$ cells is not universal among elite controllers of the virus. Roles for the innate immune response and neutralizing antibodies may also be important. Studies performed on natural killer (NK) cells indicate that variations in killer immunoglobulin receptors (KIRs) on the surface of NK cells influence signal transduction and efficiency of the innate immune response to HIV-1 infection [92]. Particular alleles (e.g., KIR3DL1 and KIR3DS1) are associated with inhibition of viral replication and attenuated disease progression when found together with the HLA-Bw4-80Ile alleles in infected individuals [93,94]. HLA-A and HLA-B molecules present HIV-1 peptides to KIRs. Therefore, variation in the components of the innate immunostimulatory mechanism understandably influences inhibition of HIV-1 replication by NK cells [95].

Demonstration that some individuals develop bNAbs in response to HIV-1 infection was an interesting and significant finding. These bNAbs are found in ~20% of HIV-1-infected people. They recognize variants of the virus from different subtypes, and evolution of the bNAbs by somatic hypermutation is a characteristic [96–98]. The breadth and potency of the bNAbs is varied, but their presence in infected individuals does not influence disease progression [99,100]. Although bNAbs usually appear years after initial exposure to the virus, they have recently been shown to develop early in life in babies who are infected perinatally, in utero, or during breast feeding [101]. The mechanisms for generating bNAbs are unclear. Simultaneous co-evolution of the founder virus and bNAbs may occur [102], although other studies suggest that bNAbs develop after superinfection of individuals with another viral strain [103,104]. The presence or absence of glycans at specific positions of the viral envelope proteins may also influence the emergence and efficacy of bNAbs [105]. The broad range of the neutralizing effects is potentially useful for the development of vaccines against HIV-1; therefore, a focus of research has been on identifying immunogens that are capable of inducing a potent bNAb response (Chapter 11). Targets on the viral envelope proteins that have been most studied are the V1V2 and V3 glycan-dependent sites of gp120, the membrane proximal external region of gp41, and CD4 binding sites of gp120 [106]. However. there may be other epitopes that are important for neutralization of the virus. Recent characterization of antibodies isolated from an HIV-1-infected individual who achieved control of the infection provided valuable insights [107]. These V1V2-directed antibodies contain protruding antigen-binding loops within the complementary determining region (CDR) of the heavy chain (H) 3. A parental antibody emerged 30–38 weeks after initial exposure to the virus and evolved to improve breadth and potency of virus neutralization. The projecting anionic antigen-binding domains of the CDR H3 antibodies are likely to provide improved access to viral epitopes that are shielded from access by more common antibodies with shorter CDRs. Understanding the mechanism of action and defining the sequences encoding the CDR H3 may have application in prophylactic gene therapy for HIV-1 infection. Delivery and sustained expression of the HIV-1-targeting CDR H3 antibody sequences may be useful for vectored immunoprophylaxis to prevent HIV-1 infection (Chapter 11) [108,109].

8.6 CURRENT AND NEW TREATMENTS OF HIV-1 INFECTION

Therapy for HIV-1 infection has rapidly evolved over the past 30 years [88]. Initially, supportive care and treatment of opportunistic infections were used to manage HIV-1-infected individuals. This approach was largely ineffective, and the mortality caused by the virus was extremely high. The discovery that azidothymidine (AZT), a nucleoside analog previously used for cancer treatment, improved the prognosis of HIV-1-infected individuals was a significant breakthrough [110] (Figure 8.1). The effect of AZT was to cause premature termination of DNA synthesis during reverse transcription. The drug is a competitive inhibitor of RT, and the azide group at the 3′ position of the deoxyribose moiety prevents formation of a phosphodiester bond during viral DNA synthesis. A lack of proofreading by the viral RT precludes removal of the drug after incorporation into the growing DNA strand. Although a promising development, it soon became clear that AZT did not provide a durable barrier to the virus. Escape mutants were rapidly selected, and the error-prone nature of DNA replication by the RT predisposes to mutability of the virus. Development of new antiretrovirals has been the focus of concerted research efforts, and this has resulted in the licensing of many more drugs that are effective against HIV-1. There are now 25 drugs, with varied viral targets, that have been licensed for use against HIV-1 [3]. During the 1990s, cART was successfully developed for treating HIV-1 [111,112]. Essentially, the principle underlying the approach is that blocking replication by targeting different viral sites provides a more effective barrier to viral escape. Treatment regimens currently advocate use of a combination of two nucleoside RT inhibitors—a non-nucleoside (noncompetitive) RT inhibitor together with a PR or IN inhibitor. Improvement in HIV-1 therapy has been impressive, and prognosis is now considerably better in HIV-1-infected individuals. A study performed in South Africa demonstrated this clearly: life expectancy improved by 11 years in the adult population after the implementation of cART in the public sector [113]. Despite the effectiveness of cART, treatment of individuals with HIV-1 infection often requires adaptation of the drug regimen to optimize efficacy. Avoiding side effects and preventing emergence of resistance are typically used to guide selection of drug combinations.

The World Health Organization now recommends that therapy for HIV-1 be initiated when the $CD4^+$ count reaches 500 cells or less per microliter [114]. Earlier timing of the start of treatment is beneficial and diminishes the transmission rates of the virus. Hopes of curative early treatment were recently raised after early cART treatment of a baby who was born to an HIV-1-infected woman [115]. After discontinuing treatment at 18 months of age, markers of viral replication were undetectable until the child was 30 months old. Unfortunately, later follow-up showed that the remission was not permanent (http://www.niaid.nih.gov/news/newsreleases/2014/pages/mississippibabyhiv.aspx). There is some variability in the response of patients to therapy, and viral suppression or immune reconstitution may predominate. To enhance immune function, the use of interleukins (IL)-2 and -7 has been assessed [116,117]. Although counts of $CD4^+$ cells may be increased, benefits of including these cytokines in treatment regimens are not established. A drawback to the approach may be that by increasing $CD4^+$ counts, the reservoir of latently infected cells is also be augmented.

Although advances in therapy of HIV-1 infection have been highly impressive, the treatment is not curative. Persistence of the provirus and reservoirs of latently infected cells are the major problems. Therefore, advancing approaches that are curative of the infection have been the major focus of research, but achieving a cure is a complicated endeavor. Successful cure in the Berlin patient has been particularly significant (see below, section 8.8.3). Insights gained from the experience have been helpful to devise strategies that use small-molecule drugs (e.g., Maraviroc [118]) and gene therapy to simulate the natural resistance of CCR5-defective $CD4^+$ cells.

8.7 MODELS OF HIV-1 REPLICATION

Various models have been developed for testing of anti-HIV drugs. These have included simulations of the viral infection in cultured cells and in animals. Earliest investigations entailed use of primary peripheral blood mononuclear cells (PBMCs), which were cultured before infection with the virus. These cells are still widely used, and protocols typically involve activation of the cells with phytohemagglutinin to render them susceptible to HIV-1 infection. The use of stably transduced lines such as Ghost [119,120], HeLa-CD4-CCR5 [121], C8166-R5 cells [122], microglial CHME5 [123], and the widely used TZM-bl reporter cells [124,125] may be more convenient than the use of PBMCs. TZM-bl cells are derived from the HeLa line and are susceptible to HIV-1 infection because they express CD4, CCR5, and CXCR4. The cells also have stably integrated *luciferase* and *β-galactosidase* reporter cassettes, which are under transcriptional control of the HIV-1 LTR. When infected with HIV-1, the activity of the reporter gene is governed by Tat, which in turn is derived from infecting viral particles. Therefore, measurement of reporter gene activity can be used conveniently as an indicator of the number of infecting viruses, which in turn may be dependent on the efficacy of an antiviral agent acting on virus-producing cells.

As with most drug development strategies, it is essential to evaluate anti-HIV-1 agents in vivo in animals before use in humans. Transgenic technologies have enabled generation of mice [126], rats [127], and rabbits [128] that express receptors and co-receptors required for cells to be infectable with HIV-1. However, these animals often do not provide all of the co-factors needed to support replication of the virus as it occurs in humans [129]. In addition, restriction factors may limit replication. To address these concerns, the focus has been on the "humanizing" of immunocompromised mice, which allows for engrafting of these animals with human cells that are infectable with HIV-1.

Several different humanized murine models have been developed to simulate human infection with HIV-1 [129,130]. Mice with the severe combined immunodeficiency (*scid*) phenotype have defective T- and B-cell function, which results from a mutation in the gene encoding the catalytic domain of DNA-dependent protein kinase [131]. These animals may be engrafted with thymus and liver or peripheral blood lymphocytes (PBLs) to generate *scid*-hu-Thy/Liv and *scid*-hu-PBL mice, respectively [129]. Subsequent infection with HIV-1 is possible and may be used to assess the efficacy of antiviral drugs. However, drawbacks are that these mice are unsuited to modeling mucosal transmission of the virus and engraftment efficiency is variable.

Nonobese diabetic (NOD) mice have defective NK cell function, which results from multiple genetic deficiencies. NOD and *scid* lineages have been crossed to generate NOD/*scid* mice that may be efficiently engrafted with human cells from various sources, such as cord blood and bone marrow. A further improvement of the model, particularly by improving grafting with heterologous cells and increasing the lifespan of NOD *scid* mice, has been achieved by making further crosses with the IL2rγ$^{-/-}$ line. The IL2rγ$^{-/-}$ phenotype results from mutations that affect signaling [132] or intracytoplasmic domains [133] of the interleukin-2 receptor common γ chain (IL2Rγ). These animals with the NOD/*scid*/IL2rγ$^{-/-}$ phenotype are called NOG mice. Engraftment of NOG mice with human cells of various sources is efficient, and they are readily disseminated throughout the body. A high viremia develops after infection with HIV-1, and the animals have been used to assess disease pathogenesis and antiviral drug efficacy [129]. A further improvement in humanized mice for study of HIV-1 has been through development of BLT (bone marrow-liver-thymus) mice [134]. These NOD-*scid* animals are engrafted with human liver and thymus, and they are then subjected to sublethal irradiation and bone marrow transplantation. BLT mice develop a high viremia after exposure to HIV-1. Depletion of CD4^{+} cells and adaptive immune responses simulate occurrences after natural HIV-1 infection [129,135]. In addition, the animals are susceptible to transmucosal infection.

Mice deficient in the V(D)J *recombination-activating gene 2* (*Rag2*), Rag2$^{-/-}$, have also been used in humanized murine models of HIV-1 infection [136]. Rag2 is essential for recombination of immunoglobulin and TCR genes during B and T cells' maturation (Chapter 1). Crossing Rag2$^{-/-}$ and IL2rγ$^{-/-}$ mice generates progeny that have defective NK cells and are deficient in humoral and cell-mediated immunity. These animals may be xenografted with human hematopoietic progenitor cells, particularly of the CD34^{+} CD38^{-} lineage, to humanize the immune system of the mice and make them susceptible to HIV-1 infection. The virus persists in these animals for up to a year, and the model is useful for assessment of efficacy of antivirals [137]. A drawback of the Rag2$^{-/-}$ IL2r$\gamma^{-/-}$ mice is that a humoral immune response to HIV-1 is usually not detectable.

Although HIV-1 and HIV-2 have their origins in African monkeys and apes, these nonhuman primates do not develop AIDS-like diseases after infection with SIVs. Conversely, SIVs do not normally infect Asian primates, and when infection is performed experimentally the animals rapidly develop an AIDS-like illness. The reason for the lack of disease pathogenesis in the African animals relates to their expression of restriction factors. These factors, which have evolved over several thousands of years, differ from their human or Asian primate orthologs in subtle ways to cause effects such as TRIM5α-mediated inhibition of capsid uncoating (Table 8.1 and see above, section 8.4). Because infection of Asian macaques with SIVs results in pathogenesis of a disease that closely resembles AIDS, these animals have been commonly used as models for analysis of pathogenesis, drug development, and advancing measures to prevent infection of humans with HIV-1. There are three main species of Asian macaques that are commonly used as models of HIV-1 infection: rhesus macaques, pig-tailed macaques, and cynomolgus macaques [129].

Despite Asian macaques being useful primate models, there are limitations to using of SIVs to simulate human infection with HIV-1. The viruses differ from each other in subtle ways that influence the entry into cells and mechanisms of replication. As a result, the efficacy of drugs and vaccines intended for use in humans may not be accurately assessed in macaques. To address this problem, chimeric SIV and HIV-1 (simian/HIVs, or SHIVs) have been used. These hybrid viruses retain essential features of SIV and are capable of infecting macaques. However, particular characteristics of HIV-1 are conferred by substituting sequences of SIV with equivalent parts of the HIV-1 genome. An example is the substitution of *env* from SIV with the equivalent sequences from HIV-1 [138], which has been used to test efficacy of vaccines [139] and pre-exposure prophylaxis [140].

8.8 GENE THERAPY FOR HIV-1 INFECTION

As discussed above, infection with HIV-1 is characterized by persistence, destruction of the immune system, and a high rate of mutation of viral sequences. Therefore, new treatments for the infection have been guided by the need to achieve durable inhibition of viral replication, disabling the provirus, restoration of immune function, and combating emergence of viral escape mutants. Although cART has unquestionably been highly successful, there are shortcomings to the available therapeutic regimens. Side effects of the drugs, failure to cure the infection, and necessity for life-long adherence to complex treatment regimens are drawbacks. Therefore, the focus of research on improving treatment has been on attempting to achieve durable curative suppression of the virus by using protocols that require administration over a limited period of time. Advances in gene therapy have been particularly important for attaining these goals. Successful anti-HIV-1 technologies have included use of RNAi-based silencing, gene editing, decoys, and dominant negative mutants [141,142]. Approaches that have been used to counter HIV-1 replication are summarized in Table 8.2 and Figure 8.5, and they are discussed in more detail below. Systemic administration of antiviral sequences and treatment ex vivo of hematopoietic stem cells (HSCs) or induced pluripotent stem cells (iPSCs) have both been used. Ex vivo treatment to render progenitor cells resistant to HIV-1 has significant advantages (see below, section 8.8.3).

8.8.1 Using RNAi against HIV-1 Infection

Many studies have reported effective silencing of HIV-1 by RNAi activators (reviewed in refs [141,142]). The variety of synthetic and expressed RNAi activators that has been used against HIV-1 is wide. Antiviral gene silencers include most of those depicted in Figures 2.2 and 2.4 of Chapter 2 and are discussed in more detail below. When silencing is used for HIV-1 therapy, the mRNA targets should be well conserved, present in all viral transcripts, and preferably within sequences that encode early proteins such as Tat, Rev, and Nef. By targeting sequences that are produced early during the infection, the viral replication is

Table 8.2 Examples of methods of using gene therapy to inhibit HIV-1 replication (see Figure 8.5)

Stage of HIV-1 Replication	Example	References
1 Binding of virion to CD4 and co-receptor (CCR5 or CXCR4)	Editing *ccr5* with ZFNs.	[239]
	Aptamers targeting gp120.	[218]
	Expression of a fusion inhibitor peptide.	[252]
2 Transcription of viral genes	Transcriptional gene silencing with RNA.	[197–199]
	Direct editing of viral genes with CRISPR/Cas.	[245,246]
	Inhibiting Tat activity with a decoy simulating TAR.	[254]
3 Inhibition of expression of genes encoding host factors	Editing *ccr5* with ZFNs.	[239]
4 Export of viral RNA	Expression of dominant negative mutant form of rev.	[260]
	Inhibition of rev with decoy simulating RRE.	[261]
5 Translation of viral proteins	RNAi activators	[150,172,173]
	Ribozymes	[258]
	Antisense	[262]
6 Facilitate immune response to HIV-1 infection	Modified TCRs containing CD4 that binds gp120.	[249–251]

Abbreviations: ZFN: Zinc finger Nuclease; CRISPR/Cas: clustered regularly interspaced short palindromic repeats (CRISPR) and CRSIPR associated (Cas); RRE: Rev response element; TCR: T-cell receptor; RNAi: RNA interference; TAR: Trans-activa*tion response element.*

more effectively impeded. Differing efficacy of RNAi against incoming genomic RNA has been reported. One study provided evidence that the genomic RNA is susceptible to RNAi-mediated degradation upon release from the capsid [143]. However, others have shown that the same RNA is inaccessible to the cellular RNAi machinery [144,145].

A recent interesting approach harnessed noncanonical processing of short hairpin RNAs (shRNAs) by Ago2, instead of Dicer, to generate guides targeting *pol* of HIV-1 [146]. The design of the so-called AgoshRNAs is modeled on the natural processing of miR-451 [147,148]. By shortening the stem length of AgoshRNAs to a minimum of 17 bp, Dicer processing is avoided, and the alternative Ago-2-mediated processing is favored [146]. In addition to the requirement of a short duplex region, weak G:U or U:G pairing at the apex of the duplex of the hairpin is important [149]. An advantage of the AgoshRNAs is that a single guide strand is produced, and problems of nonspecific effects caused by the passenger strand are obviated.

Initial work on silencing HIV-1 replication using RNAi employed standard Pol III promoters to regulate transcription of single antiviral shRNAs [150]. Predictably, it soon became clear that HIV-1 is capable of evading the inhibitory effects of single antiviral sequences [151–153]. Therefore, use of combinatorial RNAi against multiple viral targets and silencing of host factors was developed to limit the chances of viral escape and to provide maximal protection against different viral strains [154]. The principle underlying use of gene silencers in combination is similar to that on which cART is based; that is, by simultaneously disabling multiple viral targets, the ability of the virus to escape is significantly curtailed. The number of HIV-1 sites that need to be targeted at the same time to provide a good barrier to escape has been calculated to be four [155,156]. This figure was arrived at using mathematical modeling that took into account the (1) error rate of HIV-1 RT, (2) the risk of escape from single-gene silencing sequences, (3) the size and diversity of the population of viruses infecting an individual, and (4) the exponential decline in the likelihood of viral escape that occurs when the number of gene silencers is increased. Therefore, it is understandable that research has focused on development of RNAi activators that are capable of simultaneously interacting with multiple targets . Although targeting multiple sites of HIV-1 has been shown to provide a high barrier to resistance, a challenge to efficacy of combinatorial RNAi is that selection of mutations in nontargeted sequences may also cause resistance to gene silencing. Mutations to flanking sequences that alter RNA structure and render the intended target less accessible to RNAi activators is one proposed mechanism [153,157]. Selection for increased transcription caused by mutations in the U3 promoter sequences has also been proposed to overcome the silencing efficacy of RNAi activators [158], but this mechanism may not avoid efficacy of potent anti-HIV-1 gene silencers [159]. In addition to decreasing viral escape, targeting HIV-1 gene expression with multiple RNAi activators should increase the overall silencing efficacy. As a further enhancement of efficacy of combinatorial RNAi, use of licensed drugs together with RNAi activators has been assessed [160]. Impressive additive and synergistic antiviral effects were observed. Although these approaches show promise, an important consideration is that off-target effects are also increased when a larger number of virus-silencing sequences is used.

One of the first combinatorial approaches entailed the use of cassettes that encoded long hairpin RNAs (lhRNAs). The elongated stem duplex of the lhRNAs was designed to generate multiple HIV-1-targeting gene silencers [161–166]. Although theoretically a good approach, the method was complicated by variable efficiency of formation of each of the siRNAs that were generated across the span of the duplex of the lhRNAs. To generate more than one small interfering RNA (siRNA), a minimal stem length of 43 bp was required, but there is also a limit to the number of siRNAs that may be derived from a single lhRNA [166]. Formation of more than two sites from a single lhRNA is difficult to achieve. Generally, siRNAs encoded at the stem base were formed more efficiently than those located toward the loop side of the lhRNA. Varying the spacing between the intended siRNAs in the lhRNA duplex may be used to influence processing [163], but devising rules for the design of lhRNAs has been difficult. Empirical characterization to verify appropriate bias of the guide strands and assessment of silencing efficacy remain important. To increase the number of siRNAs that may be generated from one cassette containing lhRNA sequences, an alternative approach that incorporated two lhRNAs into a single cassette was used [167]. Each stem of the two lhRNAs comprised a duplex of 47 bp, and together they could be used to generate four siRNAs. However, silencing and processing efficiency from different cassettes was again variable and difficult to predict.

An alternative to using lhRNA expression cassettes has been the use of multiple independent HIV-1-targeting shRNA expression cassettes that are arranged in tandem [155,168]. However, this configuration may be prone to recombination, which is particularly problematic when using recombinant lentiviruses to deliver the sequences [155]. Thus, different transcriptional regulatory elements, such as U6, H1, and 7SK Pol III promoters and the U1 Pol II promoter, have been successfully used to prevent recombination [155]. The finding that tRNA promoters may generate artificial antiviral RNAi activators was an important development. This provided a much wider range of transcriptional regulatory elements that may be suitable for incorporation into RNAi expression cassettes [169]. Corroboration of the merits of the approach came from demonstration that expression of three antiviral shRNA sequences in a stably transduced

T cell line effectively prevented viral escape when replication occurred in mixed populations comprising protected and naïve cells [170].

Combining gene silencers is not the only configuration that may be used to prevent the emergence of viral escape. An alternative that was successfully used entailed use of a Pol II promoter to express antiviral shRNA together with a *trans* dominant negative mutant of Rev [171]. Transcription of the antiviral cassette was designed such that upstream expression of the *trans* dominant Rev mutant was coupled to a minimal polyadenylation signal. Read through from the transcription termination signal generated a *rev*-targeting shRNA. After transfection of cultured cells with this cassette, potent long-term inhibition of HIV-1 replication was achieved. Moreover, emergence of resistant viral populations was suppressed.

An improvement to combinatorial RNAi against HIV-1 was achieved through the use of Pol II cassettes that mimic natural primary miR clusters. By substituting anti-HIV-1 sequences for the partially complementary duplexes found in natural polycistronic miRs, it is possible to engineer artificial cassettes that produce multiple sequences that target the virus. This approach was used to engineer derivatives of the polycistronic miR-17-92 [172] and miR-106b [173] clusters. Naturally, miR-17-92 produces six mature miRs—miR-17, miR-18a, miR-19a, miR-20a, miR-19b, and miR-92a-1—and their overexpression is implicated in the malignant transformation of cells [174]. Liu and colleagues produced an artificial variant by using the backbone of the sequences encoding five of the miRNAs—miR-17, miR-18a, miR-19a, miR-20a, and miR-19b—to target five independent sites of the HIV-1 genome [172]. Expression from a cytomegalovirus (CMV) promoter and optimization of the positioning of the intended guide sequence led to efficient silencing of HIV-1 gene expression. The miR-106b cluster, naturally embedded within an intron, encodes miR-106b, miR-93, and miR-25. To generate HIV-targeting sequences, Aagard et al. used the backbone of this natural polycistron to generate an antiviral cassette, termed MCM7, which targeted *tat*- and *rev*-containing viral transcripts [173]. Inhibitory efficacy of the cassette was augmented by incorporating a TAR decoy into an intronic sequence of the MCM7 cassette. Results from both of the studies, using the miR-17-92 [172] or miR-106b [173] clusters, revealed that maintenance of the RNA secondary structures and flanking sequences were critical for efficient processing and silencing. Pol II expression cassettes have been developed using other miR scaffolds, such as those of pre-miR-30a and pre-miR-155 [175], to silence HIV-1 replication safely.

Future application of RNAi-based therapy for treatment of HIV-1 may involve systemic administration of the formulations containing the antiviral sequences. Given the difficulties with efficient delivery of RNAi activators to all HIV-1-infected cells, it is likely that in an individual treated using such a regimen there will be many cells that are not protected against infection with the virus. When infected, these cells are likely to provide a viral reservoir and source of mutants that may have an effect on the efficacy of antivirals. Demonstration that HIV-1 infection could be cured after bone marrow ablation, and then transplanted with cells from a donor who was deficient in CCR5 [176], prompted investigating use of RNAi to interrupt HIV-1 replication by simulating similar conditions (see discussion below of the Berlin patient, section 8.8.3). In addition to CCR5, many other host dependency factors identified using large-scale genome-wide screens [177–179] are candidate targets for similar treatment approaches. Verification of the utility of the approach came from an early study showing that silencing of *ccr5* and *cxcr4* with shRNAs, delivered using recombinant lentiviruses, rendered cultured cells resistant to HIV-1 infection [180]. Other studies have provided further support of the potential of silencing host factors as a means of treating HIV-1 infection. Inhibiting function of human transcription elongation factors, such as SPT5 [181] and cyclin T1 [182], diminished viral replication. Autophagy is required for HIV-1 replication, and disruption of factors required for this cellular process in T cell lines inhibited viral replication without evidence of toxicity [183]. Silencing of members of a panel of 30 viral host factors demonstrated variable inhibition of viral replication [184]. When silenced, three of the factors, endosomal sorting protein ALIX, transcription factor TRBP, and autophagy-related ATG16, caused sustained inhibition of viral replication. Although promising, toxicity may result from silencing of endogenous genes. Therefore, before use as a therapy, it will be critically important to exclude harmful effects of inhibiting normal endogenous genes. However, inhibition of *ccr5* does not appear to be problematic because patients with homozygous deficiency of the gene are asymptomatic.

A significant advance in the use of RNAi against HIV-1 infection was in fact based on silencing of herpes simplex virus (HSV)-2 [185]. Transmission of HIV-1 is considerably higher in HSV-2-infected individuals (Chapter 11) [186,187]. The genital ulcers that are a feature of HSV infection promote infectability and infectiousness with HIV-1. Formulation of synthetic siRNAs targeting HSV-2 within lipoplex vectors, then intravaginal administration to mice, provided good protection against transmission of the herpes virus in mice [185].

8.8.1.1 Using Lentiviral Vectors to Deliver Anti HIV-1 miR Expression Cassettes

Recombinant lentiviruses achieve sustained expression of transgenes (Chapter 4) and are capable of infecting various cells, including HSCs, which makes them useful for delivery of HIV-1-targeting RNAi activators. However, the derivation of recombinant lentiviruses from HIV-1 may complicate their use to treat HIV-1 infection. Mechanisms by which expressed RNAi activators may reduce vector titers are varied [188,189]. Attenuation of vector titers may result from (1) silencing of sequences that are common to the vector and HIV-1, (2) binding of vector genomic RNA by therapeutic RNAi sequences, (3) processing of vector-embedded primary miR sequences by Drosha/DGCR8, and (4) interference between promoters of the RNAi activator cassettes and regulatory elements of the vector genome. To optimize titers, customized alteration to the vectors' designs may be necessary. Factors such as orientation of the cassettes relative to sense or antisense polarity of the viral genome, promoter selection for expressing the RNAi activators, and use of shRNA or primary miR-like sequences are important. When using shRNA expression cassettes, inactivation of vectors by self-targeting may or not be problematic. It has been reported that stability of the hairpin within the vectors' genomic RNA limits accessibility to the silencing machinery [190]. However, contrary evidence of self-cleavage has also been reported [160,163]. Nevertheless, titers of vectors could be rescued by saturating the pathway and inhibition of RNAi using Nodamura-virus B2 protein or Adenovirus VAI RNA (Chapter 4). When incorporating artificial primary miRs into lentiviral vectors, inclusion of the expression cassette in an antisense orientation is preferable [188]. The presence of primary miR in the sense configuration within the vector genome is susceptible to processing by Drosha/DGCR8, but it only causes a minor decrease in vector titers. The selection of promoters to control RNAi activators also requires careful consideration. The commonly used CMV promoter causes a major decrease in vector titers [188]. The mechanism involves interference with the transcription of vector genomic sequences and is independent of squelching or orientation of insertion into the vector. Use of alternative transcriptional regulatory elements, especially inducible promoters, should be considered when generating lentiviral vectors to deliver RNAi activators.

8.8.1.2 Transcriptional Gene Silencing of Viral Promoter Elements

Use of RNA to inhibit transcription of genes of HIV-1 is an intriguing topic with considerable therapeutic potential. Recent emergence of long noncoding RNAs (lncRNAs) as important natural regulators of transcription has had a bearing on harnessing nucleic acids for silencing of target promoters [191]. The actions of lncRNAs are brought about by recruitment of enzymes that are responsible for chromatin modeling. These include DNA methyltransferase-3A and histone deacetylases. Through interaction with chromatin modeling complexes, epigenetic modification is brought about at sites with homology to the lncRNA. Derangement in the effects of lncRNAs on promoter activity is implicated in the etiology of certain cancers [192] and neurodegenerative diseases [193], among others. It is interesting to note that HIV-1 has recently been shown to express lncRNA that regulates transcription of the LTR [194]. The effects of epigenetic modeling are durable and may be inherited by cellular progeny, which makes them particularly appealing for use against HIV-1. The predominant effect of lncRNA is to cause heterochromatin formation and suppression of gene expression. Thus, de-repression of promoter function may be achieved by disrupting lncRNA effects, and it may be caused by using siRNA or oligonucleotides that target the lncRNAs [195]. On the other hand, it is possible to effect suppression of promoters by introducing exogenous mimics of lncRNAs, and an algorithm has been devised to aid with the design of transcription-inhibiting RNAs [196].

Successful transcriptional silencing of the LTR of HIV-1 has been demonstrated, which has significant implications for countering the virus [197–199]. This transcriptional repression has been achieved when using synthetic single-stranded RNA [199], duplex siRNAs [198], and more recently with expressed shRNAs [197]. When the NFκB binding region within the LTR was targeted, transcriptional inhibition was markedly attenuated by two centrally positioned mismatches in the 21-bp duplex [197]. This observation indicates that the effect is specific and dependent on guide and target hybdridization. To test efficacy of the approach in vivo, NOD/SCID/Janus kinase knockout mice were grafted with human PBMCs that had been transduced with lentiviral vectors that encoded the promoter-targeting shRNAs [197]. After infection of the mice with HIV-1, viral loads were significantly lower in the animals that had been engrafted with cells expressing the promoter-targeting sequences, and the $CD4^+$ cell counts were also maintained in these animals. The effect was not observed in mice that had received the control lentiviral vectors that expressed the mutant shRNA. The efficacy also correlated with the amount of antisense sequence derived from the shRNAs, and the transcriptional silencing effect could be reversed by administration of phorbol myristate acetate to cells cultured ex vivo. Effectively suppressing HIV-1 transcription could achieve a functional cure of the infection. By inducing a state of sustained transcriptional inactivity, the virus replication could be effectively inhibited. This is an exciting prospect, but success of the technology is dependent on efficient delivery to target cells, ensuring that silencing effects are specific to viral promoters and limiting the span of the epigenetic modifications.

8.8.2 Applying Aptamers and Decoys to Treatment of HIV-1 Infection

The three dimensional shapes of aptamers and decoys are central to the efficacy of these nucleic acids against HIV-1. Decoys are designed to simulate structures of functionally important motifs within the RNA of HIV-1. Examples are mimics of TAR and RRE. TAR normally interacts with Tat to enhance transcription of viral mRNA. Engineering sequences to resemble this structure may be used to sequester TAR and other essential factors required for proliferation of HIV-1 [200–202]. Likewise, sequences that sequester Rev by mimicking the RRE have been used successfully as decoys [203]. Co-localization of the decoy and equivalent viral elements within the same subcellular organelle are important, and coupling small nuclear RNA motifs to the decoys has been used successfully to direct TAR-like decoys to the nucleoli [201].

Aptamers are single-stranded RNA or DNA sequences that fold into a three-dimensional shape to have high affinity for particular target molecules. They are selected from libraries of nucleic acids using the process of systemic evolution of ligands by exponential enrichment (SELEX) [204,205]. In its simplest form, the procedure entails coupling of a molecule to a solid phase, which is then used to bind and select aptamers from a starting library comprising 10^{14}–10^{15} different sequences. Repeated cycles of amplification and transcription in vitro, for selection of RNA-based aptamers, are used to enrich sequences with intended binding affinities. Steps that increase stringency of selection are typically included to eliminate nonspecific binding and select for the aptamers with the highest target affinity. A more recent and technically demanding SELEX procedure entails the use of whole cells to purify aptamers [206]. Aptamers may be selected with high-binding affinities, and K_d values in the low nanomolar range have been generated against many different targets. As with applying RNAi activators for therapeutic use, aptamers may be administered as synthetic sequences or encoded within expression cassettes that are delivered to target tissues.

Aptamers that bind various proteins of HIV-1 have been described [207]. Targets include viral structural proteins, regulatory proteins, and enzymes. RT has been a popular target for selection of HIV-1-disabling aptamers. The positive charge of the template-binding cleft of RT is well suited to aptamer binding, and favored interaction with this site is supported by the observation that RT-targeting aptamers have similar structural features. RNA and DNA aptamers have been generated, which function efficiently as enzyme inhibitors that also diminish viral replication in cells. As with other therapies against HIV-1, emergence of escape mutants may be problematic. However, because binding of nucleic acids is fundamental to the function of RT, mutations that disrupt binding to aptamers are expected to compromise RT function severely. This notion was corroborated by studies showing that RT mutants resistant to aptamer binding had defects of processivity, and replication of viruses containing these sequences was attenuated [208]. IN has also been used as the target of potentially therapeutic aptamers, and studies have shown promising effects [209,210]. Inhibition of HIV-1 replication in cultured cells and using reporter assays has been demonstrated.

Aptamers that compete for binding to regulatory proteins have also been used successfully to counter HIV-1 in cultured cells. Detailed structural analyses have enabled identification of essential structural features of the RRE of HIV-1 [211,212]. It is interesting to note that the RRE takes on a structure that is influenced by noncanonical base pairing. Together with known sequences of RRE variants, this information was used to generate a library from which Rev-targeting aptamers were isolated. Because the RRE is embedded in the *env* sequence (Figure 8.3), flexibility of RRE is constrained. Consequently, it is possible that selected aptamers have a higher binding affinity to Rev than those of the natural RREs [213,214]. The utility of Rev aptamers and decoys has been demonstrated in models of HIV-1 infection [215,216]. The sequences have also been used in combination to augment effects of other anti-HIV-1 agents.

Potentially therapeutic aptamers that bind to gp120 with high affinity have been described [217,218]. The sequences were capable of inhibiting HIV-1 replication in cultured cells with an efficiency that was equivalent to that of neutralizing antibodies. Although promising, probability of evasion by the virus as a result of selection of escape mutants is likely to be high. In addition to targeting gp120, inhibitory aptamers that bind to the viral CD4 receptor have been described [219]. Inhibiting HIV-1 entry into cells by targeting the host receptor may be useful to counter viral escape.

8.8.2.1 Using Aptamers to Deliver HIV-1-Targeting RNAi Activators

Facilitating delivery of RNAi activators by coupling of siRNAs to cell-targeting aptamers has been an innovative method of advancing gene therapy for treatment of HIV-1 infection. Binding between an aptamer and a synthetic siRNA may be achieved through covalent linkage or through formation of "sticking" bridges (Figure 8.6; reviewed in refs [220,221]). With covalent linkage, the aptamer sequence is contiguous with one of the strands of the siRNA, and the duplex is formed through hybridization to a separate oligonucleotide. The stick arrangement is based on the use of linkers and a bridging oligonucleotide that binds the aptamer to the siRNA [221–223]. This configuration provides convenient versatility, which is useful for interchange of siRNAs and combinatorial approaches in which one aptamer may be adapted to deliver different siRNAs.

Aptamers within the chimeric antivirals have been targeted to gp120, which is expressed on the surface of HIV-1-infected cells [222,223], and also to CD4 [224]. Although the precise mechanism of internalization into the cells expressing gp120 or CD4 is not conclusively established, it appears to be based on receptor-mediated endocytosis [220]. After endosomal escape and release into the cytoplasm, the aptamer-siRNA complex is processed by Dicer to uncouple the two components. Thereafter, the siRNA is available for incorporation into the RNA-induced silencing complex (RISC) and guide strand selection to effect silencing. Use of aptamer-siRNA complexes silenced targets more effectively than siRNAs alone. The mechanism is thought to result from interaction of Dicer with the aptamer-siRNA complex and functional enhancement of downstream Argonaute proteins within RISC. Aptamer-siRNA complexes have been extensively studied and good efficacy has been observed when coupling gp120- and CD4-binding aptamers to *tat/rev*-targeting siRNAs [224,225]. Testing in a humanized murine model of HIV-1 infection revealed that replication of HIV-1 was effectively inhibited and the decline of $CD4^+$ cells in these animals is arrested [225]. In addition to serving as a targeting agent, the aptamers were

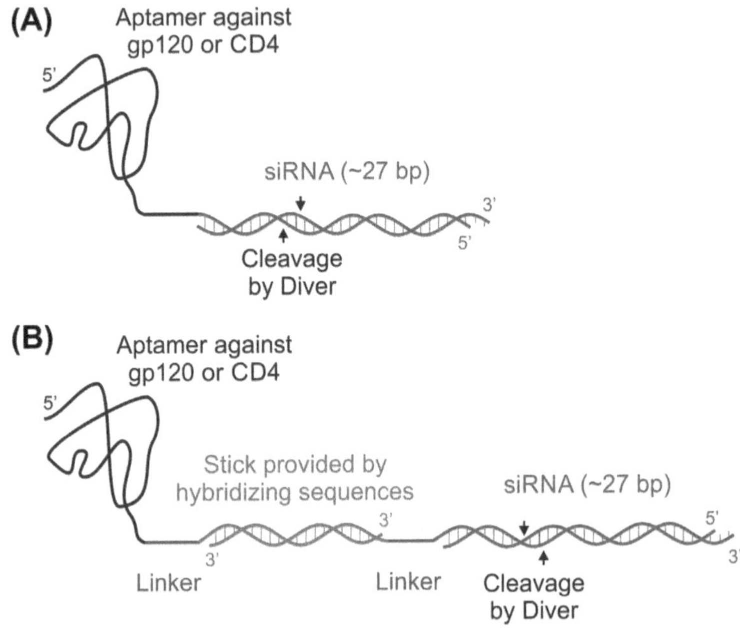

FIGURE 8.6 Chimeric molecules comprising cell-targeting aptamers coupled to anti-HIV-1 RNAi activators.
(A) The aptamer may be contiguous and covalently linked to one of the strands of the duplex Dicer substrate. To provide flexibility, a linker may be inserted between dsRNA and aptamer components. The complex with dsRNA is processed by Dicer, which results in separation of aptamers and silencing molecules. The duplex RNA is then available for incorporation into RISC and selection of the silencing guide. (B) In an alternative arrangement, the aptamer is linked to the RNA duplex through a stick sequence that is constituted by complementary hybridizing sequences. Linker sequences flank the stick region, and the siRNA precursor comprising 27 bp is processed by Dicer to release the silencing molecule. dsRNA, double-stranded RNA.

also capable of neutralizing the virus. Efficacy of the combination of aptamer with siRNA was more lasting and superior to that of either of the antiviral components alone. Moreover, the aptamer-siRNA agent was not toxic to animals receiving the treatment.

8.8.3 Modifying Stem Cells Ex vivo to Counter HIV-1 Infection

A significant influence on the HIV-1 gene therapy field has come from apparent cure of the Berlin patient from infection with the virus [176,227]. This HIV-1-infected individual developed acute myeloid leukemia and required bone marrow transplantation as part of his treatment. After myeloablation, high-dose chemotherapy, and treatment to prevent graft versus host (GVH) disease, the patient received allogeneic bone marrow cells from a matched individual who was homozygous for the *ccr5Δ32* mutation. Thus, the donor cells were

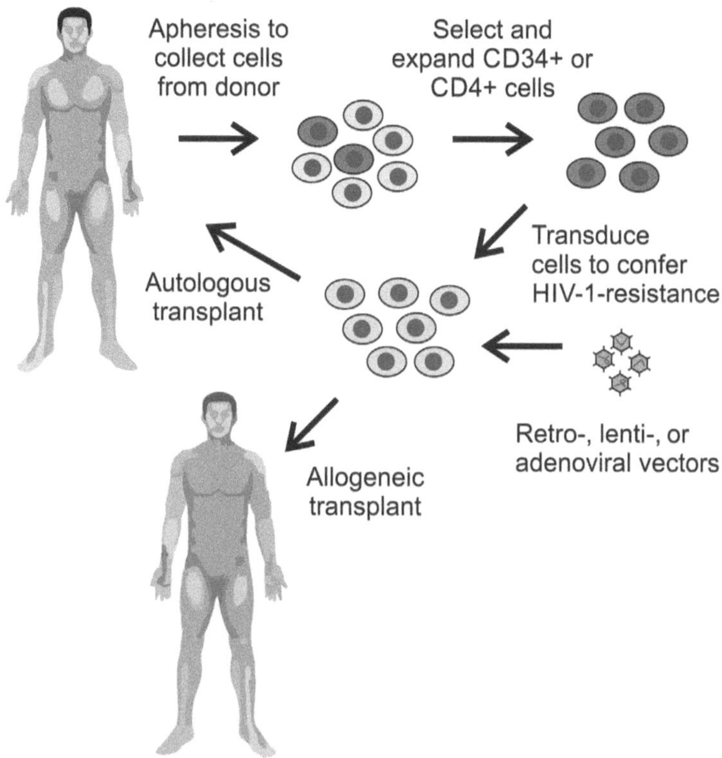

FIGURE 8.7 Modification of cells ex vivo to render them resistant to HIV-1.
Cells collected from the donor are selected and expanded in culture. After transduction with vectors encoding sequences that confer resistance to HIV-1 (Table 8.3), the selected cells may be re-infused into the original donor (autologous transplant) or a different patient (allogeneic transplant). Use of $CD34^+$ hematopoietic stem cells (HSCs) has the advantage of enabling formation of self-renewing precursor cells that generate progeny with the HIV-1 resistance phenotype. $CD4^+$ cells have the disadvantage of a limited life span.

resistant to HIV-1 infection, and since receiving the transplant, the patient has been free from HIV-1 without the need for continuation of cART. Although the myeloablation, chemotherapy, and GVH prevention measures are likely to have contributed to eliminating the reservoirs of HIV-1, repopulation with CCR5-deficient and HIV-1-resistant cells was important for the long-term suppression of viral replication. In an attempt to replicate this success, a large body of research has been directed at using gene therapy, performed ex vivo, to render $CD4^+$ cells resistant to HIV-1 [227,228] (Figure 8.7). The main objective of using this approach has been to accomplish long-term control of viral replication after a single procedure. The favored method is to modify HSCs ex vivo by introducing anti-HIV-1 genetic elements. The CD34 marker is commonly used to verify "stemness" and select these target cells [229]. Optimization of high-speed cell sorting protocols and development of methods to expand the cells in culture is a priority that is key to success of the methodology [227]. In addition to use of HSCs, iPSCs may be used in a variation of the protocol [230]. After modification ex vivo, stem cells are administered to patients. The principle underlying the procedures is that stem cells have the capacity for self-renewal and will pass the HIV-1-resistance phenotype on to progeny cells. In addition, the cells have the potential to replace a patient's immune system with cells that are resistant to viral infection, and by preventing viral proliferation viral loads would be decreased. Reservoirs of actively replicating virus and pools of latently infected cells should also be diminished. Various gene therapy-based antiviral elements have been developed. These include direct targeting of viral sequences and disabling of host factors, especially using gene editing to mutate viral co-receptors, to make cells resistant to HIV-1. Methods of delivery of the therapeutic sequences ex vivo have also been varied. They have included use of lentiviral vectors, recombinant adenoviruses, transfection of mRNA, and transposons (Chapters 4 and 5) [231].

8.8.4 Gene Editing to Counter HIV-1 Replication

Rapid advances in gene editing technology (Chapter 3) have provided impetus for implementing new ways of disabling HIV-1 replication. Methods of inhibiting viral replication may be achieved by direct targeting of viral sequences as well as by disabling host factors that are required by HIV-1. Zinc finger nucleases (ZFNs), transcription activator-like effector nucleases (TALENs), and the clustered regularly interspaced short palindromic repeats (CRISPR) with CRISPR-associated (Cas) system have all been used to inhibit HIV-1 replication. Repeated cleavage of the target and repair by error-prone nonhomologous end joining introduces mutations and permanent disruption at the targets. A particularly popular strategy has been to use sequence-specific endonucleases to target the gene encoding CCR5 [232,233]. Methodology has been devised with the aim of recreating the phenotype of the grafted cells that were successfully used to treat the Berlin patient. A difficulty with using cells with the *ccr5Δ32/ccr5Δ32* genotype is that individuals with the homozygous mutation are rare; therefore, matching them to patients for HIV-1 treatment would be difficult. By using gene editing on patients' own cells, methods are being developed by which matched cells are generated that may be used for autologous transplant. Such a transplant procedure would obviate the need for myeloablative treatment and remove a requirement for prevention of GVH disease. The advantage of gene editing over RNAi-mediated gene silencing of *ccr5* is that transient expression of the gene editor may be used to achieve permanent mutation of the target. Although RNAi-based approaches have been used successfully to inhibit *ccr5* expression [234,235], a sustained silencing effect is required to produce the desired phenotype.

Engineered ZFNs were the first to be used successfully in a gene editing approach to mutate CCR5-encoding sequences [236]. Perez and colleagues generated ZFNs made up of dimers of four-fingered ZFPs with coupled FokI endonuclease components. Cognates for these engineered proteins were in the DNA encoding the first transmembrane domain of the CCR5 protein. This region was immediately upstream of the *ccr5Δ32* mutation, and it was reasoned that the structure of the CCR5 is particularly sensitive to mutation at this site. The lead ZFN efficiently disrupted genomic DNA. HIV-1 infection of cultured Ghost cells and primary lymphocytes was significantly inhibited. Feasibility of the approach was then tested in humanized mice of the NOG line. Human $CD4^+$ cells were transduced using ZFN-expressing Ads and then infused into the mice. After challenge with HIV-1, the animals receiving cells that had been treated with the *ccr5*-targeting ZFNs showed lower viremia and enrichment of $CD4^+$ cells. In a follow-up study using the same lead ZFN, Holt and colleagues assessed efficacy after transient expression of the ZFNs in $CD34^+$ HSCs after using plasmid nucleofection [237]. The advantage of modifying progenitor cells is that long-term resistance to HIV-1 infection is potentially achievable. Transient expression of the ZFNs in $CD34^+$ cells generated CCR5-negative cells, which were capable of producing progeny $CD4^+$ cells with the same phenotype. NOD/SCID/$IL2\gamma^{-/-}$ mice were engrafted with the treated HSCs and then infected with HIV-1. The modified $CD34^+$ cells retained their ability to generate multiple lineages. CCR5-negative cells were resistant to HIV-1 infection in the xenografted animals and were rapidly selected after challenge with the virus. A similar study aimed at optimizing conditions for *ccr5* disruption used recombinant Ads to transduce $CD34^+$ cells with cassettes expressing ZFNs [238]. After testing a panel of cytokines, chemokines, and synthetic compounds, it was found that protein kinase C activators markedly improved editing of *ccr5*.

The technology has now reached an advanced stage of development, and evaluation of gene editors in a clinical trial has recently been reported [239] (Table 8.3). The primary objective of this early-phase clinical trial was to assess safety of using ZFNs to treat HIV-1 infection. Secondary objectives were the determination of the effects on circulating $CD4^+$ cells, viral loads, persistence of modified cells, and their migration to the gut. Twelve patients who were on cART and were aviremic at the time of initiating the study were recruited for the study. Autologous $CD4^+$ T cells were transduced with a recombinant adenovirus expressing a CCR5-targeting ZFN, SB-728-T developed by Sangamo Biosciences, then re-infused into the patients. An adverse effect was observed in one of the patients, and the cause was ascribed to transfusion-induced effects. The half-life of the modified T cells was estimated to be 48 days. After treatment interruption with associated viremia, the decline in CCR5-negative cells was significantly lower than that of the unmodified cells. DNA of HIV-1 in the blood of most patients decreased. In one of the patients, the increase in viral

Table 8.3 Selected clinical trials performed to evaluate the efficacy of autologous transplant for the treatment of HIV-1 infection

Antiviral Therapy and HIV-1 Target	Name	Mechanism of Delivery	Transduced and Transplanted Cells	References
ZFNs targeting *ccr5*	SB-728-T	Adenovirus	$CD4^+$ T cells	[239]
Immunostimulatory modified TCR comprising extracellular domain of CD4 coupled to zeta (ζ) chain of TCR of $CD3^+$ cells	CD4ζ TCR	Retrovirus	$CD4^+$ and $CD8^+$ cells	[249–251]
Sequence-encoding inhibitor of HIV-1 fusion derived from gp41	maC46	Retrovirus	$CD4^+$ T cells	[252]
ccr5-targeting ribozyme, shRNA against *tat/rev*, and TAR decoy	*ccr5* ribozyme, *tat/rev* shRNA and TAR decoy	Lentivirus	$CD34^+$ cells	[254]
Sequence encoding dominant negative mutant form of the rev protein	huM10	Retrovirus	$CD4^+$ T cells	[260]
Expressed ribozymes that target *tat/vpr*	OZ1	Retrovirus	$CD34^+$ cells	[258]
Decoy of RRE	RRE decoy	Retrovirus	$CD34^+$ cells	[261]
Antisense against *env*	VRX496	Conditionally replicating lentivirus	$CD4^+$ T cells	[262]

Abbreviations: ZFN: Zinc Finger Nuclease; RRE: Rev response element; TCR: T-cell receptor; shRNA: short hairpin RNA; maC46: membrane-anchored C peptide 46; huM10: humanized revM10; OZ1: tat-vpr-specific anti-HIV ribozyme; VRX496: VIRxSYS Corporation env antisense 496.

load was delayed until week 6 of the treatment interruption, but the circulating viral particle equivalents decreased thereafter. It is interesting to note that this individual was also heterozygous for the *ccr5 Δ32* deletion.

These results are promising and augur well for the future of using gene editing to treat HIV-1 infection; however, the issue of specificity remains vexed. Off-target cleavage of sites within the genome is potentially genotoxic, and limiting these effects is crucial to the ultimate success of gene editing in a clinical setting. Availability of improved sequencing technology provides the means for defining the frequency of off-target mutations, and genome-wide analysis has demonstrated that *ccr5*-targeting ZFNs do sometimes cleave nonspecifically [240,241]. A study performed to compare the specificity of TALENs and ZFNs provided some insight into comparative specificity of the nucleases [242]. Established lines and primary fibroblasts were transfected with sequences encoding *ccr5*-targeting nucleases. The toxicity of ZFNs was found to be higher than that of TALENs and is thought to be a result of higher nonspecific mutagenesis caused by ZFNs. This notion was supported by detailed analysis of mutations introduced at the *ccr5* gene and homologous *ccr2* locus. Both nucleases had equivalent efficiency at on-target sites, but nonspecific cleavage at the *ccr2* gene was significantly higher in the cells treated with the ZFN-expressing cassettes. To assist with minimizing off-target cleavage at homologous sites, algorithms have been devised to guide the design of TALENs [243,244]. Apart from properties of individual sequences themselves, chromatin structure and methylation status of the targets may also play a role in influencing nuclease cleavage, but transcription activity does not appear to be particularly important [242].

The excitement that accompanied discovery of the CRISPR/Cas system and its application to gene editing predictably led to investigating the use of the technology for treating HIV-1 infection. Two studies have reported on the efficacy of CRISPR/Cas against HIV-1 [245,246]. With both investigations, use of the nucleases to mutate viral sequences directly, rather than to disable host factors required by the virus, was described. Ebina and colleagues showed that guides complementary to the LTRs resulted in a significant inhibition of viral replication [245]. Detailed sequencing analysis verified cleavage at the intended target. Hu et al. also targeted the LTRs using single and multiplex CRISPR/Cas systems [246]. Good efficacy was demonstrated in microglial, promonocytic and T cells. The effect was observed without evidence of toxicity. Both studies applying CRISPR/Cas to target the LTRs demonstrated that in some cases proviral sequences were excised from the host genome [245,246]. Cells expressing Cas9 and multiple HIV-1-targeting guides were resistant to HIV-1 infection. The efficiency and convenience of engineering the CRISPR/Cas system make it very attractive for therapeutic use against HIV-1 infection. The technology may provide the means for directly disabling proviral sequences and thus overcoming the problematic latent infection. However, specificity of the target cleavage needs to be improved. Off-target cleavage by the CRISPR/Cas system is currently higher than it is with ZFNs and TALENs.

8.8.5 Clinical Trials Using Gene Therapy to Treat HIV-1 Infection

Several clinical trials aimed at developing gene therapy for the treatment of HIV-1 infection are in progress or have been completed. Gene-based prophylactic or therapeutic immunostimulations are discussed in Chapter 11, and clinical trials aimed at direct HIV-1 treatment are considered here. Approaches that modify autologous cells to make them resistant to HIV-1 have been favored [141,247,248]. The trial using ZFNs (see above, section 8.8.4) and selected other examples are listed in Table 8.3 and described below. There is a wide range of antiviral agents that have been developed. These include expressed ZFNs, RNAi activators, antisense molecules, ribozymes, a viral entry inhibitor peptide, and Tat and Rev decoys. Candidate therapeutics have also comprised antivirals used alone or in combination.

In one of the first trials, T cells were transduced with a chimeric TCR gene using a recombinant retroviral vector [249]. The sequence encoded extracellular and transmembrane domains of CD4 and was linked to the zeta (ζ) chain of the TCR of CD3 cells. The underlying principle of the design was that on binding of the CD4 to gp120 of HIV-1, the ζ chain would activate T cells to mediate antiviral immune activation. $CD4^+$ and $CD8^+$ T cells were modified ex vivo with a retroviral vector encoding the CD4ζ TCR and then expanded in culture for a period of 2 weeks. After re-infusion, CD4ζ-expressing cells were detectable in 1–3% of PBMCs at 8 weeks, and their counts decreased to 0.1% at 1 year. Although there was no significant change in circulating viral RNA and DNA, the modified cells migrated to the rectal tissue, where they exerted an antiviral effect. In a study involving syngeneic transplant, survival of the modified $CD4^+$ and $CD8^+$ cells was observed after stimulation with beads coated with antibodies to CD3 and CD28 [250]. A phase II trial using the same modified cells reported that the treatment did not cause a significant decrease in the viral reservoir. However, in some of the patients on the trial there was a decrease from the baseline viral burden [251].

An alternative approach used retroviral vectors to transduce T cells with a gene encoding the maC46 peptide that inhibits HIV-1 entry into cells [252]. The peptide is derived from gp41 and exerts its effects by interacting with heptad repeats of the protein [253]. As a result, gp41 is locked in a fusion-incompetent state that prevents viral entry. After transduction of cells ex vivo, autologous transplant was performed in patients who were infected with multidrug resistant virus [252]. Modified cells persisted in the blood, bone marrow, and lymph nodes for a period of a year. Although no significant decrease in viral load was observed during the first 4 months of the study, patients whose treatment regimens were changed thereafter responded with marked declines in their viral loads.

A combinatorial approach was used in a trial that used a *ccr5*-targeting ribozyme, a shRNA against *tat/rev*, and a TAR decoy [254]. Lentiviral vectors were engineered to express the three antiviral elements and then used to transduce patients' HSCs ex vivo. The four patients who were selected for full participation in the trial were undergoing autologous bone marrow transplant for treatment of non-Hodgkin lymphoma. Transduction did not affect hematopoietic potential of the $CD34^+$ cells, and the patients tolerated the re-infusion well. Markers of the modified cells persisted for 2 years, albeit at low levels during the later stages of the follow-up period. To comply with factors of ethical concern, cART was continued in the patients and the subjects also received unmodified cells during the re-infusion. Although no therapeutic benefit was discernable, the results indicated that the approach is safe and potentially a platform for RNA-based treatment of HIV-1 infection. HIV-1-targeting ribozymes, either alone or in combination with other therapeutics, have been assessed in several clinical trials [255–259]. The designs of the ribozymes have been based on different parental types of enzymatic RNA, (e.g., hairpin and hammerhead ribozymes) and have been targeted to different viral sequences. Although some efficacy has been observed, the technology recently appears to have fallen out of favor. Emergence of other methods, particularly gene silencing and gene editing technologies, have accounted for this.

Utility of a dominant negative mutant form of the Rev protein, huM10, was also investigated in a study performed in children [260]. Recombinant retroviral vectors were used to transduce $CD4^+$ cells ex vivo, and the cells were then re-infused into patients. Counts of the modified cells diminished rapidly after re-infusion. However, in one of the patients who did not adhere to her/his cART regimen, there was a spike in viral loads during this period that also coincided with a rise on the counts of huM10-positive cells. This observation suggested that there was a selective advantage in the modified cells and that expression of huM10 provided protection against HIV-1.

Two studies have reported on the use of ribozymes that target *tat/vpr* sequences [257,258]. The study reported by Mitsuyasu and colleagues was the first phase II double-blind placebo controlled study to assess efficacy of the gene therapy approach for treating HIV-1 infection [258]. Seventy-four patients were enrolled in the study to receive the ribozyme, called OZ1. Autologous $CD34^+$ cells were treated with a recombinant retrovirus expressing the ribozyme or control vector. No adverse events occurred after the treatment, and markers of the modification persisted for ~1 year. Viral loads at 47 or 48 weeks after the re-infusion were similar in control and treated groups, but the time-weighted areas under curves of the viral load plots were lower in the treated group. There was also a significant increase in $CD4^+$ counts in patients who had received the ribozyme-expressing vector.

Outcomes from other clinical trials using different antivirals, such as decoys that mimic the RRE [261] and antisense sequences expressed from a conditionally replicating lentiviral vector [262], have been similar. Overall, data from clinical trials have not yet provided conclusive proof of the efficacy of the technology for treating HIV-1 infection. Nevertheless, it is likely that with progress to more advanced phases of clinical testing, along with the availability of improved technologies, use of gene therapy for HIV-1 infection is entering an exciting phase.

8.9 CONCLUSIONS

Since the first presentation of HIV-1-infected individuals in the early 1980s, and later emergence of the global pandemic, enormous research efforts have been devoted to studying the virus. As a result, HIV-1 is now very well characterized, and this has contributed to improved management of the infection. During early stages of the pandemic, treatment was ineffective and infection with the virus was fatal. However, development of improved antivirals and their use in combination has transformed HIV-1 infection from being a deadly disease to a treatable chronic illness. Advances were derived from multidisciplinary research that provided the insights and technical means required for developing novel rational approaches to countering the infection. Nevertheless, despite the remarkable successes, HIV-1 continues to impose a major disease burden, particularly in parts of Africa where resources are limited. Lack of an effective vaccine against HIV-1 and difficulties with achieving curative therapy have been frustrating.

Applying nucleic acid transfer to treating HIV-1 infection has made impressive progress, and technology is emerging that is potentially curative or capable of providing immunoprotection (see Chapter 12). Several different approaches that use rational design to engineer nucleic acids to inhibit HIV-1 replication directly or indirectly have been developed. RNAi-based gene silencing, particularly inhibition of viral sequences using combinatorial methods, gene editing, transcriptional gene silencing, and use of aptamers and decoys all show promise. Recent progress with gene editing technology has been particularly exciting. The ability to disrupt *ccr5* to confer cellular resistance to HIV-1 infection is a significant development. Coupling gene editing to methodology that enables efficient transduction of autologous progenitor cells ex vivo suggests that using gene therapy to achieve a functional cure of the infection is feasible. In addition to making cells resistant to HIV-1 infection, gene editing could be used to eliminate the provirus from reservoirs of latently infected cells [245,246]. Successfully deploying the technology will be a great step toward achieving the ultimate goal of complete cure from the infection. As with most strategies developing the use of gene therapy to treat viral infections, there are impediments to clinical implementation. Dealing with generic issues such as ensuring specificity of the therapy and optimizing delivery to target cells are important. In addition, because most HIV-1-infected individuals come from environments where resources are limited, devising simple and affordable regimens will be essential. Nevertheless, an expectation that gene therapy will make an impact on the disease burden caused by the viral infection is realistic.

REFERENCES

[1] Sharp PM, Hahn BH. Origins of HIV and the AIDS pandemic. Cold Spring Harb Perspect Med 2011;1(1):a006841.

[2] Chitnis A, Rawls D, Moore J. Origin of HIV type 1 in colonial French Equatorial Africa? AIDS Res Hum Retroviruses 2000;16(1):5–8.

[3] Maartens G, Celum C, Lewin SR. HIV infection: epidemiology, pathogenesis, treatment, and prevention. Lancet 2014;384(9939):258–71.

[4] Hemelaar J, Gouws E, Ghys PD, Osmanov S. Isolation W-UNfH, characterisation. Global trends in molecular epidemiology of HIV-1 during 2000–2007. Aids 2011;25(5):679–89.

[5] Gottlieb MS, Schroff R, Schanker HM, Weisman JD, Fan PT, Wolf RA, et al. Pneumocystis carinii pneumonia and mucosal candidiasis in previously healthy homosexual men: evidence of a new acquired cellular immunodeficiency. N Engl J Med 1981;305(24):1425–31.

[6] Barre-Sinoussi F, Chermann JC, Rey F, Nugeyre MT, Chamaret S, Gruest J, et al. Isolation of a T-lymphotropic retrovirus from a patient at risk for acquired immune deficiency syndrome (AIDS). Science 1983;220(4599):868–71.

[7] HIV/AIDS JUNPo. Global Report: UNAIDS Report on the global AIDS Epidemic 2013. 2013. http://wwwunaidsorg/en/media/unaids/contentassets/documents/epidemiology/2013/gr2013/201309_epi_core_enpdf. [accessed 16.06.14].

[8] Quinn TC, Wawer MJ, Sewankambo N, Serwadda D, Li C, Wabwire-Mangen F, et al. Viral load and heterosexual transmission of human immunodeficiency virus type 1. Rakai Project Study Group. N Engl J Med 2000;342(13):921–9.

[9] Mugo NR, Heffron R, Donnell D, Wald A, Were EO, Rees H, et al. Increased risk of HIV-1 transmission in pregnancy: a prospective study among African HIV-1-serodiscordant couples. Aids 2011;25(15):1887–95.

[10] Baggaley RF, White RG, Boily MC. HIV transmission risk through anal intercourse: systematic review, meta-analysis and implications for HIV prevention. Int J Epidemiol 2010;39(4):1048–63.

[11] Tanser F, Barnighausen T, Grapsa E, Zaidi J, Newell ML. High coverage of ART associated with decline in risk of HIV acquisition in rural KwaZulu-Natal, South Africa. Science 2013;339(6122):966–71.

[12] Auvert B, Taljaard D, Lagarde E, Sobngwi-Tambekou J, Sitta R, Puren A. Randomized, controlled intervention trial of male circumcision for reduction of HIV infection risk: the ANRS 1265 Trial. PLoS Med 2005;2(11):e298.

[13] Wills NM, Gesteland RF, Atkins JF. Pseudoknot-dependent read-through of retroviral gag termination codons: importance of sequences in the spacer and loop 2. EMBO J 1994;13(17):4137–44.

[14] Debouck C, Gorniak JG, Strickler JE, Meek TD, Metcalf BW, Rosenberg M. Human immunodeficiency virus protease expressed in *Escherichia coli* exhibits autoprocessing and specific maturation of the gag precursor. Proc Natl Acad Sci USA 1987;84(24):8903–6.

[15] Strickler JE, Gorniak J, Dayton B, Meek T, Moore M, Magaard V, et al. Characterization and autoprocessing of precursor and mature forms of human immunodeficiency virus type 1 (HIV 1) protease purified from *Escherichia coli*. Proteins 1989;6(2):139–54.

[16] Briggs JA, Riches JD, Glass B, Bartonova V, Zanetti G, Krausslich HG. Structure and assembly of immature HIV. Proc Natl Acad Sci USA 2009;106(27):11090–5.

[17] Kido H, Kamoshita K, Fukutomi A, Katunuma N. Processing protease for gp160 human immunodeficiency virus type I envelope glycoprotein precursor in human T4+ lymphocytes. Purification and characterization. J Biol Chem 1993;268(18):13406–13.

[18] Chan DC, Kim PS. HIV entry and its inhibition. Cell 1998;93(5):681–4.

[19] Trono D. When accessories turn out to be essential. Nat Med 1998;4(12):1368–9.

[20] Guyader M, Emerman M, Sonigo P, Clavel F, Montagnier L, Alizon M. Genome organization and transactivation of the human immunodeficiency virus type 2. Nature 1987;326(6114):662–9.

[21] Yu Q, König R. Vif and the role of antiviral cytidine deaminases in HIV-1 replication. HIV Seq Compend. In: Leitner T, Foley B, Hahn BH, Marx PA, McCutchan FE, Mellors JW, et al., editors. HIV Sequence Compendium. Los Alamos, NM, USA: Theoretical Biology and Biophysics Group, Los Alamos National Laboratory; 2003:2–13.

[22] Sheehy AM, Gaddis NC, Choi JD, Malim MH. Isolation of a human gene that inhibits HIV-1 infection and is suppressed by the viral Vif protein. Nature 2002;418(6898):646–50.

[23] Neil SJ, Zang T, Bieniasz PD. Tetherin inhibits retrovirus release and is antagonized by HIV-1 Vpu. Nature 2008;451(7177):425–30.

[24] Dube M, Roy BB, Guiot-Guillain P, Mercier J, Binette J, Leung G, et al. Suppression of Tetherin-restricting activity upon human immunodeficiency virus type 1 particle release correlates with localization of Vpu in the trans-Golgi network. J Virol 2009;83(9):4574–90.

[25] Douglas JL, Viswanathan K, McCarroll MN, Gustin JK, Fruh K, Moses AV. Vpu directs the degradation of the human immunodeficiency virus restriction factor BST-2/Tetherin via a {beta}TrCP-dependent mechanism. J Virol 2009;83(16):7931–47.

[26] McNatt MW, Zang T, Bieniasz PD. Vpu binds directly to tetherin and displaces it from nascent virions. PLoS Pathog 2013;9(4):e1003299.

[27] Hrecka K, Hao C, Gierszewska M, Swanson SK, Kesik-Brodacka M, Srivastava S, et al. Vpx relieves inhibition of HIV-1 infection of macrophages mediated by the SAMHD1 protein. Nature 2011;474(7353):658–61.

[28] Laguette N, Sobhian B, Casartelli N, Ringeard M, Chable-Bessia C, Segeral E, et al. SAMHD1 is the dendritic- and myeloid-cell-specific HIV-1 restriction factor counteracted by Vpx. Nature 2011;474(7353):654–7.

[29] Ryoo J, Choi J, Oh C, Kim S, Seo M, Kim SY, et al. The ribonuclease activity of SAMHD1 is required for HIV-1 restriction. Nat Med 2014;20(8):936–41.

[30] Markle TJ, Philip M, Brockman MA. HIV-1 Nef and T-cell activation: a history of contradictions. Future Virol 2013;8(4).

[31] Doores KJ, Bonomelli C, Harvey DJ, Vasiljevic S, Dwek RA, Burton DR, et al. Envelope glycans of immunodeficiency virions are almost entirely oligomannose antigens. Proc Natl Acad Sci USA 2010;107(31):13800–5.

[32] Rossmann MG. The canyon hypothesis. Hiding the host cell receptor attachment site on a viral surface from immune surveillance. J Biol Chem 1989;264(25):14587–90.

[33] Weiss RA. Thirty years on: HIV receptor gymnastics and the prevention of infection. BMC Biol 2013;11:57.

[34] Dalgleish AG, Beverley PC, Clapham PR, Crawford DH, Greaves MF, Weiss RA. The CD4 (T4) antigen is an essential component of the receptor for the AIDS retrovirus. Nature 1984;312(5996):763–7.

[35] Klatzmann D, Champagne E, Chamaret S, Gruest J, Guetard D, Hercend T, et al. T-lymphocyte T4 molecule behaves as the receptor for human retrovirus LAV. Nature 1984;312(5996):767–8.

[36] Cocchi F, DeVico AL, Garzino-Demo A, Arya SK, Gallo RC, Lusso P. Identification of RANTES, MIP-1 alpha, and MIP-1 beta as the major HIV-suppressive factors produced by CD8+ T cells. Science 1995;270(5243):1811–5.

[37] Feng Y, Broder CC, Kennedy PE, Berger EA. HIV-1 entry cofactor: functional cDNA cloning of a seven-transmembrane, G protein-coupled receptor. Science 1996;272(5263):872–7.

[38] Liu R, Paxton WA, Choe S, Ceradini D, Martin SR, Horuk R, et al. Homozygous defect in HIV-1 coreceptor accounts for resistance of some multiply-exposed individuals to HIV-1 infection. Cell 1996;86(3):367–77.

[39] Doms RW. Beyond receptor expression: the influence of receptor conformation, density, and affinity in HIV-1 infection. Virology 2000;276(2):229–37.

[40] Warrilow D, Tachedjian G, Harrich D. Maturation of the HIV reverse transcription complex: putting the jigsaw together. Rev Med Virol 2009;19(6):324–37.

[41] Grutter MG, Luban J. TRIM5 structure, HIV-1 capsid recognition, and innate immune signaling. Curr Opin Virology 2012;2(2):142–50.

[42] Stremlau M, Owens CM, Perron MJ, Kiessling M, Autissier P, Sodroski J. The cytoplasmic body component TRIM5alpha restricts HIV-1 infection in Old World monkeys. Nature 2004;427(6977):848–53.

[43] Kleiman L. tRNA(Lys3): the primer tRNA for reverse transcription in HIV-1. IUBMB life 2002;53(2):107–14.

[44] Coffin J, Hughes S, Varmus H. Retroviruses. Plainview, NY: Cold Spring Harbor Laboratory Press; 1997. [xv].

[45] Sarafianos SG, Marchand B, Das K, Himme DM, Parniak MA, Hughes SH, et al. Structure and function of HIV-1 reverse transcriptase: molecular mechanisms of polymerization and inhibition. J Mol Biol 2009;385(3):693–713.

[46] Yan N, Regalado-Magdos AD, Stiggelbout B, Lee-Kirsch MA, Lieberman J. The cytosolic exonuclease TREX1 inhibits the innate immune response to human immunodeficiency virus type 1. Nat Immunol 2010;11(11):1005–13.

[47] Arhel N. Revisiting HIV-1 uncoating. Retrovirology 2010;7:96.

[48] Ciuffi A, Bushman FD. Retroviral DNA integration: HIV and the role of LEDGF/p75. Trends Genet 2006;22(7):388–95.

[49] Barr SD, Leipzig J, Shinn P, Ecker JR, Bushman FD. Integration targeting by avian sarcoma-leukosis virus and human immunodeficiency virus in the chicken genome. J Virol 2005;79(18):12035–44.

[50] Schroder AR, Shinn P, Chen H, Berry C, Ecker JR, Bushman F. HIV-1 integration in the human genome favors active genes and local hotspots. Cell 2002;110(4):521–9.

[51] Busschots K, Vercammen J, Emiliani S, Benarous R, Engelborghs Y, Christ F, et al. The interaction of LEDGF/p75 with integrase is lentivirus-specific and promotes DNA binding. J Biol Chem 2005;280(18):17841–7.

[52] Emiliani S, Mousnier A, Busschots K, Maroun M, Van Maele B, Tempe D, et al. Integrase mutants defective for interaction with LEDGF/p75 are impaired in chromosome tethering and HIV-1 replication. J Biol Chem 2005;280(27):25517–23.

[53] Singh DP, Ohguro N, Chylack Jr LT, Shinohara T. Lens epithelium-derived growth factor: increased resistance to thermal and oxidative stresses. Investigative Ophthalmol Vis Sci 1999;40(7):1444–51.

[54] Ge H, Si Y, Roeder RG. Isolation of cDNAs encoding novel transcription coactivators p52 and p75 reveals an alternate regulatory mechanism of transcriptional activation. EMBO J 1998;17(22):6723–9.

[55] Ciuffi A, Llano M, Poeschla E, Hoffmann C, Leipzig J, Shinn P, et al. A role for LEDGF/p75 in targeting HIV DNA integration. Nat Med 2005;11(12):1287–9.

[56] Wang W-K, Chen M-Y, Chuang C-Y, Jeang K-T, Huang L-M. Molecular biology of human immunodeficiency virus type 1. J Microbiol Immunol Infect 2000;33(3):131–40.

[57] Kingsman SM, Kingsman AJ. The regulation of human immunodeficiency virus type-1 gene expression. Eur J Biochem/FEBS 1996;240(3):491–507.

[58] Brown PH, Tiley LS, Cullen BR. Efficient polyadenylation within the human immunodeficiency virus type 1 long terminal repeat requires flanking U3-specific sequences. J Virol 1991;65(6):3340–3.

[59] Romani B, Engelbrecht S, Glashoff RH. Functions of Tat: the versatile protein of human immunodeficiency virus type 1. J Gen Virol 2010;91(Pt 1):1–12.

[60] Kuzembayeva M, Dilley K, Sardo L, Hu WS. Life of psi: how full-length HIV-1 RNAs become packaged genomes in the viral particles. Virology 2014;454–455:362–70.

[61] Clever J, Sassetti C, Parslow TG. RNA secondary structure and binding sites for gag gene products in the 5′ packaging signal of human immunodeficiency virus type 1. J Virol 1995;69(4):2101–9.

[62] De Guzman RN, Wu ZR, Stalling CC, Pappalardo L, Borer PN, Summers MF. Structure of the HIV-1 nucleocapsid protein bound to the SL3 psi-RNA recognition element. Science 1998;279(5349):384–8.

[63] McBride MS, Schwartz MD, Panganiban AT. Efficient encapsidation of human immunodeficiency virus type 1 vectors and further characterization of *cis* elements required for encapsidation. J Virol 1997;71(6):4544–54.

[64] Clever JL, Wong ML, Parslow TG. Requirements for kissing-loop-mediated dimerization of human immunodeficiency virus RNA. J Virol 1996;70(9):5902–8.

[65] Fowler L, Saksena NK. Micro-RNA: new players in HIV-pathogenesis, diagnosis, prognosis and antiviral therapy. AIDS Rev 2013;15(1):3–14.

[66] Harwig A, Das AT, Berkhout B. Retroviral microRNAs. Curr Opin Virology 2014;7C:47–54.

[67] Omoto S, Fujii YR. Regulation of human immunodeficiency virus 1 transcription by nef microRNA. J Gen Virol 2005;86(Pt 3):751–5.

[68] Bennasser Y, Le SY, Benkirane M, Jeang KT. Evidence that HIV-1 encodes an siRNA and a suppressor of RNA silencing. Immunity 2005;22(5):607–19.

[69] Ouellet DL, Plante I, Landry P, Barat C, Janelle ME, Flamand L, et al. Identification of functional microRNAs released through asymmetrical processing of HIV-1 TAR element. Nucleic Acids Res 2008;36(7):2353–65.

[70] Ouellet DL, Vigneault-Edwards J, Letourneau K, Gobeil LA, Plante I, Burnett JC, et al. Regulation of host gene expression by HIV-1 TAR microRNAs. Retrovirology 2013;10:86.

[71] Zhang Y, Fan M, Geng G, Liu B, Huang Z, Luo H, et al. A novel HIV-1-encoded microRNA enhances its viral replication by targeting the TATA box region. Retrovirology 2014;11:23.

[72] Schopman NC, Willemsen M, Liu YP, Bradley T, van Kampen A, Baas F, et al. Deep sequencing of virus-infected cells reveals HIV-encoded small RNAs. Nucleic Acids Res 2012;40(1):414–27.

[73] Klase Z, Kale P, Winograd R, Gupta MV, Heydarian M, Berro R, et al. HIV-1 TAR element is processed by Dicer to yield a viral micro-RNA involved in chromatin remodeling of the viral LTR. BMC Mol Biol 2007;8:63.

[74] Klase Z, Winograd R, Davis J, Carpio L, Hildreth R, Heydarian M, et al. HIV-1 TAR miRNA protects against apoptosis by altering cellular gene expression. Retrovirology 2009;6:18.

[75] Narayanan A, Iordanskiy S, Das R, Van Duyne R, Santos S, Jaworski E, et al. Exosomes derived from HIV-1-infected cells contain trans-activation response element RNA. J Biol Chem 2013;288(27):20014–33.

[76] Ahluwalia JK, Khan SZ, Soni K, Rawat P, Gupta A, Hariharan M, et al. Human cellular microRNA hsa-miR-29a interferes with viral nef protein expression and HIV-1 replication. Retrovirology 2008;5:117.

[77] Sung TL, Rice AP. miR-198 inhibits HIV-1 gene expression and replication in monocytes and its mechanism of action appears to involve repression of cyclin T1. PLoS Pathog 2009;5(1):e1000263.

[78] Huang J, Wang F, Argyris E, Chen K, Liang Z, Tian H, et al. Cellular microRNAs contribute to HIV-1 latency in resting primary CD4+ T lymphocytes. Nat Med 2007;13(10):1241–7.

[79] Reynoso R, Laufer N, Hackl M, Skalicky S, Monteforte R, Turk G, et al. MicroRNAs differentially present in the plasma of HIV elite controllers reduce HIV infection in vitro. Sci Reports 2014:4; http://dx.doi.org/10.1038/srep05915.

[80] International HIVCS, Pereyra F, Jia X, McLaren PJ, Telenti A, de Bakker PI, Walker BD, et al. The major genetic determinants of HIV-1 control affect HLA class I peptide presentation. Science 2010;330(6010):1551–7.

[81] Carrington M, Nelson GW, Martin MP, Kissner T, Vlahov D, Goedert JJ, et al. HLA and HIV-1: heterozygote advantage and B*35-Cw*04 disadvantage. Science 1999;283(5408):1748–52.

[82] Magierowska M, Theodorou I, Debre P, Sanson F, Autran B, Riviere Y, et al. Combined genotypes of CCR5, CCR2, SDF1, and HLA genes can predict the long-term nonprogressor status in human immunodeficiency virus-1-infected individuals. Blood 1999;93(3):936–41.

[83] Okulicz JF. Elite controllers and long-term nonprogressors: models for HIV vaccine development. J AIDS Clin Res 2012;S8:005. http://dx.doi.org/10.4172/2155-6113.1000e105.

[84] Julg B, Pereyra F, Buzon MJ, Piechocka-Trocha A, Clark MJ, Baker BM, et al. Infrequent recovery of HIV from but robust exogenous infection of activated CD4(+) T cells in HIV elite controllers. Clin Infect Dis 2010;51(2):233–8.

[85] Blankson JN, Bailey JR, Thayil S, Yang HC, Lassen K, Lai J, et al. Isolation and characterization of replication-competent human immunodeficiency virus type 1 from a subset of elite suppressors. J Virol 2007;81(5):2508–18.

[86] Pantaleo G, Menzo S, Vaccarezza M, Graziosi C, Cohen OJ, Demarest JF, et al. Studies in subjects with long-term nonprogressive human immunodeficiency virus infection. N Engl J Med 1995;332(4):209–16.

[87] Kiepiela P, Ngumbela K, Thobakgale C, Ramduth D, Honeyborne I, Moodley E, et al. CD8+ T-cell responses to different HIV proteins have discordant associations with viral load. Nat Med 2007;13(1):46–53.

[88] Barre-Sinoussi F, Ross AL, Delfraissy JF. Past, present and future: 30 years of HIV research. Nat Rev Microbiol 2013;11(12):877–83.

[89] Almeida JR, Price DA, Papagno L, Arkoub ZA, Sauce D, Bornstein E, et al. Superior control of HIV-1 replication by CD8+ T cells is reflected by their avidity, polyfunctionality, and clonal turnover. J Exp Med 2007;204(10):2473–85.

[90] Migueles SA, Sabbaghian MS, Shupert WL, Bettinotti MP, Marincola FM, Martino L, et al. HLA B*5701 is highly associated with restriction of virus replication in a subgroup of HIV-infected long term nonprogressors. Proc Natl Acad Sci USA 2000;97(6):2709–14.

[91] Saah AJ, Hoover DR, Weng S, Carrington M, Mellors J, Rinaldo Jr CR, et al. Association of HLA profiles with early plasma viral load, CD4+ cell count and rate of progression to AIDS following acute HIV-1 infection. Multicenter AIDS Cohort Study. Aids 1998;12(16):2107–13.

[92] Bashirova AA, Thomas R, Carrington M. HLA/KIR restraint of HIV: surviving the fittest. Annu Rev Immunol 2011;29:295–317.

[93] Alter G, Martin MP, Teigen N, Carr WH, Suscovich TJ, Schneidewind A, et al. Differential natural killer cell-mediated inhibition of HIV-1 replication based on distinct KIR/HLA subtypes. J Exp Med 2007;204(12):3027–36.

[94] Martin MP, Gao X, Lee JH, Nelson GW, Detels R, Goedert JJ, et al. Epistatic interaction between KIR3DS1 and HLA-B delays the progression to AIDS. Nat Genet 2002;31(4):429–34.

[95] Alter G, Heckerman D, Schneidewind A, Fadda L, Kadie CM, Carlson JM, et al. HIV-1 adaptation to NK-cell-mediated immune pressure. Nature 2011;476(7358):96–100.

[96] Sather DN, Armann J, Ching LK, Mavrantoni A, Sellhorn G, Caldwell Z, et al. Factors associated with the development of cross-reactive neutralizing antibodies during human immunodeficiency virus type 1 infection. J Virol 2009;83(2):757–69.

[97] Gray ES, Madiga MC, Hermanus T, Moore PL, Wibmer CK, Tumba NL, et al. The neutralization breadth of HIV-1 develops incrementally over four years and is associated with CD4+ T cell decline and high viral load during acute infection. J Virol 2011;85(10):4828–40.

[98] Mikell I, Sather DN, Kalams SA, Altfeld M, Alter G, Stamatatos L. Characteristics of the earliest cross-neutralizing antibody response to HIV-1. PLoS Pathog 2011;7(1):e1001251.

[99] Piantadosi A, Panteleeff D, Blish CA, Baeten JM, Jaoko W, McClelland RS, et al. Breadth of neutralizing antibody response to human immunodeficiency virus type 1 is affected by factors early in infection but does not influence disease progression. J Virol 2009;83(19):10269–74.

[100] van Gils MJ, Euler Z, Schweighardt B, Wrin T, Schuitemaker H. Prevalence of cross-reactive HIV-1-neutralizing activity in HIV-1-infected patients with rapid or slow disease progression. Aids 2009;23(18):2405–14.

[101] Goo L, Chohan V, Nduati R, Overbaugh J. Early development of broadly neutralizing antibodies in HIV-1-infected infants. Nat Med 2014;20(6):655–8.

[102] Liao HX, Lynch R, Zhou T, Gao F, Alam SM, Boyd SD, et al. Co-evolution of a broadly neutralizing HIV-1 antibody and founder virus. Nature 2013;496(7446):469–76.

[103] Cortez V, Odem-Davis K, McClelland RS, Jaoko W, Overbaugh J. HIV-1 superinfection in women broadens and strengthens the neutralizing antibody response. PLoS Pathog 2012;8(3):e1002611.

[104] Powell RL, Kinge T, Nyambi PN. Infection by discordant strains of HIV-1 markedly enhances the neutralizing antibody response against heterologous virus. J Virol 2010;84(18):9415–26.

[105] Moore PL, Gray ES, Wibmer CK, Bhiman JN, Nonyane M, Sheward DJ, et al. Evolution of an HIV glycan-dependent broadly neutralizing antibody epitope through immune escape. Nat Med 2012;18(11):1688–92.

[106] Tomaras GD, Haynes BF. Lessons from babies: inducing HIV-1 broadly neutralizing antibodies. Nat Med 2014;20(6):583–5.

[107] Doria-Rose NA, Schramm CA, Gorman J, Moore PL, Bhiman JN, DeKosky BJ, et al. Developmental pathway for potent V1V2-directed HIV-neutralizing antibodies. Nature 2014;509(7498):55–62.

[108] Balazs AB, Chen J, Hong CM, Rao DS, Yang L, Baltimore D. Antibody-based protection against HIV infection by vectored immunoprophylaxis. Nature 2012;481(7379):81–4.

[109] Balazs AB, Ouyang Y, Hong CM, Chen J, Nguyen SM, Rao DS, et al. Vectored immunoprophylaxis protects humanized mice from mucosal HIV transmission. Nat Med 2014;20(3):296–300.

[110] Fischl MA, Richman DD, Grieco MH, FischlLeedom JM, Groopman JE, Mildvan D, et al. The efficacy of azidothymidine (AZT) in the treatment of patients with AIDS and AIDS-related complex. A double-blind, placebo-controlled trial. N Engl J Med 1987;317(4):185–91.

[111] Hammer SM, Squires KE, Hughes MD, Grimes JM, Demeter LM, Currier JS, et al. A controlled trial of two nucleoside analogues plus indinavir in persons with human immunodeficiency virus infection and CD4 cell counts of 200 per cubic millimeter or less. AIDS Clinical Trials Group 320 Study Team. N Engl J Med 1997;337(11):725–33.

[112] Palella Jr FJ, Delaney KM, Moorman AC, Loveless MO, Fuhrer J, Satten GA, et al. Declining morbidity and mortality among patients with advanced human immunodeficiency virus infection. HIV Outpatient Study Investigators. N Engl J Med 1998;338(13):853–60.

[113] Bor J, Herbst AJ, Newell ML, Barnighausen T. Increases in adult life expectancy in rural South Africa: valuing the scale-up of HIV treatment. Science 2013;339(6122):961–5.

[114] Hirnschall G, Harries AD, Easterbrook PJ, Doherty MC, Ball A. The next generation of the World Health Organization's global antiretroviral guidance. J Int AIDS Soc 2013;16(1).

[115] Persaud D, Gay H, Ziemniak C, Chen YH, Piatak Jr M, Chun TW, et al. Absence of detectable HIV-1 viremia after treatment cessation in an infant. N Engl J Med 2013;369(19):1828–35.

[116] Group I-ES, Committee SS, Abrams D, Levy Y, Losso MH, Babiker A, et al. Interleukin-2 therapy in patients with HIV infection. N Engl J Med 2009;361(16):1548–59.

[117] Sereti I, Dunham RM, Spritzler J, Aga E, Proschan MA, Medvik K, et al. IL-7 administration drives T cell-cycle entry and expansion in HIV-1 infection. Blood 2009;113(25):6304–14.

[118] Hardy WD, Gulick RM, Mayer H, Fatkenheuer G, Nelson M, Heera J, et al. Two-year safety and virologic efficacy of maraviroc in treatment-experienced patients with CCR5-tropic HIV-1 infection: 96-week combined analysis of MOTIVATE 1 and 2. J Acquir Immune Defic Syndr 2010;55(5):558–64.

[119] Trkola A, Ketas T, Kewalramani VN, Endorf F, Binley JM, Katinger H, et al. Neutralization sensitivity of human immunodeficiency virus type 1 primary isolates to antibodies and CD4-based reagents is independent of coreceptor usage. J Virol 1998;72(3):1876–85.

[120] Trkola A, Matthews J, Gordon C, Ketas T, Moore JP. A cell line-based neutralization assay for primary human immunodeficiency virus type 1 isolates that use either the CCR5 or the CXCR4 coreceptor. J Virol 1999;73(11):8966–74.

[121] Kimpton J, Emerman M. Detection of replication-competent and pseudotyped human immunodeficiency virus with a sensitive cell line on the basis of activation of an integrated beta-galactosidase gene. J Virol 1992;66(4):2232–9.

[122] Krowicka H, Robinson JE, Clark R, Hager S, Broyles S, Pincus SH. Use of tissue culture cell lines to evaluate HIV antiviral resistance. AIDS Res Hum Retroviruses 2008;24(7):957–67.

[123] Wires ES, Alvarez D, Dobrowolski C, Wang Y, Morales M, Karn J, et al. Methamphetamine activates nuclear factor kappa-light-chain-enhancer of activated B cells (NF-kappaB) and induces human immunodeficiency virus (HIV) transcription in human microglial cells. J neurovirology 2012;18(5):400–10.

[124] Platt EJ, Bilska M, Kozak SL, Kabat D, Montefiori DC. Evidence that ecotropic murine leukemia virus contamination in TZM-bl cells does not affect the outcome of neutralizing antibody assays with human immunodeficiency virus type 1. J Virol 2009;83(16):8289–92.

[125] Platt EJ, Wehrly K, Kuhmann SE, Chesebro B, Kabat D. Effects of CCR5 and CD4 cell surface concentrations on infections by macrophagetropic isolates of human immunodeficiency virus type 1. J virology 1998;72(4):2855–64.

[126] Browning J, Horner JW, Pettoello-Mantovani M, Raker C, Yurasov S, DePinho RA, et al. Mice transgenic for human CD4 and CCR5 are susceptible to HIV infection. Proc Natl Acad Sci USA 1997;94(26):14637–41.

[127] Keppler OT, Welte FJ, Ngo TA, Chin PS, Patton KS, Tsou CL, et al. Progress toward a human CD4/CCR5 transgenic rat model for de novo infection by human immunodeficiency virus type 1. J Exp Med 2002;195(6):719–36.

[128] Dunn CS, Mehtali M, Houdebine LM, Gut JP, Kirn A, Aubertin AM. Human immunodeficiency virus type 1 infection of human CD4-transgenic rabbits. J Gen Virol 1995;76(Pt 6): 1327–36.

[129] Hatziioannou T, Evans DT. Animal models for HIV/AIDS research. Nat Rev Microbiol 2012;10(12):852–67.

[130] Shultz LD, Ishikawa F, Greiner DL. Humanized mice in translational biomedical research. Nat Rev Immunol 2007;7(2):118–30.

[131] Blunt T, Gell D, Fox M, Taccioli GE, Lehmann AR, Jackson SP, et al. Identification of a nonsense mutation in the carboxyl-terminal region of DNA-dependent protein kinase catalytic subunit in the scid mouse. Proc Natl Acad Sci USA 1996;93(19):10285–90.

[132] Ito M, Hiramatsu H, Kobayashi K, Suzue K, Kawahata M, Hioki K, et al. NOD/SCID/gamma(c)(null) mouse: an excellent recipient mouse model for engraftment of human cells. Blood 2002;100(9):3175–82.

[133] Shultz LD, Lyons BL, Burzenski LM, Gott B, Chen X, Chaleff S, et al. Human lymphoid and myeloid cell development in NOD/LtSz-scid IL2R gamma null mice engrafted with mobilized human hemopoietic stem cells. J Immunol 2005;174(10):6477–89.

[134] Melkus MW, Estes JD, Padgett-Thomas A, Gatlin J, Denton PW, Othieno FA, et al. Humanized mice mount specific adaptive and innate immune responses to EBV and TSST-1. Nat Med 2006;12(11):1316–22.

[135] Wege AK, Melkus MW, Denton PW, Estes JD, Garcia JV. Functional and phenotypic characterization of the humanized BLT mouse model. Curr Top Microbiol Immunol 2008;324:149–65.

[136] Goldman JP, Blundell MP, Lopes L, Kinnon C, Di Santo JP, Thrasher AJ. Enhanced human cell engraftment in mice deficient in RAG2 and the common cytokine receptor gamma chain. Br J Haematol 1998;103(2):335–42.

[137] Baenziger S, Tussiwand R, Schlaepfer E, Mazzucchelli L, Heikenwalder M, Kurrer MO, et al. Disseminated and sustained HIV infection in CD34+ cord blood cell-transplanted Rag2-/-gamma c-/- mice. Proc Natl Acad Sci USA 2006;103(43):15951–6.

[138] Harouse JM, Gettie A, Tan RC, Blanchard J, Cheng-Mayer C. Distinct pathogenic sequela in rhesus macaques infected with CCR5 or CXCR4 utilizing SHIVs. Science 1999;284(5415):816–9.

[139] Barouch DH, Stephenson KE, Borducchi EN, Smith K, Stanley K, McNally AG, et al. Protective efficacy of a global HIV-1 mosaic vaccine against heterologous SHIV challenges in rhesus monkeys. Cell 2013;155(3):531–9.

[140] Zheng Q, Ruone S, Switzer WM, Heneine W, Garcia-Lerma JG. Limited SHIV env diversification in macaques failing oral antiretroviral pre-exposure prophylaxis. Retrovirology 2012;9:40.

[141] Burnett JC, Zaia JA, Rossi JJ. Creating genetic resistance to HIV. Curr Opin Immunol 2012;24(5):625–32.

[142] Berkhout B, ter Brake O. Towards a durable RNAi gene therapy for HIV-AIDS. Expert Opin Biol Ther 2009;9(2):161–70.

[143] Jacque JM, Triques K, Stevenson M. Modulation of HIV-1 replication by RNA interference. Nature 2002;418(6896):435–8.

[144] Hu WY, Myers CP, Kilzer JM, Pfaff SL, Bushman FD. Inhibition of retroviral pathogenesis by RNA interference. Curr Biol 2002;12(15):1301–11.

[145] Waterhouse PM, Wang MB, Lough T. Gene silencing as an adaptive defence against viruses. Nature 2001;411(6839):834–42.

[146] Liu YP, Schopman NC, Berkhout B. Dicer-independent processing of short hairpin RNAs. Nucleic acids Res 2013;41(6):3723–33.

[147] Cheloufi S, Dos Santos CO, Chong MM, Hannon GJ. A Dicer-independent miRNA biogenesis pathway that requires Ago catalysis. Nature;465(7298): 584–589.

[148] Yang JS, Maurin T, Robine N, Rasmussen KD, Jeffrey KL, Chandwani R, et al. Conserved vertebrate mir-451 provides a platform for Dicer-independent, Ago2-mediated microRNA biogenesis. Proc Natl Acad Sci USA 2010;107(34):15163–8.

[149] Herrera-Carrillo E, Harwig A, Liu YP, Berkhout B. Probing the shRNA characteristics that hinder Dicer recognition and consequently allow Ago-mediated processing and AgoshRNA activity. RNA 2014;20(9):1410–8.

[150] Boden D, Pusch O, Lee F, Tucker L, Shank PR, Ramratnam B. Promoter choice affects the potency of HIV-1 specific RNA interference. Nucleic Acids Res 2003;31(17):5033–8.

[151] Boden D, Pusch O, Lee F, Tucker L, Ramratnam B. Human immunodeficiency virus type 1 escape from RNA interference. J Virol 2003;77(21):11531–5.

[152] Das AT, Brummelkamp TR, Westerhout EM, Vink M, Madiredjo M, Bernards R, et al. Human immunodeficiency virus type 1 escapes from RNA interference-mediated inhibition. J Virol 2004;78(5):2601–5.

[153] Westerhout EM, Ooms M, Vink M, Das AT, Berkhout B. HIV-1 can escape from RNA interference by evolving an alternative structure in its RNA genome. Nucleic Acids Res 2005;33(2):796–804.

[154] McIntyre GJ, Groneman JL, Yu YH, Tran A, Applegate TL. Multiple shRNA combinations for near-complete coverage of all HIV-1 strains. AIDS Res Ther 2011;8(1):1.

[155] ter Brake O, t Hooft K, Liu YP, Centlivre M, von Eije KJ, Berkhout B. Lentiviral vector design for multiple shRNA expression and durable HIV-1 inhibition. Mol Ther: J Am Soc Gene Ther 2008;16(3):557–64.

[156] Berkhout B. Toward a durable anti-HIV gene therapy based on RNA interference. Ann N. Y Acad Sci 2009;1175:3–14.

[157] Westerhout EM, Berkhout B. A systematic analysis of the effect of target RNA structure on RNA interference. Nucleic Acids Res 2007;35(13):4322–30.

[158] Shah PS, Pham NP, Schaffer DV. HIV develops indirect cross-resistance to combinatorial RNAi targeting two distinct and spatially distant sites. Mol Ther: J Am Soc Gene Ther 2012;20(4):840–8.

[159] Berkhout B, Das AT. HIV-1 escape from RNAi antivirals: yet another Houdini action? Mol Ther Nucleic Acids 2012;1:e26.

[160] Boutimah F, Eekels JJ, Liu YP, Berkhout B. Antiviral strategies combining antiretroviral drugs with RNAi-mediated attack on HIV-1 and cellular co-factors. Antivir Res 2013;98(1):121–9.

[161] Barichievy S, Saayman S, von Eije KJ, Morris KV, Arbuthnot P, Weinberg MS. The inhibitory efficacy of RNA POL III-expressed long hairpin RNAs targeted to untranslated regions of the HIV-1 5′ long terminal repeat. Oligonucleotides 2007;17(4):419–31.

[162] Konstantinova P, ter Brake O, Haasnoot J, de Haan P, Berkhout B. Trans-inhibition of HIV-1 by a long hairpin RNA expressed within the viral genome. Retrovirology 2007;4:15.

[163] Liu YP, Haasnoot J, Berkhout B. Design of extended short hairpin RNAs for HIV-1 inhibition. Nucleic Acids Res 2007;35(17):5683–93.

[164] Nishitsuji H, Kohara M, Kannagi M, Masuda T. Effective suppression of human immunodeficiency virus type 1 through a combination of short- or long-hairpin RNAs targeting essential sequences for retroviral integration. J Virol 2006;80(15):7658–66.

[165] Saayman S, Barichievy S, Capovilla A, Morris KV, Arbuthnot P, Weinberg MS. The efficacy of generating three independent anti-HIV-1 siRNAs from a single U6 RNA Pol III-expressed long hairpin RNA. PloS one 2008;3(7):e2602.

[166] Liu YP, von Eije KJ, Schopman NC, Westerink JT, ter Brake O, Haasnoot J, et al. Combinatorial RNAi against HIV-1 using extended short hairpin RNAs. Mol Ther: J Am Soc Gene Ther 2009;17(10):1712–23.

[167] Saayman S, Arbuthnot P, Weinberg MS. Deriving four functional anti-HIV siRNAs from a single Pol III-generated transcript comprising two adjacent long hairpin RNA precursors. Nucleic Acids Res 2010;38(19):6652–63.

[168] ter Brake O, Konstantinova P, Ceylan M, Berkhout B. Silencing of HIV-1 with RNA interference: a multiple shRNA approach. Mol Ther: J Am Soc Gene Ther 2006;14(6):883–92.

[169] Scherer LJ, Frank R, Rossi JJ. Optimization and characterization of tRNA-shRNA expression constructs. Nucleic Acids Res 2007;35(8):2620–8.

[170] Herrera-Carrillo E, Liu YP, Berkhout B. The impact of unprotected T cells in RNAi-based gene therapy for HIV-AIDS. Mol Ther: J Am Soc Gene Ther 2014;22(3):596–606.

[171] Unwalla HJ, Li HT, Bahner I, Li MJ, Kohn D, Rossi JJ. Novel Pol II fusion promoter directs human immunodeficiency virus type 1-inducible coexpression of a short hairpin RNA and protein. J Virol 2006;80(4):1863–73.

[172] Liu YP, Gruber J, Haasnoot J, Konstantinova P, Berkhout B. RNAi-mediated inhibition of HIV-1 by targeting partially complementary viral sequences. Nucleic Acids Res 2009;37(18):6194–204.

[173] Aagaard LA, Zhang J, von Eije KJ, Li H, Saetrom P, Amarzguioui M, et al. Engineering and optimization of the miR-106b cluster for ectopic expression of multiplexed anti-HIV RNAs. Gene Ther 2008;15(23):1536–49.

[174] He L, Thomson JM, Hemann MT, Hernando-Monge E, Mu D, Goodson S, et al. A microRNA polycistron as a potential human oncogene. Nature 2005;435(7043):828–33.

[175] Zhang T, Cheng T, Wei L, Cai Y, Yeo AE, Han J, et al. Efficient inhibition of HIV-1 replication by an artificial polycistronic miRNA construct. Virology J 2012;9:118.

[176] Hutter G, Nowak D, Mossner M, Ganepola S, Mussig A, Allers K, et al. Long-term control of HIV by CCR5 Delta32/Delta32 stem-cell transplantation. N Engl J Med 2009;360(7):692–8.

[177] Brass AL, Dykxhoorn DM, Benita Y, Yan N, Engelman A, Xavier RJ, et al. Identification of host proteins required for HIV infection through a functional genomic screen. Science 2008;319(5865):921–6.

[178] Bushman FD, Malani N, Fernandes J, D'Orso I, Cagney G, Diamond TL, et al. Host cell factors in HIV replication: meta-analysis of genome-wide studies. PLoS Pathog 2009;5(5):e1000437.

[179] Konig R, Zhou Y, Elleder D, Diamond TL, Bonamy GM, Irelan JT, et al. Global analysis of host-pathogen interactions that regulate early-stage HIV-1 replication. Cell 2008;135(1):49–60.

[180] Anderson J, Akkina R. HIV-1 resistance conferred by siRNA cosuppression of CXCR4 and CCR5 coreceptors by a bispecific lentiviral vector. AIDS Res Ther 2005;2(1):1.

[181] Ping YH, Chu CY, Cao H, Jacque JM, Stevenson M, Rana TM. Modulating HIV-1 replication by RNA interference directed against human transcription elongation factor SPT5. Retrovirology 2004;1:46.

[182] Chiu YL, Cao H, Jacque JM, Stevenson M, Rana TM. Inhibition of human immunodeficiency virus type 1 replication by RNA interference directed against human transcription elongation factor P-TEFb (CDK9/CyclinT1). J Virol 2004;78(5):2517–29.

[183] Eekels JJ, Sagnier S, Geerts D, Jeeninga RE, Biard-Piechaczyk M, Berkhout B. Inhibition of HIV-1 replication with stable RNAi-mediated knockdown of autophagy factors. Virology J 2012;9:69.

[184] Eekels JJ, Geerts D, Jeeninga RE, Berkhout B. Long-term inhibition of HIV-1 replication with RNA interference against cellular co-factors. Antivir Res 2011;89(1):43–53.

[185] Palliser D, Chowdhury D, Wang QY, Lee SJ, Bronson RT, Knipe DM, et al. An siRNA-based microbicide protects mice from lethal herpes simplex virus 2 infection. Nature 2006;439(7072):89–94.

[186] Freeman EE, Weiss HA, Glynn JR, Cross PL, Whitworth JA, Hayes RJ. Herpes simplex virus 2 infection increases HIV acquisition in men and women: systematic review and meta-analysis of longitudinal studies. Aids 2006;20(1):73–83.

[187] Reynolds SJ, Risbud AR, Shepherd ME, Zenilman JM, Brookmeyer RS, Paranjape RS, et al. Recent herpes simplex virus type 2 infection and the risk of human immunodeficiency virus type 1 acquisition in India. J Infect Dis 2003;187(10):1513–21.

[188] Liu YP, Vink MA, Westerink JT, Ramirez de Arellano E, Konstantinova P, Ter Brake O, et al. Titers of lentiviral vectors encoding shRNAs and miRNAs are reduced by different mechanisms that require distinct repair strategies. RNA 2010;16(7):1328–39.

[189] ter Brake O, Westerink JT, Berkhout B. Lentiviral vector engineering for anti-HIV RNAi gene therapy. Methods Mol Biol;614: 201–213.

[190] Zhou D, Zhang J, Wang C, Bliesath JR, He Q, Yu D, et al. A method for detecting and preventing negative RNA interference in preparation of lentiviral vectors for siRNA delivery. RNA 2009;15(4):732–40.

[191] Morris KV. Non-coding RNAs and the epigenetic control of gene expression. In: Arbuthnot P, Weinberg MS, editors. Applied RNAi: from fundamental research to therapeutic applications. Norfolk, UK: Caister Academic Press; 2014. p. 43–50.

[192] Tahira AC, Kubrusly MS, Faria MF, Dazzani B, Fonseca RS, Maracaja-Coutinho V, et al. Long noncoding intronic RNAs are differentially expressed in primary and metastatic pancreatic cancer. Mol Cancer 2011;10:141.

[193] Lipovich L, Dachet F, Cai J, Bagla S, Balan K, Jia H, et al. Activity-dependent human brain coding/noncoding gene regulatory networks. Genetics 2012;192(3):1133–48.

[194] Saayman S, Ackley A, Turner AM, Famiglietti M, Bosque A, Clemson M, et al. An HIV-encoded antisense long noncoding RNA epigenetically regulates viral transcription. Mol Ther: J Am Soc Gene Ther 2014;22(6):1164–75.

[195] Modarresi F, Faghihi MA, Lopez-Toledano MA, Fatemi RP, Magistri M, Brothers SP, et al. Inhibition of natural antisense transcripts in vivo results in gene-specific transcriptional upregulation. Nat Biotechnol 2012;30(5):453–9.

[196] Ackley A, Lenox A, Stapleton K, Knowling S, Lu T, Sabir KS, et al. An algorithm for generating small RNAs capable of epigenetically modulating transcriptional gene silencing and activation in human cells. Mol Ther Nucleic Acids 2013;2:e104.

[197] Suzuki K, Hattori S, Marks K, Ahlenstiel C, Maeda Y, Ishida T, et al. Promoter targeting shRNA suppresses HIV-1 infection in vivo through transcriptional gene silencing. Mol Ther Nucleic Acids 2013;2:e137.

[198] Suzuki K, Juelich T, Lim H, Ishida T, Watanebe T, Cooper DA, et al. Closed chromatin architecture is induced by an RNA duplex targeting the HIV-1 promoter region. J Biol Chem 2008;283(34):23353–63.

[199] Weinberg MS, Villeneuve LM, Ehsani A, Amarzguioui M, Aagaard L, Chen ZX, et al. The antisense strand of small interfering RNAs directs histone methylation and transcriptional gene silencing in human cells. RNA 2006;12(2):256–62.

[200] Li MJ, Bauer G, Michienzi A, Yee JK, Lee NS, Kim J, et al. Inhibition of HIV-1 infection by lentiviral vectors expressing Pol III-promoted anti-HIV RNAs. Mol Ther: J Am Soc Gene Ther 2003;8(2):196–206.

[201] Michienzi A, Li S, Zaia JA, Rossi JJ. A nucleolar TAR decoy inhibitor of HIV-1 replication. Proc Natl Acad Sci USA 2002;99(22):14047–52.

[202] Sullenger BA, Gallardo HF, Ungers GE, Gilboa E. Overexpression of TAR sequences renders cells resistant to human immunodeficiency virus replication. Cell 1990;63(3):601–8.

[203] Lee SW, Gallardo HF, Gilboa E, Smith C. Inhibition of human immunodeficiency virus type 1 in human T cells by a potent rev response element decoy consisting of the 13-nucleotide minimal rev-binding domain. J Virol 1994;68(12):8254–64.

[204] Gold L, Brown D, Y-y He, Shtatland T, Singer BS, Wu Y. From oligonucleotide shapes to genomic SELEX: novel biological regulatory loops. Proc Natl Acad Sci 1997;94(1):59–64.

[205] Gold L, Polisky B, Uhlenbeck O, Yarus M. Diversity of oligonucleotide functions. Annu Rev Biochem 1995;64(1):763–97.

[206] Guo KT, Ziemer G, Paul A, Wendel HP. CELL-SELEX: novel perspectives of aptamer-based therapeutics. Int J Mol Sci 2008;9(4):668–78.

[207] Held DM, Kissel JD, Patterson JT, Nickens DG, Burke DH. HIV-1 inactivation by nucleic acid aptamers. Front Biosci 2006;11:89–112.

[208] Fisher TS, Joshi P, Prasad VR. Mutations that confer resistance to template-analog inhibitors of human immunodeficiency virus (HIV) type 1 reverse transcriptase lead to severe defects in HIV replication. J Virol 2002;76(8):4068–72.

[209] Ojwang JO, Buckheit RW, Pommier Y, Mazumder A, De Vreese K, Este JA, et al. T30177, an oligonucleotide stabilized by an intramolecular guanosine octet, is a potent inhibitor of laboratory strains and clinical isolates of human immunodeficiency virus type 1. Antimicrob Agents Chemother 1995;39(11):2426–35.

[210] Mazumder A, Neamati N, Ojwang JO, Sunder S, Rando RF, Pommier Y. Inhibition of the human immunodeficiency virus type 1 integrase by guanosine quartet structures. Biochemistry 1996;35(43):13762–71.

[211] Bartel DP, Zapp ML, Green MR, Szostak JW. HIV-1 rev regulation involves recognition of non-Watson-Crick base pairs in viral RNA. Cell 1991;67(3):529–36.

[212] Hung LW, Holbrook EL, Holbrook SR. The crystal structure of the rev binding element of HIV-1 reveals novel base pairing and conformational variability. Proc Natl Acad Sci USA 2000;97(10):5107–12.

[213] Giver L, Bartel DP, Zapp ML, Green MR, Ellington AD. Selection and design of high-affinity RNA ligands for HIV-1. Rev Gene 1993;137(1):19–24.

[214] Tuerk C, MacDougal-Waugh S. In vitro evolution of functional nucleic acids: high-affinity RNA ligands of HIV-1 proteins. Gene 1993;137(1):33–9.

[215] Good PD, Krikos AJ, Li SX, Bertrand E, Lee NS, Giver L, et al. Expression of small, therapeutic RNAs in human cell nuclei. Gene Ther 1997;4(1):45–54.

[216] Konopka K, Lee NS, Rossi J, Duzgunes N. Rev-binding aptamer and CMV promoter act as decoys to inhibit HIV replication. Gene 2000;255(2):235–44.

[217] Cohen C, Forzan M, Sproat B, Pantophlet R, McGowan I, Burton D, et al. An aptamer that neutralizes R5 strains of HIV-1 binds to core residues of gp120 in the CCR5 binding site. Virology 2008;381(1):46–54.

[218] Khati M, Schuman M, Ibrahim J, Sattentau Q, Gordon S, James W. Neutralization of infectivity of diverse R5 clinical isolates of human immunodeficiency virus type 1 by gp120-binding 2′F-RNA aptamers. J Virol 2003;77(23):12692–8.

[219] Kraus E, James W, Barclay AN. Cutting edge: novel RNA ligands able to bind CD4 antigen and inhibit CD4+ T lymphocyte function. J Immunol 1998;160(11):5209–12.

[220] Zhou J, Rossi JJ. Therapeutic potential of aptamer-siRNA conjugates for treatment of HIV-1. BioDrugs: Clin Immunother Biopharm Gene Ther 2012;26(6):393–400.

[221] Zhou J, Neff CP, Swiderski P, Li H, Smith DD, Aboellail T, et al. Functional in vivo delivery of multiplexed anti-HIV-1 siRNAs via a chemically synthesized aptamer with a sticky bridge. Mol Ther: J Am Soc Gene Ther 2013;21(1):192–200.

[222] Zhou J, Li H, Li S, Zaia J, Rossi JJ. Novel dual inhibitory function aptamer-siRNA delivery system for HIV-1 therapy. Mol Ther: J Am Soc Gene Ther 2008;16(8):1481–9.

[223] Zhou J, Swiderski P, Li H, Zhang J, Neff CP, Akkina R, et al. Selection, characterization and application of new RNA HIV gp 120 aptamers for facile delivery of Dicer substrate siRNAs into HIV infected cells. Nucleic Acids Res 2009;37(9):3094–109.

[224] Wheeler LA, Trifonova R, Vrbanac V, Basar E, McKernan S, Xu Z, et al. Inhibition of HIV transmission in human cervicovaginal explants and humanized mice using CD4 aptamer-siRNA chimeras. J Clin Invest 2011;121(6):2401–12.

[225] Neff CP, Zhou J, Remling L, Kuruvilla J, Zhang J, Li H, et al. An aptamer-siRNA chimera suppresses HIV-1 viral loads and protects from helper CD4(+) T cell decline in humanized mice. Sci Transl Med 2011;3(66):66ra6.

[226] Hutter G, Thiel E. Allogeneic transplantation of CCR5-deficient progenitor cells in a patient with HIV infection: an update after 3 years and the search for patient no. 2. Aids 2011;25(2):273–4.

[227] Kiem HP, Jerome KR, Deeks SG, McCune JM. Hematopoietic-stem-cell-based gene therapy for HIV disease. Cell Stem Cell 2012;10(2):137–47.

[228] Kitchen SG, Shimizu S, An DS. Stem cell-based anti-HIV gene therapy. Virology 2011;411(2):260–72.

[229] Berenson RJ, Andrews RG, Bensinger WI, Kalamasz D, Knitter G, Buckner CD, et al. Antigen CD34+ marrow cells engraft lethally irradiated baboons. J Clin Invest 1988;81(3):951–5.

[230] Kambal A, Mitchell G, Cary W, Gruenloh W, Jung Y, Kalomoiris S, et al. Generation of HIV-1 resistant and functional macrophages from hematopoietic stem cell-derived induced pluripotent stem cells. Mol Ther:J Am Soc Gene Ther 2011;19(3):584–93.

[231] Bauer G, Anderson JS. Gene therapy vectors. Gene therapy for HIV. Springer; 2014. 27–33.

[232] Hofer U, Henley JE, Exline CM, Mulhern O, Lopez E, Cannon PM. Pre-clinical modeling of CCR5 knockout in human hematopoietic stem cells by zinc finger nucleases using humanized mice. J Infect Dis 2013;208(Suppl 2):S160–4.

[233] Cannon P, June C. Chemokine receptor 5 knockout strategies. Curr Opin Hiv AIDS 2011;6(1):74–9.

[234] Kim SS, Peer D, Kumar P, Subramanya S, Wu H, Asthana D, et al. RNAi-mediated CCR5 silencing by LFA-1-targeted nanoparticles prevents HIV infection in BLT mice. Mol Ther: J Am Soc Gene Ther 2010;18(2):370–6.

[235] Kumar P, Ban HS, Kim SS, Wu H, Pearson T, Greiner DL, et al. T cell-specific siRNA delivery suppresses HIV-1 infection in humanized mice. Cell 2008;134(4):577–86.

[236] Perez EE, Wang J, Miller JC, Jouvenot Y, Kim KA, Liu O, et al. Establishment of HIV-1 resistance in CD4+ T cells by genome editing using zinc-finger nucleases. Nat Biotechnol 2008;26(7):808–16.

[237] Holt N, Wang J, Kim K, Friedman G, Wang X, Taupin V, et al. Human hematopoietic stem/progenitor cells modified by zinc-finger nucleases targeted to CCR5 control HIV-1 in vivo. Nat Biotechnol 2010;28(8):839–47.

[238] Li L, Krymskaya L, Wang J, Henley J, Rao A, Cao LF, et al. Genomic editing of the HIV-1 coreceptor CCR5 in adult hematopoietic stem and progenitor cells using zinc finger nucleases. Mol Ther: J Am Soc Gene Ther 2013;21(6):1259–69.

[239] Tebas P, Stein D, Tang WW, Frank I, Wang SQ, Lee G, et al. Gene editing of CCR5 in autologous CD4 T cells of persons infected with HIV. N Engl J Med 2014;370(10):901–10.

[240] Gabriel R, Lombardo A, Arens A, Miller JC, Genovese P, Kaeppel C, et al. An unbiased genome-wide analysis of zinc-finger nuclease specificity. Nat Biotechnol 2011;29(9):816–23.

[241] Pattanayak V, Ramirez CL, Joung JK, Liu DR. Revealing off-target cleavage specificities of zinc-finger nucleases by in vitro selection. Nat Methods 2011;8(9):765–70.

[242] Mussolino C, Alzubi J, Fine EJ, Morbitzer R, Cradick TJ, Lahaye T, et al. TALENs facilitate targeted genome editing in human cells with high specificity and low cytotoxicity. Nucleic Acids Res 2014;42(10):6762–73.

[243] Doyle EL, Booher NJ, Standage DS, Voytas DF, Brendel VP, Vandyk JK, et al. TAL Effector-Nucleotide Targeter (TALE-NT) 2.0: tools for TAL effector design and target prediction. Nucleic Acids Res 2012;40(Web Server issue):W117–22.

[244] Fine EJ, Cradick TJ, Zhao CL, Lin Y, Bao G. An online bioinformatics tool predicts zinc finger and TALE nuclease off-target cleavage. Nucleic Acids Res 2014;42(6):e42.

[245] Ebina H, Misawa N, Kanemura Y, Koyanagi Y. Harnessing the CRISPR/Cas9 system to disrupt latent HIV-1 provirus. Sci Reports 2013;3:2510.

[246] Hu W, Kaminski R, Yang F, Zhang Y, Cosentino L, Li F, et al. RNA-directed gene editing specifically eradicates latent and prevents new HIV-1 infection. Proc Natl Acad Sci USA 2014;111(31):11461–6.

[247] Didigu C, Doms R. Gene therapy targeting HIV entry. Viruses 2014;6(3):1395–409.

[248] Younan P, Kowalski J, Kiem HP. Genetically modified hematopoietic stem cell transplantation for HIV-1-infected patients: can we achieve a cure? Mol Ther: J Am Soc Gene Ther 2014;22(2):257–64.

[249] Mitsuyasu RT, Anton PA, Deeks SG, Scadden DT, Connick E, Downs MT, et al. Prolonged survival and tissue trafficking following adoptive transfer of CD4zeta gene-modified autologous CD4(+) and CD8(+) T cells in human immunodeficiency virus-infected subjects. Blood 2000;96(3):785–93.

[250] Walker RE, Bechtel CM, Natarajan V, Baseler M, Hege KM, Metcalf JA, et al. Long-term in vivo survival of receptor-modified syngeneic T cells in patients with human immunodeficiency virus infection. Blood 2000;96(2):467–74.

[251] Deeks SG, Wagner B, Anton PA, Mitsuyasu RT, Scadden DT, Huang C, et al. A phase II randomized study of HIV-specific T-cell gene therapy in subjects with undetectable plasma viremia on combination antiretroviral therapy. Mol Ther: J Am Soc Gene Ther 2002;5(6):788–97.

[252] van Lunzen J, Glaunsinger T, Stahmer I, von Baehr V, Baum C, Schilz A, et al. Transfer of autologous gene-modified T cells in HIV-infected patients with advanced immunodeficiency and drug-resistant virus. Mol Ther: J Am Soc Gene Ther 2007;15(5):1024–33.

[253] Egelhofer M, Brandenburg G, Martinius H, Schult-Dietrich P, Melikyan G, Kunert R, et al. Inhibition of human immunodeficiency virus type 1 entry in cells expressing gp41-derived peptides. J Virol 2004;78(2):568–75.

[254] DiGiusto DL, Krishnan A, Li L, Li H, Li S, Rao A, et al. RNA-based gene therapy for HIV with lentiviral vector-modified CD34(+) cells in patients undergoing transplantation for AIDS-related lymphoma. Sci Transl Med; 2(36): 36ra43.

[255] Amado RG, Mitsuyasu RT, Rosenblatt JD, Ngok FK, Bakker A, Cole S, et al. Anti-human immunodeficiency virus hematopoietic progenitor cell-delivered ribozyme in a phase I study: myeloid and lymphoid reconstitution in human immunodeficiency virus type-1-infected patients. Hum Gene Ther 2004;15(3):251–62.

[256] Feng Y, Leavitt M, Tritz R, Duarte E, Kang D, Mamounas M, et al. Inhibition of CCR5-dependent HIV-1 infection by hairpin ribozyme gene therapy against CC-chemokine receptor 5. Virology 2000;276(2):271–8.

[257] Macpherson JL, Boyd MP, Arndt AJ, Todd AV, Fanning GC, Ely JA, et al. Long-term survival and concomitant gene expression of ribozyme-transduced CD4+ T-lymphocytes in HIV-infected patients. J Gene Med 2005;7(5):552–64.

[258] Mitsuyasu RT, Merigan TC, Carr A, Zack JA, Winters MA, Workman C, et al. Phase 2 gene therapy trial of an anti-HIV ribozyme in autologous CD34+ cells. Nat Med 2009;15(3):285–92.

[259] Wong-Staal F, Poeschla EM, Looney DJ. A controlled, Phase 1 clinical trial to evaluate the safety and effects in HIV-1 infected humans of autologous lymphocytes transduced with a ribozyme that cleaves HIV-1 RNA. Hum Gene Ther 1998;9(16):2407–25.

[260] Podsakoff GM, Engel BC, Carbonaro DA, Choi C, Smogorzewska EM, Bauer G, et al. Selective survival of peripheral blood lymphocytes in children with HIV-1 following delivery of an anti-HIV gene to bone marrow CD34(+) cells. Mol Ther: J Am Soc Gene Ther 2005;12(1):77–86.

[261] Kohn DB, Bauer G, Rice CR, Rothschild JC, Carbonaro DA, Valdez P, et al. A clinical trial of retroviral-mediated transfer of a rev-responsive element decoy gene into CD34(+) cells from the bone marrow of human immunodeficiency virus-1-infected children. Blood 1999;94(1):368–71.

[262] Levine BL, Humeau LM, Boyer J, MacGregor RR, Rebello T, Lu X, et al. Gene transfer in humans using a conditionally replicating lentiviral vector. Proc Natl Acad Sci USA 2006;103(46):17372–7.

CHAPTER 9

Gene Therapy for Respiratory Viral Infections

9.1 INTRODUCTION

Many viruses have a tropism for cells of the respiratory tract. Infection with these viruses is the cause of sporadic and sometimes serious outbreaks that may lead to significant morbidity and mortality. Transmission of the viruses to humans may occur as a result of cross-species spread from animals, such as swines or birds in the case of influenza virus A. Spread of the viruses between individuals is usually as a result of inhalation of virus-containing droplets or through contact with contaminated objects. Efficient transmission makes control of the infections difficult to achieve, especially when containment of outbreaks has been delayed. Spread of respiratory viruses may be rapid in areas of dense population or when an infected individual has direct close contact with many people. This was the case at the initiation of the outbreak of the pandemic caused by infection with the severe acute respiratory syndrome (SARS) coronavirus (CoV) during 2002/2003 [1,2]. A health-care worker who treated infected individuals was in turn responsible for wider spread of the virus to people of the Gaundong province of China.

Hypermutability of the viruses leads to significant changes in their properties, and sometimes new viral strains emerge that have serious pathological effects. High rates of mutation may also account for ability of the viruses to spread from one species to another. Variability in viral gene sequences and their expression further complicates management by thwarting development of effective vaccines and treatments. Therefore, the availability of expertise that may be rapidly and rationally deployed to counter epidemics of respiratory viral infections would be valuable, and gene therapy provides resources for such purposes. Because the viruses that cause respiratory infections have genomes that comprise RNA, gene silencing and antisense approaches have been favored to disable viral replication. Delivery of antiviral nucleic acids to the infected tissues is facilitated by the anatomy of the pulmonary system. Intranasal and intratracheal administration of the therapeutics may conveniently be used to distribute candidate drugs to infected cells. Significant progress has now been made with use of nucleic acids to treat viral infections, and gene therapy is well positioned for advancement to clinical use in the treatment of respiratory viral infections.

Gene Therapy for Viral Infections. http://dx.doi.org/10.1016/B978-0-12-410518-8.00009-0

9.2 RESPIRATORY SYNCYTIAL VIRUS

Respiratory syncytial virus (RSV) is a member of the family of *Paramyxoviridae* and *Pneumovirus* genus (reviewed in refs [3,4]). The virions have a variable spherical shape that ranges from 100 to 350 nm in diameter. Viral capsids are enveloped during the budding process that releases viral particles from infected cells. The name of the virus is derived from the observation that infected cells become fused to each other and form syncytia. There are two antigenic subtypes of the virus: A and B. Infection with subtype B is more common and associated with milder symptoms. Infections with RSV are acquired through close contact with respiratory secretions of an infected person and may also be transmitted by contact with contaminated objects. RSV infection is prevalent throughout the world and is a major cause of seasonal lower respiratory infections in young children [5,6]. It is estimated that almost all children have been infected with RSV by the time that they reach 2 years of age [7]. Severity of the infection is usually dependent on age and is most serious in infants and young children. The inherent narrowness of the airways of young children, which is exacerbated by mucosal swelling that is associated with the infection, contributes to particularly restricted air flow in the tracheobronchial tree of the younger age group [3]. Elderly patients [8] and adult bone marrow transplant recipients [9] infected with RSV may also experience significant morbidity.

Evidence indicates that RSV infection typically occurs in children who were previously healthy [4]. However, prophylactic treatments may be implemented to prevent complications in certain groups of young children who are thought to be at high risk for a serious clinical course, such as those with congenital heart defects. After exposure, the infection usually manifests after 2–4 days and has an overall duration of approximately 2 weeks. In a small proportion of infected children, bronchiolitis may develop and carries a risk for asthma later in life. Immunity to the virus is not durable; therefore, individuals are susceptible to multiple infections during a lifetime. When occurring in older children and adults, RSV infection is typically restricted to the upper respiratory tract. Progress has been made with vaccine development [10], but a widely available prophylactic immunogen is not currently available. Treatment of infected individuals is mainly supportive and is aimed at improving pulmonary air flow.

The RSV genome comprises an RNA strand of approximately 15.2 kb with negative strand polarity kilobases [11] (Figure 9.1). Eleven proteins are encoded, which include structural and nonstructural sequences. The virion contains three surface glycoproteins, the attachment (G), fusion (F), and SH proteins, which act in concert to fuse the particle to the respiratory epithelial membranes and deliver the RNA genome before replication and expression of viral proteins. The helical capsid is formed by the nucleocapsid (N) protein and binds and protects the RNA genome. Nonstructural (NS) proteins include

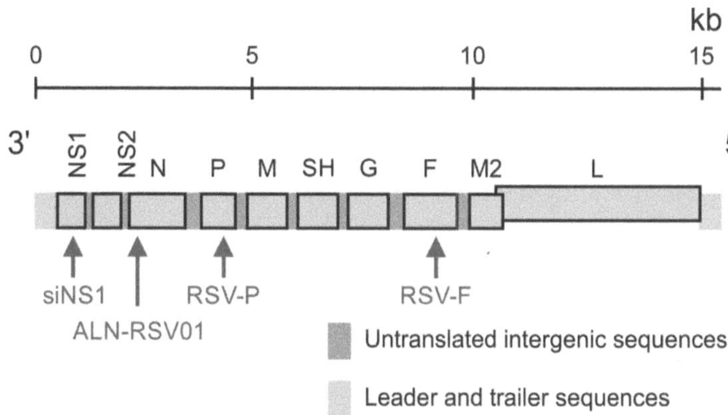

FIGURE 9.1 Illustration of the genome of RSV with sites that have been targeted using RNAi. The single-stranded RNA of antisense polarity comprises ~15.2 kb. Plus strands derived from the genome are used as translation templates for the synthesis of the indicated viral proteins. Intergenic sequences between the viral ORFs and leader and trailer sequences at the 3′ and 5′ ends of the genome are not translated. Sites within the genome that have been successfully targeted using RNAi activators are indicated. Efficacy of siNS1 was reported by Zhang and colleagues [16], ALN-RSV01 was characterized by DeVincenzo et al. [17,18], and the RSV-P and RSV-F silencers were developed by Bitko et al. [14,15].

NS1, NS2, the large polymerase subunit (L), and phosphoprotein (P). NS1 and NS2 are responsible for inhibiting the host's innate immune response to infection with the virus. Together with other nonstructural proteins, L and P components enable replication of the viral genome and regulate expression of viral genes. RSV proliferation is restricted to respiratory epithelial cells [12], which is useful for delivery of therapeutic nucleic acids that target the virus.

9.2.1 Gene Therapy for RSV Infection

Several studies have reported on the utility of RNA interference (RNAi) activators that effectively inhibit RSV replication [13]. The first study performed on a cell culture model of RSV infection demonstrated successful inhibition of viral replication by small interfering RNAs (siRNAs) that targeted *P* and *F* sequences (Figure 9.1) [14]. A follow-up investigation confirmed efficacy in vivo when using mice to simulate human RSV infection [15]. The siRNAs were effective when administered as naked nucleic acids and when formulated with a transfection reagent. Markers of viral replication were suppressed, and indicators of bronchoconstriction were diminished without evidence for nonspecific activation of the interferon response. Another study reported on effective inhibition of RSV replication in mice after administration of *NS1*-targeting expression cassettes that were administered intranasally within chitosan-containing nanoparticles [16] (Chapter 5). Antiviral efficacy was accompanied by diminished inflammation and reactivity of the airways.

Advancing gene therapy for treatment of RSV infection has now progressed to a stage of testing in clinical trials. Alnylam Pharmaceuticals developed use of an N-targeting lead synthetic siRNA, called ALN-RSV01, that was administered to patients in a nasal spray [17,18]. The treatment was well tolerated without evidence of side effects. Antiviral efficacy of ALN-RSV01was demonstrated in healthy adult volunteers who received an inoculum of the virus [18]. Follow-up studies were performed on RSV-infected patients who had received lung transplants (http://www.alnylam.com/capella/presentations/complete-results-of-our-aln-rsv01-phase-iib-study/). Although the primary endpoint of reducing bronchiolitis obliterans syndrome was not met in an intent-to-treat analysis, a therapeutic effect of ALN-RSV01 was observed.

9.3 INFLUENZA VIRUS

Influenza is an acute respiratory infection that is caused by viruses that belong to the *Orthomymyxoviridae* family. There are three genera of the family: *Influenzavirus A*, *Influenzavirus B*, and *Influenzavirus C*. The *Influenzavirus A* genus has one species, which is *influenzavirus A*. This virus naturally infects a range of species, especially migratory aquatic birds, and may be transmitted to animals and humans. Influenza virus A is also the most virulent and has been a cause of sporadic pandemic outbreaks with significant mortality and morbidity (recently reviewed in ref. [19]).

There are several serotypes of influenza virus A, which are based on the presence of particular epitopes within the hemagglutinin (HA) or neuraminidase (NA) proteins that are located on the surface of the viral particle (Figure 9.2).

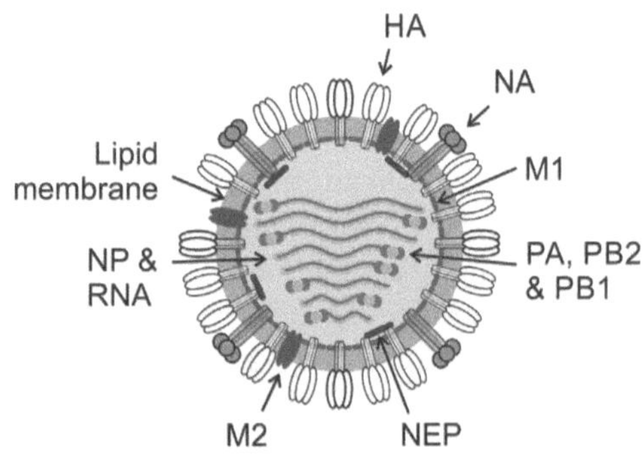

FIGURE 9.2 Schematic of the structure of the virion of influenza virus A.
The eight RNA segments of the influenza virus A, comprising strands of negative polarity, are complexed to the viral nucleoprotein (NP). Proteins of the polymerase complex—PA, P2, and P1—are also associated with the viral genome. M1 and NEP are located within the lipid envelope. Abundant HA and NA proteins are exposed on the surface of the viral particle. The ion channel protein (M2), which plays a role in mediating endosomal escape during viral entry into cells, is also embedded in the lipid envelope.

The serotypes are named according to the particular epitopes that characterize the pathogenic viruses. For example, the H1N1 serotype of the virus was responsible for the outbreak of Spanish flu in 1918 and swine flu in 2009, and the Asian flu pandemic of 1957 was caused by the H2N2 serotype [20–22]. The virus may be transmitted by inhalation of influenza virus A-containing aerosols or through direct contact with virus-contaminated surfaces. However, influenza virus A is inactivated by standard hygienic washing practices and by direct exposure to sunlight [23,24]. Only approximately 67% of individuals exposed to the virus become symptomatic [25]. When symptoms occur, they are nonspecific and include headache, fever, general body pains, fatigue, nasal congestion, and other upper respiratory tract symptoms.

The viral particles of influenza virus A are typically spherical and enveloped (reviewed in ref. [26]; Figure 9.2). The genome, which is segmented, comprises eight single-stranded RNA molecules of negative polarity (Figure 9.3). Viral sequences are prone to mutation and to genetic changes resulting from reassortment of the genomic segments [27]. These factors contribute to variability in the viral proteins, antigenic drift, and sometimes a propensity of the virus to cause cross-species infections with outbreaks of pandemics [28–30]. Reassortment of the influenza genome occurs when more than one virus infects a cell, and segments are packaged into individual viral particles in different combinations. After packaging into the virions, the viral particles acquire new properties that may influence host range and disease pathogenesis.

The eight segments of the influenza A genome, from largest (~2.3 kb) to smallest (~900 nt), encode polymerase PB2, polymerase PB1, polymerase PA, HA, nucleoprotein (NP), NA, matrix (M1 and M2), and nonstructural (NS1 and nuclear export protein (NEP)) proteins [26] (Figure 9.3). The total size of the genome is approximately 13.5 kb. After infection of cells, the negative strands are converted to positive stranded cRNA/mRNA by the viral RNA polymerases. Cap snatching from cellular mRNA and "stuttering" of the viral polymerases at a stretch of U residues on the 5′ ends of the genomic segments result in modification of the positive strands to resemble eukaryotic mRNAs. Alternative splicing and reading frame shifts on the viral mRNAs increase the number of proteins that may be expressed by the influenza virus A genome. For example, the M2 protein is translated from an alternatively spliced mRNA that is derived from segment 7 of the viral genome [31,32]. Depending on the strain, 12–14 proteins may be expressed by the virus.

Uptake of influenza virus A into cells is mediated by interaction of HA of the viral particle with sialic acid receptors on the cell surface [33]. Clathrin-mediated endocytosis, release from the endosome, and nuclear transport follow [34]. On reaching the endosome, the M2 protein functions as a pH-gated proton channel [35]. It causes acidification of the interior of the virion to

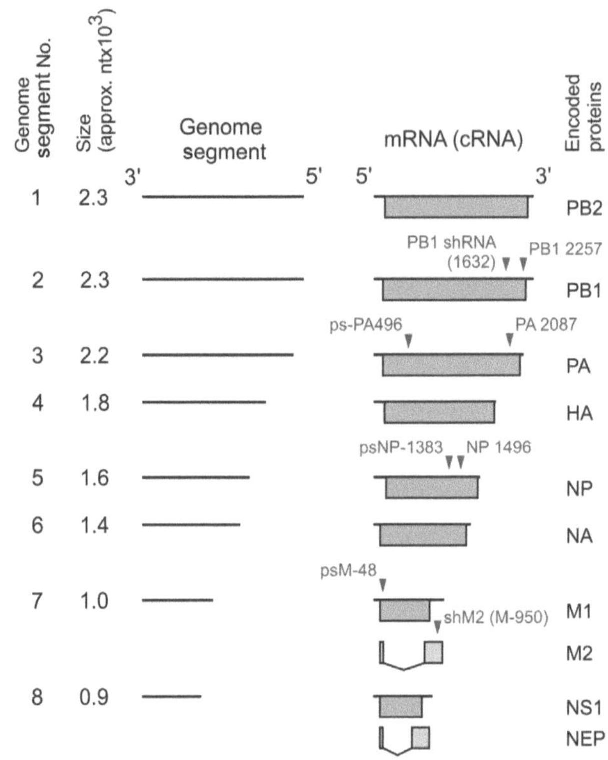

FIGURE 9.3 Eight segments that collectively make up the genome of influenza virus A together with sites that have been targeted using gene silencing.

The segments, reverse complements that serve as mRNAs, their sizes, and encoded proteins are indicated. Additional proteins, not shown in this illustration, may also be generated from segments 2 and 3 as a result of alternative translational initiation or shifts in the reading frames. Sites that have been targeted by studies aimed at developing use of gene therapy to inhibit replication of the virus are indicated by arrowheads. siRNAs targeted to the *PB1* (PB1 2257), *PA* (PA 2087), and *NP* (NP 1496) sequences were described by Ge et al. [29,43], Tompkins and colleagues [46], and Stewart et al. [50]. The PB1 shRNA antiviral RNAi activator was described by Li et al. [44]. ps-PA496 [47], psNP-1383 [49], and psM-48 [49] provided protection of mice against lethal exposure to influenza virus A. The expressed sequence, shM2 (M-950), was effective against viral replication and when delivered with a recombinant lentiviral vector [45]. In each case, the numbers refer to the nucleotide coordinates on the viral segments that are targeted by the RNAi activators.

result in dissociation of the ribonucleoprotein complexes before transport of the genomic segments into the nucleus. Replication of the viral genome and formation of translational templates occurs in the nucleus. Accumulation of the matrix protein and ribonucleoprotein complexes leads to their export to the cytoplasm, which is facilitated by the NEP. Thereafter, budding occurs to release the newly formed viral particles. NA is required for maturation of the virions and functions by cleaving the cellular sialic acid receptor molecules from the viral particles to facilitate release of the newly formed virions. Deficiency of NA restricts replication of the virus, and the enzyme is the target of an important class of antiviral drugs.

Vaccination against influenza virus A remains the preferred mode of protecting against illness from the virus [36,37]. Components of the vaccines are regularly derived from most recently isolated strains of the virus, and vaccination is recommended for administration to people who are older than 6 months of age [36]. Licensed drugs for treatment of influenza A infection fall into two broad categories: inhibitors of viral NA and inhibitors of M2 proteins (reviewed in ref. [37]). The NA inhibitors act by interfering with budding of the maturing virions from the infected cells (see section 9.3). Currently licensed NA inhibitors are oseltamivir and zanamivir. Inhibitors of M2, derivatives of adamantane that are amantadine and rimantadine, function by preventing endosomal release of the virions from the infected cell. M2 inhibitors are associated with toxicity and emergence of mutants that escape effects of the drugs, but NA inhibitors provide a high barrier to viral resistance and are generally well tolerated. Although there is evidence for beneficial effects of treatment when administered early during infection with influenza virus A, some studies indicate that the favorable effects of treatment are minimal [38].

9.3.1 Gene Therapy for Infection with Influenza Virus A

Evidence indicates that the RNAi pathway serves as an antiviral mechanism of host cells against influenza virus A [39,40]. However, the virus may counter the antiviral effects of host gene silencing through effects of the NS1 protein [41], although this effect has been disputed [28], and the effect of NS1 appears to vary depending on the strain of virus [42]. Despite the possible inhibitory effect of influenza virus A on RNAi, several studies have shown that gene silencing is effective against influenza virus A. In 2003, Ge and colleagues reported on inhibition of viral gene expression when using synthetic siRNAs [43]. Several different sites were targeted, and the greatest efficacy was shown with siRNAs with cognates within the *PB1*, *PA*, and *NP* regions (Figure 9.3). Efficacy was demonstrated when Madin–Darby canine kidney (MDCK) cells were transfected with the siRNAs before infection with isolates of influenza virus A. Good inhibition of viral replication was also observed after infection of embryonated chicken eggs. An interesting observation was that the siRNAs were particularly effective against the cRNA/mRNA sequences, with little effects against the negative stranded genomic segments. The specificity for particular viral RNA strands may result from bias occurring during the processing of the duplex siRNA. In addition, presence of the cRNA/mRNA in the cytoplasm or diminished access of the NP-protected genomic segments to the RNAi machinery may contribute to strand selectivity. Li et al. recently confirmed that the *PB1* sequence is a good target site for inhibition of influenza virus A replication when using an expressed RNAi activator (PB1 short hairpin RNA (shRNA)) [44]. A follow-up study of Ge and colleagues evaluated efficacy of siRNAs in virus-infected mice [29]. siRNAs were complexed to polyethyline imine and intravenously administered. A significant antiviral effect was observed when the gene silencers were administered before exposure and when siRNAs were given after the viral infection was established in the mice. The same study showed that expressed RNAi activators were also active against viral replication in cultured cells when they were delivered with recombinant lentiviral vectors. Use of lentiviral vectors to deliver influenza virus A-targeting shRNAs was described in other studies [30,45]. Silencing of *M1/M2* sequences effectively inhibited replication of the virus in MDCK cells. There was no evidence of escape mutants emerging after long-term suppression [45]. In addition, the gene silencing was effective against both H1N1 and avian H5N1 strains of the virus.

Tompkins et al. utilized siRNAs that targeted sites in *PA* and *NP*, but they administered the gene silencers as formulations with a commercially available transfection reagent, Oligofectamine™ [46]. The antiviral sequences were active against different viral isolates and protected mice from lethal challenge with highly pathogenic strains of influenza virus A. Suppression of viral replication in cultured cells and in mice was also demonstrated in a study that described use of siRNAs that targeted the *PA* sequence of the virus [47]. A particular siRNA, ps-PA496, provided protection against the virus. Suzuki and colleagues used bicistronic expression cassettes to target *NP* sequences of influenza A and influenza B viruses [48]. Inhibition of replication of both viruses was demonstrated in cultured cells. The cassettes were also effective when using recombinant baculovirus vectors. *M2* and *NP* sequences were targeted by Zhou and colleagues in a study performed on MDCK cells and in mice [49]. Effective suppression of H1N1, H5N1, and H9N2 isolates was demonstrated. Activation of the innate immune response by duplex RNA has also been investigated as a means of inhibiting influenza virus A. Stewart et al. showed that positioning the 5′ UGUGU 3′ sequence at the 5′ end of the sense strand of a siRNA targeting *PB1* (PB1-2257) augmented antiviral efficacy [50]. Although this is an interesting approach, and utility has been verified in cultured cells, avoiding inadvertent unintended effects in vivo may be more difficult to achieve.

Because influenza virus A is predisposed to mutation, developing gene silencing methodology that prevents escape is important. In addition to using combinatorial RNAi, silencing of host factors that are required by the virus to replicate may also be used. A high-throughput screening study reported by Karlas and colleagues enabled identification of such human host factors [51]. However, investigating use of silencing of host factors for therapeutic application has not yet been thoroughly explored.

Using antisense-based methodology, inhibition of replication of the highly pathogenic avian strain of influenza virus A, H5N1, has been demonstrated in chickens [52]. RNA oligonucleotides targeting the *NS1* and *HA* sequences were formulated with a commercially available transfection reagent (Lipofectamine™) and administered to the airways of the birds. Three doses of the antisense formulations effectively protected the chickens from the virus. Recently, positively charged phosphorodiamidate morpholino oligonucleotides have also been used to treat influenza virus infection, and a candidate drug, AVI-7100, is now at a clinical stage of evaluation by Sarepta Therapeutics (http://www.sarepta.com/pipeline/).

Overall, good efficacy of various influenza virus A-targeting nucleic acids has been demonstrated in cultured cells and in vivo. However, most studies have not yet progressed beyond evaluation in small animal models.

9.4 SARS CoV

The importance of implementing prompt therapeutic intervention after emergence of a pathogen was emphasized by the outbreak of SARS in 2002–2003 [53,54]. After cases were initially reported in Asia, there was rapid spread of the infection to other parts of the world. Approximately 8000 people were affected, and there were approximately 800 deaths that resulted from the epidemic. Mortality was high in people over the age of 50 years, but for unexplained reasons mortality was low in children younger than 12 years [55]. The causative agent was rapidly identified as a CoV and named the SARS-CoV [56]. The species is a member of the *Coronaviridae* family and *Coronavirus* genus. Viruses of the genus are transmissible across species and may infect many different animals to cause disease (reviewed in ref. [1]). The pleomorphic virion is enveloped (Figure 9.4) and contains a single-stranded RNA genome of positive polarity that comprises

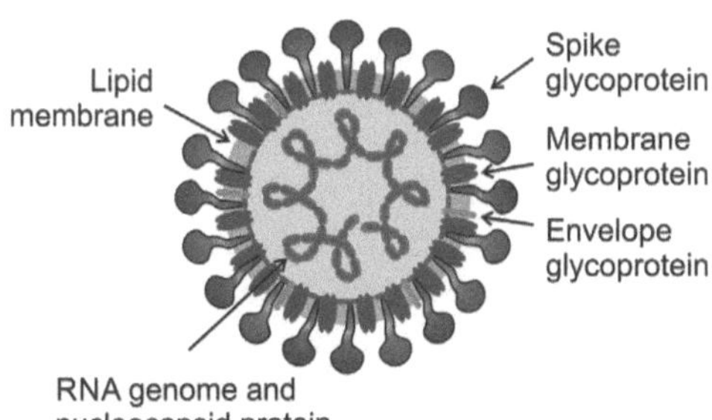

FIGURE 9.4 Schematically illustrated structure of the viral particle of SARS-CoV.
The single-stranded RNA of the genome is complexed to the N protein, which in turn is located within the membranous envelope. E, M, and S glycoproteins are embedded in the envelope. The circular structure in cross-section and protruding S proteins give the virion its characteristic crown-like appearance when viewed using transmission electron microscopy.

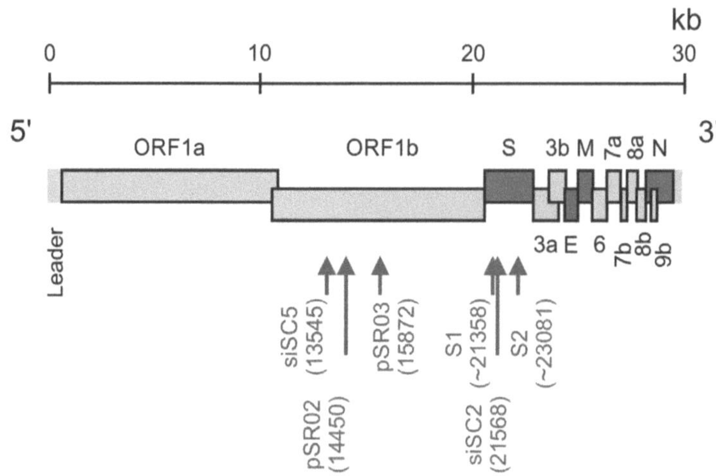

FIGURE 9.5 Genome of SARS-CoV with sites targeted by RNAi activators.
The single-stranded RNA of sense polarity comprises ~30 kb. The sequence is flanked by 5′ and 3′ UTRs and the intervening sequence encodes the viral structural and nonstructural proteins. The large *ORFs*, *1a* and *1b*, encode polyproteins that are processed to generate nonstructural proteins required for replication of the viral genome. The four structural genes, *S*, *E*, *M*, and *N*, encode the spike, envelope, membrane, and nucleocapsid proteins, respectively. Eight accessory proteins are encoded by *ORFs*, some of which are overlapping, within the 3′ quarter of the genome. Selected sites within the genome that have been successfully targeted using RNAi activators are indicated. Efficacy of siSC5 and siSC2 were reported by Li and colleagues [66], pSR02 and pSR03 were characterized by Wang et al. [63], and the S1 and S2 silencers were developed by Zhang et al. [65]. Approximate coordinates of the first nucleotide of the cognates of the gene silencing guides are indicated in parentheses. UTR, untranslated region.

approximately 30 kb (Figure 9.5) [57]. After its discovery, the genome of the virus was characterized with remarkable speed. This provided a basis for rational design of gene therapy-based approaches to countering SARS-CoV.

As a CoV, SARS-CoV has a crown-like (coronal) appearance when viewed using transmission electron microscopy [58]. The spike (S) glycoprotein is embedded in the viral envelope and forms protrusions that give the characteristic appearance to the virions (Figure 9.4). At the initiation of entry of the virus into cells to be infected, S interacts with the human angiotensin converting enzyme-2 [59]. Two other glycoproteins are also positioned in the envelope: the membrane (M) and envelope (E) proteins. The nucleocapsid protein (N) is responsible for packaging of the viral genome in a helical arrangement. The *1a* and *1b* viral *open reading frames (ORFs)* encode polyproteins, referred to as pp1a and pp1b, which function as viral replicases. These sequences comprise two-thirds of the genome (Figure 9.5) [1]. They are translated from the genomic RNA and rapidly convert the genomic RNA into minus strands, which in turn lead to formation of "nested" subgenomic mRNAs that serve as the translation templates for synthesis of other viral proteins. Synthesis of pp1b occurs after a translational reading frame shift of the *ORF 1a* sequence. Maturation of pp1a and pp1b follows proteolytic cleavage that is performed by viral chymotrypsin-like protease and papain-like protease [1,60]. There are eight accessory proteins encoded by the virus, which are not essential for replication of the virus in cultured cells (reviewed in refs [1,58]). Their functions include regulation of programmed cell death, cell signaling control, innate immune response suppression, and release of proinflammatory cytokines.

As the name indicates, infection with SARS-CoV is characterized by rapid onset of severe respiratory symptoms. Individuals who are infected with the virus are only contagious when the virus affects the lower respiratory tract [1]. This property facilitated restriction of spread of SARS-CoV through identification of infectious individuals and their placement in quarantine. Transmission of the virus occurs as a result of droplet spread. Although the respiratory tract is the primary site of infection, the virus may also spread to other organs such as the gastrointestinal tract [61]. Pathogenesis of the disease primarily occurs as a result of the host's immune response to the infection and correlates with the observation that symptoms become worse before clearance of the virus [62]. The epidemic that arose in 2002/2003 is likely to have originated from cross-species spread from palm civets to handlers working at markets selling the animals [2]. However, Chinese horseshoe bats are implicated as the primary reservoir of the virus [13,14], and these animals are likely to have transmitted the virus to other mammalian species before infecting humans.

9.4.1 Gene Therapy for SARS-CoV

Because there is no effective treatment for SARS-CoV infection, and there is no good vaccine, harnessing gene therapy to counter the infection was a logical line of investigation. Swift detailed characterization of the SARS-CoV genome enabled rapid development of potentially therapeutic anti SARS-CoV RNAi activators (Figure 9.5). Although none of the investigations progressed to clinical trial, proof of principle was demonstrated using various cell culture models that simulated the human infection [63–65]. In addition, the speed with which candidate nucleic acid therapeutics were identified reinforced the notion that rational design using gene therapy may be a powerful means of rapidly advancing treatment for emerging pathogens. Wang and colleagues initially reported on the use of expression cassettes that generated six antiviral RNAi activators that targeted various sites within the genome [63]. The lead antivirals, pSR02 and pSR03, had cognates in the *ORF 1b* sequence (Figure 9.5). Inhibition of viral replication and attenuation of the cytopathic effects were demonstrated in cultured Vero cells derived from African green monkeys. A subsequent study, also performed on Vero cells, demonstrated efficacy of synthetic siRNAs against the virus [64]. Particularly good inhibition of viral replication was demonstrated by siRNAs targeting the *S* sequence. The *S* gene was also shown to be a good target for inhibiting SARS-CoV in cultured cells by expressed RNAi activators [65]. The most advanced study that used RNAi against SARS-CoV assessed efficacy of synthetic siRNAs against SARS-CoV in nonhuman primates [66,67]. RNAi activators with cognates in the *ORF 1b* and *S* regions of the viral genome were developed as the candidate drugs. The short duplexes had dTdT overhangs at their 3′ ends but were otherwise unmodified. Initial assessment in a murine model of the infection was followed by evaluation in *Rhesus macaques*. The siRNAs, administered in a dextrose solution, were given intranasally and took advantage of the anatomy of the respiratory airways to distribute the siRNAs to SARS-CoV-infected cells. The treatment effectively inhibited replication of SARS-CoV and alleviated pathogenesis of virus-related disease.

The epidemic of SARS-CoV of 2002/2003 was effectively contained and deaths resulting from the infection have not been reported since 2004 (http://www.cdc.gov/sars/). However, despite this success, an important lesson from the outbreak has been that CoVs have the potential to be serious pathogens [1,68]. The ability of the viruses to adapt and achieve cross-species infections is particularly concerning, and there may be further serious outbreaks of infections caused by viruses of the *Coronaviridae* family. Indeed, emergence of the Middle East respiratory syndrome (MERS), which is also caused by a CoV, supports this notion [69]. Problems related to infection with the MERS virus have not been as widespread and severe as those caused by SARS-CoV. Nevertheless, vigilance and advancing technology that would be able to deal with such an event is important. Insights gained from using RNAi against SARS-CoV have been valuable and will be useful to advance treatment of respiratory coronaviral infections.

9.5 GENE THERAPY FOR OTHER RESPIRATORY VIRAL INFECTIONS

The feasibility of gene therapy for treatment of other respiratory viral infections, such as measles virus [70,71] and parainfluenza virus [15], has also been explored. Harnessing RNAi has been a popular strategy to inhibit expression of viral genes and to silence host dependency factors that the viruses require for their replication (reviewed in ref. [72]). An example is with the development of RNAi-based treatment of measles virus infections. Despite widespread vaccination programs that effectively prevent infection with the virus, disease related to transmission of measles virus remains an important global problem. The modest efficacy of available drugs, such as ribavirin, has prompted investigation of gene therapy to treat the infection. Synthetic and expressed RNAi activators that target viral mRNA containing the *L* sequences, which are responsible for encoding the essential viral RNA-dependent RNA polymerase, have been used to inhibit viral replication in cultured cells [70]. The effect was observed without evidence of toxicity. A more recent investigation verified that good silencing of viral replication could be achieved with gene silencers that had cognates in other viral sequences [73]. The measles virus receptor (signaling lymphocyte activation molecule) [74] and Rab9 GTPase [71,73], essential host factors required for replication of the virus, have also been effectively silenced to counter measles virus replication. In addition, viral genes and host factor-encoding genes have been silenced together in an attempt to augment antiviral efficacy [73]. Although these approaches have demonstrated impressive results that show potential for therapeutic application, feasibility of implementing gene therapy for measles virus infection is likely to be faced with several challenges. Moreover, the availability of an effective vaccine and the acute nature of measles virus infection suggest that gene therapy is not a priority for treating the disease.

9.6 CONCLUSIONS

Viruses that cause infections of the respiratory tracts are common and highly varied in their origins. Because presentation with pathognomonic signs and symptoms is unusual, molecular diagnosis and immunoassays play an important role in diagnosis. In settings where resources are limited, this may retard identification of the etiological agent that is responsible for the clinical presentation. To compound problems of delays in diagnosis, infection with respiratory viruses usually occurs rapidly after contact with respiratory secretions of infected individuals. Transmission may occur between individuals and in some cases as a result of interspecies infection. Moreover, respiratory viruses, such as influenza virus A, readily mutate their genomes to generate viruses that have new properties of disease pathogenesis. In addition, vaccines and available therapies for respiratory viral infections often only have modest efficacy. Collectively, the potential severity of respiratory viral infections coupled to risks for emergence of new strains of particularly pathogenic strains pose public health challenges.

Improvements in the technologies that enable rapid and thorough characterization of genomes of emergent pathogens have provided important information that is required for development of gene therapy-based treatment of respiratory viral infections. Because most respiratory viruses have RNA genomes, and some of the viruses replicate only in the cytoplasm, therapeutic strategies based on use of antisense and RNAi activators have been favored. As with gene therapy for other viral infections, delivery of antiviral nucleic acids is challenging. However, in the case of respiratory viral infections, the anatomical structure of the airways may be exploited for administration of the therapeutic nucleic acids. Inhalation of solutions or formulations containing the candidate drugs is convenient to distribute the nucleic acids along the trachea, bronchi, bronchioles, and alveoli. This approach has been used in clinical trials for the treatment of RSV infection [17,18], and the results have been useful in informing the broader development of gene therapy for infection with respiratory viruses. Airway administration of antiviral nucleic acids has also been shown to be effective for countering influenza virus A [46,47] and SARS-CoV [66,67] in animal models of the infections.

The persistent risk of emergence of highly pathogenic respiratory viruses remains concerning, and the lessons learned from the outbreak of the SARS-CoV pandemic of 2002/2003 were valuable. Although know-how that may be used as a platform for treating respiratory viral infections is still being developed, the objective of achieving rapid implementation of effective treatment seems reachable. The availability of sophisticated molecular techniques, improved methods of disabling viral gene expression, and ease of administration to the respiratory tract are particularly important to advancing gene therapy for respiratory viral infections. These developments will surely contribute significantly to more effective management of current and emergent pathogenic respiratory viral infections in the future.

REFERENCES

[1] Perlman S, Netland J. Coronaviruses post-SARS: update on replication and pathogenesis. Nat Rev Microbiol 2009;7(6):439–50.

[2] Riley S, Fraser C, Donnelly CA, Ghani AC, Abu-Raddad LJ, Hedley AJ, et al. Transmission dynamics of the etiological agent of SARS in Hong Kong: impact of public health interventions. Science 2003;300(5627):1961–6.

[3] Fraire AE, Woda BA. Respiratory syncytial virus. Viruses and the lung. Springer; 2014. p. 95–99.

[4] Hall CB, Weinberg GA, Iwane MK, Blumkin AK, Edwards KM, Staat MA, et al. The burden of respiratory syncytial virus infection in young children. N Engl J Med 2009;360(6):588–98.

[5] Ogra PL. Respiratory syncytial virus: the virus, the disease and the immune response. Paediatr Respir Rev 2004;5(Suppl. A):S119–26.

[6] Nokes JD, Cane PA. New strategies for control of respiratory syncytial virus infection. Curr Opin Infect Dis 2008;21(6):639–43.

[7] Henderson FW, Collier AM, Clyde Jr WA, Denny FW. Respiratory-syncytial-virus infections, reinfections and immunity: a prospective, longitudinal study in young children. N Engl J Med 1979;300(10):530–4.

[8] Han LL, Alexander JP, Anderson LJ. Respiratory syncytial virus pneumonia among the elderly: an assessment of disease burden. J Infect Dis 1999;179(1):25–30.

[9] Fouillard L, Mouthon L, Laporte JP, Isnard F, Stachowiak J, Aoudjhane M, et al. Severe respiratory syncytial virus pneumonia after autologous bone marrow transplantation: a report of three cases and review. Bone Marrow Transplant 1992;9(2):97–100.

[10] Durbin AP, Karron RA. Progress in the development of respiratory syncytial virus and parainfluenza virus vaccines. Clin Infect Dis 2003;37(12):1668–77.

[11] Cowton VM, McGivern DR, Fearns R. Unravelling the complexities of respiratory syncytial virus RNA synthesis. J Gen Virol 2006;87(7):1805–21.

[12] Zhang L, Peeples ME, Boucher RC, Collins PL, Pickles RJ. Respiratory syncytial virus infection of human airway epithelial cells is polarized, specific to ciliated cells, and without obvious cytopathology. J Virol 2002;76(11):5654–66.

[13] Barik S. RNAi applications to defeat respiratory viral infections. RNA Interf Viruses: Curr Innovations Future Trends 2010;187.

[14] Bitko V, Barik S. Phenotypic silencing of cytoplasmic genes using sequence-specific double-stranded short interfering RNA and its application in the reverse genetics of wild type negative-strand RNA viruses. BMC Microbiol 2001;1:34.

[15] Bitko V, Musiyenko A, Shulyayeva O, Barik S. Inhibition of respiratory viruses by nasally administered siRNA. Nat Med 2005;11(1):50–5.

[16] Zhang W, Yang H, Kong X, Mohapatra S, San Juan-Vergara H, Hellermann G, et al. Inhibition of respiratory syncytial virus infection with intranasal siRNA nanoparticles targeting the viral NS1 gene. Nat Med 2005;11(1):56–62.

[17] DeVincenzo J, Cehelsky JE, Alvarez R, Elbashir S, Harborth J, Toudjarska I, et al. Evaluation of the safety, tolerability and pharmacokinetics of ALN-RSV01, a novel RNAi antiviral therapeutic directed against respiratory syncytial virus (RSV). Antivir Res 2008;77(3):225–31.

[18] DeVincenzo J, Lambkin-Williams R, Wilkinson T, Cehelsky J, Nochur S, Walsh E, et al. A randomized, double-blind, placebo-controlled study of an RNAi-based therapy directed against respiratory syncytial virus. Proc Natl Acad Sci USA 2010;107(19):8800–5.

[19] Urbaniak K, Kowalczyk A, Markowska-Daniel I. Influenza A viruses of avian origin circulating in pigs and other mammals. Acta Biochim Pol 2014;61(3):433–9.

[20] Patterson KD, Pyle GF. The geography and mortality of the 1918 influenza pandemic. Bull Hist Med 1991;65(1):4–21.

[21] Trifonov V, Khiabanian H, Rabadan R. Geographic dependence, surveillance, and origins of the 2009 influenza A (H1N1) virus. N Engl J Med 2009;361(2):115–9.

[22] Kilbourne ED. Influenza pandemics of the 20th century. Emerg Infect Dis 2006;12(1):9.

[23] Brankston G, Gitterman L, Hirji Z, Lemieux C, Gardam M. Transmission of influenza A in human beings. Lancet Infect Dis 2007;7(4):257–65.

[24] Suarez DL, Spackman E, Senne DA, Bulaga L, Welsch AC, Froberg K. The effect of various disinfectants on detection of avian influenza virus by real time RT-PCR. Avian Dis 2003;47 (Suppl. 3):1091–5.

[25] Carrat F, Vergu E, Ferguson NM, Lemaitre M, Cauchemez S, Leach S, et al. Time lines of infection and disease in human influenza: a review of volunteer challenge studies. Am J Epidemiol 2008;167(7):775–85.

[26] Lamb RA, Choppin PW. The gene structure and replication of influenza virus. Annu Rev Biochem 1983;52:467–506.

[27] Hay AJ, Gregory V, Douglas AR, Lin YP. The evolution of human influenza viruses. Philos Trans R Soc Lond Ser B Biol Sci 2001;356(1416):1861–70.

[28] Kok KH, Jin DY. Influenza A virus NS1 protein does not suppress RNA interference in mammalian cells. J Gen Virol 2006;87(Pt 9):2639–44.

[29] Ge Q, Filip L, Bai A, Nguyen T, Eisen HN, Chen J. Inhibition of influenza virus production in virus-infected mice by RNA interference. Proc Natl Acad Sci USA 2004;101(23):8676–81.

[30] Hui EK, Yap EM, An DS, Chen IS, Nayak DP. Inhibition of influenza virus matrix (M1) protein expression and virus replication by U6 promoter-driven and lentivirus-mediated delivery of siRNA. J Gen Virol 2004;85(Pt 7):1877–84.

[31] Alonso-Caplen FV, Nemeroff ME, Qiu Y, Krug RM. Nucleocytoplasmic transport: the influenza virus NS1 protein regulates the transport of spliced NS2 mRNA and its precursor NS1 mRNA. Genes Dev 1992;6(2):255–67.

[32] Shih S-R, Nemeroff ME, Krug RM. The choice of alternative 5′ splice sites in influenza virus M1 mRNA is regulated by the viral polymerase complex. Proc Natl Acad Sci 1995;92(14):6324–8.

[33] Weis W, Brown J, Cusack S, Paulson I, Skehel I, Wiley D. Structure of the influenza virus haemagglutinin complexed with its receptor, sialic acid. Nature 1988;333:426–31.

[34] Lakadamyali M, Rust MJ, Zhuang X. Endocytosis of influenza viruses. Microbes Infect 2004;6(10):929–36.

[35] Helenius A. Unpacking the incoming influenza virus. Cell 1992;69(4):577–8.

[36] Grohskopf LA, Olsen SJ, Sokolow LZ, Bresee JS, Cox NJ, Broder KR, et al. Prevention and control of seasonal influenza with vaccines: recommendations of the advisory committee on immunization practices (ACIP)–United States, 2014–15 influenza season. MMWR Morb Mortal Wkly Rep 2014;63(32):691–7.

[37] Moscona A. Neuraminidase inhibitors for influenza. N Engl J Med 2005;353(13):1363–73.

[38] Michiels B, Van Puyenbroeck K, Verhoeven V, Vermeire E, Coenen S. The value of neuraminidase inhibitors for the prevention and treatment of seasonal influenza: a systematic review of systematic reviews. PloS One 2013;8(4):e60348.

[39] Matskevich AA, Moelling K. Dicer is involved in protection against influenza A virus infection. J Gen Virol 2007;88(Pt 10):2627–35.

[40] Van Stry M, Oguin 3rd TH, Cheloufi S, Vogel P, Watanabe M, Pillai MR, et al. Enhanced susceptibility of Ago1/3 double-null mice to influenza A virus infection. J Virol 2012;86(8):4151–7.

[41] Li WX, Li H, Lu R, Li F, Dus M, Atkinson P, et al. Interferon antagonist proteins of influenza and vaccinia viruses are suppressors of RNA silencing. Proc Natl Acad Sci USA 2004;101(5):1350–5.

[42] de Vries W, Haasnoot J, Fouchier R, de Haan P, Berkhout B. Differential RNA silencing suppression activity of NS1 proteins from different influenza A virus strains. J Gen Virol 2009;90(Pt 8):1916–22.

[43] Ge Q, McManus MT, Nguyen T, Shen CH, Sharp PA, Eisen HN, et al. RNA interference of influenza virus production by directly targeting mRNA for degradation and indirectly inhibiting all viral RNA transcription. Proc Natl Acad Sci USA 2003;100(5):2718–23.

[44] Li W, Yang X, Jiang Y, Wang B, Yang Y, Jiang Z, et al. Inhibition of influenza A virus replication by RNA interference targeted against the PB1 subunit of the RNA polymerase gene. Arch Virol 2011;156(11):1979–87.

[45] Sui HY, Zhao GY, Huang JD, Jin DY, Yuen KY, Zheng BJ. Small interfering RNA targeting m2 gene induces effective and long term inhibition of influenza A virus replication. PloS One 2009;4(5):e5671.

[46] Tompkins SM, Lo CY, Tumpey TM, Epstein SL. Protection against lethal influenza virus challenge by RNA interference in vivo. Proc Natl Acad Sci USA 2004;101(23):8682–6.

[47] Zhang W, Wang CY, Yang ST, Qin C, Hu JL, Xia XZ. Inhibition of highly pathogenic avian influenza virus H5N1 replication by the small interfering RNA targeting polymerase A gene. Biochem Biophys Res Commun 2009;390(3):421–6.

[48] Suzuki H, Saitoh H, Suzuki T, Takaku H. Inhibition of influenza virus by baculovirus-mediated shRNA. Nucleic Acids Symp Ser 2009;53:287–8.

[49] Zhou H, Jin M, Yu Z, Xu X, Peng Y, Wu H, et al. Effective small interfering RNAs targeting matrix and nucleocapsid protein gene inhibit influenza A virus replication in cells and mice. Antivir Res 2007;76(2):186–93.

[50] Stewart CR, Karpala AJ, Lowther S, Lowenthal JW, Bean AG. Immunostimulatory motifs enhance antiviral siRNAs targeting highly pathogenic avian influenza H5N1. PloS One 2011;6(7):e21552.

[51] Karlas A, Machuy N, Shin Y, Pleissner KP, Artarini A, Heuer D, et al. Genome-wide RNAi screen identifies human host factors crucial for influenza virus replication. Nature 2010;463(7282):818–22.

[52] Wu Y, Zhang G, Li Y, Jin Y, Dale R, Sun LQ, et al. Inhibition of highly pathogenic avian H5N1 influenza virus replication by RNA oligonucleotides targeting NS1 gene. Biochem Biophys Res Commun 2008;365(2):369–74.

[53] Rosling L, Rosling M. Pneumonia causes panic in Guangdong province. BMJ 2003;326(7386):416.

[54] Suresh MR, Bhatnagar PK, Das D. Molecular targets for diagnostics and therapeutics of severe acute respiratory syndrome (SARS-CoV). J Pharm Pharm Sci 2008;11(2):1s–13s.

[55] Leung CW, Kwan YW, Ko PW, Chiu SS, Loung PY, Fong NC, et al. Severe acute respiratory syndrome among children. Pediatrics 2004;113(6):e535–43.

[56] Drosten C, Gunther S, Preiser W, van der Werf S, Brodt HR, Becker S, et al. Identification of a novel coronavirus in patients with severe acute respiratory syndrome. N Engl J Med 2003;348(20):1967–76.

[57] Marra MA, Jones SJ, Astell CR, Holt RA, Brooks-Wilson A, Butterfield YS, et al. The genome sequence of the SARS-associated coronavirus. Science 2003;300(5624):1399–404.

[58] McBride R, Fielding BC. The role of severe acute respiratory syndrome (SARS)-coronavirus accessory proteins in virus pathogenesis. Viruses 2012;4(11):2902–23.

[59] Li W, Moore MJ, Vasilieva N, Sui J, Wong SK, Berne MA, et al. Angiotensin-converting enzyme 2 is a functional receptor for the SARS coronavirus. Nature 2003;426(6965):450–4.

[60] Masters PS. The molecular biology of coronaviruses. Adv Virus Res 2006;66:193–292.

[61] Gu J, Gong E, Zhang B, Zheng J, Gao Z, Zhong Y, et al. Multiple organ infection and the pathogenesis of SARS. J Exp Med 2005;202(3):415–24.

[62] Perlman S, Dandekar AA. Immunopathogenesis of coronavirus infections: implications for SARS. Nat Rev Immunol 2005;5(12):917–27.

[63] Wang Z, Ren L, Zhao X, Hung T, Meng A, Wang J, et al. Inhibition of severe acute respiratory syndrome virus replication by small interfering RNAs in mammalian cells. J Virol 2004;78(14):7523–7.

[64] Wu CJ, Huang HW, Liu CY, Hong CF, Chan YL. Inhibition of SARS-CoV replication by siRNA. Antivir Res 2005;65(1):45–8.

[65] Zhang Y, Li T, Fu L, Yu C, Li Y, Xu X, et al. Silencing SARS-CoV spike protein expression in cultured cells by RNA interference. FEBS Lett 2004;560(1–3):141–6.

[66] Li BJ, Tang Q, Cheng D, Qin C, Xie FY, Wei Q, et al. Using siRNA in prophylactic and therapeutic regimens against SARS coronavirus in *Rhesus macaque*. Nat Med 2005;11(9): 944–51.

[67] Tang Q, Li B, Woodle M, Lu PY. Application of siRNA against SARS in the *Rhesus macaque* model. Methods Mol Biol 2008;442:139–58.

[68] Graham RL, Donaldson EF, Baric RS. A decade after SARS: strategies for controlling emerging coronaviruses. Nat Rev Microbiol 2013;11(12):836–48.

[69] Zaki AM, van Boheemen S, Bestebroer TM, Osterhaus AD, Fouchier RA. Isolation of a novel coronavirus from a man with pneumonia in Saudi Arabia. N Engl J Med 2012;367(19): 1814–20.

[70] Otaki M, Sada K, Kadoya H, Nagano-Fujii M, Hotta H. Inhibition of measles virus and subacute sclerosing panencephalitis virus by RNA interference. Antivir Res 2006;70(3):105–11.

[71] Shi JY, Liu B, Wang ML, Luo EJ. Short hairpin RNA-mediated inhibition of measles virus replication in vitro. Can J Microbiol 2010;56(1):77–80.

[72] Blasquez L, Fortes P. Harnessing RNAi for the treatment of viral infections. In: Arbuthnot P, Weinberg MS, editors. Applied RNAi: from fundamental research to therapeutic applications. Norfolk, UK: Caister Academic Press; 2014. p. 151–80.

[73] Shi J, Wang M, Wang J, Wang S, Luo E. Comparison of inhibitory efficacy of short interfering RNAs targeting different genes on measles virus replication. J basic Microbiol 2012;52(3): 332–9.

[74] Hu L, Wang Z, Hu C, Liu X, Yao L, Li W, et al. Inhibition of measles virus multiplication in cell culture by RNA interference. Acta Virol 2005;49(4):227–34.

CHAPTER 10

Gene Therapy for Infection with Hemorrhagic Fever Viruses

10.1 INTRODUCTION

Hemorrhagic fever is a very serious complication of viral infection. The condition is characterized by sudden onset of constitutional symptoms that rapidly progress to bleeding diatheses, shock, and organ failure with high mortality. Various RNA viruses may cause hemorrhagic fevers. There are differences in their modes of transmission, geographical distribution, and clinical course, which in turn are dependent on the nature of the etiological agent. Infection with some viruses, such as Ebola virus and Marburg virus, is frequently complicated by hemorrhagic fever and a grave prognosis. However, Dengue virus (DENV) is rarely associated with hemorrhagic fever, but mortality may be high when the serious complication arises. Infection with Ebola virus and Marburg virus occurs as a result of human–human or animal–human contact with contaminated secretions or blood. On the other hand, many of the hemorrhagic fever viruses (e.g., DENV and West Nile virus [WNV]) are arboviruses and are transmitted by mosquito vectors. Therefore, measures to prevent infection in humans may be directed at the insects. Given the often lethal nature of the infection with hemorrhagic fever viruses and their potential global impact, developing vaccines and effective antivirals is understandably an active field of research.

A difficulty with management of viral hemorrhagic fevers is that the presenting symptoms are often not pathognomonic of the infection and early diagnoses are not easily made. Symptoms of headache, fever, and malaise are also often associated with infections such as malaria, which may be endemic to regions where the viral hemorrhagic fevers occur. Treatment of viral hemorrhagic fevers is very limited, and containment of the infections is often the only means available to prevent spread. Implementing measures to restrict transmission of viruses is often also complicated by occurrence of the infections in resource-poor environments. Therefore, development of effective means to counter the serious problems caused by the viruses is a global priority. Successful treatment of viral hemorrhagic fevers is not only important from a humanitarian perspective but also to negate the possible threat of bioterrorism that the viruses pose.

Gene Therapy for Viral Infections. http://dx.doi.org/10.1016/B978-0-12-410518-8.00010-7

The rational design that is a feature of gene therapy is particularly useful to development of therapy for hemorrhagic fever viruses. This appealing property has led to many investigations that aim to advance various gene therapy-based methods to disable the viruses that cause hemorrhagic fevers.

10.2 FILOVIRIDAE

There are three genera of the family of *Filoviridae*: *Ebolavirus*, *Marburgvirus*, and *Cuevavirus* [1]. The species of virus from the *Cuevavirus* genus was recently isolated from the carcasses of cave bats in Europe and was associated with high death rates in these animals [2]. Viruses of the genus were also provisionally named *Lloviu*, after the place in Spain where the isolates were detected. Evidence to date indicates that these viruses do not cause disease in humans. On the other hand, infection with species from the *Ebolavirus* and *Marburgvirus* genera may lead to hemorrhagic fevers that are associated with high mortality in humans.

The virions are filamentous in structure and are enveloped after budding from the previously infected cells (Figure 10.1; reviewed in ref. [3]). Particles are typically approximately 80 nm in diameter, but the length of the virions is variable and may be as long as 1.4 μm [4]. Their genomes comprise single-stranded nonsegmented RNA of antisense polarity. Conversion of the negative stranded genome to positive polarity, which is essential for replication of the viruses, requires the viral L polymerase protein. The genomes of the *Filoviridae* family have similar structure. They comprise approximately 19 kb and encode seven proteins. Sequences from the 3′ end of the genome (5′ end of the reverse complement that is the translational template) encode nucleoprotein (NP), viral protein (VP) 35 (VP35), VP40, glycoprotein (GP), VP30,

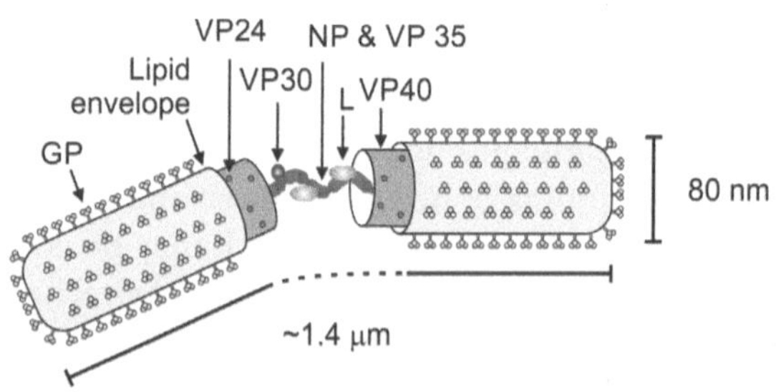

FIGURE 10.1 Schematic illustration of the structure of the Ebola viral particle.
The enveloped virion is filamentous and has a variable shape and length. Trimers of the viral glycoprotein are embedded in the envelope, which surrounds the capsid. VP40 and VP24 constitute the capsid, which contains the viral RNA genome. The NP and VP35 proteins are complexed to the genomic RNA. VP30 and L proteins required for replication of the genome are also enclosed in the capsid. VP, viral protein; NP, nucleoprotein.

VP24, and the L protein [5] (Figure 10.2). Intergenic noncoding sequences are located between viral open reading frames. In the filamentous viral particles, the GP proteins are embedded into the envelope as spiked trimeric structures (Figure 10.1) and are important for the cellular internalization of the virions. Several proteins have been implicated in uptake of virions of the *Filoviridae* family, and these include dendritic cell-specific ICAM-3 grabbing nonintegrin (DC-SIGN) [6], its homolog DC-SIGNR, and β1 integrins [7]. However, none has been proven to be essential for the process. Internalization of the viral particles is primarily by macropinocytosis and clathrin-mediated endocytosis [4]. An interesting recent finding has been demonstration that the Niemann-Pick C1 (NPC1) protein, a cholesterol transporter, facilitates uptake of Marburg viruses and Ebola viruses [8]. The host factor was identified after genome-wide screening showed that resistance to Ebola virus

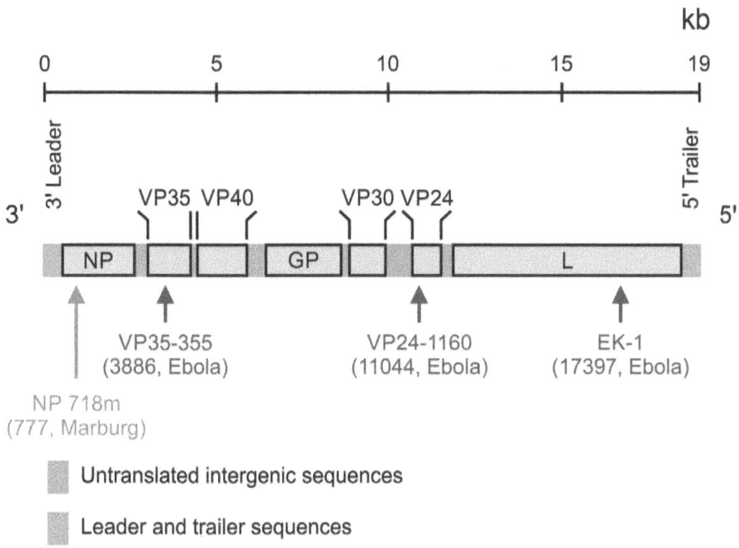

FIGURE 10.2 Genome of Ebola virus with sites that have been targeted using RNAi. The single-stranded RNA of antisense polarity comprises approximately 19 kb. Plus strands derived from the genome are used as translation templates for the synthesis of the indicated structural and nonstructural proteins. Leader and trailer sequences at the 3′ and 5′ ends of the genome do not encode sequences that are translated. There are also untranslated intergenic sequences located between the viral open reading frames. Sites within the genome that have been successfully targeted using synthetic siRNAs formulated in SNALPs are indicated, and the nucleotide coordinates of these gene silencers are shown in parentheses. Nomenclature is that used by Geisbert et al. in their study on preclinical evaluation of RNAi-based therapy in nonhuman primates [18]. The genome of Marburg virus has a similar structure to that of Ebola virus. The target site of the siRNA that has been successfully used against Marburg virus infection of nonhuman primates is also shown [26]. RNAi, RNA interference; siRNA, small interfering RNAs; SNALP, stable nucleic acid–lipid particles.

and Marburg virus infection was conferred by mutations that disrupted the homotypic fusion and vacuole protein-sorting (HOPS) complex. The mechanism by which HOPS facilitates viral uptake is through enabling fusion of the virus to the endosomal membranes, and the process is independent of the role of HOPS as a cholesterol transporter. Elucidation of the role of NPC1 in cellular entry of filoviruses generated enthusiasm for targeting this protein in the development of new antiviral strategies.

10.2.1 Ebola Virus

The first reports of known outbreaks of Ebola virus disease (EVD) were made in 1976 [1]. At the time, cases of the infection occurred in two African countries: Sudan and Zaire (now the Democratic Republic of the Congo [DRC]). The Ebola River in the DRC was nearby the village where the first outbreak happened and gave its name to the virus. Since initial reports, outbreaks of EVD have been sporadic but regular. The most recent epidemic, which commenced in December 2013 in Guinea then spread to Liberia and Sierra Leone, has been particularly severe. By March 2015, the infection had claimed the lives of approximately 10,000 people in West Africa.

There are five species in the *Ebolavirus* genus: Bundibugyo ebolavirus (BDBV), Zaire ebolavirus (EBOV), Reston ebolavirus, Sudan ebolavirus (SUDV), and Taï Forest ebolavirus [1]. BDBV, EBOV, and SUDV are the cause of severe illness and death in humans, but viruses from the other two species of the genus generally do not cause human mortality. Transmission of the viruses occurs as a result of human-to-human or animal-to-human contact. The infection may also be severe in primates, and outbreaks have been responsible for death in chimpanzees and gorillas. Other animals such as antelopes and bats may also transmit the infection to humans. The natural hosts of Ebola viruses are thought to be certain species of African fruit bats, and the distribution of the virus correlates with the location of these animals. Practices related to the consumption of bushmeat are implicated in infection with the virus. Transmission typically occurs as a result of contact through secretions from mucosal membranes or through blood. Certain traditional practices involving close contact between mourners and corpses of individuals who have died from EVD is another important mode of infection. It has been reported that to date 60% of cases of EVD in Guinea have resulted from such traditional burial practices [9]. Of course, health-care workers who are in close contact with patients with EVD are also at high risk.

After infection, the onset of symptoms and signs is rapid and dramatic. They manifest between 2 and 21 days from the time of exposure and are initially characterized by headache, fever, weakness, and malaise [1]. Rash, gastrointestinal disturbance, liver derangements, bleeding, and kidney failure may follow. Mortality has been reported to range from 50% to 90% in epidemics where there are more than 10 cases of the infection. Preventative vaccination and curative treatment of EVD are largely ineffective. Unfortunately, the outbreaks also typically occur in settings where access to human and material resources is severely constrained. As a result, containment is particularly difficult to achieve.

Several different human cell types are permissive for infection by Ebola virus and Marburg virus [3]. These include lymphocytes, monocytes, macrophages, dendritic cells, hepatocytes, fibroblasts, and endothelial cells. Infections are characterized by profound release of proinflammatory cytokines. However, the viruses are capable of evading the antiviral effects of activation of the interferon (IFN) response (reviewed in ref. [10]). Particularly, VP24 and VP35 inhibit production of IFNs and downstream signaling by these mediators of the innate antiviral response. Suppressing the host's innate immune response through a mechanism that involves the viral GPs may also play a role [11]. The cascade of events that culminates in the manifestation of the hemorrhagic fever is complex [3]. It is initiated by an overwhelming effect that the virus has on the host's immune system. Severe lymphopenia, induction of apoptosis in infected and bystanding cells, dysregulation of immunostimulation by dendritic cells, release of tissue factor, and induction of the coagulation cascade all contribute to the striking signs and symptoms.

10.2.2 Marburg Virus

The genus *Marburgvirus* derives its name from the city in Germany that was one of the sites where infection with the virus was initially identified (reviewed in ref. [12]). The first cases of Marburg virus disease (MVD) were reported in 1967. At the time, similarly afflicted patients were identified in Belgrade and Frankfurt. However, the virus did not originate from Europe, and the outbreak was caused by infection from African green monkeys that had been imported from Uganda. Transmission of the infection occurred during postmortem removal of the kidneys from the animals for the purposes of propagating vaccine strains of poliovirus. Isolation of the infectious agent was remarkably rapid and was reported in three papers published in 1968 [13–15].

MVD and EVD follow similar clinical courses and may have equally devastating consequences. As members of the *Filovirus* genus, their genome organizations are similar and there is considerable sequence homology [5,16]. However, despite the close relationship, the viral proteins are antigenically distinct. Epidemics caused by MVD have been sporadic and the mortality variable, ranging from 25% to more than 80% [12]. Transmission from reservoirs of the virus and pathogenesis of the infections are similar. Both diseases are characterized by profound lymphopenia, disruption of immune responses, proinflammatory cytokine release, and disseminated intravascular coagulation. As with EVD, there is no effective treatment and containment is used to limit spread of the virus.

10.2.3 Gene Therapy for Infections with Ebola Virus and Marburg Virus

Given the seriousness of EVD, it is not surprising that the quest for a cure has been vigorous. Because Ebola viruses have an RNA genome and replicate in the cytoplasm, exploiting RNA interference (RNAi) to develop novel therapy was logical. However, a feature of the viruses that is particularly relevant for gene silencing-based therapy is that the viral VP35 protein has suppressive effects on RNAi [17]. One method that may overcome this effect is by silencing of the *VP35* sequence itself. *VP* has indeed been targeted by a cocktail of small interfering RNAs (siRNAs) that was used successfully to counter Ebola virus infection in nonhuman primates [18] (see section 10.2.3).

The first study that used RNAi to counter Ebola virus replication utilized a combination of four synthetic siRNAs that targeted the *L* gene of EBOV [19]. Initial evaluation was performed in cell culture and verified that effective and specific silencing of the viral gene could be achieved to result in inhibition of viral replication. To assess the RNAi-based approach in vivo, efficacy of the antiviral sequences was assessed in guinea pigs. These animals were selected as the model of EVD because they simulate the human state of infection better than does infection of murine species [20,21]. The siRNAs were delivered using two different nonviral vector formulations: stable nucleic acid–lipid particles (SNALPs) and complexes to polyethylene imine (PEI). SNALPs had previously been used to achieve effective inhibition of hepatitis B virus replication [22] (Chapters 5 and 6). Selection of this vector was also appropriate because the liver is the primary site of EBOV infection in some animal models [3,23]. One hour after lethal exposure to Ebola virus, siRNA-SNALP formulations were administered intraperitoneally and then daily for 6 days. Significant decreases in circulating viral particle equivalents were observed in the animals that had received the *L* gene-targeting siRNAs. Control animals that were injected with scrambled sequences died or were euthanized by 12 days after the challenge. Comparison of the efficacy of the two methods of delivering the antivirals demonstrated that the SNALPs were more effective than the complexes to PEI. Specificity of the effect was further corroborated by demonstration that the SNALP formulations did not cause activation of the innate immune response. An important finding was that one of the siRNAs, EK1 (Figure 10.2), was particularly effective against EBOV and was as effective as the combination of four sequences.

The highly effective *L*-targeting EK-1 siRNA was taken forward for testing in rhesus macaques [18]. Two additional siRNAs, each with a cognate in the *VP24* and *VP35*-containing sequences of EBOV, were also tested (Figure 10.2).

Chemical modification of both strands of the siRNAs, by incorporating G and U nucleotides with 2′-*O*-methyl residues, was included to minimize immunostimulation and improve stability (Chapter 2). The cocktail of siRNAs was formulated in SNALPs and then given to the macaques as four or seven intravenous administrations. The gene silencers were administered at a dose of 2 mg/kg and with both treatment regimens the first injection was given 30 min after exposure to a normally lethal dose of the virus. One of the three animals receiving the four injections succumbed, but the other two survived. It is important to note that all four of the macaques that were given the seven injections of the siRNAs were protected against exposure to EBOV. Adverse effects were observed in some of the animals, but these were ascribed to the EBOV infection rather than administration of the SNALP/siRNA formulations. Unfortunately, despite promising results from evaluations performed in nonhuman primates, progress with implementing clinical use of RNAi-based therapy for treatment of EBOV has been slow [24]. After concerns about adverse effects of the formulations, a hold was placed on an early-stage clinical trial. This has subsequently been lifted and the outcome of the evaluations is awaited with interest.

Initial evaluation of using RNAi against Marburg virus replication involved transfection of cultured cells with siRNAs and plasmids expressing individual sequences targeting the viral genes [25]. Efficient inhibition of protein expression was observed without evidence of nonspecific induction of the innate immune response by the RNAi activators. Infection of cultured cells confirmed that the siRNAs acted against replication of the virus. Since this first report of efficacy against Marburg virus, gene silencing for treatment of Marburg virus infection has now reached an advanced stage [26,27]. The technology used to counter Marburg virus infection has been similar to that used to treat Ebola virus infection. Investigations performed on guinea pigs entailed administration of synthetic siRNAs targeted to the viral *NP* sequence [27]. After identifying the lead gene silencers from results of evaluation in cultured cells, siRNAs were formulated in liposomal nonviral vectors and tested in vivo. The most effective siRNA, NP-718m (Figure 10.2), provided 100% protection against challenge with the Angola strain of the virus. This study was followed up with an assessment of efficacy in Rhesus macaques [26]. The NP-718m-containing lipoplexes were administered to the monkeys 30–45 min and 1, 2, and 3 days after exposure to a lethal dose of Marburg virus. All of the animals that received the antiviral gene silencer were protected against the challenge whereas the control animals all succumbed to the infection. Protection afforded at day 3 after exposure to Marburg virus is significant because symptoms typically begin to manifest at or earlier than this time point during natural infection. In addition to being effective against the virus, the treatment did not appear to cause any adverse effects.

Using antisense-based technology, Warren and colleagues demonstrated that protection against lethal exposure to Ebola virus and Marburg virus could also be achieved in macaques [28]. The positively charged phosphorodiamidate morpholino oligonucleotides (PMOs) targeted *VP24* and *VP35* sequences of Ebola virus and *NP* and *VP24* of Marburg virus. This study was a refinement of previous work that showed efficacy of different versions of the antiviral sequences in vivo [29,30]. The *VP24*- and *VP35*-targeting oligonucleotides were modified to make them positively charged and to enhance efficacy they included phosphoramidite and morphilino residues. More than 60% of animals were protected against Ebola virus exposure. Monkeys receiving Marburg virus-targeting antisense molecules were completely protected against the virus. The candidate drug targeting Marburg virus, AVI-7288, is being developed by Serepta Therapeutics and is now in early stage clinical trial testing (http://www.sarepta.com/pipeline/avi-7288-marburg).

Aptamers targeting the VP35 protein of EBOV have been isolated after systematic evolution of ligands by exponential enrichment to select oligonucleotides that bind to the protein [31]. Analysis revealed good binding affinity to the viral target and functional inhibition of VP35 through disruption of interaction with NP. The inhibitors are at an early stage of their development and indications are that the approach could be therapeutically applicable.

10.3 DENGUE VIRUS

Infection with DENV is common, particularly in the tropics, and is a significant cause of public health problems [32]. The origin of the name is not clear. Possible derivations from Swahili or Spanish words describing the gait of people infected with DENV have been proposed [33]. Up until recently, four types of DNEV were known: DENV1-4 [34], and a fifth, DENV-5, was reported in 2013 [35]. As a member of the *Flaviviridae* family, and *Flavivirus* genus, the virus is enveloped and has a single-stranded positive sense RNA genome of approximately 11 kb (Figure 10.3). The encoded polyprotein is processed by cellular and viral proteases to form three structural proteins and seven nonstructural proteins [36]. The structural components are the capsid (C), envelope (E), and membrane (M) proteins. The seven nonstructural proteins are NS1, NS2a, NS2b, NS3, NS4a, NS4b, and NS5. The capsid, comprising the C protein, is surrounded by the cell-derived envelope. M and E proteins, together with a peptide derived from the precursor M protein, are embedded in the envelope. They are exposed on the surface of the virion to enable binding to cells during infection. NS2b and NS3 have proteolytic function; NS5 is required for viral genome replication; and NS2a, NS4a, and NS4b inhibit the innate immune response to facilitate viral replication. Uptake of DENV into cells is by a clathrin-dependent endocytosis [37]. Detailed analysis using single particle tracking showed that the virions diffuse across the cell surface until they encounter a

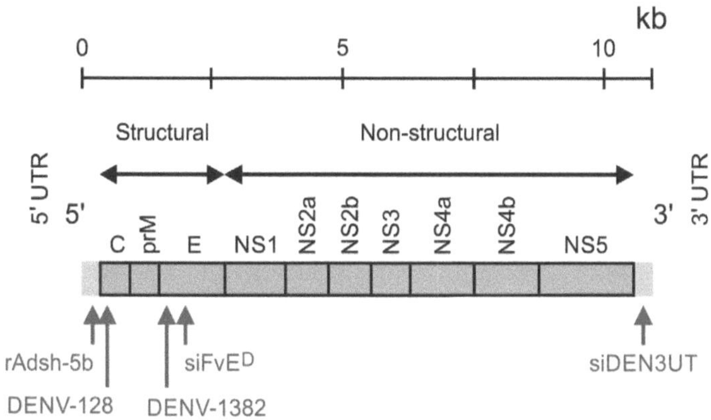

FIGURE 10.3 Genome of DENV with sites that have been targeted using RNAi.
The single-stranded RNA of sense polarity comprises approximately 11 kb. This sequence also serves as a translational template for the formation of structural and nonstructural proteins of the virus. UTRs located at the 5′ and 3′ ends contain *cis* elements that are important for the regulation of viral replication and protein synthesis. Approximate targeting sites of reported lead anti-DENV RNAi activators are indicated. Efficacy of rAdsh-5b was reported by Korrapati et al. [41], DENV-128 and DENV-1382 were characterized by Xie et al. [42], the synthetic siFvED was developed by Subramanya et al. [43], and siDEN3UT was described by Zhang et al. [40]. UTR, untranslated region. DENV, Dengue virus; UTR, Untranslated regions.

clathrin-coated pit. After cellular internalization, the viral particle releases the genome into the cytoplasm when viral and endosomal membranes fuse.

The virus is an arbovirus and is transmitted by biting of mosquitoes of the *Aedes aegypti* species [32]. Following transmission to humans, the virus replicates in the reticuloendothelial system. After a viremia is established, DENV may be transmitted back to mosquitoes during their taking of blood meals. Clinical presentation is with sudden onset of headache, fever, rash, myalgia, and arthralgia. The infection is usually self-limiting and resolves within 4–7 days. Uncommonly, DENV infection may be complicated by vasculopathy with hypovolemia, hemorrhage, and shock. A complication of the management of DENV infection is that the peak in viremia occurs before the onset of symptoms. Thus, patients seek medical care at a stage when the viremia is subsiding. Present treatment is aimed at providing supportive care and implementation of mosquito vector control strategies.

Immunity to DENV is serotype-specific, which has been problematic for development of an effective vaccine against the virus [32]. Moreover, when immunity is induced to one serotype, antibody-mediated augmentation of disease pathogenesis may occur when exposed to another serotype [38]. In such cases of a repeat infection, hemorrhagic fever may be precipitated. This feature of DENV infection and the variability of the viral types have thwarted development of a vaccine.

10.3.1 Gene Therapy for Infection with DENV

Given the global importance of DENV infection, research on developing new approaches to treating the disease is a priority. The most effective future therapies are likely to entail use of a combination of prophylactic vaccination, vector control, and direct treatment of infected individuals [39]. RNAi-based silencing of DENV has the potential to address two of the three arms: treatment of the virus in individuals and limiting of viral spread in mosquitoes. Initial studies on the use of RNAi reported on silencing of DENV in mammalian cells when using recombinant adeno-associated viral vectors [40]. Subsequently, adenoviral [41], lentiviral [42], and nonviral vectors [43] were also used to deliver DENV-targeting gene silencers (Figure 10.3). To provide a greater barrier to viral escape, Xie et al. also used a combination of two artificial DENV-targeting primary microRNA sequences expressed in a bicistronic cassette [42]. Most studies were performed using simulations of viral replication in cells in culture, and some have also entailed assessment of efficacy in murine models of DENV infection [44]. In addition to direct targeting of the viral genes, host factors that are required for viral replication have also been assessed as suitable targets. To inhibit receptor engagement and endocytosis-mediated uptake of the virus, Alhoot and colleagues used synthetic siRNAs to silence expression of CD-14-associated molecule, clathrin heavy polypeptide, and Dynamin-2 in monocytes [45]. These proteins are important for viral attachment and the clathrin-mediated cellular internalization of DENV. Their results showed that markers of viral replication and the numbers of infected cells were decreased as a result of the treatment. A follow-up study performed on cultured liver-derived HepG2 cells entailed silencing of the clathrin-mediated endocytosis pathway and a cellular surface receptor for DENV, GRP78, on these cells [46]. Markers of viral replication and numbers of infected cells were again diminished by the treatment. Although encouraging, therapeutic use of RNAi-based drugs in patients remains some way from clinical application. Antisense technologies based on use of PMOs against the virus show promise and are in early preclinical development by Sarepta Therapeutics (http://www.sarepta.com/pipeline/). Developing of the appropriate technology for delivering of the sequences to virus-infected cells and the utility of treatment in symptomatic patients when the peak of viremia has passed (see section 10.3) remain challenging.

As an alternative to targeting of DENV replication in patients, controlling virus transmission by replacing wild-type mosquitoes with DENV-resistant transgenic strains of the vector has been investigated (Figure 10.4). By expressing antiviral hairpin sequences, it is possible to harness RNAi to render mosquito cells resistant to DENV infection [47–49]. Initial studies entailed use of a transposon system to generate the transgenic insects [49]. Constitutively active baculovirus immediate early and *Drosophila* polyubiquitin promoters were used to drive expression of DENV-targeting long hairpin RNA (lhRNA) sequences.

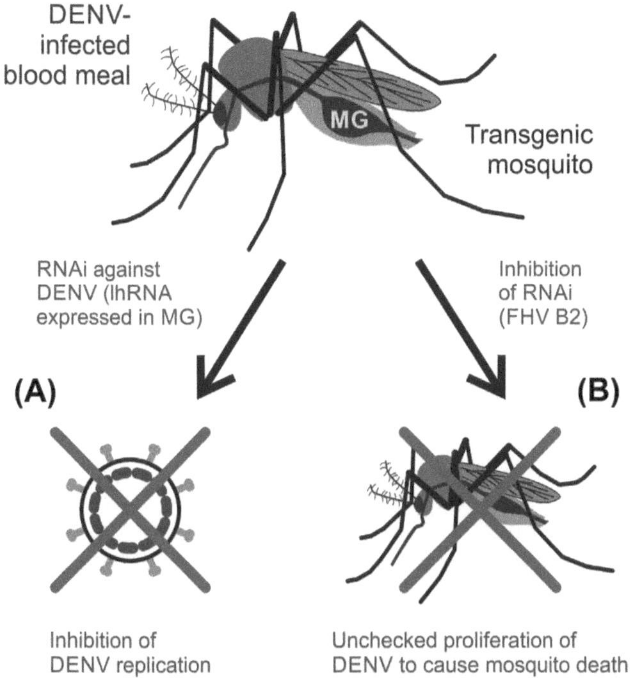

FIGURE 10.4 Schematic illustration of the two approaches that have been used to counter DENV infection of humans by targeting the mosquito vector.
Transgenic mosquitoes have been engineered to express antiviral lhRNAs in the MG of the insects. Resultant silencing of viral genes may lead to inhibition of DENV replication in the vector (A). Alternatively, by inhibiting the normal antiviral RNAi function in the mosquitoes with the FHV B2 protein, enhanced replication of DENV may exacerbate pathogenesis of the infection in mosquitoes (B). MG, midgut; FHV, Flock house virus; DENV, Dengue virus; RNAi, RNA interference; lhRNA, long hairpin RNA.

Although the antiviral siRNAs were generated, their production in the mosquito midgut was found to be inadequate for effective silencing in vivo. This was a significant drawback because DENV infection of mosquitoes is initiated in the midgut, and the early stages of infection of mosquitoes is also when the virus is at its most vulnerable. In a subsequent study, the promoter of the *carboxypeptidase* gene of the *A. aegypti* species was used to drive expression of the DENV-targeting lhRNA [48]. This transcriptional control element is activated in the midgut of mosquitoes at 8–48 h after the insects' ingestion of blood meals [50], which also corresponds to the predicted timing of highest susceptibility of DENV to RNAi. The *mariner Mos1* transposon system was used to propagate the transgenic mosquitoes harboring the DENV-targeting RNAi expression cassettes. DENV replication in the midgut of transgenic mosquitoes was effectively inhibited [48]. However, a limitation was that silencing of DENV was not complete and replication occurred in some of the transgenic insects. Variation in expression and efficacy of antiviral sequences, as well as suppression of RNAi-based gene silencing by the virus, are likely to be causative. It was considered unlikely that the virus was capable of escaping the silencing effects of the several siRNAs derived from the lhRNAs.

Another approach to targeting the mosquito carrier of DENV entailed inhibition of RNAi in the vector [51] (Figure 10.4). Silencing mechanisms normally form an important part of the insect's immune response to infection by the virus. The rationale for the investigation aimed at inhibiting RNAi in the mosquitoes was based on the hypothesis that proliferation of the virus would be facilitated and reach levels that would be pathogenic to the vector. To inhibit RNAi, transgenic mosquitoes were generated by incorporation of sequences encoding the B2 protein of Flock house virus. To enable precise integration, site-directed recombination with the *ΦC31* system was used to introduce the transgene, which was under control of the poly-ubiquitin promoter of the mosquitoes. Analysis revealed that proliferation of DENV was markedly augmented in the midgut of transgenic insects. However, despite high proliferation, dissemination of the virus was not increased. A similar effect was observed when using a recombinant Sindbis virus to infect transgenic mosquitoes. Thus, although RNAi is important to inhibit viral replication in insects, it is not the only antiviral mechanism. Mosquitoes of the *A. aegypti* species are also vectors for other viruses such as yellow fever virus (YFV) and chikungunya virus [52]. Thus, targeting the vector as a means of controlling spread of DENV and other viruses that the mosquito may carry is an intriguing possibility. With more research on the topic, feasible strategies may emerge to prevent transmission of the virus. Success will also depend on safe and effective replacement of mosquito populations with genetically engineered variants.

10.4 GENE THERAPY FOR OTHER HEMORRHAGIC FEVER VIRUSES

Antiviral efficacy of gene therapy approaches has been demonstrated against other viruses that may cause hemorrhagic fevers. These viruses include YFV, Rift Valley fever virus (RVFV), and WNV.

YFV is a member of the *Flaviviridae* family, as is DENV. The mode of transmission, genome structure, and pathogenesis of diseases caused by the two viruses are similar. YFV is endemic to parts of Africa and Latin America, where the annual incidence of infection is approximately 200,000 [53]. Of the cases of YFV infection, 30,000 are fatal and the number of fatalities appears to be increasing. Although infection with YFV may be serious, an effective vaccine is available. Because DENV and YFV share many common features, similar technologies aimed at countering YFV or DENV may be used. That is, the mosquito vector and the viral genes themselves could be targets for countering the viruses. One study has reported on efficacy of and RNAi-based strategy to disable YFV replication [54]. Pacca and colleagues showed that targeting *NS1* and *E* sequences resulted in good silencing. This was verified in cultured cells and in mice. In addition to gene silencing, antisense and aptamer technologies have also been used against YFV (reviewed in refs [55,56]).

WNV is another mosquito-borne virus of the family of *Flavividiae*. The first reported cases of infection with the virus occurred in the West Nile region of Uganda, and this location gave the virus its name. Most infections with WNV are asymptomatic, but when symptoms occur, patients complain of headache, anorexia, nausea, and malaise. Rarely, the virus may cause lethal encephalitis. Since its initial discovery in Africa, the virus has spread widely to other parts of the world. There is no effective vaccine against WNV, and treatment is largely supportive. To advance gene therapy against WNV, several groups have developed RNAi-activating gene silencers (reviewed in ref. [57]). These have been shown to be effective in cultured cells and in murine models of WNV infection [58,59]. In addition to direct silencing of viral genes, host factors required by WNV may also be suitable targets for inhibition of WNV replication. Several host factors that may be suitable for inhibiting WNV replication have been identified in a genome-wide screen [60]. As with DENV, the mosquito vector may also be the focus of RNAi-mediated gene silencing to disable WNV replication [61].

RVFV is a member of the *Buyaviridae* family, which may cause outbreaks of disease in livestock and humans. It is particularly common in Africa, although cases were also reported in the Yemen and Saudi Arabia [62–67]. When infected with the virus, humans typically develop mild febrile illness, which rarely develops into a hemorrhagic fever. Humans usually acquire the virus as a result of contact with contaminated blood of infected animals or after mosquito-borne transmission. The RNA genome of RVFV comprises three segments: the ambisense small (S) strand, the medium (M), and large (L) negative sense components [68–70]. The infectious particle is an enveloped spherical structure with diameter of 12–80 nm (reviewed in ref. [71]). Gn and Gc glycoproteins of the virion surround the envelope in an icosadedral lattice. The N proteins of the S segment arrange themselves into hexameric structures to which the viral genomic RNA is bound. The nonstructural protein of the S segment (NSs) performs immunomodulatory effects and is capable of global inhibition of transcription in infected cells. Disruption of NSs markedly attenuates pathogenesis of RVFV-associated disease and has been targeted using RNAi [72]. Scott and colleagues screened a panel of expressed antiviral sequences that targeted *N* and *NSs* sequences. Lead antiviral sequences were capable of inhibiting expression of viral proteins, attenuating cytotoxicity caused by NSs and diminished viral replication in cultured cells.

10.5 CONCLUSIONS

Given the potentially devastating consequences of spread of infections with viruses that cause hemorrhagic fevers and the paucity of available treatment options, development of innovative new prophylactic and therapeutic measures is essential. The importance is for humanitarian reasons and to counter possible threats of deliberately releasing the viruses into the environment. Significant progress has been made in advancing gene therapy for the treatment of viral infections that cause hemorrhagic fevers. Viral species of the group have RNA genomes that replicate in the cytoplasm. Therefore, most studies to date have applied RNAi-based gene silencing, but antisense and aptamer-based strategies have also been successfully used. Countering arboviruses that cause hemorrhagic fever through the interruption of transmission to mosquitoes has been attempted and is an interesting approach. Inhibiting viral replication in the insect vectors and augmenting pathological effects of the viruses in mosquitoes are the two approaches that have been attempted to date.

Chemically modified synthetic siRNAs, formulated in lipoplex vectors, have been used successfully to protect nonhuman primates from lethal exposure to Ebola virus [18] and Marburg virus [26]. Although promising, the technology has not yet reached a stage of treating patients infected with the viruses. Given that Ebola virus continues to be a life-threatening pathogen, and that the most serious recorded outbreak of the viral infection is currently in West Africa, rapid advancement is critical. Mortality from the epidemic has been high, and potential for spread was confirmed by isolated cases occurring in the United States and Spain among travelers and aid workers returning from West Africa (http://www.cdc.gov/vhf/ebola/outbreaks/2014-west-africa/). The gravity of the situation has galvanized researchers into making concerted efforts to develop effective treatment. Several different novel therapies against Ebola virus show promise. These include small-molecule compounds and the plant-derived antibody-based cocktail antiviral, ZMapp [73]. Given the serious nature of infections with Ebola virus and Marburg virus, conventional clinical trials for candidate therapeutics are impractical. Consequently, the U.S. Food and Drug Administration devised the so-called "Animal Rule" (reviewed in ref. [74]), which enables use in humans of drugs that show good efficacy in animal studies. Because nucleic acid-based treatment of Ebola virus and Marburg virus has been convincingly demonstrated in nonhuman primates, a strong case may be made for initiating gene therapy for the hemorrhagic virus infections. Although this is commendable, practical issues relating to rapid economic implementation of large-scale synthesis of antivirals of acceptable quality for roll out in affected populations may be problematic. Nevertheless, technical advances have gained significant momentum and hopefully transition to deployment of effective countermeasures is imminent.

REFERENCES

[1] World Health Organization. Ebola virus disease Fact sheet No 103. 2014.

[2] Negredo A, Palacios G, Vazquez-Moron S, Gonzalez F, Dopazo H, Molero F, et al. Discovery of an ebolavirus-like filovirus in Europe. PLoS Pathog 2011;7(10):e1002304.

[3] Geisbert TW, Hensley LE. Ebola virus: new insights into disease aetiopathology and possible therapeutic interventions. Expert Rev Mol Med 2004;6(20):1–24.

[4] Aleksandrowicz P, Marzi A, Biedenkopf N, Beimforde N, Becker S, Hoenen T, et al. Ebola virus enters host cells by macropinocytosis and clathrin-mediated endocytosis. J Infect Dis 2011;204(Suppl. 3):S957–67.

[5] Sanchez A, Kiley MP, Holloway BP, Auperin DD. Sequence analysis of the Ebola virus genome: organization, genetic elements, and comparison with the genome of Marburg virus. Virus Res 1993;29(3):215–40.

[6] Alvarez CP, Lasala F, Carrillo J, Muniz O, Corbi AL, Delgado R. C-type lectins DC-SIGN and L-SIGN mediate cellular entry by Ebola virus in cis and in trans. J Virol 2002;76(13):6841–4.

[7] Takada A, Watanabe S, Ito H, Okazaki K, Kida H, Kawaoka Y. Downregulation of beta1 integrins by Ebola virus glycoprotein: implication for virus entry. Virology 2000;278(1):20–6.

[8] Carette JE, Raaben M, Wong AC, Herbert AS, Obernosterer G, Mulherkar N, et al. Ebola virus entry requires the cholesterol transporter Niemann-Pick C1. Nature 2011;477(7364):340–3.

[9] Chan M. Ebola virus disease in west africa - no early end to the outbreak. N Engl J Med 2014;371(13):1183–5.

[10] Basler CF, Amarasinghe GK. Evasion of interferon responses by Ebola and Marburg viruses. J Interferon Cytokine Res 2009;29(9):511–20.

[11] Cook JD, Lee JE. The secret life of viral entry glycoproteins: moonlighting in immune evasion. PLoS Pathog 2013;9(5):e1003258.

[12] Brauburger K, Hume AJ, Muhlberger E, Olejnik J. Forty-five years of Marburg virus research. Viruses 2012;4(10):1878–927.

[13] Kissling RE, Robinson RQ, Murphy FA, Whitfield SG. Agent of disease contracted from green monkeys. Science 1968;160(3830):888–90.

[14] Kunz C, Hofmann H, Kovac W, Stockinger L. Biological and morphological characteristics of the "hemorrhagic fever" virus occurring in Germany. Wien klin Wochenschr 1968;80(9): 161–2. passim.

[15] Siegert R, Shu HL, Slenczka HL, Peters D, Muller G. The aetiology of an unknown human infection transmitted by monkeys (preliminary communication). Ger Med Mon 1968;13(1):1–2.

[16] Sanchez A, Kiley MP, Klenk HD, Feldmann H. Sequence analysis of the Marburg virus nucleoprotein gene: comparison to Ebola virus and other non-segmented negative-strand RNA viruses. J Gen Virol 1992;73(Pt 2):347–57.

[17] Haasnoot J, de Vries W, Geutjes EJ, Prins M, de Haan P, Berkhout B. The Ebola virus VP35 protein is a suppressor of RNA silencing. PLoS Pathog 2007;3(6):e86.

[18] Geisbert TW, Lee AC, Robbins M, Geisbert JB, Honko AN, Sood V, et al. Postexposure protection of non-human primates against a lethal Ebola virus challenge with RNA interference: a proof-of-concept study. Lancet 2010;375(9729):1896–905.

[19] Geisbert TW, Hensley LE, Kagan E, Yu EZ, Geisbert JB, Daddario-DiCaprio K, et al. Postexposure protection of guinea pigs against a lethal ebola virus challenge is conferred by RNA interference. J Infect Dis 2006;193(12):1650–7.

[20] Bray M, Hatfill S, Hensley L, Huggins JW. Haematological, biochemical and coagulation changes in mice, guinea-pigs and monkeys infected with a mouse-adapted variant of Ebola Zaire virus. J Comp Pathol 2001;125(4):243–53.

[21] Geisbert TW, Pushko P, Anderson K, Smith J, Davis KJ, Jahrling PB. Evaluation in nonhuman primates of vaccines against Ebola virus. Emerg Infect Dis 2002;8(5):503–7.

[22] Morrissey DV, Lockridge JA, Shaw L, Blanchard K, Jensen K, Breen W, et al. Potent and persistent in vivo anti-HBV activity of chemically modified siRNAs. Nat Biotechnol 2005;23(8):1002–7.

[23] Geisbert TW, Hensley LE, Larsen T, Young HA, Reed DS, Geisbert JB, et al. Pathogenesis of Ebola hemorrhagic fever in cynomolgus macaques: evidence that dendritic cells are early and sustained targets of infection. Am J Pathol 2003;163(6):2347–70.

[24] Strauss S. Biotech drugs too little, too late for Ebola outbreak. Nat Biotechnol 2014;32(9):849–50.

[25] Fowler T, Bamberg S, Moller P, Klenk HD, Meyer TF, Becker S, et al. Inhibition of Marburg virus protein expression and viral release by RNA interference. J Gen Virol 2005;86(Pt 4):1181–8.

[26] Thi EP, Mire CE, Ursic-Bedoya R, Geisbert JB, Lee AC, Agans KN, et al. Marburg virus infection in nonhuman primates: therapeutic treatment by lipid-encapsulated siRNA. Sci Transl Med 2014;6(250):250ra116.

[27] Ursic-Bedoya R, Mire CE, Robbins M, Geisbert JB, Judge A, MacLachlan I, et al. Protection against lethal Marburg virus infection mediated by lipid encapsulated small interfering RNA. J Infect Dis 2014;209(4):562–70.

[28] Warren TK, Warfield KL, Wells J, Swenson DL, Donner KS, Van Tongeren SA, et al. Advanced antisense therapies for postexposure protection against lethal filovirus infections. Nat Med 2010;16(9):991–4.

[29] Warfield KL, Panchal RG, Aman MJ, Bavari S. Antisense treatments for biothreat agents. Curr Opin Mol Ther 2006;8(2):93–103.

[30] Warfield KL, Swenson DL, Olinger GG, Nichols DK, Pratt WD, Blouch R, et al. Gene-specific countermeasures against Ebola virus based on antisense phosphorodiamidate morpholino oligomers. PLoS Pathog 2006;2(1):e1.

[31] Binning JM, Wang T, Luthra P, Shabman RS, Borek DM, Liu G, et al. Development of RNA aptamers targeting Ebola virus VP35. Biochemistry 2013;52(47):8406–19.

[32] Swaminathan S, Khanna N. Dengue: recent advances in biology and current status of translational research. Curr Mol Med 2009;9(2):152–73.

[33] Dengue virus net. 2014. http://www.denguevirusnet.com/.

[34] Vasilakis N, Cardosa J, Hanley KA, Holmes EC, Weaver SC. Fever from the forest: prospects for the continued emergence of sylvatic dengue virus and its impact on public health. Nat Rev Microbiol 2011;9(7):532–41.

[35] Vasilakis N. The daemon in the forest-emergence of a new dengue serotype in SouthEast Asia. In: Third international conference on dengue and dengue haemorrhagic fever Bangkok, Thailand. 2013.

[36] Bartenschlager R, Miller S. Molecular aspects of Dengue virus replication. Future Microbiol 2008;3:155–65.

[37] van der Schaar HM, Rust MJ, Chen C, van der Ende-Metselaar H, Wilschut J, Zhuang X, et al. Dissecting the cell entry pathway of dengue virus by single-particle tracking in living cells. PLoS Pathog 2008;4(12):e1000244.

[38] Dejnirattisai W, Jumnainsong A, Onsirisakul N, Fitton P, Vasanawathana S, Limpitikul W, et al. Cross-reacting antibodies enhance dengue virus infection in humans. Science 2010; 328(5979):745–8.

[39] Whitehorn J, Yacoub S, Anders KL, Macareo LR, Cassetti MC, Nguyen Van VC, et al. Dengue therapeutics, chemoprophylaxis, and allied tools: state of the art and future directions. PLoS Negl Trop Dis 2014;8(8):e3025.

[40] Zhang W, Singam R, Hellermann G, Kong X, Juan HS, Lockey RF, et al. Attenuation of dengue virus infection by adeno-associated virus-mediated siRNA delivery. Genet Vaccines Ther 2004;2(1). http://dx.doi.org/10.1186/479-0556-2-8.

[41] Korrapati AB, Swaminathan G, Singh A, Khanna N, Swaminathan S. Adenovirus delivered short hairpin RNA targeting a conserved site in the 5′ non-translated region inhibits all four serotypes of dengue viruses. PLoS Negl Trop Dis 2012;6(7):e1735.

[42] Xie PW, Xie Y, Zhang XJ, Huang H, He LN, Wang XJ, et al. Inhibition of Dengue virus 2 replication by artificial micrornas targeting the conserved regions. Nucleic Acid Ther 2013;23(4):244–52.

[43] Subramanya S, Kim SS, Abraham S, Yao J, Kumar M, Kumar P, et al. Targeted delivery of small interfering RNA to human dendritic cells to suppress dengue virus infection and associated proinflammatory cytokine production. J Virol 2010;84(5):2490–501.

[44] Stein DA, Perry ST, Buck MD, Oehmen CS, Fischer MA, Poore E, et al. Inhibition of dengue virus infections in cell cultures and in AG129 mice by a small interfering RNA targeting a highly conserved sequence. J Virol 2011;85(19):10154–66.

[45] Alhoot MA, Wang SM, Sekaran SD. Inhibition of dengue virus entry and multiplication into monocytes using RNA interference. PLoS Negl Trop Dis 2011;5(11):e1410.

[46] Alhoot MA, Wang SM, Sekaran SD. RNA interference mediated inhibition of dengue virus multiplication and entry in HepG2 cells. PloS One 2012;7(3):e34060.

[47] Adelman ZN, Sanchez-Vargas I, Travanty EA, Carlson JO, Beaty BJ, Blair CD, et al. RNA silencing of dengue virus type 2 replication in transformed C6/36 mosquito cells transcribing an inverted-repeat RNA derived from the virus genome. J Virol 2002;76(24):12925–33.

[48] Franz AW, Sanchez-Vargas I, Adelman ZN, Blair CD, Beaty BJ, James AA, et al. Engineering RNA interference-based resistance to dengue virus type 2 in genetically modified *Aedes aegypti*. Proc Natl Acad Sci USA 2006;103(11):4198–203.

[49] Travanty EA, Adelman ZN, Franz AW, Keene KM, Beaty BJ, Blair CD, et al. Using RNA interference to develop dengue virus resistance in genetically modified *Aedes aegypti*. Insect Biochem Mol Biol 2004;34(7):607–13.

[50] Moreira LA, Edwards MJ, Adhami F, Jasinskiene N, James AA, Jacobs-Lorena M. Robust gut-specific gene expression in transgenic *Aedes aegypti* mosquitoes. Proc Natl Acad Sci USA 2000;97(20):10895–8.

[51] Khoo CC, Doty JB, Heersink MS, Olson KE, Franz AW. Transgene-mediated suppression of the RNA interference pathway in *Aedes aegypti* interferes with gene silencing and enhances Sindbis virus and dengue virus type 2 replication. Insect Mol Biol 2013;22(1):104–14.

[52] Weaver SC, Reisen WK. Present and future arboviral threats. Antivir Res 2010;85(2):328–45.

[53] Yellow fever fact sheet No 100. World Health Organization; 2014.

[54] Pacca CC, Severino AA, Mondini A, Rahal P, D'Avila SG, Cordeiro JA, et al. RNA interference inhibits yellow fever virus replication in vitro and in vivo. Virus Genes 2009;38(2):224–31.

[55] Stein DA, Shi PY. Nucleic acid-based inhibition of flavivirus infections. Front Biosci 2008; 13:1385–95.

[56] Bruno JG, Carrillo MP, Richarte AM, Phillips T, Andrews C, Lee JS. Development, screening, and analysis of DNA aptamer libraries potentially useful for diagnosis and passive immunity of arboviruses. BMC Res Notes 2012;5:633.

[57] Fortes P, Blazquez L. Harnessing RNAi for the treatment of viral infections. In: Arbuthnot P, Weinberg MS, editors. Applied RNAi. Norfolk, UK: Caister Academic Press; 2014. p. 151–80.

[58] Bai F, Wang T, Pal U, Bao F, Gould LH, Fikrig E. Use of RNA interference to prevent lethal murine West Nile virus infection. J Infect Dis 2005;191(7):1148–54.

[59] Kumar P, Lee SK, Shankar P, Manjunath N. A single siRNA suppresses fatal encephalitis induced by two different flaviviruses. PLoS Med 2006;3(4):e96.

[60] Krishnan MN, Ng A, Sukumaran B, Gilfoy FD, Uchil PD, Sultana H, et al. RNA interference screen for human genes associated with West Nile virus infection. Nature 2008;455(7210):242–5.

[61] Brackney DE, Beane JE, Ebel GD. RNAi targeting of West Nile virus in mosquito midguts promotes virus diversification. PLoS Pathog 2009;5(7):e1000502.

[62] Peters C, Liu C, Anderson GJ, Morrill J, Jahrling P. Pathogenesis of viral hemorrhagic fevers: Rift Valley fever and Lassa fever contrasted. Rev Infect Dis 1989;11:743–9.

[63] Gerdes GH. Rift valley fever. Vet Clin North Am Food Anim Pract 2002;18(3):549–55.

[64] Miller BR, Godsey MS, Crabtree MB, Savage HM, Al-Mazrao Y, Al-Jeffri MH, et al. Isolation and genetic characterization of Rift Valley fever virus from Aedes vexans arabiensis, Kingdom of Saudi Arabia. Emerg Infect Dis 2002;8(12):1492–4.

[65] Jup PG, Kemp A, Grobbelaar A, Lema P, Burt FJ, Alahmed AM, et al. The 2000 epidemic of Rift Valley fever in Saudi Arabia: mosquito vector studies. Med Vet Entomol 2002;16(3):245–52.

[66] Shoemaker T, Boulianne C, Vincent MJ, Pezzanite L, Al-Qahtani MM, Al-Mazrou Y, et al. Genetic analysis of viruses associated with emergence of Rift Valley fever in Saudi Arabia and Yemen, 2000–01. Emerg Infect Dis 2002;8(12):1415–20.

[67] Fagbo SF. The evolving transmission pattern of Rift Valley fever in the Arabian Peninsula. Ann NY Acad Sci 2002;969:201–4.

[68] Muller R, Poch O, Delarue M, Bishop DHL, Bouloy M. Rift valley fever virus L segment: correction of the sequence and possible functional role of newly identified regions conserved in RNA-dependent polymerases. J Gen Virol 1994;75(6):1345–52.

[69] Collett M. Messenger RNA of the M segment RNA of Rift Valley fever virus. Virology 1986;151(1):151–6.

[70] Giorgi C, Accardi L, Nicoletti L, Gro MC, Takehara K, Hilditch C, et al. Sequences and coding strategies of the S RNAs of Toscana and Rift Valley fever viruses compared to those of Punta Toro, Sicilian Sandfly fever, and Uukuniemi viruses. Virology 1991;180(2):738–53.

[71] Ikegami T. Molecular biology and genetic diversity of Rift Valley fever virus. Antivir Res 2012;95(3):293–310.

[72] Scott T, Paweska JT, Arbuthnot P, Weinberg MS. Pathogenic effects of Rift Valley fever virus NSs gene are alleviated in cultured cells by expressed antiviral short hairpin RNAs. Antivir Ther 2012;17(4):643–56.

[73] Qiu X, Wong G, Audet J, Bello A, Fernando L, Alimonti JB, et al. Reversion of advanced Ebola virus disease in nonhuman primates with ZMapp. Nature 2014;514(7520):47–53.

[74] Snoy PJ. Establishing efficacy of human products using animals: the US food and drug administration's "animal rule". Veterinary Pathol 2010;47(5):774–8.

CHAPTER 11

Gene Transfer for Prophylaxis and Therapy of Viral Infections

11.1 INTRODUCTION

Protection against viruses through the use of vaccination has been fundamental to global prevention of mortality and morbidity. Administration of vaccines to limit transmission of measles, mumps, rubella, polio, and hepatitis B virus (HBV) is now widely used in the expanded programs of immunization in many countries. To date, most vaccines used against viruses have comprised live attenuated viruses, killed viruses, or recombinant proteins. Essentially, the rationale for administration of these compounds is based on an attempt to induce memorized immunity to the infectious agent without risk of disease. Recently, DNA and recombinant vectors that express immunogens have been used in an attempt to improve antiviral prophylaxis and therapeutic efficacy.

Edward Jenner, an English physician who lived from 1749 to 1823, is generally credited as being the pioneer of vaccination (http://www.historyofvaccines.org/). At the time that he lived, smallpox infection was a major cause of illness and death all over the world. The well-known line of inquiry that Jenner pursued was to investigate whether infection with cowpox virus prevented infection with smallpox virus. Proof of efficacy was provided after collecting pustular material from the skin of a dairymaid. This material was rubbed into scratches made on the skin of a subject who was then protected against subsequent challenge with smallpox virus (http://www.jenner.ac.uk/edwardjenner). This initial proof was performed on the son of Jenner's gardener and was followed up with studies performed on many other subjects. Before Jenner's work on cowpox virus, the procedure of variolation was used in an attempt to protect against smallpox infection. The method, which Jenner himself used, entailed transdermal administration or inhalation of material from scabs of patients who had been infected with smallpox virus. It is interesting to note that the earliest evidence for variolation comes from Chinese documents of approximately 1000 years ago. The procedure was also practiced in Turkey and was brought to the West during the early eighteenth century by Lady Mary Montagu [1]. A problem with variolation was that it was not entirely safe. There was a risk for full-blown smallpox infection after administration of the virus-containing scab-derived material. Therefore, Jenner's use of cowpox virus was significant

Gene Therapy for Viral Infections. http://dx.doi.org/10.1016/B978-0-12-410518-8.00011-9

because infection after the vaccination was very mild and therefore safe. The term "vaccination" was coined by Jenner and makes reference to the source of his discovery. The Latin word for cow is *vacca*, and *vaccinus* is an adjective that means "relating to cows."

Most commonly, vaccination is implemented to cause active immunity in an immunized individual. That is, the immunogen is administered to a patient to induce the patient's own immune system to respond to the pathogen. When the immunogen comprises a viral surface protein, such as is the case with the HBV vaccine, a humoral immune response is induced and the antibodies provide protection against the pathogen. Vaccines that comprise inactivated whole viral particles (e.g., the Salk vaccine that provides protection against poliovirus) also induce humoral immunity to the pathogen. Administration of a vaccine that is derived from an attenuated infectious viral particle results in activation of humoral and cell-mediated arms of the immune system. Attenuated vaccines (e.g., measles and mumps vaccines) should replicate in the recipient to be effective. Therefore, they simulate the natural infection and are able to achieve good protection by activating both arms of the adaptive immune response. Vaccines that comprise naked DNA or recombinant viral vectors, such as that are derived from adenoviruses and certain pox viruses (see section 11.1.3 and Chapter 4), also are capable of inducing cell-mediated and humoral immunity to a pathogen. This is an important advantage of this new class of immunogens and is a reason for their gaining popularity.

Passive immunity to a virus is induced when the immunity to the pathogen is provided by an exogenous source and does not involve induction of the recipient's own immune response. An example is the use of hyperimmune globulin, which may be used to prevent transmission of HBV from a highly replicative mother to her child during birth [2]. The method is not widely used, and the immunity is of short duration. However, adaptation of the passive immunity approach has recently generated interest with the discovery and characterization of broadly neutralizing antibodies (bNAbs) that are effective against a range of isolates of human immunodeficiency virus-1 (HIV-1) [3]. So-called vectored immunoprophylaxis (VIP) entails administration of recombinant viral vectors that express HIV-1-targeting immunoadhesins or derivatives of bNAbs, which are capable of disabling HIV-1 before the pathogen establishes an infection (see section 11.2.3 and Chapter 8) [4].

11.1.1 Principles Underlying Use of Gene Transfer for Immunotherapy and Prophylaxis

The idea that DNA encoding immunogenic proteins could be used for vaccination was supported directly and indirectly by several studies that were published during the 1980s [5,6] and 1990s [7]. The seminal investigation by Wolff et al. demonstrated that reporter genes encoded by plasmids could be expressed in myocytes after intramuscular injection of the naked DNA. The efficiency of delivery was modest, which suggested that utility of the approach was not suited to applications requiring transduction of many cells. However, it was realized that highly efficient transduction of many cells is not an absolute requirement for an immunostimulatory effect, and investigation of the vaccine-like properties of injected DNA followed soon thereafter [8]. Tang and colleagues showed that introducing plasmid DNA encoding human growth hormone into dermal cells of mice could elicit an antibody response to the human protein. In addition to use of naked DNA, recombinant viral vectors may be used to deliver immunogenic transgenes [9].

Using gene transfer to elicit a protective antiviral immune response to encoded proteins has many important advantages [10]:

1. DNA is inherently stable, and requirements for maintenance of the cold chain are not necessary with DNA-based vaccines.
2. DNA is safe and not toxic.
3. The technology for manipulating DNA sequences to enable expression of particular protein sequences is advanced. Procedures are inexpensive, easy to implement, and widely used. Moreover, large-scale preparation of DNA, especially plasmid DNA, is convenient.
4. To optimize an immune response to a particular encoded protein, it is possible to include additional open reading frames (ORFs) that augment or bias immunogenicity.
5. The presence of unmethylated CpG islands found in plasmid DNA activates the innate immune response; therefore, they may serve as an adjuvant for the intended immunogen.
6. Although CpG islands may stimulate an innate immune response, DNA itself is typically not antigenic. Therefore, it is possible to administer several doses of a DNA vaccine without attenuating its immunostimulatory effect.
7. Expression of a transgene from DNA, as is the case with live attenuated vaccines but not with protein vaccines, may resemble viral gene expression in an infected cell. Resultant simulation of antigen presentation and mounting of humoral and cell-mediated immune responses may cause more effective neutralization of the pathogen.

Despite these significant advantages, there are some difficulties with use of DNA-based vaccines. Ensuring good antigen expression to result in a protective immune response, adequate delivery of the immunogenic DNA, and manipulation of the bias of the immune response to achieve optimal antiviral efficacy are all important.

Mechanisms of activation of innate and adaptive immune responses after typical viral infections have been summarized in Chapter 1. There are many similarities to the ways in which gene transfer after administration of naked DNA or viral vectors act as immunostimulatants [11]. When administered intramuscularly or intradermally, different cell types may be transfected or transduced (Figure 11.1). These include myocytes, dendritic cells, and monocytes [10,12,13]. Major histocompatibility complex (MHC) class I and MHC class II are activated during an adaptive immune response. Immunogens expressed in myocytes may be shed and displayed on MHC class I molecules. Transduction of dendritic cells leads to MHC class I presentation, and uptake of immunogenic proteins released from myocytes may result in presentation on MHC

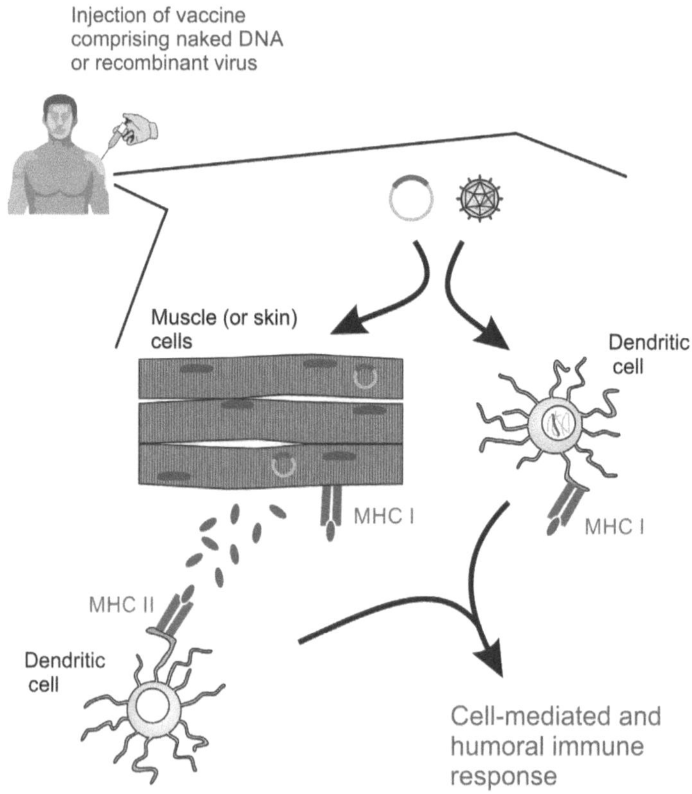

FIGURE 11.1 Immunostimulation after transduction of cells with plasmids or recombinant viral vectors.
The immunogenic transgene cassette is inserted into a plasmid or recombinant virus, such as poxvirus, and then administered to a patient. Transduction of cells at the site of injection, which may be muscle, skin, or dendritic cells, leads to expression of the transgene. Subsequently, epitopes are processed in these cells and presented on major histocompatibility complex class I (MHC I) proteins. Antigenic proteins that are secreted from the transduced cells, or present in remnants of dead cells, may be processed by dendritic cells and presented on MHC class II molecules. These mechanisms lead to lymphocyte activation and stimulation of humoral and cell-mediated immunity against the exogenous transgene-derived antigen.

class II molecules of dendritic cells. Subsequent stimulation of cytotoxic T cells and B cell proliferation takes place by mechanisms that occur during natural viral infection.

11.1.2 Optimization of Immunostimulation by Expression Cassettes

Optimization of the design of sequences encoding immunogens plays an important role in maximizing the efficacy of DNA-based vaccines (reviewed in ref. [14]). Many variables may be modified to improve safety and immunogenicity of the vaccines.

1. **Codon optimization**: The preponderance of tRNAs within cells of different organisms varies. Codon usage within the reading frames of an expression cassette should match the abundance of the tRNAs within the cells. Changing of the protein coding sequence may also facilitate the removal of repressive or destabilizing RNA sequences that impede translation of the proteins (see section 11.2.1) [15].
2. **Expression cassette design**: This should be devised to ensure optimal production of the immunogenic proteins. Some of the considerations are (1) inclusion of an intron within the cassette, (2) placement of suitable elements in the 5′ untranslated region (UTR) and 3′ UTR, (3) positioning of optimal Kozak sequences at the translational start site, (4) elimination of cryptic splice donor and acceptor sites in the cassette, and (5) avoidance of sequences that are conducive to formation of strong secondary structures that may interfere with ribosomal interaction with the mRNA. Of course, 5′ UTRs should not contain additional ORFs that may interfere with expression of a transgene.
3. **Fusing sequences to the transgene**: This may improve secretion and immunogenicity. An example includes use of plasminogen activator to facilitate overall expression and secretion of HIV-1 Gag antigens from transfected cells [16]. In addition, the immunostimulatory effects of exogenous immunogens may be enhanced by their fusion to the monocyte chemoattractant protein-3, which targets antigen-presenting cells [17].
4. **Promoter elements and transcriptional termination signals**: These should be incorporated into the cassettes to control initiation and polyadenylation of the mRNA transcribed from the transgene. The cytomegalovirus immediate/early promoter/enhancer element is a commonly used transcriptional regulatory element that is active in most tissues. Sequences from the rabbit β-globin gene are widely used to regulate termination of transcription and mRNA polyadenylation.
5. **Cryptic eukaryotic promoters**: Cryptic eukaryotic promoters that occur within plasmid sequences have been described [18]. Identifying and

altering such elements is important to avoid unintended production of duplex RNA, which may inhibit translation by activating innate immunity or RNA interference (RNAi).

6. **Antibiotic resistance genes**: Antibiotic resistance genes, particularly *β-lactamase*, which confers ampicillin resistance, should be excluded from plasmids encoding immunogenic transgenes. This is important to prevent inadvertent dissemination of the antibiotic resistance gene in bacterial flora. In addition, residual antibiotics in the DNA may evoke an allergic response in susceptible individuals. To avoid use of antibiotic resistance genes, various antibiotic-free selection markers have been developed [14]. A promising example of RNA-based selection is the RNA-OUT technology [19]. It is based on production of a short RNA sequence (<150 nt) that is antisense to mRNA encoding the toxic enzyme, levansucrase. In the presence of sucrose, bacteria only survive when the plasmid-encoded antisense RNA sequence is present. Plasmids containing the RNA-OUT selection mechanism also have the advantage of being small and easy to prepare with minimal effects of shearing.
7. **Adjuvants**: Adjuvants may be administered together with the immunogenic transgene to augment immunogenicity. In addition to the traditional adjuvants such as aluminum salts and oil-based compounds, additional expression cassettes may be incorporated into DNA vaccines to improve efficacy in a defined manner [20]. These immunomodulators may be various cytokines and co-stimulatory molecules. For example, Th-2-mediated humoral immunity may be selectively enhanced by co-administration of interleukin (IL)-4, IL-5, and IL-10 [21]. By contrast, inclusion of sequences encoding IL-12, tumor necrosis factor-α, and IL-15, among others, enhances a cytotoxic response [20,21]. IL-12 stimulates $CD4^+$ cells to produce cytokines that favor a Th-1-type immune response. An additional effect of IL-12 on $CD8^+$ cells is to enhance their function as memory cells to cause an overall enhancement of the cytotoxic T lymphocyte (CTL) response. IL-15 also has an augmenting effect on memory T cell function, and IL-12 and IL-15 have both been incorporated into candidate vaccines that have been aimed at countering HIV-1 infection [22].
8. **The method and route of delivery of DNA expression cassettes**: This has an influence on the efficacy of immunostimulation [20]. The efficiency of transfection after injection of naked DNA is typically low, and the resultant immunogenicity is poor. To enhance utility, different techniques of delivering DNA have been used. These include use of small ballistic devices, ultrasound, recombinant viral particles (Chapter 4 and see section 11.1.3), nonviral vectors (Chapter 5), and electroporation (EP). EP is a simple procedure that conveniently increases transfection of cells to improve immunogenicity of antiviral vaccines [23,24].

The underlying principle is based on induction of a transient increase in the permeability of cell membranes after application of pulses of electric current to cells at the site of DNA injection. By improving transfection efficiency, lower doses of DNA are required to achieve an effect. Although EP boosts transfection efficiency, the duration of transgene expression is not prolonged and the risk of integration into host cell DNA is not increased [25,26]. The procedure itself leads to localized inflammation at the injection site [27]. Cellular infiltration and recruitment of antigen-presenting cells also contribute to an augmented immune response.

11.1.3 Recombinant Viral Vectors for Vaccination

The main types of recombinant viral vectors that have been developed for use in vaccination are derived from poxviruses and adenoviruses (Ads). Both types are now well characterized, and knowledge about how to manipulate their genomes to incorporate immunogenic antiviral sequences is well established. Ads, discussed in Chapter 4, have several useful properties that may be exploited for use as vaccines. They have a broad tissue tropism and transduce cells with high efficiency. The recombinant vectors are replication defective but are capable of inducing powerful CTL responses. However, because community-acquired immunity to the Ad5 serotype is high, generation of vectors derived from other viral strains, such as Ad35, may be necessary to augment the vectors' ability to transduce cells [28]. A particularly serious complication of immune stimulation by Ads was highlighted by the Step vaccination trial, which entailed administration of three doses of a recombinant Ad expressing *gag*, *pol*, and *nef* of HIV-1 [29]. Compared with patients receiving the placebo, a slightly increased risk for infection by HIV-1 was observed in individuals given the vaccination. The effect may be a consequence of the predilection of HIV-1 for infecting T cells activated by the vaccine (Chapter 8).

Smallpox disease was eradicated in 1980 through global implementation of programs that entailed use of vaccinia virus (VV), a poxvirus, as a vaccine [30]. Since this landmark development, poxviruses have gained widespread popularity as vectors for delivery of immunogenic protein-coding sequences (reviewed in refs [9,31–33]). The viral genome comprises 130–300 kb of linear duplex DNA, and it is possible to alter the viral sequences significantly without compromising infectivity of the vectors. Transgenes comprising as much as 25 kb of DNA may be incorporated into recombinant poxvirus vectors, which is convenient for engineering the vectors to express a cocktail of immunogenic proteins. Attenuation of the virulence of parental strains may be achieved through serial passage in unnatural host cells or by deliberate deletion of genetic elements. Examples of poxviruses that have been manipulated in this way are the New York Vaccinia derived from Copenhagen and Modified Vaccinia Virus Ankara (MVA) strains. Members of the *Avipoxvirus* genus have also been popular in vaccine development and include the canarypox virus (e.g., ALVAC) and fowlpox virus (e.g., TROVAC). These viral vaccines do not replicate in mammalian cells; therefore, they are safe to use in immunocompromised individuals. Although they are replication defective, the ability to infect cells and express immunogenic proteins is retained. Moreover, poxviruses replicate exclusively in the cytoplasm of infected cells; therefore, risk of mutagenesis is minimal. Improved insights into the molecular biology of poxviruses have enabled development of platform technologies for the manipulation of viral genomes to incorporate immunogenic transgenes.

To generate recombinant poxvirus vectors, homologous recombination is typically used to engineer the viral sequences. The technology was developed in the 1980s [34,35], and has since been in widespread use. The procedures typically entail transfection of plasmids containing the donor sequences and infection of the same cells with the parental target virus (Figure 11.2). To add to the versatility

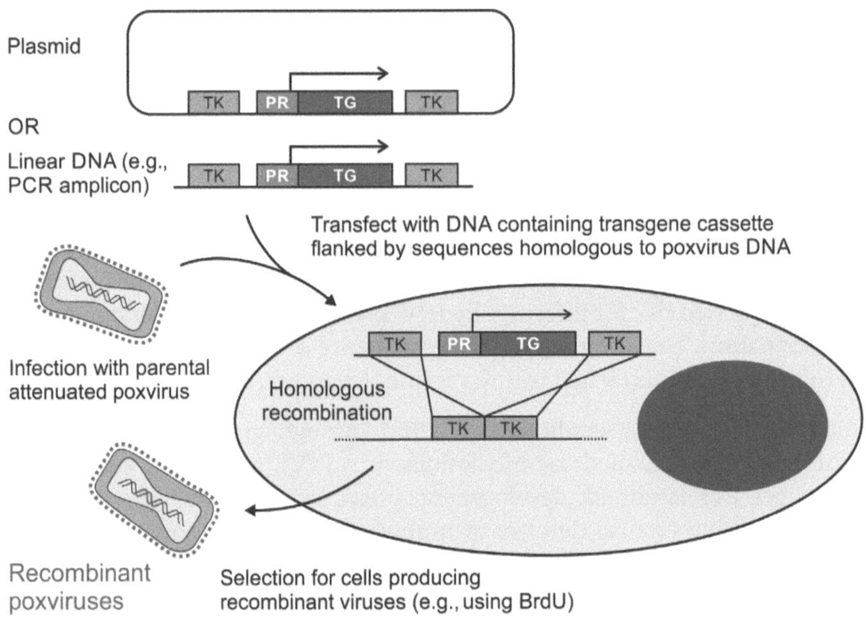

FIGURE 11.2 Outline of methods used to generate recombinant poxvirus vectors.
Plasmid or linear DNA is used to transfect cells that are co-infected with an attenuated parental poxvirus. The transfected DNA contains the immunogenic cassette, with PR and TG sequences, which are flanked by elements that are included to direct homologous recombination. In the illustrated example, sequences from the viral *TK* gene are used. Homologous recombination within the cytoplasm of infected and transfected cells leads to insertion of the transgene at the target site. Disruption of the target gene enables selection for the recombinant viral vector and purification of recombinant viral particles containing the transgene. PR, promoter; TG, transgene; TK, thymidine kinase; BrdU, bromodeoxyuridine.

of the methodology, linear DNA such as a polymerase chain reaction product, may be used instead of plasmid DNA. DNA that is transfected is engineered to encode a transgene cassette and has flanking sequences with homology to the poxvirus target. Intended homologous recombination is usually rare, and selection of recombinant clones is required to purify the vectors. Several selection procedures have been developed, which include sorting on the basis of thymidine kinase positive or negative status [34,35] and antibiotic resistance [36,37].

11.1.3.1 Heterologous Prime-Boosting

The efficacy of recombinant viral vectors may be diminished by the host's immunity to the recombinant viruses. As a result, reuse of vectors is limited and the desired immune response to a transgene-encoded protein may also be diminished. A strategy that has successfully been used to overcome these problems entails use of heterologous prime-boosting. A priming event, which may be caused by previous exposure to a pathogen, administration of an antigen, recombinant virus, or DNA-based vaccine, is followed by administration of a recombinant vector such as a derivative of MVA [31,38]. The approach overcomes the difficulties associated with repeat administration of the same vaccines. Moreover, the immune response is markedly augmented, which probably results from dissimilar immunostimulatory mechanisms of the different types of vaccines. Several studies have shown improved efficacy of using the heterologous prime-boost approach to augment cell-mediated and humoral immunity of vaccines against tuberculosis [39], malaria [40,41], and viruses [42–46].

11.2 GENE-BASED VACCINATION FOR HIV-1 INFECTION

Developing an effective vaccine against HIV-1 has been a priority of research aimed at countering the significant global health problems caused by the virus. Unfortunately, there has been little success, and results from clinical trials performed to date have largely been disappointing. Consequently, development of HIV-1 vaccination continues to be a vigorous field of research. Ideally, broad cross-clade vaccination should be attained to protect individuals. In addition, vaccination should achieve activation of humoral and cell-mediated arms of the adaptive immune response to prevent or treat the infection. Naturally, antibodies are typically generated to the readily accessible surface epitopes of the virus and do not effectively prevent HIV-1 entry into cells [3]. Epitopes that make the virus vulnerable may only be exposed transiently during conformational changes when the HIV-1 particle fuses with target cells. In addition, ideal vaccine epitopes may be concealed by the viral glycoproteins and protected by the glycan shield to limit access by antibodies. Because the humoral immune response does not typically bias selection of antibodies to these epitopes, the formation of effective neutralizing antibodies is infrequent. Another complicating factor is that the virus is prone to mutation, and alterations in sequences of the viral proteins enable HIV-1 to evade neutralizing antibodies. Collectively, these properties pose serious obstacles to achieving effective vaccination against HIV-1 infection.

Recent insights that have been gained into the mechanisms of action of bNAbs against HIV-1 have been helpful in developing HIV-1 vaccination strategy [3]. bNAbs are typically only detectable 2–4 years after initial infection and are directed to different epitopes of the viral envelope proteins [47]. The long time course of bNAb evolution suggests that their appearance is a consequence of a selective process [3]. Evidence indicates that there is a particularly high rate of somatic mutation of sequences encoding the complementarity-determining regions (CDRs) of the bNAbs. This occurs in response to viral mutations and results in selection of the broadly neutralizing capabilities. In addition, post-translational modifications, such as tyrosine sulfation, may contribute to function of the bNAbs [48]. A difficulty of analyzing and applying information based on the evolution of bNAbs is that the originally infecting viruses are not easily identified. Moreover, the antecedent Abs that are initially induced, and from which the bNAbs are derived, often appear to have poor affinity for the viral envelope proteins [47,49].

Many bNAbs have now been described, and the specific epitopes that they target are varied (reviewed in refs [3,50]). The mechanisms by which bNAbs interact with the virus to neutralize infection are better understood. First-generation bNAbs, such as b12 [51] and 447-52D [52], have limited potency and breadth of neutralization. However, second-generation bNAbs that have been described more recently have a better antiviral effect [3,50]. Examples of mechanisms by which bNAbs counter HIV-1 are through formation of sequences with long heavy-chain CDR3 domains that are capable of penetrating the glycan shield to access vulnerable targets of the envelope protein [49] or binding to heptad repeats of gp41 [53].

There are two main ways in which an understanding of the mechanisms of action of bNAbs may facilitate development of gene transfer for immunoprophylaxis or therapy:

1. Identification of viral epitopes recognized by the bNAbs informs the design of most suitable antigens that may be used as protective immunogens.
2. Characterization of the antibodies themselves enables the design of expression cassettes encoding antigen-binding proteins that may have utility for therapy or prophylaxis.

Proof of the utility of bNAbs for protection against HIV-1 has come from clinical and preclinical studies. HIV-1 infection of neonates born to mothers who are positive for the virus is lower when maternal bNAbs are present [54]. Several studies performed in nonhuman primates exposed to HIV-1 or simian-human immunodeficiency virus, confirmed that the animals could be protected after challenge following different routes of exposure [55–58].

However, dosage of the bNAbs that is required to achieve protection was variable, and there appears to be a poor correlation between efficacy in cultured cells and in vivo [3]. It seems that the discrepancies may be accounted for by variability in the abilities of different antibodies to activate particular antiviral functions, such as binding of the Fc receptor [59]. Nevertheless, the data indicate that passive immunization with bNAbs or eliciting a bNAb response in patients may be protective against the virus. To overcome concerns about the practicality of repeated administration of bNAbs for passive immunization protocols, VIP has been developed. This entails administration of vectors that are capable of expressing HIV-1-neutralizing antibodies or antibody-like molecules over a prolonged period (Figure 11.3) [4,60].

11.2.1 Transfer of Genes Encoding Immunogenic HIV-1 Proteins

The versatility of DNA-based vaccines is particularly attractive for countering HIV-1 infection [15]. As with all such antiviral approaches, induction of a high level of transgene expression is vital to inducing an effective immune response. Achieving this goal was initially hampered by complicating effects of

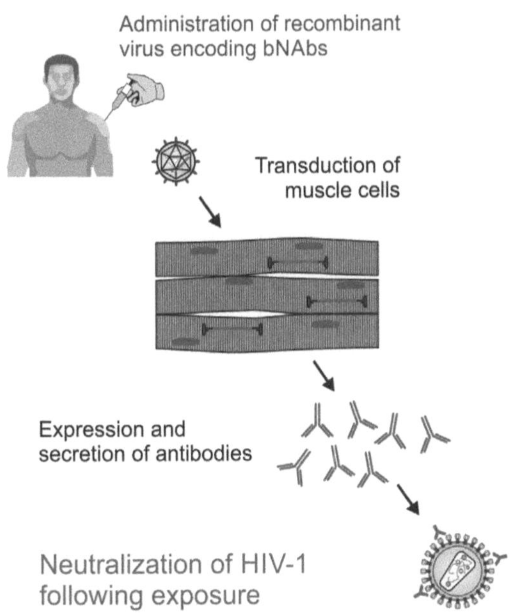

FIGURE 11.3 Vectored immunoprophylaxis to protect against HIV-1 infection.
A recombinant viral vector, such as one derived from an adeno-associated virus, is engineered to contain an expression cassette that encodes derivatives of an HIV-1-targeting broadly neutralizing antibody (bNAb). After intramuscular injection, the antibody or antibody-like protein is expressed, secreted, and is then available to neutralize exposure to the virus.

viral sequences on protein translation [15]. One such factor was the influence of interaction between Rev and the Rev response element (RRE) on regulating the export of mRNA from the nucleus to cytoplasm (Chapter 8). Instability sequences (INSs) [61] and *cis*-acting repressive signals (CRSs) [62] have also been shown to inhibit translation of HIV-1 proteins. Therefore, to produce good HIV-1 immunogenic protein expression, strategies have been devised to counter the effects of these elements [15]. Codon optimization has been used to improve translation and change INS, CRS, and RRE elements without altering the amino acid sequences of the immunogenic proteins [61,63]. Importantly, the effects of INS, CRS, and RRE sequences are dependent on their interaction with nuclear proteins. By delivering expression cassettes with poxvirus vectors, which replicate in the cytoplasm, the inhibitory effects are bypassed. Use of Ads for delivery of immunogenic HIV-1 proteins involves a nuclear stage of transgene expression; thus, elimination of INS and CRS sequences is important when using these vectors in vaccination regimens. To facilitate nuclear export of HIV-1 protein-encoding mRNAs, other export systems instead of Rev/RRE may be used (reviewed in ref. [64]).

Augmenting immunity to HIV-1 transgenes has involved use of the generic methods described above for the enhancement of transgenes' immunity. These include addition of sequences that improve secretion of the immunogenic proteins [16], administration of adjuvants [65], prime-boost, EP, and coupling of sequences that improve activation of antigen-presenting cells [17].

11.2.2 DNA-Based Vaccines in Nonhuman Primates and in Clinical Trial

Activation of cellular and humoral immune responses to HIV-1 has been demonstrated in several preclinical trials performed in nonhuman primates [15]. As with other applications of DNA-based vaccination, EP has been shown to improve efficacy. Administration of plasmid DNA encoding various antigens of simian immunodeficiency virus (SIV) caused a broad and sustained immune response, which was augmented with each repeat vaccination [66]. In other examples, the response to EP of DNA encoding gp120 was followed up by intramuscular administration of the protein [67,68]. The immune response lasted for more than a year and was accompanied by induction of a Th1 cytokine response. Antiviral responses were also detectable in mucosal secretions and in blood.

Many clinical trials have been undertaken to assess efficacy of transgenes as immunogens to protect against HIV-1 infection [15,20,69]. The Step trial (discussed briefly in section 11.1.3) was a double-blind study that involved administration of an Ad5 vaccine vector that expressed *gag/pol/nef* sequences [29]. Three thousand seronegative participants were enrolled and were assigned randomly to receive three doses of the Ad5 vaccine or a placebo. The study was unexpectedly terminated early because it "met prespecified futility boundaries at the first interim analysis." It was of some concern that mucosal T cell activation increased risk for HIV-1 infection among those individuals who had received the Ad vector. A subsequent study entailed detailed analysis of HIV-1 sequences derived from the patients who were infected while on the trial [70]. Evidence indicated that the vaccination elicited T cell responses that applied some pressure on the virus, and in particular on *gag* sequences encompassed by the vaccine. However, the effects were weak, not protective, and easily evaded by the virus. Another study used Ad5 as a vaccination boost after three priming doses of plasmid DNA [71]. The plasmids encoded *gag, pol, nef,* and *env* sequences, whereas the Ad5 vector transduced cells with *gag/pol* fusion and *env* sequences. The study was performed on 2504 transgender women and men who have sex with men. Although the vaccination regimen caused limited side effects, efficacy against HIV-1 was not demonstrated. The risk of acquiring the infection and the viral set points after infection were not altered by the vaccination protocol.

A community-based double-blind randomized trial performed in Thailand demonstrated some benefit of vaccination against HIV-1 [72]. The vaccination regimen of this RV144 trial comprised four primary administrations of a *gag/env/pol*-expressing canarypox viral vector, which was followed by two booster injections of recombinant gp120. The volunteers for the trial were mainly individuals who were at risk for heterosexual infection with the virus. Analysis of the subjects was performed for 3 years after completion of the vaccination program. Results showed that there was modest but significant protection of approximately 31%. However, vaccination did not affect the viral loads or $CD4^+$ counts in those participants who became infected. In addition, efficacy of the vaccine appeared to wane with time [15,72]. Detailed analysis showed that an antibody response to linear epitopes of the V1 and V2 regions of the viral envelope protein played an important role in providing protection against the infection [73–75]. Two vaccination signatures were identified in the V2 sequence, which were at positions 169 and 181 [74]. There was some correlation between the matches that occurred between the vaccine and viral sequences at these points and the efficacy of the prophylaxis. Another study aimed at defining the role of specific antibody types in response to the vaccine, perhaps surprisingly, showed that a high concentration of immunoglobulin (Ig)-A directed to gp120 attenuated vaccine efficacy [76]. Collectively, results from the RV144 study have been encouraging. Although the regimen does not provide complete protection against HIV-1 infection, a basis is provided for further improvement of immunostimulatory prophylactic and therapeutic strategies.

11.2.3 VIP for Treatment of HIV-1 Infection

Characterization of bNAbs has led to investigating therapeutic use of vector-mediated transfer of genes encoding derivatives of these antibodies. The rationale for this VIP approach is that sustained production of an exogenous

HIV-1-binding Ab may provide protection to individuals. Although an attractive idea, ensuring sustained expression and protection regardless of the mode of transmission is challenging. VIP has been used to prevent transmission of other pathogens, such as malaria [77], and the principle is summarized in Figure 11.3.

An early demonstration of the utility of VIP was provided in a study performed in mice that received adeno-associated viruses (AAVs) (Chapter 4) expressing sequences encoding b12, a first-generation bNAb [78]. After a single intramuscular injection of the recombinant AAV, Abs were detectable in serum samples that were collected for 6 months. Moreover, HIV-1-neutralizing activity was detectable in the serum samples over the same time period. In a subsequent investigation performed by the same group, Johnson et al. generated AAVs that expressed antibody-like molecules, immunoadhesins, which had specificity for epitopes of SIV [60]. The steady-state concentrations of immunoadhesins in blood were better than those achieved when using similar technology to produce single-chain variable fragments of antibodies (scFVs) or complete Abs. Essentially, the immunoadhesins comprised coupled variable domains from light and heavy chains that were fused to a sequence encoding Fc of IgG. The variable sequences were derived from previously characterized macaque Abs that had broad SIV-neutralizing properties. When tested in monkeys, six of nine animals that received the recombinant vectors were protected against intravenous administration of SIV, and none of the nine developed acquired immune deficiency syndrome. Conversely, the control animals that did not receive the VIP treatment were all infected, and most developed lethal complications of SIV transmission. Although encouraging, the efficacy of the immunoadhesins was later compromised by an immune response mounted by the macaques to the antiviral molecules.

Sequences encoding a panel of bNAbs were used by Balazs et al. in a study that aimed to assess efficacy of VIP in humanized mice [79]. After intramuscular administration of the AAVs to nonobese diabetic, severe combined immunodeficient and interleukin 2 receptor gamma chain deficient (NSG) mice (Chapter 8), they were adoptively transplanted with human peripheral blood mononuclear cells. Good expression of the exogenous antibodies was achieved. Of the different expressed bNAbs, b12 provided the best protection against HIV-1 infection after intravenous injection of the AAVs. Efficacy was associated with minimal depletion of transplanted human $CD4^+$ cells. A later study, also by Balazs and colleagues, extended evaluation of the utility of VIP for HIV-1 therapy [4]. Humanized mice were again treated with recombinant AAVs expressing bNAbs. In addition to b12, second-generation VCR01 and VCR07 bNAbs were tested. VCR01 and VCR07 have broader neutralizing properties and may be more effective against the few founder viruses that are responsible for establishing HIV-1 infection. As demonstrated before, protection against intravenous challenge was demonstrated after administration of the virus to grafted NSG mice. In an extension of the study, the bone marrow-liver-thymus (BLT) murine model was used to assess protection against mucosal transmission. The reason for choice of this model is that grafted human cells colonize mucosal surfaces more efficiently in BLT mice. Administration of AAVs that expressed VCR01 and VCR07 protected against transmucosal transmission after repeated intravaginal challenge of the mice with HIV-1. This also correlated with detectable gp-120 binding activity in the vaginal washings of experimental animals. The observation is particularly important because transmucosal infection with HIV-1 as a result of heterosexual contact is the most common mode of human HIV-1 infection. Recently an interesting study reported on the use of HIV-1-binding molecules comprising IgG coupled to CD4 and CCR5 derivatives [80]. These chimeric sequences were generally capable of binding a wider range of viral isolates than bNAbs. Incorporation into recombinant AAVs provided protection to rhesus macaques following challenge with SIV. Also the immune response to the immunoadhesins was minimal. These results from studies using VIP are encouraging. Durability of the prophylaxis; ability to prevent infection with different quasispecies; and the influence of immunoadhesins or antibodies on binding of exogenous HIV-1 to cells will be important next steps of the evaluations [3].

11.3 GENE-BASED IMMUNOPROPHYLAXIS AND IMMUNOTHERAPY FOR HBV INFECTION

Despite the availability of an effective protein-based vaccine against HBV, persistent infection with the virus remains an important global health problem (Chapter 6). Infection with HBV that occurs during early life is particularly problematic. The immune response to the exposure is attenuated and often leads to chronicity of the infection. This is unlike the case with adult infection with HBV, in which a vigorous innate, humoral, and cell-mediated attack on the virus is mounted. When exposed to HBV during early life, T cell responses to the virus may be restricted or undetectable. Exposure over prolonged periods to the viral antigens, particularly HBV surface and e antigens (HBsAg and HBeAg, respectively), leads to exhaustion of T cell effector functions with attenuated release of Th1 cytokines after stimulation of T cells (reviewed in ref. [81]). The mechanism is thought to involve the expression of inhibitory molecules such as programmed death 1 (PD-1), signaling lymphocyte activation molecule, and CTL-associated protein-4 on the surface of virus-specific T cells [82]. Therefore, some efforts have focused on reducing the inhibitory effects of these proteins to achieve a therapeutic effect against HBV [83–84]. In addition to employing this approach to augment T cell function, treatment with licensed antivirals to diminish "exhausting" effects of HBsAg and HBeAg has been used. Recent studies have demonstrated that adjuvant effects alone may partially reverse the inhibitory effects on exhausted T cells [85]. The prime-boost strategy of transgene-mediated induction of immunity also potentially provides a mechanism for attenuating PD-1-mediated T cell exhaustion [86]. An important aspect of devising

HBV-vaccination strategies is that the compact nature of the HBV genome limits sequence flexibility of the viral genes and vaccine escape is unusual (Chapter 6). These features, along with the versatility inherent in DNA manipulation and the ability to augment immune responses (see section 11.1.2), have provoked an interest in the use of DNA-based vaccines to counter HBV.

The utility of DNA-based vaccination for the prevention and treatment of HBV infection has been addressed in several studies [87]. Initial investigations were performed in the mid-1990s and entailed assessment of immunogenic properties of injected plasmids encoding viral surface antigens [88–92]. Evidence for induction of humoral and cell-mediated immunity was demonstrated. Subsequent verification was performed in chimpanzees [93,94], woodchucks [95], and ducks [96,97]. Overall, good humoral and cell-mediated immunity to viral antigens was reported in most studies, which was interpreted as indicating that the approach has clinical utility. Interestingly, maternal immunity to the duck hepatitis B virus (DHBV) induced in ducks could be transferred to hatchlings [98]. An antibody response to viral protein expression in transgenic mice could also be produced after DNA vaccination [99] or adoptive transfer of $CD4^+$ and $CD8^+$ cells from vaccinated mice to transgenic animals [100]. Although these early studies were encouraging, their translation to clinical utility has been disappointing (discussed below, this section).

To augment the efficacy of anti-HBV DNA-based vaccines prime-boost, EP and co-expression of cytokines have been used. When tested in a chimpanzee, DNA vaccination followed by administration of a canarypox viral vector achieved a durable therapeutic effect [101]. Circulating HBV DNA diminished by more than 400-fold over a period of 186 weeks, but HBsAg decrease was only transient. Administration of plasmids encoding woodchuck hepatitis virus (WHV) *core* and *interferon (INF)-γ* sequences increased antiviral immunity, although a sterilizing effect was not attained [102,103]. Likewise, efficacy of bicistronic vectors was demonstrated in ducks infected with DHBV [104,105]. As with other applications of DNA vaccines, EP enhanced the antiviral immune response and this was demonstrated in woodchucks [106], ducks [107], and mice [108].

Recently, a series of studies using improved immunization protocols was performed on woodchucks [109,110]. In the first of two reports by Kosinska and colleagues, a prime-boost strategy was used to augment the virus-specific T cell response [109]. The immunization regimen entailed sequential administration of optimized plasmids and Ads encoding the core protein of WHV. Preliminary treatment of mice demonstrated that a vigorous $CD8^+$ T cell response was induced. Naïve woodchucks, after they had received the immunogenic plasmids and Ads, were protected against challenge with WHV. A follow-up study by the same group aimed to assess the therapeutic utility of the prime-boost treatment [110]. Administration of the *core*-expressing plasmids and Ads together with a licensed antiviral (entecavir) was very effective. Strong $CD4^+$

and CD8+ T cell responses to WHV surface and core antigens were observed. The antiviral effect was durable, and half of the woodchucks tested remained negative for markers of viral replication after withdrawal of entecavir. A study performed by a separate group investigated the antiviral effects of a combination of DNA vaccination, PD-1 blockade, and entecavir treatment [111]. To inhibit PD-1 interaction with its ligand, antibodies were generated against PD-1 of woodchucks, and these were administered to the animals. The function of T cells targeting the virus was markedly enhanced as a result of the treatment. This activity was associated with suppression of viral replication and complete clearance of the virus occurred in some of the woodchucks.

The recently described therapeutic utility of using transgenes to induce an immune response to HBV in humans has been disappointing [112,113]. This accords with earlier observations that although a CTL activation could be elicited with DNA vaccination of humans, the magnitude of the response was considered to be inadequate for clearance of the virus [114]. In the phase I/II clinical study reported by Godon et al. [113], HBeAg-negative chronic carriers of the virus were treated with a combination of DNA vaccine and standard licensed HBV therapy. Although vaccination resulted in activation of polyfunctional CD4+ cells, there was no significant efficacy against the virus after withdrawal of treatment. A follow-up study performed by the same group confirmed these findings [112]. Five intramuscular injections of a DNA vaccine were given over a period of 44 weeks. The patients had been receiving treatment for the infection with nucleoside/nucleotide analogues. Compared with the control group, vaccination did not restore the immune response to HBV. Moreover, although well tolerated, the risk for relapse was not diminished by treatment with the DNA-based vaccination. On the basis of the information of preclinical studies, particularly the results from prime-boost vaccination of woodchucks in combination with entecavir [109–111], the protocols used in humans may not have been ideal. Future adaptation of immunization regimens that have been proven to be effective in woodchucks may well provide better protection and possible translation to therapy for chronic HBV infection.

11.3.1 Engineering Anti-HBV T Cell Receptors and Chimeric Antigen Receptors

In addition to using transgenes to vaccinate against HBV, the engineering of T cell receptors (TCRs) or chimeric antigen receptors (CARs) has interesting therapeutic potential [82,115,116]. Essentially, the procedures entail transduction of T cells with sequences encoding TCRs or CARs that target specific epitopes of HBV. CARs typically comprise scFVs that are fused to signaling domains, which are capable of activating cytotoxicity, T cell proliferation, and survival (Figure 11.4). Antiviral immunity is augmented in patients after re-infusion of the modified cells (Figure 11.5). The modified T cells are activated by interaction with a defined epitope and an advantage over engineered TCRs is that they

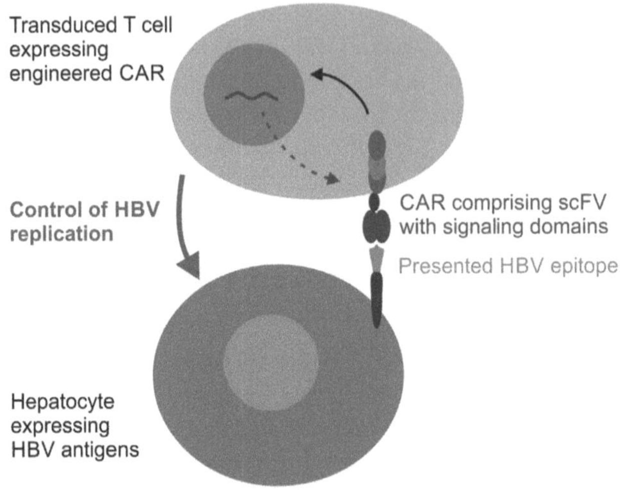

FIGURE 11.4 Engineering T cells with modified chimeric antigen receptors (CARs) to target HBV.
T cells are transduced with sequences encoding the single-chain variable fragments (scFVs) and intracellular signaling domains that constitute complete CARs. Extracellular antigen recognition by the CAR, performed by the scFV component, follows interaction with epitopes of HBV proteins presented on infected hepatocytes. Stimulation of pathways by the signaling domains of the CARs results in T cell activation and control of HBV replication in the infected liver cells.

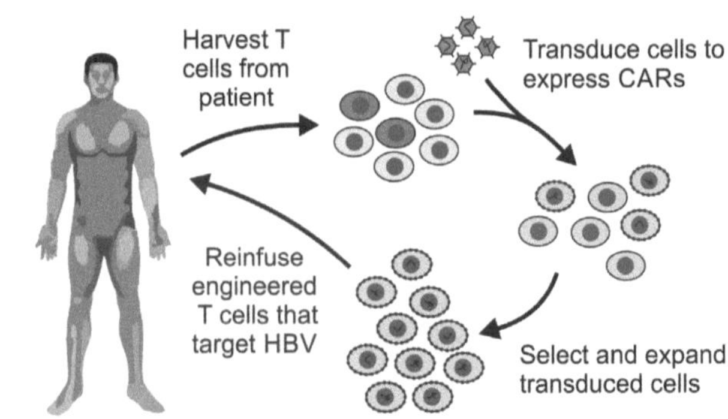

FIGURE 11.5 Procedure for ex vivo engineering of T cells to incorporate sequences expressing HBV-targeting chimeric antigen receptors (CARs).
T cells are collected from the patient then transduced with sequences encoding HBV-targeting CARs. The CAR-encoding sequences may be present within recombinant lentiviral vectors or mRNA for stable or transient expression, respectively. Transduced cells are selected and expanded before re-infusion into the patient. Colonization of the liver by the engineered T cells should result in augmented immunity to HBV.

are effective without dependence on associated specific human leukocyte antigen (HLA) molecules. Initial feasibility of the approach was demonstrated by Gehring and colleagues [115]. T cells from patients who were chronic carriers of HBV or who had hepatocellular carcinoma were transduced with sequences encoding an HBV-targeting HLA-A2-restricted TCR. Good expression of the TCRs and lysis of a cell line expressing HBV genes was demonstrated. In a subsequent study, Krebs et al. demonstrated the effectiveness of CARs that were engineered to target HBV surface proteins in mice [116]. Efficient engraftment by the T cells and colonization of the livers of HBV transgenic mice were observed. Replication of HBV was also controlled, and evidence of toxicity was minimal. Importantly, function of the modified T cells was not compromised by circulating viral antigens and the approach is not dependent on HLA type.

A concern of engineering TCRs and CARs to target HBV epitopes lies with the possibility that the T cells may cause immune-mediated hepatotoxicity. To address this, a recently described approach entailed use of EP with TCR-expressing mRNA [117]. The exogenous TCRs were only transiently expressed and obviated potential complications of toxicity. Highly functional HBV-targeting T cells could be generated, and they were also of a grade that would be suitable for clinical use. The methodology may be clinically applicable to HBV infection and has been used successfully to target liver cancer cells expressing HBsAg.

11.4 PREVENTION AND TREATMENT OF HEPATITIS C VIRUS INFECTION USING IMMUNOSTIMULATORY GENE TRANSFER

Prevention of transmission of hepatitis C virus (HCV) infection through the use of gene-based vaccines is an active research field. Infection with the virus is common and causes serious complications of cirrhosis and liver cancer (Chapter 7). Although strides have been made in improving treatment of HCV infection, the availability of an effective vaccine would contribute significantly to alleviating problems caused by the virus. As with HIV-1, HCV is prone to mutations and a concern is that vaccination should provide an adequately broad immunity to prevent infection. Feasibility of vaccination to prevent HCV infection has been endorsed by recent demonstration that an early humoral immune response to HCV is associated with protection against the virus [118]. Further support comes from demonstrations that recombinant envelope glycoproteins may cause neutralizing immune responses [119,120]. In addition to an antibody-mediated response to the virus, induction of a CTL response to HCV infection is also important. Therefore, gene transfer is well suited to developing vaccination against HCV. DNA and recombinant viral vectors have been used to develop HCV vaccines, and the progress to date is promising. Initial testing in small animals has been followed by testing in nonhuman primates and then in clinical trials.

An important early study aimed at developing HCV vaccination used naked DNA and Ads derived from the rare human serotype 6 (Ad6) to transduce cells of chimpanzees [121]. Viral sequences encoding nonstructural (NS) proteins of HCV—NS3, NS4A, NS4B, NS5A, and NS5B—were included in the vectors

(Figure 7.3 of Chapter 7). Robust Ab-induced immunity was elicited, and $CD4^+$ responses were augmented after EP with plasmid DNA that served as a booster. The vaccines were tolerated well, and the induced responses were multispecific and effective against challenge with heterologous strains of HCV. Cell-mediated immunity against at least two different epitopes of the NS proteins of HCV was induced. $CD8^+$ T cells within the liver were activated, which is particularly useful for control of the hepatotropic HCV. However, one of the five animals mounted a weak $CD8^+$ T cell immunity, and challenge resulted in emergence of a viral escape variant. Interestingly, this chimpanzee also did not develop antibodies to the Ad6 vector. Thus, protection was incomplete in all chimpanzees, but the vaccines provided good protection in most animals.

To overcome problems of preexisting immunity to Ads, vectors derived from chimpanzees were used to generate vaccines that avoided immunity to the common Ad5 strain [122]. One of the chimpanzee adenoviral strains, ChAd3, and Ad6 were used to generate vectors that transduce cells with cassettes expressing the NS3, NS4A, NS4B, NS5A, and NS5B proteins of HCV genotype 1B [44]. In a phase I clinical trial performed on healthy volunteers, vigorous polyfunctional CTL responses to multiple epitopes of the NS proteins were induced. The immunostimulatory effect was boosted by administration of a heterologous vector, which resulted in durable activation of T cells of both $CD4^+$ and $CD8^+$ subsets. The data indicated that the immune response elicited by the vaccination was of sufficient scale to confer protection to the virus.

In an investigation aimed at broadening the immunity to HCV, recombinant Ad6-derived vectors were engineered to express complete E1/E2 or truncated E2 viral glycoproteins [123]. These vectors were administered to rodents in a prime-boost regimen that entailed intramuscular injection of recombinant E1/E2 proteins after animals had received the Ads. Thus, the protocol was developed to maximize antibody-based and cell-mediated immunity to the virus. The prime-boost protocol elicited a potent CTL response, which was more effective than when the Ads were administered alone. In addition, a Th1-mediated IgG response was elicited, and sera from immunized animals were capable of neutralizing infection by pseudoviral particles of cells in culture.

One clinical trial that uses EP to administer HCV sequences is in progress, but results from the study have not yet been posted (http://clinicaltrials.gov/ct2/show/NCT00563173). Safety and potential therapeutic efficacy of DNA-based vaccination against HCV has been reported in two other studies performed on patients who had not responded to treatment with IFN and ribavirin [124,125]. The vaccine, termed CIGB-230, comprised DNA encoding viral structural proteins together with recombinant core proteins, and it was administered 5 times in four-weekly intervals. After receiving the vaccinations, the 15 patients showed only minor adverse effects [125]. Detailed analysis demonstrated a good neutralizing antibody response with all subjects producing IgG against the viral core protein. Improved HCV-specific activation of $CD4^+$ T cells and IFN-γ secretion was shown in most patients who had received the vaccination regimen. Moreover, the liver histology was stabilized or improved in 40% of patients after the vaccination [124].

Collectively, the results from studies performed on animals and humans augur well for the prospects of gene-based immunoprophylaxis and perhaps therapy for HCV infection. However, the ultimate success of vaccination against HCV will also be dependent on ensuring that a strong and broad immune response to the virus is generated. Several different parts of the viral genome (e.g., *NS* sequences, *E1/E2*, and *core*) have been shown to elicit an immune response that may be protective against HCV. A combination of vaccines that target different epitopes of HCV will be valuable to counter the diversity of viral strains responsible for HCV infection.

11.5 GENE TRANSFER TO PROTECT AGAINST HUMAN PAPILLOMAVIRUS INFECTION

Human papillomaviruses (HPV) are the etiological agents that are responsible for cervical carcinoma in women. DNA from the virus has been detected in almost all tumor samples (reviewed in ref. [126]). The numbered species of HPV belong to the *Alphapapillomavirus* genus and *Papillomaviridae* family. The virus is transmitted sexually, and risk for development of cervical cancer is associated with sexual activity at a young age and multiple sexual partners. Widespread screening to detect early premalignant epithelial changes, alterations to sexual behavior, and availability of effective prophylactic vaccines is contributing to a decrease in incidence of cervical carcinoma in developed countries. However, cervical carcinoma remains an important problem in developing countries where the malignancy is the major gynecological cancer.

Papillomaviruses have been isolated from many animals, which include species of birds [127] and reptiles [128]. The viruses infect cells of the skin and mucous membranes. Infection with HPV is typically subclinical and is cleared within 2 years in most cases [126]. When the virus persists, risk for development of precancerous lesions is high. HPV is nonenveloped and has a circular double-stranded DNA genome that comprises approximately 7000–8000 bp (reviewed in ref. [129]). ORFs are categorized as early (E) or late (L) according to the stage of the viral replication cycle when the genes are expressed. The early genes are *E1*, *E2*, *E4*, *E5*, *E6*, and *E7*. There are two late genes: *L1* and *L2*. Early proteins are responsible for controlling replication of the virus whereas L1 and L2 constitute the major and minor structural capsid proteins, respectively. Viral DNA is frequently integrated into the host genome. E6 and E7 are implicated in the transformation process through the disruptive effects of these proteins on tumor suppressor genes: E6 binds to p53 whereas E7 interacts with Rb. Novel approaches to countering the infection remain a priority. Gene-based strategies for immunostimulation and disabling the viral sequences have potential utility. As a DNA virus, HPV is susceptible to inactivation by mutagenic sequence-specific gene editing (Chapter 3). This approach has recently been shown to be feasible [130]. Hu et al. reported on successful targeting of *E6* and *E7* sequences of the virus to restore tumor suppressor function of p53 and Rb in host cells. Using a murine model, it was shown that a transfection reagent could be used to deliver the transcription activator-like effector nuclease-expressing DNA to target cells after topical administration of the formulation.

11.5.1 Vaccination against HPV

There are more than 160 numbered species of HPV, which are distinguished by sequence differences of at least 10% within *L1* [131]. Infection with HPV types 16 and 18 poses a particularly high risk for cervical carcinoma and according to the World Health Organization account for approximately 70% of cases of the malignancy (http://www.who.int/biologicals/areas/human_papillomavirus/en/). Because HPV is a nonenveloped virus, prophylactic vaccination strategies have been aimed at inducing an immune response to the capsid proteins. Currently available vaccines, such as Gardasil and Cervarix, comprise virus-like particles that are formed by expression and self-assembly of L1 derived from HPV types 16 and 18 [132,133]. Administration to young girls provides good prophylaxis and HPV-targeting vaccines are now increasingly being included in immunization programs. Although the vaccines are good, activation of cell-mediated immunity could improve elimination of HPV from infected epithelial cells. Therefore, use of gene-based immunotherapy may have added benefits of inducing a broader antiviral immune response for prophylaxis and treatment of the infection. Properties of several of the DNA-based vaccines that have been tested in clinical studies are summarized in Table 11.1.

Garcia and colleagues were the first to evaluate clinical utility of a plasmid encoding viral proteins to counter HPV [134]. The vaccine, termed ZYC101a, was designed to express E6 and E7 proteins of HPV-16 and HPV-18. After administration to women with confirmed cervical intraepithelial neoplasia, the most impressive effect was demonstrated in women younger than 25 years of age. In this group, the vaccination regimen promoted resolution of the premalignant lesions. Subsequently, Kim et al. also developed a plasmid-based vaccination strategy, which was tested in mice [135]. The engineered transgenes described in this study encoded E6 and E7 together with L2. An effective immune response was demonstrated in mice. Activation of a CTL effect against cells expressing the viral antigens and neutralization of infection of cultured cells by pseudoviral

Table 11.1 Gene-Based Vaccines That Have Been Used against Human Papillomavirus in Clinical Trials

Vaccine name and references	Composition and mode of administration	Encoded proteins	Results of efficacy assessments
ZYC101a [134]	Plasmid administered IM	E6 and E7 proteins of HPV-16 and HPV-18	Resolution of premalignant lesions in women younger than 25 years of age.
pNGVL4a-hCRTE6E7L2 [136]	IM administration of plasmid with electroporation	E6 and E7 with L2 fused to calreticulin	$CD8^+$ activation against cells expressing HPV proteins despite depletion of $CD4^+$ cells.
TA-HPV [137]	Recombinant vaccinia virus	E6 and E7 proteins of HPV-16 and HPV-18	Immune response elicited to the virus with evidence of decreased viral replication. Some patients showed a decrease in size of HPV-related lesions.
Lm-LLO-E7 [138]	DNA delivered with attenuated *Listeria monocytogenes*	Sequences encoding HPV-16-derived E7	Good safety, but efficacy was not conclusively established.

Abbreviations: IM, intramuscularly; HPV, human papillomavirus.

particles was demonstrated. Follow-up studies using the same expression cassettes have shown enhanced efficacy when EP was used to facilitate delivery of the plasmid DNA [136]. Interestingly, the procedure resulted in powerful stimulation of $CD8^+$ cells when $CD4^+$ cells were depleted. This property may be of use in the vaccination of HIV-1-infected immunocompromised patients.

Vectors have also been used to deliver HPV-targeting immunogens. In one of the first such studies, a recombinant VV was engineered to encode E6 and E7 of HPV-16 and HPV-18 [137]. Eighteen women with vulval intraepithelial neoplasia were enrolled in the study. Induction of an immune response to the viral antigens, decreases in viral loads, and improvements in symptoms were observed in many but not all of the patients. Utility of attenuated *Listeria monocytogenes* as a vector to express HPV-16-derived E7 protein has also been investigated in a clinical setting [138]. Although the vaccine was tolerated well, efficacy was not demonstrated and the strategy does not appear to have been developed further.

A more recent clinical trial to develop prophylactic and therapeutic vaccination entailed intramuscular administration of plasmid DNA, with EP, to 18 patients with premalignant cervical lesions [139]. The plasmid DNA encoded E6 and E7 and was given in escalating doses. The vaccination protocol did not cause significant adverse effects, and vigorous, dose-dependent, HPV-directed $CD8^+$ T cell responses were observed. The results are encouraging, but the therapeutic utility has not yet been established.

11.6 NUCLEIC ACID-BASED IMMUNOPROTECTION AGAINST INFECTION WITH HERPES SIMPLEX VIRUSES

Herpes simplex viruses (HSVs) belong to the *Herpesviridae* family, *Alphaherpesvirinae* subfamily, and *Simplexvirus* genus. Their genomes comprise linear double-stranded DNA of approximately 150 kbp (reviewed in refs [140,141]). HSV-1 and HSV-2 are the two species of *Alphaherpesvirinae* that infect humans. Globally, infection with HSV-2, which occurs as a result of sexual transmission, is very common. The virus is estimated to be present in 500 million people, and approximately 23 million new infections occur each year [142]. Analysis of serum markers of the infection indicates that prevalence occurs twice as frequently in women [143,144] and is particularly common in sub-Saharan Africa, where markers of the infection may be found in more than 80% of some populations [145]. Symptoms of the infection are typically blistering of the skin or mucosal membranes, but an infectious individual may be asymptomatic. Several antivirals are available for treatment of HSV infections, but the curative efficacy is low. Although the drugs may inhibit viral replication, shedding of the virus is not eliminated and risk for transmission persists on treatment. In addition, there is no vaccine available to prevent HSV infection. Barrier methods to preventing infection, such as the use of condoms, are effective. Although HSV infection per se may not be serious, HSV-2 increases risk for transmission of HIV-1 [146,147]. Acquisition of HIV-1 is increased approximately threefold in both men and women and is even higher when infection with HSV-2 has recently occurred. Genital lesions caused by HSV-2 may increase susceptibility to and infection by HIV-1. Ulcers are enriched with HIV-1-infectable activated T cells, HIV-1 is present in the lesions, and disruption of mucosal integrity all contribute to increased transmission of HIV-1 caused by HSV-2 infection. Thus, developing effective methods of countering the HSV-2 infection is an important global priority. In a seminal study, silencing of HSV-2 sequences after murine intravaginal administration of synthetic RNAi activators protected against lethal infection with the virus (Chapter 8) [148]. Gene editing of the viral sequences is potentially useful as treatment against HSV infection [149]. Although feasible, the approach has not yet reached an advanced stage of development, as is the case for gene-based immunotherapy for HSV infections.

The genome homology of HSV-1 and HSV-2 is approximately 50%, and the two viruses share many clinical features. The unique long (UL) and unique short (US) sequences comprise approximately 108 and 13 kbp of DNA, respectively. Terminal and internal repeat sequences flank the UL and US regions of the genome. These sequences are referred to as the terminal repeat long, internal repeat long (IRL), internal repeat short (IRS), and terminal repeat short. In addition, the HSVs' genomes contain many short sequence repeats, which are particularly abundant in the IRL and IRS elements [150–153]. The entire HSV genome comprises approximately 150 kbp. Fifty-six and 12 of the 74 viral

ORFs are located in UL and US, respectively. The genes of HSV-1 and HSV-2 are numbered and named with a prefix to indicate their location in the genome (e.g., *UL1*, *US1*, etc.). Viral DNA is tightly wound in a spool-like arrangement within the icosahedral capsid of the viral particle. A tegument, comprising several viral proteins, surrounds the capsid, and the complete virion has an outer lipid envelope with embedded viral glycoproteins.

Entry of HSV-2 into cells occurs as a result of interaction of some of the surface glycoproteins with cellular receptors [154]. For example the glycoprotein D (gD) binds to cellular herpesvirus entry mediator. After entry into cells, the viral DNA is delivered to the nucleus by the capsid. At initiation of replication of the viral genome, the linear sequence is circularized by the host DNA ligase IV/XRCC4 [155]. After circularization, replication initially proceeds by a theta replication mechanism, which involves duplication of the circular DNA structures [141]. Thereafter, rolling circle replication occurs, which leads to the formation of linear concatemers of the viral DNA. Genes of the HSVs are categorized as being immediate early (IE), early (E), or late (L). Proteins encoded by *IE* genes are regulatory and those of *E* sequences are enzymatic; together, they perform functions required for replication of the viral genome. The *L* genes, which are expressed after the *IE* and *E* genes, encode structural proteins that are responsible for the formation of the viral particles.

An important and interesting feature of HSV infection is that initial lytic replication may enter a dormant latent phase, which in turn may again be activated to become lytic (Figure 11.6). HSV-1 and HSV-2 typically infect cells of the skin and mucosal membranes. During transition to the latent phase, the virus enters innervating sensory neuronal cells and travels to the dorsal root ganglion where the DNA remains dormant. Reactivation may occur later, and the virus spreads along neurons to establish infection on the skin or mucosal membranes. Molecular triggers for emergence from the latent state are not well understood, but they are associated with factors such as an immunocompromised state and exposure to ultraviolet light [156].

11.6.1 Prophylactic and Therapeutic Vaccination against HSV

Naturally, the immune response to HSV-2 infection is inadequate, and the virus persists for life after latency is established. Understanding and identifying the factors that induce immunity to the infection have been difficult, but insights to enable development of appropriate vaccination strategies are slowly emerging. The presence of memory T cells in the genital mucosa plays an important role in preventing viral transmission [157,158]. Evidence also indicates that $CD4^+$ T cells have a significant role. Rapid immunostimulation at the site of infection, especially leading to the release of IFN-γ, is particularly important. An antibody response caused by B cells is secondary, but it initiates and augments the combined immune response to the virus [157,159]. Antiviral antibodies may prevent

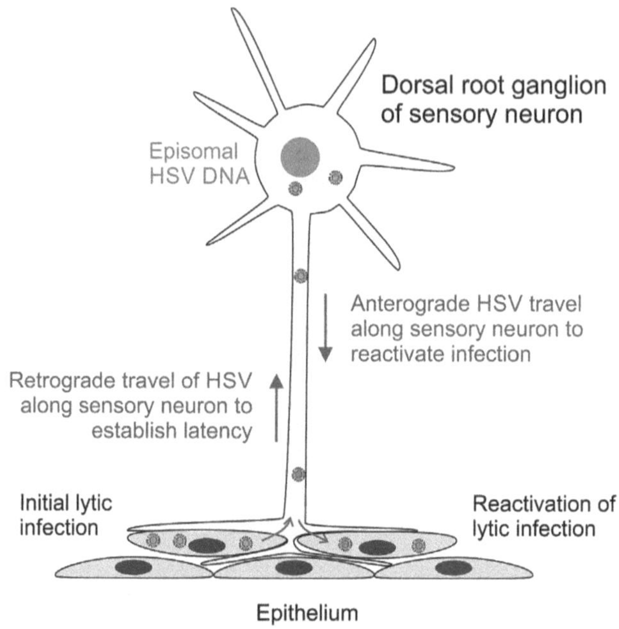

FIGURE 11.6 Lytic and latent states of HSV infection.
Initial infection at the epithelial surface results in a lytic infection with active viral replication. Latency is established after retrograde tracking of the virus along sensory neurons to the dorsal root ganglion in the spinal cord. Dormant HSV DNA may exist episomally in the latent state of viral replication. Reactivation of lytic infection may occur after anterograde passage of the virus along the sensory neuron and then proliferation in the innervated epithelial cells.

infection by HSV-2, but they are only protective when present in sufficient quantities at the exposed mucosal surfaces. Therefore, a challenge of developing effective vaccination regimens to prevent HSV-2 infection has been to ensure that a strong immune response at the site of exposure is elicited. To date, several different vaccination strategies have been used (reviewed in refs [158,160,161]). Vaccination has included use of attenuated viruses and recombinant viral envelope proteins, such as gD and glycoprotein B (gB). In an attempt to augment mucosal immunity, an interesting approach has harnessed the so-called "prime-and-pull" method [162]. Essentially, the vaccination regimen entails initial systemic administration of an HSV-2 immunogen. Topical administration (e.g., intravaginally) of chemokines results in redirecting of the immune response to the likely site of HSV-2 exposure.

Different approaches to vaccination using expression cassettes have been investigated as a mode of achieving prophylaxis against HSV infection. Chiuppesi and colleagues investigated the utility of lentiviral vectors, derived from the feline immunodeficiency virus, encoding the gB protein of HSV-1 [163]. After three intradermal injections into the foot pad and tail base at 0, 7, and 21 days, the mice were subjected to intravaginal challenge with HSV-1 or HSV-2. Analysis showed that good cross-neutralizing cell-mediated and humoral immunity was elicited. Severe disease caused by HSV-1 was eliminated in all vaccinated mice. Although efficacy of prophylaxis against HSV-2 infection was lower, severe disease was prevented in animals exposed to this species of virus. Capacity of the lentiviral vectors for inclusion of additional immunogenic HSV sequences may be useful to enhance prophylaxis.

Plasmid DNA containing codon-optimized cassettes encoding gD and tegument proteins of HSV-2 was used in another study aimed at HSV vaccine development [164]. The DNA was formulated with a lipid-based adjuvant and used for intramuscular administration to guinea pigs. The animals received three doses of the vaccine, which were given in twice-weekly intervals. Prophylactic and therapeutic efficacy of the vaccination regimen were evaluated in the animals. After vaginal challenge, detectable viral loads were significantly diminished in the immunized animals. The load of HSV-2 DNA was also diminished in dorsal root ganglia, indicating that viral latency was also reduced. Therapeutic efficacy was evident from lessening of viral load shedding and recurrence of genital lesions.

Albeit slow, progress is being made in developing effective prophylactic and therapeutic vaccination against HSV infections [158,160,161]. Results from studies performed using immunogenic gene transfer in mice and guinea pigs are encouraging. However, translating these findings to clinical use will be challenging. This is supported by the observation that administration of plasmid DNA encoding gD2 to human subjects elicited modest immune responses to the virus [165]. The DNA was administered in four doses over a period of 24 weeks using a needle-free device that was applied to the deltoid muscles of the participants. Cell-mediated immune response was observed in some of the patients who received the highest dose of the vaccine, but an antibody response to the viral antigen was not induced at all.

11.7 CONCLUSIONS

Use of gene transfer to enhance immunity against viral infections of public health importance has made impressive progress during the past 20 years. Importantly, the tested regimens have largely been safe and well tolerated. However, the goal of achieving sufficiently potent immunostimulatory effects to neutralize chronic viral infections has largely not yet been met. Use of gene therapy to produce antiviral immune responses to viral infections is not limited to activation of the endogenous humoral and cell-mediated immune responses. VIP and engineering of transgene cassettes that express TCRs and CARs have also shown promise.

Convenience of engineering DNA sequences to encode multiple immunogens, together with adjuvants and activation of cell-mediated immunity after transgene expression have been particularly important advantages of this vaccination approach. After initial studies, it soon became apparent that increased expression of immunogenic transgenes is crucially important for optimal activation of a prophylactic or therapeutic immune response. Consequently, a major focus has been on developing methods that may be used to augment immunogenic transgene expression. Insights gained from this research have enabled the technology and enhanced antiviral efficacy. Expression of transgenes has been improved through development of strategies that use methods of EP, engineering viral vectors, and co-administration or co-expression of adjuvant molecules. Augmenting immune responses by use of prime-boosting has been important and is now widely used in vaccination regimens. Among other techniques, codon optimization has been useful to ensure that reading frames are efficiently translated by the most abundant tRNAs within human cells. Such alteration of transgenes' sequences without changing the amino acid order is also useful to remove repressive or destabilizing elements from mRNA, and it has been exploited to optimize immunogenic cassettes encoding HIV-1 immunogens.

Gene transfer as an immune-based antiviral strategy is likely to benefit from improved insights that are gained from broader advances in gene therapy and basic molecular biology, virology, and immunology. For example, one of the factors that has hampered vaccination against HIV-1 has been the incomplete knowledge about the structure of the trimeric gp120 protein on the surface of the viral particle. Recent solving of the structure of this important viral envelope component [166] is likely to facilitate design of optimized immunogenic expression cassettes. Better understanding of the mechanisms by which viruses evade host immune responses and characterization of epitopes that are vulnerable for the viruses will aid with the design of suitable immunogenic transgenes. Collectively, use of gene-based strategies for protection against viral infections is at an interesting and promising stage. Coupling advances in immunotherapy to new direct antiviral strategies, which include other gene therapy-based antivirals, are likely to contribute to improved efficacy. Knowledge gained from these developments may well have generic application and are likely to contribute to treatment of other viral infections such as those caused by hemorrhagic fever viruses.

REFERENCES

[1] Dinc G, Ulman YI. The introduction of variolation "A La Turca" to the West by Lady Mary Montagu and Turkey's contribution to this. Vaccine 2007;25(21):4261–5.

[2] Beasley RP, Hwang LY, Stevens CE, Lin C-C, Hsieh F-J, Wang K-Y, et al. Efficacy of hepatitis B immune globulin for prevention of perinatal transmission of the hepatitis B virus carrier state: final report of a randomized double-blind, placebo-controlled trial. Hepatology 1983;3(2):135–41.

[3] Corti D, Lanzavecchia A. Broadly neutralizing antiviral antibodies. Annu Rev Immunol 2013;31:705–42.

[4] Balazs AB, Ouyang Y, Hong CM, Chen J, Nguyen SM, Rao DS, et al. Vectored immunoprophylaxis protects humanized mice from mucosal HIV transmission. Nat Med 2014;20(3):296–300.

[5] Benvenisty N, Reshef L. Direct introduction of genes into rats and expression of the genes. Proc Natl Acad Sci USA 1986;83(24):9551–5.

[6] Will H, Cattaneo R, Koch HG, Darai G, Schaller H, Schellekens H, et al. Cloned HBV DNA causes hepatitis in chimpanzees. Nature 1982;299(5885):740–2.

[7] Wolff JA, Malone RW, Williams P, Chong W, Acsadi G, Jani A, et al. Direct gene transfer into mouse muscle in vivo. Science 1990;247(4949 Pt 1):1465–8.

[8] Tang DC, DeVit M, Johnston SA. Genetic immunization is a simple method for eliciting an immune response. Nature 1992;356(6365):152–4.

[9] Rocha CD, Caetano BC, Machado AV, Bruna-Romero O. Recombinant viruses as tools to induce protective cellular immunity against infectious diseases. Int Microbiol 2004;7(2):83–94.

[10] Abdulhaqq SA, Weiner DB. DNA vaccines: developing new strategies to enhance immune responses. Immunol Res 2008;42(1–3):219–32.

[11] Kutzler MA, Weiner DB. DNA vaccines: ready for prime time? Nat Rev Genet 2008;9(10):776–88.

[12] Chattergoon MA, Kim JJ, Yang JS, Robinson TM, Lee DJ, Dentchev T, et al. Targeted antigen delivery to antigen-presenting cells including dendritic cells by engineered Fas-mediated apoptosis. Nat Biotechnol 2000;18(9):974–9.

[13] Dupuis M, Denis-Mize K, Woo C, Goldbeck C, Selby MJ, Chen M, et al. Distribution of DNA vaccines determines their immunogenicity after intramuscular injection in mice. J Immunol 2000;165(5):2850–8.

[14] Williams JA. Improving DNA vaccine performance through vector design. Curr Gene Ther 2014;14(3):170–89.

[15] Felber BK, Valentin A, Rosati M, Bergamaschi C, Pavlakis GN. HIV DNA vaccine: stepwise improvements make a difference. Vaccines 2014;2(2):354–79.

[16] Wallace A, West K, Rothman AL, Ennis FA, Lu S, Wang S. Post-translational intracellular trafficking determines the type of immune response elicited by DNA vaccines expressing Gag antigen of human immunodeficiency virus type 1 (HIV-1). Hum Vaccin Immunother 2013;9(10):2095–102.

[17] Kulkarni V, Jalah R, Ganneru B, Bergamaschi C, Alicea C, von Gegerfelt A, et al. Comparison of immune responses generated by optimized DNA vaccination against SIV antigens in mice and macaques. Vaccine 2011;29(39):6742–54.

[18] Lemp NA, Hiraoka K, Kasahara N, Logg CR. Cryptic transcripts from a ubiquitous plasmid origin of replication confound tests for cis-regulatory function. Nucleic Acids Res 2012;40(15):7280–90.

[19] Luke J, Carnes AE, Hodgson CP, Williams JA. Improved antibiotic-free DNA vaccine vectors utilizing a novel RNA based plasmid selection system. Vaccine 2009;27(46):6454–9.

[20] Villarreal LP, DeFilippis VR. A hypothesis for DNA viruses as the origin of eukaryotic replication proteins. J Virol 2000;74(15):7079–84.

[21] Kim JJ, Trivedi NN, Nottingham LK, Morrison L, Tsai A, Hu Y, et al. Modulation of amplitude and direction of in vivo immune responses by co-administration of cytokine gene expression cassettes with DNA immunogens. Eur J Immunol 1998;28(3):1089–103.

[22] Halwani R, Boyer JD, Yassine-Diab B, Haddad EK, Robinson TM, Kumar S, et al. Therapeutic vaccination with simian immunodeficiency virus (SIV)-DNA + IL-12 or IL-15 induces distinct CD8 memory subsets in SIV-infected macaques. J Immunol 2008;180(12):7969–79.

[23] Hirao LA, Wu L, Khan AS, Hokey DA, Yan J, Dai A, et al. Combined effects of IL-12 and electroporation enhances the potency of DNA vaccination in macaques. Vaccine 2008;26(25):3112–20.

[24] Hirao LA, Wu L, Khan AS, Satishchandran A, Draghia-Akli R, Weiner DB. Intradermal/ subcutaneous immunization by electroporation improves plasmid vaccine delivery and potency in pigs and rhesus macaques. Vaccine 2008;26(3):440–8.

[25] Brave A, Gudmundsdotter L, Sandstrom E, Haller BK, Hallengard D, Maltais AK, et al. Biodistribution, persistence and lack of integration of a multigene HIV vaccine delivered by needle-free intradermal injection and electroporation. Vaccine 2010;28(51): 8203–9.

[26] Roos AK, Eriksson F, Timmons JA, Gerhardt J, Nyman U, Gudmundsdotter L, et al. Skin electroporation: effects on transgene expression, DNA persistence and local tissue environment. PloS One 2009;4(9):e7226.

[27] Liu J, Kjeken R, Mathiesen I, Barouch DH. Recruitment of antigen-presenting cells to the site of inoculation and augmentation of human immunodeficiency virus type 1 DNA vaccine immunogenicity by in vivo electroporation. J Virol 2008;82(11):5643–9.

[28] Nwanegbo E, Vardas E, Gao W, Whittle H, Sun H, Rowe D, et al. Prevalence of neutralizing antibodies to adenoviral serotypes 5 and 35 in the adult populations of the Gambia, South Africa, and the United States. Clin Diagn Lab Immunol 2004;11(2):351–7.

[29] Buchbinder SP, Mehrotra DV, Duerr A, Fitzgerald DW, Mogg R, Li D, et al. Efficacy assessment of a cell-mediated immunity HIV-1 vaccine (the step study): a double-blind, randomised, placebo-controlled, test-of-concept trial. Lancet 2008;372(9653):1881–93.

[30] Fenner F. Global eradication of smallpox. Clin Infect Dis 1982;4(5): 916–30.

[31] Cottingham MG, Carroll MW. Recombinant MVA vaccines: dispelling the myths. Vaccine 2013;31(39):4247–51.

[32] Gilbert SC. Clinical development of modified vaccinia virus Ankara vaccines. Vaccine 2013;31(39):4241–6.

[33] Volz A, Sutter G. Protective efficacy of modified vaccinia virus Ankara in preclinical studies. Vaccine 2013;31(39):4235–40.

[34] Mackett M, Smith GL, Moss B. General method for production and selection of infectious vaccinia virus recombinants expressing foreign genes. J Virol 1984;49(3):857–64.

[35] Weir JP, Bajszar G, Moss B. Mapping of the vaccinia virus thymidine kinase gene by marker rescue and by cell-free translation of selected mRNA. Proc Natl Acad Sci USA 1982;79(4):1210–4.

[36] Perkus ME, Limbach K, Paoletti E. Cloning and expression of foreign genes in vaccinia virus, using a host range selection system. J Virol 1989;63(9):3829–36.

[37] White SD, Conwell K, Langland JO, Jacobs BL. Use of a negative selectable marker for rapid selection of recombinant vaccinia virus. Biotechniques 2011;50(5):303–9.

[38] Vuola JM, Keating S, Webster DP, Berthoud T, Dunachie S, Gilbert SC, et al. Differential immunogenicity of various heterologous prime-boost vaccine regimens using DNA and viral vectors in healthy volunteers. J Immunol 2005;174(1):449–55.

[39] Whelan KT, Pathan AA, Sander CR, Fletcher HA, Poulton I, Alder NC, et al. Safety and immunogenicity of boosting BCG vaccinated subjects with BCG: comparison with boosting with a new TB vaccine, MVA85A. PloS One 2009;4(6):e5934.

[40] Li S, Rodrigues M, Rodriguez D, Rodriguez JR, Esteban M, Palese P, et al. Priming with recombinant influenza virus followed by administration of recombinant vaccinia virus induces CD8+ T-cell-mediated protective immunity against malaria. Proc Natl Acad Sci USA 1993;90(11):5214–8.

[41] O'Hara GA, Duncan CJ, Ewer KJ, Collins KA, Elias SC, Halstead FD, et al. Clinical assessment of a recombinant simian adenovirus ChAd63: a potent new vaccine vector. J Infect Dis 2012;205(5):772–81.

[42] Casimiro DR, Chen L, Fu TM, Evans RK, Caulfield MJ, Davies ME, et al. Comparative immunogenicity in rhesus monkeys of DNA plasmid, recombinant vaccinia virus, and replication-defective adenovirus vectors expressing a human immunodeficiency virus type 1 gag gene. J Virol 2003;77(11):6305–13.

[43] Casimiro DR, Tang A, Chen L, Fu TM, Evans RK, Davies ME, et al. Vaccine-induced immunity in baboons by using DNA and replication-incompetent adenovirus type 5 vectors expressing a human immunodeficiency virus type 1 gag gene. J Virol 2003;77(13):7663–8.

[44] Barnes E, Folgori A, Capone S, Swadling L, Aston S, Kurioka A, et al. Novel adenovirus-based vaccines induce broad and sustained T cell responses to HCV in man. Sci Transl Med 2012;4(115):115ra1.

[45] Berthoud TK, Hamill M, Lillie PJ, Milicic A, Poyntz HC, Lambe T, et al. Potent CD8+ T-cell immunogenicity in humans of a novel heterosubtypic influenza A vaccine, MVA-NP+M1. Clin Infect Dis 2011;52(1):1–7.

[46] Gonzalo RM, Rodriguez D, Garcia-Sastre A, Rodriguez JR, Palese P, Esteban M. Enhanced CD8+ T cell response to HIV-1 env by combined immunization with influenza and vaccinia virus recombinants. Vaccine 1999;17(7–8):887–92.

[47] Gray ES, Madiga MC, Hermanus T, Moore PL, Wibmer CK, Tumba NL, et al. The neutralization breadth of HIV-1 develops incrementally over four years and is associated with CD4+ T cell decline and high viral load during acute infection. J Virol 2011;85(10):4828–40.

[48] Choe H, Li W, Wright PL, Vasilieva N, Venturi M, Huang CC, et al. Tyrosine sulfation of human antibodies contributes to recognition of the CCR5 binding region of HIV-1 gp120. Cell 2003;114(2):161–70.

[49] Doria-Rose NA, Schramm CA, Gorman J, Moore PL, Bhiman JN, DeKosky BJ, et al. Developmental pathway for potent V1V2-directed HIV-neutralizing antibodies. Nature 2014;509(7498):55–62.

[50] Klein F, Mouquet H, Dosenovic P, Scheid JF, Scharf L, Nussenzweig MC. Antibodies in HIV-1 vaccine development and therapy. Science 2013;341(6151):1199–204.

[51] Binley JM, Wrin T, Korber B, Zwick MB, Wang M, Chappey C, et al. Comprehensive cross-clade neutralization analysis of a panel of anti-human immunodeficiency virus type 1 monoclonal antibodies. J Virol 2004;78(23):13232–52.

[52] Conley AJ, Gorny MK, Kessler 2nd JA, Boots LJ, Ossorio-Castro M, Koenig S, et al. Neutralization of primary human immunodeficiency virus type 1 isolates by the broadly reactive anti-V3 monoclonal antibody, 447-52D. J Virol 1994;68(11):6994–7000.

[53] Sabin C, Corti D, Buzon V, Seaman D, Lutje Hulsik D, Hinz D, et al. Crystal structure and size-dependent neutralization properties of HK20, a human monoclonal antibody binding to the highly conserved heptad repeat 1 of gp41. PLoS Pathog 2010;6(11):e1001195.

[54] Barin F, Jourdain G, Brunet S, Ngo-Giang-Huong N, Weerawatgoompa S, Karnchanamayul W, et al. Revisiting the role of neutralizing antibodies in mother-to-child transmission of HIV-1. J Infect Dis 2006;193(11):1504–11.

[55] Baba TW, Liska V, Hofmann-Lehmann R, Vlasak J, Xu W, Ayehunie S, et al. Human neutralizing monoclonal antibodies of the IgG1 subtype protect against mucosal simian–human immunodeficiency virus infection. Nat Med 2000;6(2):200–6.

[56] Ferrantelli F, Rasmussen RA, Buckley KA, Li P-L, Wang T, Montefior DC, et al. Complete protection of neonatal rhesus macaques against oral exposure to pathogenic simian-human immunodeficiency virus by human anti-HIV monoclonal antibodies. J Infect Dis 2004;189(12):2167–73.

[57] Mascola JR, Stiegler G, VanCott TC, Katinger H, Carpenter CB, Hanson CE, et al. Protection of macaques against vaginal transmission of a pathogenic HIV-1/SIV chimeric virus by passive infusion of neutralizing antibodies. Nat Med 2000;6(2):207–10.

[58] Eichberg JW, Murthy KK, Ward RH, Prince AM. Prevention of HIV infection by passive immunization with HIVIG or CD4-IgG. AIDS Res Hum Retroviruses 1992;8(8):1515.

[59] Moldt B, Shibata-Koyama M, Rakasz EG, Schultz N, Kanda Y, Dunlop DC, et al. A nonfucosylated variant of the anti-HIV-1 monoclonal antibody b12 has enhanced FcγRIIIa-mediated antiviral activity in vitro but does not improve protection against mucosal SHIV challenge in macaques. J Virol 2012;86(11):6189–96.

[60] Johnson PR, Schnepp BC, Zhang J, Connell MJ, Greene SM, Yuste E, et al. Vector-mediated gene transfer engenders long-lived neutralizing activity and protection against SIV infection in monkeys. Nat Med 2009;15(8):901–6.

[61] Schwartz S, Felber BK, Pavlakis GN. Distinct RNA sequences in the gag region of human immunodeficiency virus type 1 decrease RNA stability and inhibit expression in the absence of Rev protein. J Virol 1992;66(1):150–9.

[62] Cochrane AW, Jones KS, Beidas S, Dillon PJ, Skalka AM, Rosen CA. Identification and characterization of intragenic sequences which repress human immunodeficiency virus structural gene expression. J Virol 1991;65(10):5305–13.

[63] Schwartz S, Campbell M, Nasioulas G, Harrison J, Felber BK, Pavlakis GN. Mutational inactivation of an inhibitory sequence in human immunodeficiency virus type 1 results in Rev-independent gag expression. J Virol 1992;66(12):7176–82.

[64] Felber BK, Zolotukhin AS, Pavlakis GN. Posttranscriptional control of HIV-1 and other retroviruses and its practical applications. Adv Pharmacol 2007;55:161–97.

[65] Boyer JD, Robinson TM, Kutzler MA, Parkinson R, Calarota SA, Sidhu MK, et al. SIV DNA vaccine co-administered with IL-12 expression plasmid enhances CD8 SIV cellular immune responses in cynomolgus macaques. J Med Primatol 2005;34(5–6):262–70.

[66] Patel V, Valentin A, Kulkarni V, Rosati M, Bergamaschi C, Jalah R, et al. Long-lasting humoral and cellular immune responses and mucosal dissemination after intramuscular DNA immunization. Vaccine 2010;28(30):4827–36.

[67] Cristillo AD, Weiss D, Hudacik L, Restrepo S, Galmin L, Suschak J, et al. Persistent antibody and T cell responses induced by HIV-1 DNA vaccine delivered by electroporation. Biochem Biophys Res Commun 2008;366(1):29–35.

[68] Jalah R, Kulkarni V, Patel V, Rosati M, Alicea C, Bear J, et al. DNA and protein co-immunization improves the magnitude and longevity of humoral immune responses in macaques. PloS One 2014;9(3):e91550.

[69] Adis IL. HIV gp120 vaccine-VaxGen: AIDSVAX, AIDSVAX B/B, AIDSVAX B/E, HIV gp120 vaccine-Genentech, HIV gp120 vaccine AIDSVAX-VaxGen, HIV vaccine AIDSVAX-VaxGen. Drugs R&D 2003;4(4):249.

[70] Rolland M, Tovanabutra S, deCamp AC, Frahm N, Gilbert PB, Sanders-Buell E, et al. Genetic impact of vaccination on breakthrough HIV-1 sequences from the STEP trial. Nat Med 2011;17(3):366–71.

[71] Hammer SM, Sobieszczyk ME, Janes H, Karuna ST, Mulligan MJ, Grove D, et al. Efficacy trial of a DNA/rAd5 HIV-1 preventive vaccine. N Engl J Med 2013;369(22):2083–92.

[72] Rerks-Ngarm S, Pitisuttithum P, Nitayaphan S, Kaewkungwal J, Chiu J, Paris R, et al. Vaccination with ALVAC and AIDSVAX to prevent HIV-1 infection in Thailand. N Engl J Med 2009;361(23):2209–20.

[73] Karasavvas N, Billings E, Rao M, Williams C, Zolla-Pazner S, Bailer RT, et al. The Thai phase III HIV type 1 vaccine trial (RV144) regimen induces antibodies that target conserved regions within the V2 loop of gp120. AIDS Res Hum Retroviruses 2012;28(11):1444–57.

[74] Rolland M, Edlefsen PT, Larsen BB, Tovanabutra S, Sanders-Buell E, Hertz T, et al. Increased HIV-1 vaccine efficacy against viruses with genetic signatures in Env V2. Nature 2012;490(7420):417–20.

[75] Zolla-Pazner S, deCamp AC, Cardozo T, Karasavvas N, Gottardo R, Williams C, et al. Analysis of V2 antibody responses induced in vaccines in the ALVAC/AIDSVAX HIV-1 vaccine efficacy trial. PloS One 2013;8(1):e53629.

[76] Haynes BF, Gilbert PB, McElrath MJ, Zolla-Pazne S, Tomaras GD, Alam SM, et al. Immune-correlates analysis of an HIV-1 vaccine efficacy trial. N Engl J Med 2012;366(14):1275–86.

[77] Rodrigues MM, Soares IS. Gene-therapy for malaria prevention. Trends Parasitol 2014;30(11):511–3.

[78] Lewis AD, Chen R, Montefiori DC, Johnson PR, Clark KR. Generation of neutralizing activity against human immunodeficiency virus type 1 in serum by antibody gene transfer. J Virol 2002;76(17):8769–75.

[79] Balazs AB, Chen J, Hong CM, Rao DS, Yang L, Baltimore D. Antibody-based protection against HIV infection by vectored immunoprophylaxis. Nature 2012;481(7379):81–4.

[80] Gardner MR, Kattenhorn LM, Kondur HR, von Schaewen M, Dorfman T, Chiang JJ, et al. AAV-expressed eCD4-Ig provides durable protection from multiple SHIV challenges. Nature 2015;519(7541):87–91.

[81] Bertoletti A, Ferrari C. Innate and adaptive immune responses in chronic hepatitis B virus infections: towards restoration of immune control of viral infection. Gut 2012;61(12):1754–64.

[82] Maini MK, Schurich A. The molecular basis of the failed immune response in chronic HBV: therapeutic implications. J Hepatol 2010;52(4):616–9.

[83] Boni C, Laccabue D, Lampertico P, Giuberti T, Vigano M, Schivazappa S, et al. Restored function of HBV-specific T cells after long-term effective therapy with nucleos(t)ide analogues. Gastroenterology 2012;143(4):963–73:e9.

[84] Schurich A, Khanna P, Lopes AR, Han KJ, Peppa D, Micco L, et al. Role of the coinhibitory receptor cytotoxic T lymphocyte antigen-4 on apoptosis-prone CD8 T cells in persistent hepatitis B virus infection. Hepatology 2011;53(5):1494–503.

[85] Gehring AJ, Haniffa M, Kennedy PT, Ho ZZ, Boni C, Shin A, et al. Mobilizing monocytes to cross-present circulating viral antigen in chronic infection. J Clin Invest 2013;123(9):3766–76.

[86] Bot A, Qiu Z, Wong R, Obrocea M, Smith KA. Programmed cell death-1 (PD-1) at the heart of heterologous prime-boost vaccines and regulation of CD8+ T cell immunity. J Transl Med 2010;8:132.

[87] Cova L. Advances and challenges in the development of therapeutic DNA vaccines against hepatitis B virus infection. Curr Gene Ther 2014;14(3):149–60.

[88] Davis HL, Mancini M, Michel ML, Whalen RG. DNA-mediated immunization to hepatitis B surface antigen: longevity of primary response and effect of boost. Vaccine 1996;14(9): 910–5.

[89] Mancini M, Davis H, Tiollais P, Michel ML. DNA-based immunization against the envelope proteins of the hepatitis B virus. J Biotechnol 1996;44(1–3):47–57.

[90] Michel ML. DNA-mediated immunization: prospects for hepatitis B vaccination. Res Virol 1995;146(4):261–5.

[91] Michel ML, Davis HL, Schleef M, Mancini M, Tiollais P, Whalen RG. DNA-mediated immunization to the hepatitis B surface antigen in mice: aspects of the humoral response mimic hepatitis B viral infection in humans. Proc Natl Acad Sci USA 1995;92(12):5307–11.

[92] Khatri K, Goyal AK, Gupta PN, Mishra N, Vyas SP. Plasmid DNA loaded chitosan nanoparticles for nasal mucosal immunization against hepatitis B. Int J Pharm 2008;354(1–2):235–41.

[93] Davis HL, McCluskie MJ, Gerin JL, Purcell RH. DNA vaccine for hepatitis B: evidence for immunogenicity in chimpanzees and comparison with other vaccines. Proc Natl Acad Sci USA 1996;93(14):7213–8.

[94] Prince AM, Whalen R, Brotman B. Successful nucleic acid based immunization of newborn chimpanzees against hepatitis B virus. Vaccine 1997;15(8):916–9.

[95] Lu M, Hilken G, Kruppenbacher J, Kemper T, Schirmbeck R, Reimann J, et al. Immunization of woodchucks with plasmids expressing woodchuck hepatitis virus (WHV) core antigen and surface antigen suppresses WHV infection. J Virol 1999;73(1):281–9.

[96] Jilbert AR, Miller DS, Scougall CA, Turnbull H, Burrell CJ. Kinetics of duck hepatitis B virus infection following low dose virus inoculation: one virus DNA genome is infectious in neonatal ducks. Virology 1996;226(2):338–45.

[97] Triyatni M, Jilbert AR, Qiao M, Miller DS, Burrell CJ. Protective efficacy of DNA vaccines against duck hepatitis B virus infection. J Virol 1998;72(1):84–94.

[98] Rollier C, Charollois C, Jamard C, Trepo C, Cova L. Early life humoral response of ducks to DNA immunization against hepadnavirus large envelope protein. Vaccine 2000;18(27):3091–6.

[99] Mancini M, Hadchouel M, Davis HL, Whalen RG, Tiollais P, Michel ML. DNA-mediated immunization in a transgenic mouse model of the hepatitis B surface antigen chronic carrier state. Proc Natl Acad Sci USA 1996;93(22):12496–501.

[100] Mancini M, Hadchouel M, Tiollais P, Michel ML. Regulation of hepatitis B virus mRNA expression in a hepatitis B surface antigen transgenic mouse model by IFN-gamma-secreting T cells after DNA-based immunization. J Immunol 1998;161(10):5564–70.

[101] Pancholi P, Lee DH, Liu Q, Tackney C, Taylor P, Perkus M, et al. DNA prime/canarypox boost-based immunotherapy of chronic hepatitis B virus infection in a chimpanzee. Hepatology 2001;33(2):448–54.

[102] Siegel F, Lu M, Roggendorf M. Coadministration of gamma interferon with DNA vaccine expressing woodchuck hepatitis virus (WHV) core antigen enhances the specific immune response and protects against WHV infection. J Virol 2001;75(11):5036–42.

[103] Wang J, Gujar SA, Cova L, Michalak TI. Bicistronic woodchuck hepatitis virus core and gamma interferon DNA vaccine can protect from hepatitis but does not elicit sterilizing antiviral immunity. J Virol 2007;81(2):903–16.

[104] Saade F, Buronfosse T, Guerret S, Pradat P, Chevallier M, Zoulim F, et al. In vivo infectivity of liver extracts after resolution of hepadnaviral infection following therapy associating DNA vaccine and cytokine genes. J Viral Hepat 2013;20(4):e56–65.

[105] Saade F, Buronfosse T, Pradat P, Abdul F, Cova L. Enhancement of neutralizing humoral response of DNA vaccine against duck hepatitis B virus envelope protein by co-delivery of cytokine genes. Vaccine 2008;26(40):5159–64.

[106] Liu KH, Ascenzi MA, Bellezza CA, Bezuidenhout AJ, Cote PJ, Gonzalez-Aseguinolaza G, et al. Electroporation enhances immunogenicity of a DNA vaccine expressing woodchuck hepatitis virus surface antigen in woodchucks. J Virol 2011;85(10):4853–62.

[107] Khawaja G, Buronfosse T, Jamard C, Abdul F, Guerret S, Zoulim F, et al. In vivo electroporation improves therapeutic potency of a DNA vaccine targeting hepadnaviral proteins. Virology 2012;433(1):192–202.

[108] Luxembourg A, Hannaman D, Ellefsen B, Nakamura G, Bernard R. Enhancement of immune responses to an HBV DNA vaccine by electroporation. Vaccine 2006;24(21):4490–3.

[109] Kosinska AD, Johrden L, Zhang E, Fiedler M, Mayer A, Wildner O, et al. DNA prime-adenovirus boost immunization induces a vigorous and multifunctional T-cell response against hepadnaviral proteins in the mouse and woodchuck model. J Virol 2012;86(17):9297–310.

[110] Kosinska AD, Zhang E, Johrden L, Liu J, Seiz PL, Zhang X, et al. Combination of DNA prime–adenovirus boost immunization with entecavir elicits sustained control of chronic hepatitis B in the woodchuck model. PLoS Pathog 2013;9(6):e1003391.

[111] Liu J, Zhang E, Ma Z, Wu W, Kosinska A, Zhang X, et al. Enhancing virus-specific immunity in vivo by combining therapeutic vaccination and PD-L1 blockade in chronic hepadnaviral infection. PLoS Pathog 2014;10(1):e1003856.

[112] Fontaine H, Kahi S, Chazallon C, Bourgine M, Varaut A, Buffet C, et al. Anti-HBV DNA vaccination does not prevent relapse after discontinuation of analogues in the treatment of chronic hepatitis B: a randomised trial–ANRS HB02 VAC-ADN. Gut 2014;64(1):139–47.

[113] Godon O, Fontaine H, Kahi S, Meritet J, Scott-Algara D, Pol S, et al. Immunological and antiviral responses after therapeutic DNA immunization in chronic hepatitis B patients efficiently treated by analogues. Mol Ther 2014;22(3):675–84.

[114] Heathcote J, McHutchison J, Lee S, Tong M, Benner K, Minuk G, et al. A pilot study of the CY-1899 T-cell vaccine in subjects chronically infected with hepatitis B virus. The CY1899 T cell vaccine study group. Hepatology 1999;30(2):531–6.

[115] Gehring AJ, Xue SA, Ho ZZ, Teoh D, Ruedl C, Chia A, et al. Engineering virus-specific T cells that target HBV infected hepatocytes and hepatocellular carcinoma cell lines. J Hepatol 2011;55(1):103–10.

[116] Krebs K, Bottinger N, Huang LR, Chmielewski M, Arzberger S, Gasteiger G, et al. T cells expressing a chimeric antigen receptor that binds hepatitis B virus envelope proteins control virus replication in mice. Gastroenterology 2013;145(2):456–65.

[117] Koh S, Shimasaki N, Suwanarusk R, Ho ZZ, Chia A, Banu N, et al. A practical approach to immunotherapy of hepatocellular carcinoma using T cells redirected against hepatitis B virus. Mol Ther Nucleic Acids 2013;2:e114.

[118] Osburn WO, Snider AE, Wells BL, Latanich R, Bailey JR, Thomas DL, et al. Clearance of hepatitis C infection is associated with the early appearance of broad neutralizing antibody responses. Hepatology 2014;59(6):2140–51.

[119] Meunier JC, Gottwein JM, Houghton M, Russell RS, Emerson SU, Bukh J, et al. Vaccine-induced cross-genotype reactive neutralizing antibodies against hepatitis C virus. J Infect Dis 2011;204(8):1186–90.

[120] Stamataki Z, Coates S, Abrignani S, Houghton M, McKeating JA. Immunization of human volunteers with hepatitis C virus envelope glycoproteins elicits antibodies that cross-neutralize heterologous virus strains. J Infect Dis 2011;204(5):811–3.

[121] Folgori A, Capone S, Ruggeri L, Meola A, Sporeno E, Ercole BB, et al. A T-cell HCV vaccine eliciting effective immunity against heterologous virus challenge in chimpanzees. Nat Med 2006;12(2):190–7.

[122] Colloca S, Barnes E, Folgori A, Ammendola V, Capone S, Cirillo A, et al. Vaccine vectors derived from a large collection of simian adenoviruses induce potent cellular immunity across multiple species. Sci Transl Med 2012;4(115):115ra2.

[123] Chmielewska AM, Naddeo M, Capone S, Ammendola V, Hu K, Meredith L, et al. Combined adenovirus vector and hepatitis C virus envelope protein prime-boost regimen elicits T cell and neutralizing antibody immune responses. J Virol 2014;88(10):5502–10.

[124] Alvarez-Lajonchere L, Shoukry NH, Gra B, Amador-Canizares Y, Helle F, Bedard N, et al. Immunogenicity of CIGB-230, a therapeutic DNA vaccine preparation, in HCV-chronically infected individuals in a phase I clinical trial. J Viral Hepat 2009;16(3):156–67.

[125] Castellanos M, Cinza Z, Dorta Z, Veliz G, Vega H, Lorenzo I, et al. Immunization with a DNA vaccine candidate in chronic hepatitis C patients is safe, well tolerated and does not impair immune response induction after anti-hepatitis B vaccination. J Gene Med 2010;12(1):107–16.

[126] Ault KA. Epidemiology and natural history of human papillomavirus infections in the female genital tract. Infect Dis Obstet Gynecol 2006;2006(Suppl.):40470.

[127] Terai M, DeSalle R, Burk RD. Lack of canonical E6 and E7 open reading frames in bird papillomaviruses: *Fringilla coelebs* papillomavirus and *Psittacus erithacus* timneh papillomavirus. J Virol 2002;76(19):10020–3.

[128] Herbst LH, Lenz J, Van Doorslaer K, Chen Z, Stacy BA, Wellehan Jr JF, et al. Genomic characterization of two novel reptilian papillomaviruses, *Chelonia mydas* papillomavirus 1 and *Caretta caretta* papillomavirus 1. Virology 2009;383(1):131–5.

[129] zur Hausen H. Papillomaviruses and cancer: from basic studies to clinical application. Nat Rev Cancer 2002;2(5):342–50.

[130] Hu Z, Ding W, Zhu D, Yu L, Jiang X, Wang X, et al. TALEN-mediated targeting of HPV oncogenes ameliorates HPV-related cervical malignancy. J Clin Invest 2014;125(1):425–36.

[131] Burk RD, Harari A, Chen Z. Human papillomavirus genome variants. Virology 2013;445(1–2):232–43.

[132] Paavonen J, Jenkins D, Bosch FX, Naud P, Salmeron J, Wheeler CM, et al. Efficacy of a prophylactic adjuvanted bivalent L1 virus-like-particle vaccine against infection with human papillomavirus types 16 and 18 in young women: an interim analysis of a phase III double-blind, randomised controlled trial. Lancet 2007;369(9580):2161–70.

[133] Villa LL, Costa RL, Petta CA, Andrade RP, Ault KA, Giuliano AR, et al. Prophylactic quadrivalent human papillomavirus (types 6, 11, 16, and 18) L1 virus-like particle vaccine in young women: a randomised double-blind placebo-controlled multicentre phase II efficacy trial. Lancet Oncol 2005;6(5):271–8.

[134] Garcia F, Petry KU, Muderspach L, Gold MA, Braly P, Crum CP, et al. ZYC101a for treatment of high-grade cervical intraepithelial neoplasia: a randomized controlled trial. Obstet Gynecol 2004;103(2):317–26.

[135] Kim D, Gambhira R, Karanam B, Monie A, Hung C-F, Roden R, et al. Generation and characterization of a preventive and therapeutic HPV DNA vaccine. Vaccine 2008;26(3):351–60.

[136] Peng S, Song L, Knoff J, Wang JW, Chang YN, Hannaman D, et al. Control of HPV-associated tumors by innovative therapeutic HPV DNA vaccine in the absence of CD4+ T cells. Cell Biosci 2014;4(1):11.

[137] Davidson EJ, Boswell CM, Sehr P, Pawlita M, Tomlinson AE, McVey RJ, et al. Immunological and clinical responses in women with vulval intraepithelial neoplasia vaccinated with a vaccinia virus encoding human papillomavirus 16/18 oncoproteins. Cancer Res 2003;63(18):6032–41.

[138] Maciag PC, Radulovic S, Rothman J. The first clinical use of a live-attenuated Listeria monocytogenes vaccine: a phase I safety study of Lm-LLO-E7 in patients with advanced carcinoma of the cervix. Vaccine 2009;27(30):3975–83.

[139] Bagarazzi ML, Yan J, Morrow MP, Shen X, Parker RL, Lee JC, et al. Immunotherapy against HPV16/18 generates potent TH1 and cytotoxic cellular immune responses. Sci Transl Med 2012;4(155):155ra38.

[140] Lehman IR, Boehmer PE. Replication of herpes simplex virus DNA. J Biol Chem 1999;274(40):28059–62.

[141] Muylaert I, Tang KW, Elias P. Replication and recombination of herpes simplex virus DNA. J Biol Chem 2011;286(18):15619–24.

[142] Looker KJ, Garnett GP, Schmid GP. An estimate of the global prevalence and incidence of herpes simplex virus type 2 infection. Bull World Health Organ 2008;86(10):805–12.

[143] Centers for Disease C, Prevention. Seroprevalence of herpes simplex virus type 2 among persons aged 14–49 years–United States, 2005–2008. MMWR Morb Mortal Wkly Rep 2010;59(15):456–9.

[144] Xu F, Sternberg MR, Kottiri BJ, McQuillan GM, Lee FK, Nahmias AJ, et al. Trends in herpes simplex virus type 1 and type 2 seroprevalence in the United States. JAMA 2006;296(8):964–73.

[145] Paz-Bailey G, Ramaswamy M, Hawkes SJ, Geretti AM. Herpes simplex virus type 2: epidemiology and management options in developing countries. Sex Transm Infect 2007;83(1):16–22.

[146] Freeman EE, Weiss HA, Glynn JR, Cross PL, Whitworth JA, Hayes RJ. Herpes simplex virus 2 infection increases HIV acquisition in men and women: systematic review and meta-analysis of longitudinal studies. Aids 2006;20(1):73–83.

[147] Reynolds SJ, Risbud AR, Shepherd ME, Zenilman JM, Brookmeyer RS, Paranjape RS, et al. Recent herpes simplex virus type 2 infection and the risk of human immunodeficiency virus type 1 acquisition in India. J Infect Dis 2003;187(10):1513–21.

[148] Palliser D, Chowdhury D, Wang QY, Lee SJ, Bronson RT, Knipe DM, et al. An siRNA-based microbicide protects mice from lethal herpes simplex virus 2 infection. Nature 2006;439(7072):89–94.

[149] Aubert M, Boyle NM, Stone D, Stensland L, Huang ML, Magaret AS, et al. In vitro inactivation of latent HSV by targeted mutagenesis using an HSV-specific homing endonuclease. Mol Ther Nucleic acids 2014;3:e146.

[150] Deback C, Boutolleau D, Depienne C, Luyt CE, Bonnafous P, Gautheret-Dejean A, et al. Utilization of microsatellite polymorphism for differentiating herpes simplex virus type 1 strains. J Clin Microbiol 2009;47(3):533–40.

[151] Deback C, Luyt CE, Lespinats S, Depienne C, Boutolleau D, Chastre J, et al. Microsatellite analysis of HSV-1 isolates: from oropharynx reactivation toward lung infection in patients undergoing mechanical ventilation. J Clin Virol 2010;47(4):313–20.

[152] Szpara ML, Gatherer D, Ochoa A, Greenbaum B, Dolan A, Bowden RJ, et al. Evolution and diversity in human herpes simplex virus genomes. J Virol 2014;88(2):1209–27.

[153] Umene K, Kawana T. Divergence of reiterated sequences in a series of genital isolates of herpes simplex virus type 1 from individual patients. J Gen Virol 2003;84(Pt 4):917–23.

[154] Spear PG, Eisenberg RJ, Cohen GH. Three classes of cell surface receptors for alphaherpesvirus entry. Virology 2000;275(1):1–8.

[155] Muylaert I, Elias P. Knockdown of DNA ligase IV/XRCC4 by RNA interference inhibits herpes simplex virus type I DNA replication. J Biol Chem 2007;282(15):10865–72.

[156] Whitley RJ. Herpesviruses. In: Baron S, editor. Medical microbiology. 4th ed. 1996. Galveston (TX).

[157] Iijima N, Linehan MM, Zamora M, Butkus D, Dunn R, Kehry MR, et al. Dendritic cells and B cells maximize mucosal Th1 memory response to herpes simplex virus. J Exp Med 2008;205(13):3041–52.

[158] Shin H, Iwasaki A. Generating protective immunity against genital herpes. Trends Immunol 2013;34(10):487–94.

[159] Harandi AM, Svennerholm B, Holmgren J, Eriksson K. Differential roles of B cells and IFN-gamma-secreting CD4(+) T cells in innate and adaptive immune control of genital herpes simplex virus type 2 infection in mice. J Gen Virol 2001;82(Pt 4):845–53.

[160] Johnston C, Koelle DM, Wald A. Current status and prospects for development of an HSV vaccine. Vaccine 2014;32(14):1553–60.

[161] Zhu XP, Muhammad ZS, Wang JG, Lin W, Guo SK, Zhang W. HSV-2 vaccine: current status and insight into factors for developing an efficient vaccine. Viruses 2014;6(2):371–90.

[162] Shin H, Iwasaki A. A vaccine strategy that protects against genital herpes by establishing local memory T cells. Nature 2012;491(7424):463–7.

[163] Chiuppesi F, Vannucci L, De Luca A, Lai M, Matteoli B, Freer G, et al. A lentiviral vector-based, herpes simplex virus 1 (HSV-1) glycoprotein B vaccine affords cross-protection against HSV-1 and HSV-2 genital infections. J Virol 2012;86(12):6563–74.

[164] Veselenak RL, Shlapobersky M, Pyles RB, Wei Q, Sullivan SM, Bourne N. A Vaxfectin (®)-adjuvanted HSV-2 plasmid DNA vaccine is effective for prophylactic and therapeutic use in the guinea pig model of genital herpes. Vaccine 2012;30(49):7046–51.

[165] Cattamanchi A, Posavad CM, Wald A, Baine Y, Moses J, Higgins TJ, et al. Phase I study of a herpes simplex virus type 2 (HSV-2) DNA vaccine administered to healthy, HSV-2-seronegative adults by a needle-free injection system. Clin Vaccine Immunol 2008;15(11):1638–43.

[166] Pancera M, Zhou T, Druz A, Georgiev IS, Soto S, Gorman J, et al. Structure and immune recognition of trimeric pre-fusion HIV-1 Env. Nature 2014;514(7523):455–61.

Antiviral Gene Therapy: Summary and Perspectives

12.1 VIRUSES: ESSENTIALS OF THEIR REPLICATION AND SUSCEPTIBILITY TO ANTIVIRAL GENE THERAPY

Viruses constitute the simplest and most abundant life form. Replication of viruses is largely dependent on delivery of their genomes to host cells and then usurping functions of target cells to produce proteins that are required to form viral progeny. The mechanisms by which viruses appropriate cellular functions predictably involve some disruption, which is responsible for induction of disease in virus-infected individuals. In addition to viral proteins, host factors are also required to facilitate replication of viruses. Typically, antiviral gene therapy aims to deploy DNA or RNA to achieve a therapeutic effect by inhibiting the expression of viral or host factor genes. A timeline of some of the significant developments and ongoing research efforts in gene therapy for viral infections is illustrated in Figure 12.1. Capacity of viruses to bypass dependence on host factors is highly constrained; therefore, inhibiting the function of host factors is useful to suppress emergence of escape by the pathogens. This has been useful in developing gene therapy for infections with HIV-1 [1] and hepatitis C virus (HCV) [2]. When using this approach, appropriate target selection is important. If targeted host factors have important cellular functions, then their silencing may cause toxicity. Another method of providing a high barrier to viral infection is through simultaneous inhibition of multiple viral targets using combinatorial approaches.

Types of diseases caused by viral infections are varied and depend on characteristics of the individual pathogens. For example, human immunodeficiency virus (HIV)-1 has a tropism for $CD4^+$ T cells of the immune system, and manifestation of the acquired immune deficiency syndrome results from insidious virus-induced depletion of the infected host T cells. By contrast, infection with Ebola virus follows a rapid and usually fulminant course. Although the prognosis of cases of Ebola virus infection is grave, the pathogen does not persist, and successful treatment is curative. This is different from the situation with HIV-1 infection, in which the stable proviral replication intermediate remains embedded in the genome of infected host cells and makes elimination of the virus very difficult.

Gene Therapy for Viral Infections. http://dx.doi.org/10.1016/B978-0-12-410518-8.00012-0

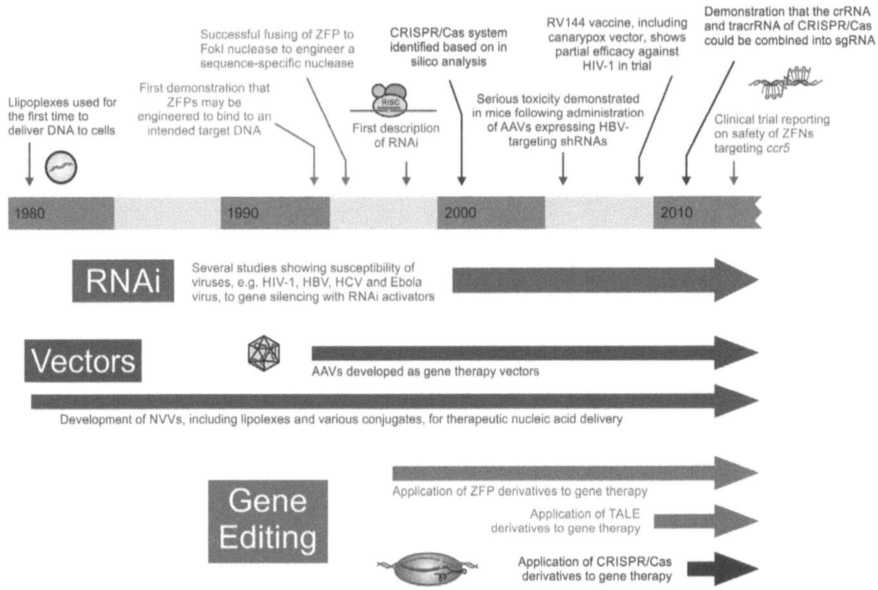

FIGURE 12.1 Some significant developments in the advancement of gene therapy for viral infections.

Milestones and major areas of ongoing technological development are indicated. AAV, adeno-associated virus; sgRNA, single guide RNA.

Although infections with viruses remain significant global causes of public health problems, advances in their treatment have been particularly impressive during the past 25 years. Several improvements in antiviral therapy have been made, and many diseases are now effectively managed. Antivirals comprising small molecules have demonstrated particularly good efficacy against a range of pathogens. Prophylactic immunization and other preventative measures, such as behavioral modification to minimize risks of transmission of viruses, have also contributed to limiting the spread of viral infections. Despite these advances, current therapeutics have inadequate curative efficacy against certain viral infections, and it is likely that disease caused by the pathogens will remain problematic for many years. Mutations may change properties of existing viruses, and new pathogens are likely to emerge. Hence, flexible technology that allows for rapid development of antivirals based on rational design is particularly important to counter the serious health problems that may result. Gene therapy has the potential to be effective against a range of existing and emergent viruses. The technology may also complement and augment efficacy of current antiviral drugs. However, gene therapy is still faced with significant hurdles before it is widely implemented for treatment of viral infections. Particularly important are the specificity of the antiviral nucleic acids for their targets and concerns about efficiency and safety of delivery, costs, and manufacturing capabilities.

In addition to direct targeting of viral genes and host factors, gene transfer for immune-based treatment and prophylaxis has potential for countering many viruses (Chapter 11). Relative ease of generating sequences expressing appropriate antigens and ability to induce humoral and cell-mediated arms of the immune response are advantages over use of protein-based vaccines. Gene transfer using DNA transfection and recombinant viral vectors, such as derivatives of poxviruses, have become popular for eliciting immune responses, and efficacy of the approach is continually improving. Using the procedures to effect immunostimulation is generally easy to implement and has already been tested in clinical trials performed in resource-poor settings. Although complex, vectored immunoprophylaxis (VIP) [3] and modification of immune cells with engineered T cell receptors (TCRs) [4] or chimeric antigen receptors (CARs) [5] are exciting new developments that have potential antiviral utility.

Advances in antiviral gene therapy have been made through steady accumulation of knowledge and as a result of significant breakthrough discoveries (Figure 12.1). Two momentous advances in the field have been development of technologies based on gene editing (Chapter 3) and harnessing the naturally occurring RNA interference (RNAi) pathway (Chapter 2). Gene editing is a process by which specific target DNA sequences are altered to inhibit viral replication. Therapy based on this technology is applicable to direct inactivation of viral DNA replication intermediates, such as are produced during proliferation of hepatitis B virus (HBV), or through inhibition of host factors required for replication of a particular virus [6]. RNAi is an endogenous gene regulatory mechanism that typically causes post-transcriptional gene silencing. Artificial RNAi activators, which comprise mimics of intermediates of the pathway and contain homology to intended targets, have been used successfully to inhibit viral gene expression. Exploiting RNAi and gene editing have both generated considerable enthusiasm, and the technology has advanced to use in clinical trials for treatment of viral infections, which include those caused by HBV and HIV-1. Other effective methods of inhibiting viral gene expression included use of antisense oligonucleotides, aptamers and RNase P substrates.

12.1.1 Exploiting RNAi for Antiviral Gene Therapy

Both expressed and synthetic RNAi activators have been utilized successfully against various viral infections. Considerable effort has gone into ensuring good specificity and potency of the silencers against their viral cognates. Use of Pol III promoters to transcribe RNA that forms short hairpin RNA (shRNA) structures, which mimic precursor microRNAs (pre-miRs), has been particularly popular. However, demonstration that overexpression of HBV-targeting shRNAs could result in fatal toxicity in mice was an important caveat to researchers in the field [7]. Since this observation, various configurations of expressed RNAi activators have been engineered to improve control of production of the antivirals. Using antiviral mimics of primary microRNAs (pri-miRs) has shown promise against HBV [8,9], HCV [10], and HIV-1 [11,12]. These RNAi activators may be transcribed from inducible and tissue-specific Pol II promoters. The artificial pri-miRs have been multimerized to result in combinatorial silencing that improves efficacy and provides a greater barrier to viral escape. As mimics of early intermediates of the RNAi pathway, functional coupling of processing of the pri-miR mimics to downstream steps of the pathway is thought to improve potency. Better silencing by antivirals allows for use of lower doses of the candidate therapeutics and diminished off-target effects. Developments with use of synthetic RNAi activators have also been impressive. Typically, synthetic RNAi activators comprise short complementary strands of RNA that constitute short interfering RNAs (siRNAs). Various chemical modifications have been used to augment silencing, diminish immunostimulation, and stabilize the gene silencers. Synthetic procedures are also adaptable for other purposes. One interesting application is the coupling of targeting aptamers to synthetic siRNAs, which enables specific delivery to virus-infected cells [13–16]. Availability of new methods for bioinformatic analysis has also been important to advance RNAi-based antiviral therapy. Because only the seed region of the guide strand of a potentially therapeutic siRNA is required to effect translational silencing [17], the likelihood of nonspecific inhibition of cellular targets may be high. However, availability of vast sequence databases and analytical algorithms have allowed for the design of antiviral RNAi activators that have minimal interaction with off-target sites.

The ultimate success of RNAi-based antiviral therapy is likely to be dependent on demonstrating superior efficiency when compared with existing therapies. In addition, adapting the silencing strategies to attain the inhibition that is required for a curative effect is important. These considerations are illustrated by the example of development of RNAi-based therapy for treatment of persistent HBV infection. A clinical trial is in progress for the use of nonviral formulations containing synthetic RNAi activators to treat HBV infection (http://www.arrowheadresearch.com/programs-overview). Suppression of viral replication that is better or as good as that of licensed drugs needs to be demonstrated. In addition, because the gene silencing mechanism is not permanent, ability of the treatment to achieve a durable effect that leads to curative elimination of all viral replication intermediates will be important. Transient RNAi-based silencing may be better suited to treatment of RNA viruses that do not have a stable DNA replication intermediate.

12.1.2 Advancing Gene Editing to Counter Viral Infections

Antiviral gene editing technology entails inactivation of DNA sequences; therefore, it is adaptable to disabling viral DNA replication intermediates and genes of host factors that are required for viral proliferation. To effect mutagenesis, engineered sequence-specific DNA-targeting proteins have been coupled to nucleases. Repeated cleavage followed by cellular error-prone nonhomologous end joining results in mutagenesis at the intended target. Earliest studies showing feasibility of gene editing used derivatives of zinc finger proteins (ZFPs). Twenty years ago, the approach was shown to inhibit expression of the *bcr-abl* sequence that is generated after the chromosomal translocation that characterizes chronic myeloid leukemia [18]. The inhibitory effect resulted from binding of the engineered ZFPs to DNA and physical disruption of transcription of the *bcr-abl* sequence.

Engineered ZFPs comprise tandem arrangements of fingers that bind to triplets of nucleotides. Zinc finger nucleases (ZFNs) are typically generated by fusing DNA-binding domains with the nuclease from the type IIS restriction endonuclease, FokI, which cuts one strand of a DNA duplex. Therefore, complete ZFNs comprise dimers, and each of the units is designed to cleave complementary strands at an intended target. Although engineered ZFNs have been the subject of many years of study, application of the technology is complicated by variability in the efficiency of target binding and cleavage. Homing endonucleases (HEs), also known as meganucleases, may cause sequence-specific cleavage of viral targets, but engineering this class of nuclease is fraught with technical difficulties. Problems associated with the dual functions of target binding and DNA cleavage of domains of the HEs are constraining.

Use of derivatives of transcription activator-like effectors (TALEs), which include TALE nucleases (TALENs) or repressor TALEs, overcomes some of the problems of ZFNs. Monomers that make up TALEs do not show the same context dependence of the fingers of ZFPs, which circumvents the need to screen large panels of the engineered proteins to identify optimally acting editors. As an alternative to engineering nucleases, sequence-specific transcriptional repressors may well find favor for therapeutic inhibition of viral gene expression. The Krüppel-associated box (KRAB) is capable of inducing epigenetic changes at specific sequences of DNA to cause long-term transcriptional suppression. Substituting a repressor for a nuclease in the engineered proteins gets around the problem of possible mutagenic effects that may arise from nonspecific cleavage of host genomic DNA.

Recent discovery of the clustered regularly interspaced short palindromic repeats (CRISPR) with CRISPR-associated (Cas) systems has also generated enthusiasm among researchers interested in developing antiviral gene editing. Engineering these nucleases is convenient because site-specific cleavage of target DNA is based on interaction of a single-stranded RNA guide with the target DNA. Therefore, the approach obviates the more complicated methodologies that entail generation of sequences encoding proteins that bind to specific DNA targets.

All classes of gene editing technologies have been investigated for efficacy against viral infections. The early-stage clinical trial that assessed *ccr5*-targeting ZFNs is auspicious [1], and the potential for rendering of autologous $CD4^+$ T cells resistant to HIV-1 is an exciting prospect. TALENs have been generated that target HBV genes [19,20] and *ccr5* [21], and their potential for antiviral therapy is promising. Engineered CRISPR-Cas systems were also shown to be effective against replication of HIV-1 [22,23] and HBV [24] in cultured cells. As with RNAi against viral infections, efficient delivery and specificity of the effectors for their intended targets are important for clinical application of gene editing.

With the exception of the smaller sequences encoding HEs, expression cassettes producing gene editors are typically larger than 3 kb in length. Moreover, in the case of ZFNs and TALENs, two cassettes are required to produce the two monomers that make up the dimeric nucleases. Likewise, typical engineered CRISPR/Cas systems comprise two cassettes: one encodes the nuclease enzyme and the other the guide RNA. Recombinant adeno-associated viral vectors, popular gene therapy vectors, have a limited capacity for accommodating transgenes. Therefore, complete dimeric nucleases may only be constituted by transduction of cells with two vectors, each expressing one of the monomers. An alternative to this complicated delivery approach is use of nonviral vectors (NVVs) to deliver mRNA encoding the gene editors' subunits. Formulation methods are not limited by the size of the mRNA that may be incorporated into the vectors, and the technique is now becoming more commonly used.

A useful feature of gene editing for viral gene therapy is that transient expression of the gene editor may be all that is required to effect potentially curative inhibition of viral replication. Once a target has been altered by introduction of mutations, a gene editing nuclease is no longer required. In fact, persistence of the effector is undesirable because the risk for off-target effects is increased. This feature is well suited to using NVVs to deliver nuclease-encoding mRNAs. Durability of gene editing that involves use of targeted transcriptional repressors, such as KRAB, to introduce epigenetic inhibitory effects is not yet well characterized. Transient expression of such repressors may be sufficient to cause the intended durable therapeutic inhibition. Avoiding unwanted effects on the host genome may be particularly useful for treatment of chronic HBV infection with transcriptional repressors. In individuals who are carriers of the virus, HBV DNA is often integrated into the hepatocyte genomes, and nuclease-mediated digestion may induce chromosomal translocations to cause unintended side effects.

12.1.3 Delivery of Antiviral Nucleic Acids to Target Cells

Optimizing delivery of nucleic acids is an objective that is common to advancing gene therapy for any disease. Location of the virus-infected cells is pertinent to devising the appropriate method of delivering the therapeutics. For example, systemic administration of antivirals may be necessary to disable HBV genes in the liver. However, for treatment of respiratory infections, administration of the therapeutic formulations to the airways is usually more effective. The two main types of delivery vehicles, viral vectors and NVVs, have been discussed in detail in Chapters 4 and 5. Each has particular features that are suited to specific applications to antiviral gene therapy. Viral vectors are efficient carriers of sequences encoding expression cassettes, whereas delivery in vivo of expression cassettes with NVV formulations is usually modest. Because viruses essentially exist as nucleic acids parasites, they have evolved efficient mechanisms of cellular delivery of the nucleic acids, and this property has been exploited to generate viral vectors. Although recombinant viruses are highly capable vectors, they have drawbacks of high costs of production, difficulties with regulating the dose of therapeutics, induction of mutations in transduced cells, and eliciting neutralizing and potentially toxic immunostimulation. Durable expression, particularly by recombinant lentiviral vectors, is useful to achieve sustained antiviral effects, but utility of these vectors for antiviral application is narrow. RNA-based antivirals are growing in popularity, and NVVs are the preferred vectors for delivery of these nucleic acids. Because NVVs do not readily cross the nuclear membrane, the cytoplasmic site of action of antiviral RNAs, which include synthetic siRNAs and mRNAs, is a convenient feature. Various NVVs are under development, and these include lipoplex formulations and polymer conjugates. Because they are based on use of synthetic compounds, NVVs may be prepared on the large scale that is required for clinical application. Moreover, RNA-containing nonviral formulations are appealing because of their limited immunostimulation and avoidance of the mutagenic effects of DNA.

To attain adequate expression of an immunogenic protein, fewer cells need to be transduced than when delivering effectors that have a direct effect on viral replication. Moreover, the cells to be transduced may be readily accessible after intramuscular or intradermal delivery and are not necessarily the virus-infected cells. Recombinant adenoviruses, various poxviruses, protein coding sequences within NVVs, and electroporation of naked DNA together with adjuvants have all been used successfully to elicit antiviral immune responses.

12.2 PROSPECTS FOR ANTIVIRAL GENE THERAPY

Predicting the long-term role of gene therapy for viral infections is difficult. Apart from the many technical challenges, there are additional factors that will influence the future general use of antiviral gene therapy. When alternative treatment or prevention of a viral infection is effective and easily administered, the importance of antiviral gene therapy may be diminished. An example of such a case could play out with developments in the treatment of HCV infection. Recent progress in HCV therapy using small molecules has been very impressive. The newly licensed combination drug, Harvoni™, produced by Gilead, is rapidly effective against HCV genotype 1 (http://www.gilead.com/news/press-releases/2014). Harvoni contains inhibitors of NS5A and NS5B proteins of HCV, and its administration to patients avoids the need for including interferon-α in the regimen. With successful implementation of the new HCV treatments, and if the management becomes widely affordable, the importance of gene therapy for HCV infection may diminish. Miravirsen, a modified oligonucleotide that is antisense to miR-122, has shown good efficacy in clinical trials [2]. Licensing and general use of this candidate host factor-disabling nucleic acid is likely to depend on its economical production and achieving antiviral efficacy that is at least as good as the recently licensed anti-HCV agents. Nevertheless, gene therapy will continue to be important for treating serious infections when cure using other approaches is not possible. Eliminating HBV and HIV-1 are examples where gene therapy is likely to play a role in curing the infections. In the case of chronic HBV infection, persistence of the stable covalently closed circular DNA is unaffected by existing therapies, and the gene editing approach potentially provides the means to disable this intractable viral transcription template. Likewise, use of gene editing to inactivate proviral sequences of HIV-1 or to render $CD4^+$ T cells resistant to infection with the virus should be useful to cure HIV-1 infection.

A serious consideration for the long-term prospects of implementing gene therapy for treatment of viral infections relates to the practicalities of using the technology in resource-poor settings. Infection with viruses that are causes of serious health problems, such as HIV-1, HBV, and Ebola virus, often occurs in financially impoverished communities. Gene therapy formulations are complicated and expensive to prepare, and it is unlikely that the necessary financial resources are currently available in many of the countries where the infections have the most serious effects. Likewise, gene therapy-based methods may be difficult to execute and require a high level of expertise to implement. An example is the procedures involved with modification of autologous progenitors of $CD4^+$ T cells ex vivo, their selection, and then re-infusion into patients to confer resistance to HIV-1 (Chapter 8). The technology is highly specialized, and a lack of expertise would limit access in some of the communities where treatment of HIV-1 infection is most needed. With improvements in methodologies and advancement of new technologies, it is expected that costs will diminish and procedures for performing gene therapy may become simpler. Ideally, this should result in future broader access of the technology to disadvantaged patients.

In the case of using gene transfer for immunostimulation to prevent or treat viral infections, the practical considerations are different. DNA-based vaccines and recombinant viral vectors that are used to induce immunity are generally administered using simpler procedures. Giving the vaccines to large numbers of people is uncomplicated as long as the necessary cold chain requirements are met. However, immunotherapy using VIP and modified T cells that express engineered TCRs or CARs is more complex. As with gene therapy to target genes of viruses or host factors, implementation is likely to require simplification and improvement to make the technologies cost-effective for wider use.

A major advantage of gene therapy is the ability to develop the antiviral agents based on application of what are becoming simple rational design principles. This is the case for antiviral gene therapies that harness RNAi-based silencing, antisense oligonucleotides, and gene editing. On the basis of knowledge of the sequences of the targets, it is theoretically possible to develop generic methods that enable very rapid design of antiviral nucleic acids. Therefore, the speed with which gene therapy could be developed against an emerging pathogen is an advantage over use of small-molecule drugs to treat these viral infections. Identifying effective small molecule antivirals is also guided by rational design, but it typically requires detailed structural analysis of viral protein targets and empirical testing with optimization before use as therapy.

With advances in the broader fields of gene therapy, virology, synthetic chemistry, and molecular biology, a better understanding of factors that influence the safety, delivery efficiency, and antiviral actions of nucleic acids is likely to be reached. This will be important to facilitate implementation of more effective therapeutic measures to address the serious health challenges posed by existing and emergent pathogens. Availability of suitable tools will enable countermeasures to be put into place, which in turn will be important to deal with outbreaks of infections such as caused by the severe acute respiratory syndrome coronavirus during 2002/2003 and infection with Ebola virus. Gene therapy for viral infections is currently poised at an exciting phase. There have been significant successes, and realization of the full potential of the powerful technology is tantalizingly close. Significant goals are likely to be met in the near future, which in turn will influence the direction of the field.

REFERENCES

[1] Tebas P, Stein D, Tang WW, Frank I, Wang SQ, Lee G, et al. Gene editing of CCR5 in autologous CD4 T cells of persons infected with HIV. N Engl J Med 2014;370(10):901–10.

[2] Janssen HL, Reesink HW, Lawitz EJ, Zeuzem S, Rodriguez-Torres M, Patel K, et al. Treatment of HCV infection by targeting microRNA. N Engl J Med 2013;368(18):1685–94.

[3] Balazs AB, Ouyang Y, Hong CM, Chen J, Nguyen SM, Rao DS, et al. Vectored immunoprophylaxis protects humanized mice from mucosal HIV transmission. Nat Med 2014;20(3):296–300.

[4] Gehring AJ, Xue SA, Ho ZZ, Teoh D, Ruedl C, Chia A, et al. Engineering virus-specific T cells that target HBV infected hepatocytes and hepatocellular carcinoma cell lines. J Hepatol 2011;55(1):103–10.

[5] Krebs K, Bottinger N, Huang LR, Chmielewski M, Arzberger S, Gasteiger G, et al. T cells expressing a chimeric antigen receptor that binds hepatitis B virus envelope proteins control virus replication in mice. Gastroenterology 2013;145(2):456–65.

[6] Schiffer JT, Swan DA, Stone D, Jerome KR. Predictors of hepatitis B cure using gene therapy to deliver DNA cleavage enzymes: a mathematical modeling approach. PLoS Comput Biol 2013;9(7):e1003131.

[7] Grimm D, Streetz KL, Jopling CL, Storm TA, Pandey K, Davis CR, et al. Fatality in mice due to oversaturation of cellular microRNA/short hairpin RNA pathways. Nature 2006;441(7092):537–41.

[8] Ely A, Naidoo T, Arbuthnot P. Efficient silencing of gene expression with modular trimeric Pol II expression cassettes comprising microRNA shuttles. Nucleic Acids Res 2009;37(13):e91.

[9] Ely A, Naidoo T, Mufamadi S, Crowther C, Arbuthnot P. Expressed anti-HBV primary microRNA shuttles inhibit viral replication efficiently in vitro and in vivo. Mol Ther J Am Soc Gene Ther 2008;16(6):1105–12.

[10] Yang X, Haurigot V, Zhou S, Luo G, Couto LB. Inhibition of hepatitis C virus replication using adeno-associated virus vector delivery of an exogenous anti-hepatitis C virus microRNA cluster. Hepatology 2010;52(6):1877–87.

[11] Liu YP, Gruber J, Haasnoot J, Konstantinova P, Berkhout B. RNAi-mediated inhibition of HIV-1 by targeting partially complementary viral sequences. Nucleic Acids Res 2009;37(18): 6194–204.

[12] Aagaard LA, Zhang J, von Eije KJ, Li H, Saetrom P, Amarzguioui M, et al. Engineering and optimization of the miR-106b cluster for ectopic expression of multiplexed anti-HIV RNAs. Gene Ther 2008;15(23):1536–49.

[13] Zhou J, Neff CP, Swiderski P, Li H, Smith DD, Aboellail T, et al. Functional in vivo delivery of multiplexed anti-HIV-1 siRNAs via a chemically synthesized aptamer with a sticky bridge. Mol Ther J Am Soc Gene Ther 2013;21(1):192–200.

[14] Zhou J, Rossi JJ. Aptamer-targeted RNAi for HIV-1 therapy. Methods Mol Biol 2011;721: 355–71.

[15] Zhou J, Rossi JJ. Cell-specific aptamer-mediated targeted drug delivery. Oligonucleotides 2011;21(1):1–10.

[16] Zhou J, Rossi JJ. Therapeutic potential of aptamer-siRNA conjugates for treatment of HIV-1. BioDrugs Clin Immunother Biopharm Gene Ther 2012;26(6):393–400.

[17] Grimson A, Farh KK, Johnston WK, Garrett-Engele P, Lim LP, Bartel DP. MicroRNA targeting specificity in mammals: determinants beyond seed pairing. Mol Cell 2007;27(1):91–105.

[18] Choo Y, Sanchez-Garcia I, Klug A. In vivo repression by a site-specific DNA-binding protein designed against an oncogenic sequence. Nature 1994;372(6507):642–5.

[19] Bloom K, Ely A, Mussolino C, Cathomen T, Arbuthnot P. Inactivation of hepatitis B virus replication in cultured cells and in vivo with engineered transcription activator-like effector nucleases. Molecular therapy. J Am Soc Gene Ther 2013;21(10):1889–97.

[20] Chen J, Zhang W, Lin J, Wang F, Wu M, Chen C, et al. An efficient antiviral strategy for targeting hepatitis B virus genome using transcription activator-like effector Nucleases. Mol Ther J Am Soc Gene Ther 2013;22(3):303–11.

[21] Mussolino C, Alzubi J, Fine EJ, Morbitzer R, Cradick TJ, Lahaye T, et al. TALENs facilitate targeted genome editing in human cells with high specificity and low cytotoxicity. Nucleic Acids Res 2014;42(10):6762–73.

[22] Ebina H, Misawa N, Kanemura Y, Koyanagi Y. Harnessing the CRISPR/Cas9 system to disrupt latent HIV-1 provirus. Sci Rep 2013;3:2510.

[23] Hu W, Kaminski R, Yang F, Zhang Y, Cosentino L, Li F, et al. RNA-directed gene editing specifically eradicates latent and prevents new HIV-1 infection. Proc Natl Acad Sci USA 2014;111(31):11461–6.

[24] Lin SR, Yang HC, Kuo YT, Liu CJ, Yang TY, Sung KC, et al. The CRISPR/Cas9 system facilitates clearance of the intrahepatic HBV templates in vivo. Mol Ther Nucleic Acids 2014;3:e186.

Index

Note: Page numbers followed by "f" and "t" indicate figures and tables, respectively.

FSC
www.fsc.org
MIX
Papier | Fördert gute Waldnutzung
FSC® C083411

Kolibri 360 GmbH
Ferdinand-Jühlke-Straße 7
99095 Erfurt, Deutschland
produktsicherheit@kolibri360.de